3. wissenschaftliche Konferenz der Gesellschaft
Deutscher Naturforscher und Ärzte
Semmering bei Wien 1965

Funktionelle und morphologische
Organisation der Zelle

Probleme der biologischen Reduplikation

Problems of Reduplication in Biology

Herausgegeben von *P. Sitte*, Heidelberg

Springer-Verlag Berlin Heidelberg New York 1966

ISBN-13: 978-3-540-03638-8   e-ISBN-13: 978-3-642-47409-5
DOI: 10.1007/978-3-642-47409-5

Titel-Nr. 1335

Vorwort des Herausgebers

Im Oktober 1965 trafen sich zum dritten Male auf Einladung der Gesellschaft Deutscher Naturforscher und Ärzte zahlreiche Wissenschaftler aus Europa und Amerika, um über bestimmte Aspekte der funktionellen und morphologischen Organisation der Zelle zu diskutieren. Diese dritte Konferenz galt der Behandlung eines der allgemeinsten und zentralsten biologischen Probleme, jenem der Reduplikation: Jedes lebende System besitzt die Fähigkeit zur Vermehrung — und umgekehrt erscheint es durch sie vor allem als Lebewesen charakterisiert.

Verständlicherweise schloß aber gerade diese zentrale Stellung des Themas, zu dem tatsächlich jeder Zweig der biologischen Forschung vielfältige Beziehung hat, eine umfassende Behandlung aus. Das Schwergewicht der Erörterungen lag — im Hinblick auf die beiden vorangegangenen, so erfolgreichen Konferenzen von Rottach-Egern und von Reinhardsbrunn — bei denjenigen Organellen, die als eigene Reduplikantensysteme innerhalb der Zelle erscheinen. Diese Diskussionen standen freilich in einem größeren Rahmen, der von der Reduplikation der informativen Makromoleküle bis zu jener der Zelle und zu den theoretischen Problemen des Vermehrungsvorganges schlechthin reichte.

Die 19 Vorträge und die überaus lebhaften Diskussionen, die mit diesem Bericht der breiteren Öffentlichkeit zugänglich gemacht werden (letztere naturgemäß in stark gestraffter Form), dienten in fruchtbarer Weise dem Austausch von Erkenntnissen, von Ansichten und Einsichten. So konnte auch diesmal wieder das vom Gastgeber gesteckte Ziel erreicht und reiche Ernte eingebracht werden. Auch sonst stand die Konferenz unter glücklichen Sternen: Die Landschaft am Semmering, wo sich in den Ernst des Gebirges noch der heitere Charakter des Wiener Beckens mischt, leuchtete im schönsten Spätherbstwetter. Und die Teilnehmer werden sich stets des wunderbaren Abends erinnern, an dem sich die gesamte Konferenz — zu Gast bei Herrn MANFRED VON MAUTNER-MARKHOF und seiner Frau Gemahlin — an vollendet gebotener Kammermusik aus dem alten Österreich erfreuen durfte.

Mein Dank gilt jenen vielen, die das Zustandekommen der Konferenz ermöglicht und zu ihrem guten Verlauf geholfen haben: allen voran der Gesellschaft Deutscher Naturforscher und Ärzte, die auch diesmal wieder eine großzügige Gastgeberin war. Besonders habe ich dem Generalsekretär der Gesellschaft, Herrn Kollegen H. J. ANTWEILER, und Herrn Präsidenten Dr. Dr. h. c. M. v. MAUTNER-MARKHOF, sowie Herrn Prok. F. KREPP zu danken. Unermüdlich hat mich Dr. HEINZ FALK unterstützt — ohne seine ständige Hilfe bei der Vorbereitung und der Nacharbeit hätte weder die Konferenz so glatt ablaufen noch der Bericht so rasch erscheinen können. Der Verlag hat in bewährter Weise das Seine getan.

Heidelberg, im Januar 1966　　　　　　　　　　　　　　　　PETER SITTE

Inhaltsverzeichnis

Teilnehmerverzeichnis

Referenten

Prof. Dr. *A. Bajer*, Dept. of Biology, University of Oregon, Eugene, Oregon 97403 (USA).

Dr. *L. Diers*, Botanisches Institut der Universität, 5 Köln-Lindenthal, Gyrhofstr. 15.

Prof. Dr. *F. Duspiva*, Zoologisches Institut der Universität, 69 Heidelberg, Tiergartenstraße.

Doz. Dr. *O. Hess*, Max-Planck-Institut für Biologie, 74 Tübingen, Spemannstr. 34.

Doz. Dr. *H.-G. Keyl*, Max-Planck-Institut für Meeresbiologie, 74 Tübingen, Melanchthonstr. 36.

Doz. Dr. *J. Klima*, Laboratorium für Elektronenmikroskopie der Universität, Innsbruck, Schöpfstr. 41 (Austria).

Dr. *D. J. L. Luck*, Rockefeller University, New York 21, N.Y. 10021 (USA).

Dr. *B. Parthier*, Institut für Allgemeine Botanik der Universität, X 401 Halle/Saale, Am Kirchtor 1 (present address: Wenner Grens Institute, Norrtullsgatan 16, Stockholm V A, Sweden).

Doz. Dr. *W. Sachsenmaier*, Institut für experimentelle Krebsforschung der Universität, 69 Heidelberg, Voßstr. 3.

Dr. *H. Schaller*, Max-Planck-Institut für Virusforschung, 74 Tübingen, Spemannstraße 35.

Doz. Dr. *E. Schnepf*, Pflanzenphysiologisches Institut der Universität, 34 Göttingen, Untere Karspüle 2.

Doz. Dr. *H. Sitte*, Institut für Biophysik der Universität, Med. Fakultät,, 665 Homburg/Saar.

Prof. Dr. *W. Stubbe*, Botanisches Institut der Universität, 4 Düsseldorf, Strümpellstraße 4 (z. Z. der Tagung noch am Botanischen Institut der Universität zu Köln).

Prof. Dr. *J. H. Taylor*, Institute of Molecular Biophysics, Florida State University, Tallahassee, Fla. 32306 (USA).

Prof. Dr. *H. Tuppy*, Institut für Biochemie der Universität, Wien IX, Wasagasse 9 (Austria).

Dr. *W. Wehrmeyer*, Botanisches Institut der Tierärztl. Hochschule, 3 Hannover-Kirchrode, Bünteweg 17 (Westfalenhof).

Prof. Dr. *N. Weissenfels*, Zoologisches Institut der Universität, 53 Bonn, Poppelsdorfer Schloß.

Prof. Dr. *W. G. Whaley*, Cell Research Institute, University of Texas, Austin, Texas 78712 (USA).

Doz. Dr. *E. Wintersberger*, Institut für Biochemie der Universität, Wien IX, Wasagasse 9 (Austria).

Prof. Dr. *K. E. Wohlfarth-Bottermann*, Institut für Cytologie und Mikromorphologie der Universität, 53 Bonn-Endenich, Gartenstr. 61 a.

Dr. *R. Wollgiehn*, Institut für Biochemie der Pflanzen (DAW), X 401 Halle/Saale, Weinbergweg.

Diskussionsteilnehmer

Dr. *W. O Abel*, Max-Planck-Institut für Pflanzengenetik, 6802 Ladenburg, Rosenhof.

Prof. Dr. *H. J. Antweiler*, Organisch-chemisches Institut der Universität, 53 Bonn, Meckenheimer Allee 168.

Dr. *F. Bartels*, Max-Planck-Institut für Züchtungsforschung, 5 Köln-Vogelsang.

Prof. Dr. *W. Beermann*, Max-Planck-Institut für Biologie, 74 Tübingen, Spemannstraße 34.

Prof. Dr. *K. Bier*, Zoologisches Institut der Universität, 44 Münster/Westf., Badestr. 9.

Dr. *D. Brdiczka*, Physiologisch-chemisches Institut der Universität, 8 München 15, Goethestr. 33.

Prof. Dr. *G. Bruns*, Institut für Mikrobiologie und experimentelle Therapie (DAW), Abt. Experimentelle Pathologie, X 69 Jena, Beuthenbergstr. 11.

Doz. Dr. *H. David*, Pathologisches Institut der Universität, X 105 Berlin N 4, Schumannstr. 20.

Dr. *P. Döbel*, Institut für Kulturpflanzenforschung (DAW), X 4325 Gatersleben, Kreis Aschersleben.

Dr. *H. Falk*, Botanisches Institut der Universität, 69 Heidelberg, Hofmeisterweg 4.

Prof. Dr. *E. Grundmann*, Institut für experimentelle Pathologie der Farbenfabriken Bayer AG., 56 Wuppertal-Elberfeld, Friedr.-Ebert-Str. 217.

Dr. *R. Hagemann*, Institut für Kulturpflanzenforschung (DAW), X 4325 Gatersleben, Kreis Aschersleben.

Dr. *Hedwige Jakob*, Laboratoire de Génétique Physiologique, C.N.R.S., Gif-sur-Yvette (Seine-&-Oise (France)).

Prof. Dr. *P. Karlson*, Physiologisch-chemisches Institut der Universität, 355 Marburg/Lahn, Deutschhaus-Str. 1.

Doz. Dr. *Renate Lettré*, Institut für experimentelle Krebsforschung der Universität, 69 Heidelberg, Voßstr. 3.

Dr. *G. F. Meyer*, Max-Planck-Institut für Biologie, 74 Tübingen, Spemannstr. 34.

Prof. Dr. *F. Miller*, Pathologisches Institut der Universität, 8 München 15, Thalkirchner Str. 36.

Dipl.-Chem. *W. Neupert*, Physiologisch-chemisches Institut der Universität, 8 München 15, Goethestr. 33.

Prof. Dr. *F. Schötz*, Botanisches Institut der Universität, 8 München 19, Menzinger Straße 67.

Prof. Dr. *P. Sitte*, Botanisches Institut der Universität, 78 Freiburg/Breisgau, Schänzlestr. 9.

Dr. *K. F. Springer*, 69 Heidelberg, Neuenheimer Landstr. 28.

Prof. Dr. *W. Stoeckenius*, Rockefeller University, New York 21, N.Y. 10021, (USA).

Prof. Dr. *Elisabeth Tschermak-Woess*, Botanisches Institut der Universität, Wien III, Rennweg 14 (Austria).

Probleme der Reduplikation von Nucleinsäuren

Von

HEINZ SCHALLER, Tübingen

Mit 2 Abbildungen

Bei der Reduplikation biologischer Systeme muß von Generation zu Generation ein gewisser Grundbestand an Information, der in den Erbanlagen niedergelegt ist, weitergegeben werden. Dies setzt vor jeder Zellteilung eine Kopierung der genetischen Information voraus. Man kann heute mit Sicherheit annehmen, daß in autonomen Systemen die DNS der einzige Träger genetischer Information ist. Es gibt zwar noch andere Makromoleküle in der Zelle, die eine Information weiterleiten können, wie z. B. die RNS oder die Proteine, aber nur die DNS gewährleistet durch den Mechanismus der identischen Reduplikation eine konstante Informationsübermittlung.

Die DNS erfüllt somit zwei Funktionen: Einmal steuert sie über die Synthese der m-RNS die Proteinsynthese und damit alle chemischen Vorgänge in der Zelle (DNS → RNS → Protein), zum anderen kann sie sich selbst vermehren (DNS → DNS). In beiden Fällen sichert der einzigartige Mechanismus der Basenpaarung [32] ein exaktes Ablesen. Dieses Schema läßt sich auf alle Organismen und in etwas abgewandelter Form auch auf die RNS-Viren anwenden.

Eine Diskussion der Reduplikation informativer Makromoleküle wird also immer mit der Frage nach der Vermehrung der DNS verbunden sein.

DNS-Replikation

Die Replikation der DNS kann nach einem der drei folgenden Modelle vor sich gehen [5]: 1. Der Elterndoppelstrang bleibt unverändert erhalten, ein Tochterdoppelstrang wird neu gebildet (konservativ). 2. Der Elterndoppelstrang teilt sich und an jeder Hälfte wird ein neuer Tochterstrang gebildet (semi-konservativ). 3. Der Elternstrang bleibt weder als Doppel- noch als Einzelstrang erhalten, sondern wird auf die Tochterstränge verteilt (dispers).

Bei Mikroorganismen gilt die semi-konservative Replikation als bewiesen [17, 18, 27]. Man nimmt an, daß dabei Strangtrennung und Replikation gleichzeitig erfolgen, was durch das Y-Modell (Abb. 1) veranschaulicht wird. In Sonderfällen sind andere Arten der DNS-Replikation gefunden worden, die aber erst später diskutiert werden.

Auch bei höheren Organismen konnte eine semi-konservative DNS-Vermehrung nachgewiesen werden [26, 29], jedoch gibt es hier bisher keinen Beweis, daß die stabile Einheit der Replikation ein DNS-Einzel-strang ist [*29*].

Über die molekularen Vorgänge bei der semi-konservativen Redupli-kation der DNS geben fast ausschließlich Experimente an Bakterien und Viren Auskunft. Deshalb wird im folgenden nur von ihnen die Rede sein.

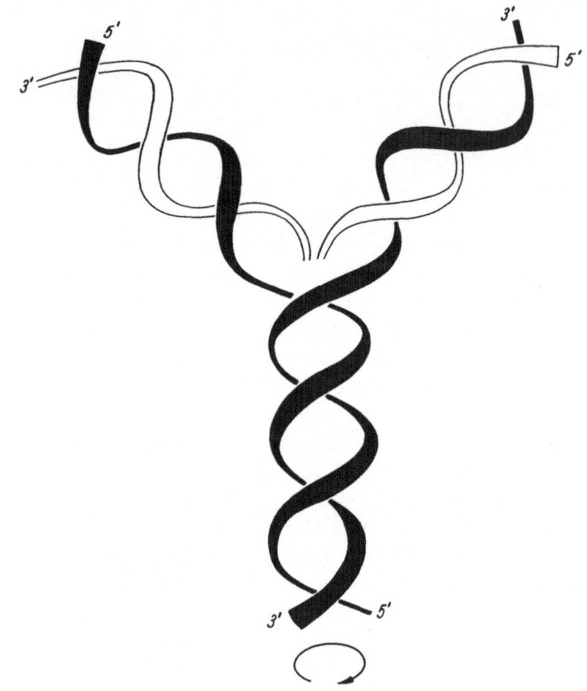

Abb. 1. Schema einer semi-konservativen *DNS*-Replikation (nach [*2*], verändert)

Semi-Konservative Vermehrung des Bakterienchromosoms

Bei dem Bacterium *Escherichia coli* liegen alle Gene auf einer cycli-schen Genkarte [*11*]. Durch Radioautographie konnte gezeigt werden, daß dieser ringförmigen Genkarte auch physikalisch eine cyclische Struktur, das Ringchromosom, entspricht [*3*]. Die quantitative Aus-wertung der Versuche ergab als physikalisches Strukturelement einen DNS-Doppelstrang von etwa 1 mm Länge. Aus den radioautographi-schen Daten und aus anderen physikalischen und genetischen Befunden [*1, 19, 28, 30*] ergibt sich folgendes Bild der semi-konservativen Repli-kation der DNS (Abb. 2):

Die Replikation beginnt nicht an mehreren Stellen gleichzeitig, son-dern nur an einem Punkt des Bakterienchromosoms. Sollte es mehrere mögliche Startstellen geben, scheint die Aktivierung einer einzigen die anderen auszuschließen. Normalerweise findet keine zweite Replikation statt, bevor die erste nicht abgeschlossen ist.

Daraus folgt, daß es auch nur einen Wachstumspunkt gibt, der sich in einer Richtung über das ganze Chromosom hinweg bewegt. Die beiden neusynthetisierten Stränge wachsen also in der gleichen Richtung. Lage der Startstelle auf der Genkarte und Wachstumsrichtung können bei verschiedenen Bakterienstämmen verschieden sein.

Die Trennung der DNS-Doppelspirale kann entweder durch Drehen des Moleküls oder durch Schneiden und „Heilen" (eines) der Stränge bewirkt werden. Im allgemeinen wird das einfachere Modell der Trennung durch Rotation bevorzugt. Auf eine Ring-DNS läßt sich dieses Modell jedoch nur durch den zusätzlichen Einbau einer oder mehrerer Drehstellen, den sogenannten "swivels" [3], anwenden. In diesem Zusammenhang wird die Frage diskutiert, ob Startpunkt der DNS-Synthese und Rotationsstelle vielleicht identisch sind. Möglicherweise sorgt ein aktiver Drehmechanismus für die Entspiralisierung der DNS am räumlich entfernten Wachstumspunkt [3].

Bei einer Generationszeit von 20 min ergibt sich auf Grund obiger Aussagen eine Replikationsgeschwindigkeit von ungefähr 10^5 Nucleotidpaaren pro Minute, das entspricht einem Aufdrehen von etwa 10^4 Windungen pro Minute. Physikalisch-chemische Überlegungen lassen dies möglich erscheinen [15]. Die Energie, die notwendig wäre, um die DNS um ihre eigene Achse rotieren zu lassen, ist klein gegenüber der bei der Synthese umgesetzten Energie. Andererseits reicht die Stabilität der

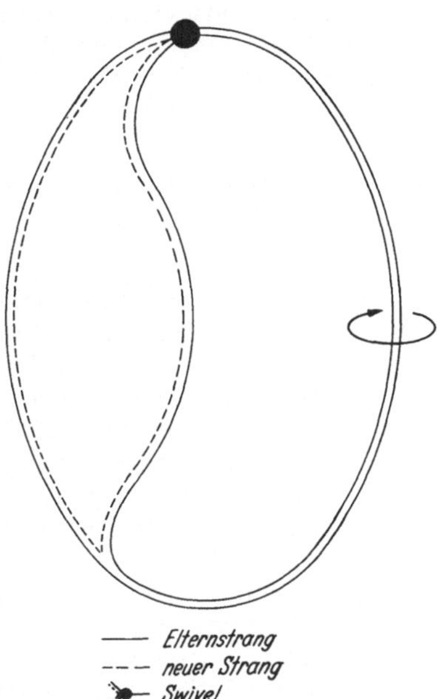

Abb. 2. Schema der Replikation des Bakterienchromosoms (nach CAIRNS [9], verändert)

chemischen Bindungen in der Kette aus, um den bei der Rotation auftretenden Torsionskräften standzuhalten.

Über den Ort der DNS-Synthese ist noch nicht sehr viel bekannt. Neuere Arbeiten deuten darauf hin, daß nascierende DNS mit der Zellwand verbunden ist [7]. In synchronisierten Bakterienkulturen enthält diese DNS-Fraktion immer die gleichen genetischen Marker [8].

Zur Frage der Regulation der DNS-Replikation [14] kann man sehr wenig Definitives sagen. DNS-Reduplikation ist nur nach vorheriger Proteinsynthese möglich. Ist sie jedoch einmal gestartet, läuft sie auch bei fehlender Proteinsynthese weiter, bis ein bestimmter Punkt auf der Genkarte, die Startstelle für die nächste Replikationsrunde, erreicht ist [16]. Nach dem heute üblichen Konzept entspricht dies einer positiven Regulation [10]. Ein spezifisches Initiatorprotein wirkt auf ein zugehöriges Operatorgen, den Replikator, und löst so den Start der DNS-Synthese aus.

In diesem Zusammenhang ist das Problem der zeitlichen Korrelation von DNS- und m-RNS-Synthese zu erwähnen. In Bakterien, bei denen im Gegensatz zu höheren Organismen ständig [23] DNS produziert wird, muß die m-RNS während der DNS-Replikationsphase synthetisiert werden. Hybridisationsexperimente mit pulsmarkierter DNS und RNS aus synchronen Kulturen [4] sprechen dafür, daß die Region der DNS-Replikation der bevorzugte Ort der m-RNS-Synthese ist.

Für eine einmalige Produktion einer bestimmten m-RNS während des Replikationscyclus sprechen Messungen von Enzymaktivitäten in synchronen Kulturen von Mikroorganismen [13]. Es sieht so aus, als ob jedes Enzym auch nach Induktion nur zu einer bestimmten Zeit synthetisiert werden kann.

Biochemische Aspekte der Replikation

Eines der Kernprobleme in der Nucleinsäureforschung ist die Frage nach dem biochemischen Mechanismus der DNS-Reduplikation. Man weiß heute, daß die Mononucleotide als Triphosphate aktiviert werden. Welche Polymerasen dann ihren Einbau in die wachsende Kette bewirken, ist noch immer ungeklärt.

Bei der Replikation nach dem Y-Modell haben die beiden wachsenden DNS-Stränge entgegengesetzte Polarität. Die Wachstumsstelle ist einmal das 5'-, zum anderen das 3'-Ende der Kette. Es ist bisher nicht bekannt, ob ein Enzym beide Funktionen erfüllt, oder ob ein ganzer Enzymkomplex dafür verantwortlich ist. Beide Möglichkeiten sind theoretisch denkbar.

Lange Zeit war man sicher, mit der E. coli-Polymerase [12] das richtige Replikationssystem gefunden zu haben. Dieses Enzym verlängert DNS-Ketten in vitro nur am 3'-Ende. Trotzdem ist es in der Lage, native zweisträngige DNS verschiedenster Herkunft um ein Vielfaches zu vermehren. Basenzusammensetzung und Dinucleotidmuster des Syntheseproduktes zeigen, daß die Basensequenz genau kopiert wird, und die neusynthetisierten Stränge gegenläufige Polarität haben. Schließlich konnte in einem Spezialfall ein semi-konservativer Replikationsmechanismus auch mit physikalischen Methoden nachgewiesen werden [31].

Diese Ergebnisse legten den Schluß nahe, daß in vivo die E. coli-Polymerase für die Replikation zuständig sei. Einige neuere Resultate lassen daran Zweifel aufkommen. Auch mit endonucleasefreien Polymerasefraktionen [22] ist es immer noch nicht gelungen, biologische Aktivität in Form von transformierender DNS durch Replikation zu vermehren [21]. Genauere physikalische Untersuchungen haben gezeigt, daß im Syntheseprodukt Verzweigungen auftreten [24]. Beides hat vermutlich seine Ursache darin, daß Kettenwachstum in vitro nur an einem Strang erfolgt und daher andauernd die Gefahr des Überwechselns des Enzyms auf den Gegenstrang gegeben ist. Hingegen vermag die E. coli-Polymerase sehr schnell und exakt Lücken in einer DNS-Kette nach der Sequenz des Komplementärstranges aufzufüllen (Reparation) [20]. Hierbei kann auch verlorene biologische Aktivität wiederhergestellt werden [*21*]. Man hat daher die Frage gestellt, ob die Hauptfunktion des Enzyms in der Zelle die Reparation und nicht die Replikation ist. Auf Grund des vorliegenden Materials kann diese Frage nicht beantwortet werden. Wahrscheinlich ist die E. coli-Polymerase, wenn überhaupt, dann nur als Teil eines ganzen Synthesesystems an der Replikation in vivo beteiligt.

Im weiteren Sinne könnte man natürlich Reparationsenzyme als Teil eines solchen Synthesekomplexes ansehen. Es gibt Enzyme, die durch Strahlungen oder chemische Agenzien verursachte Schäden im DNS-Doppelstrang reparieren können [*2, 9, 25*]. Die geschädigten Stellen werden dabei herausgeschnitten und die richtige Sequenz entsprechend dem intakten Komplementärstrang wieder hergestellt. Dieser Vorgang kann schon durch eine ungepaarte Base ausgelöst werden. Eine solche Fehlstelle ist aber auch das Charakteristikum eines Replikationsfehlers. Daher ist es durchaus möglich, daß die extrem hohe Genauigkeit der Replikation von Doppelstrang-DNS mit auf einem nachträglichen Reparationsprozeß beruht.

Andere Arten der Replikation

Bisher wurde die normale, d. h. die semi-konservative Reduplikation von Doppelstrang-DNS beschrieben. Das genetische Material wird jedoch nicht immer in Form einer Doppelstrang-DNS weitergegeben. So enthält das Tabakmosaikvirus, das klassische Objekt der Virusforschung, eine Einstrang-RNS als Träger der genetischen Information und der Bakteriophage $\Phi X 174$ eine Einstrang-DNS.

Nach unseren heutigen Kenntnissen erfolgt die Vermehrung dieser Einstrang-Nucleinsäuren immer über einen Doppelstrang, die sogenannte Replikative Form (RF). Diese besteht aus dem ursprünglichen infizierenden Viruselternstrang, dem (+)-Strang, und einem in der Wirtszelle neu gebildeten komplementären (−)-Strang.

An dieser replikativen Form sollen nun vorwiegend (+)-Stränge neu synthetisiert werden, d. h. nur die eine Hälfte des Doppelstranges soll reproduziert werden. Es sind heute zwei Mechanismen bekannt, die diesem Schema folgen:

1. Bei RNS-Viren liegt eine asymmetrische semi-konservative Vermehrung vor [*33*]. Am (−)-Strang der replikativen Form werden hauptsächlich (+)-Stränge synthetisiert. Der wachsende (+)-Strang verdrängt dabei kontinuierlich den älteren aus dem Synthesekomplex.

2. Bei DNS-Viren findet man eine konservative Vermehrung ähnlich der Synthese der m-RNS an Doppelstrang-DNS. An der replikativen Form, die als Einheit erhalten bleibt [*6*], werden die neuen (+)-Stränge synthetisiert.

Wahrscheinlich beruht der Unterschied in der Replikation der DNS- und der RNS-Phagen auf dem Unterschied in der Stabilität der Replikativen Formen. Im Gegensatz zu der RF der RNS-Phagen liegt die RF der DNS-Phagen als Doppelring vor.

Abschließend kann man fragen, warum überall in der Natur Doppelstrang-DNS Träger der genetischen Information ist und Einstrang-nucleinsäuren nur bei primitiven Organismen gefunden werden. Offensichtlich ist es wegen der Gefahr eines Informationsverlustes günstiger, wenn jede Information doppelt vorhanden ist, weil nur dann eine einmal verloren gegangene Information durch Reparatur zurückgewonnen werden kann. Dies ist umso wichtiger, je mehr Information übertragen werden muß. Aus diesen Gründen ist eine Weitergabe des genetischen Materials als Einstrang-Nucleinsäure wahrscheinlich nur auf sehr niedriger Stufe des Lebens möglich.

Summary

An outline of aspects of the reduplication of nucleic acids is given and some problems concerning this subject are discussed.

Literatur

[1] Bonhoeffer, F., and A. Gierer: J. mol. u. Biol. 7, 534 (1963).
[2] Boyce, R. P., and P. Howard-Flanders: Proc. nat. Acad. Sci. (Wash.) 51, 293 (1964).
[3] Cairns, J.: Cold Spr. Harb. Symp. quant. Biol. 28, 43 (1963).
[4] Cutler, R. G., and J. E. Evans: Abst. Biophys. Soc. Meeting, FG 2 (1965).
[5] Delbrück, M., and G. S. Stent: In „The Chemical Basis of Heredity". Baltimore: Johns Hopkins Press 1957.
[6] Denhardt, D. T., and R. L. Sinsheimer: J. molec. Biol. 12, 674 (1965).
[7] Ganesan, A. T., and J. Lederberg: Biochem. biophys. Res. Commun. 18, 824 (1965).
[8] — Persönliche Mitteilung.
[9] Hanawalt, P. C., and R. H. Haynes: Abst. Biophys. Soc. Meeting, FG 9 (1965).
[10] Jacob, F., S. Brenner, and F. Cuzin: Cold Spr. Harb. Symp. quant. Biol. 28, 329 (1963).
[11] —, and E. L. Wollman: In "Sexuality and the Genetics of Bacteria". New York: Academic Press 1961.
[12] Kornberg, A.: In "Enzymatic Synthesis of DNA". New York: John Wiley 1962.
[13] Kuempel, P. L., M. Masters, and A. B. Pardee: Biochem. biophys. Res. Commun. 18, 858 (1965).
[14] Lark, K. G.: In "Molecular Genetics". New York: Academic Press 1963.
[15] Levinthal, C., and H. R. Crane: Proc. nat. Acad. Sci. (Wash.) 42, 436 (1956).
[16] Maaløe, O.: Cold Spr. Harb. Symp. quant. Biol. 26, 45 (1961).
[17] Meselson, M., and F. Stahl: Proc. nat. Acad. Sci. (Wash.) 44, 671 (1958).
[18] —, and J. J. Weigle: Proc. nat. Acad. Sci. (Wash.) 47, 857 (1961).
[19] Nagata, T.: Proc. nat. Acad. Sci. (Wash.) 49, 551 (1963).
[20] Richardson, C. C., R. B. Inman, and A. Kornberg: J. molec. Biol. 9, 46 (1964).
[21] —, C. L. Schildkraut, H. V. Aposhian, A. Kornberg, W. Bodmer, and J. Lederberg: In "Sympos. on Informational Macromolecules", p. 13. New York: Academic Press (1963).
[22] — — — — J. biol. Chem. 239, 222 (1964).
[23] Schächter, M., M. W. Bentzon, and O. Maaløe: Nature (Lond.) 183, 1207 (1959).
[24] Schildkraut, C. L., C. C. Richardson, and A. Kornberg: J. molec. Biol. 9, 24 (1964).

[*25*] SETLOW, R. B., and W. L. CARRIER: Proc. nat. Acad. Sci. (Wash.) **51**, 227 (1964).
[*26*] SIMON, E. H.: J. molec. Biol. **3**, 101 (1961).
[*27*] SUEOKA, N.: Proc. nat. Acad. Sci. (Wash.) **46**, 83 (1960).
[*28*] —, and H. YOSHIKAWA: Cold Spr. Harb. Symp. quant. Biol. **28**, 47 (1963).
[*29*] TAYLOR, J. H.: In "Molecular Genetics". New York: Academic Press 1963.
[*30*] VIELMETTER, W., u. W. MESSER: In Vorbereitung.
[*31*] WAKE, R. G., and R. L. BALDWIN: J. molec. Biol. **5**, 201 (1962).
[*32*] WATSON, J. D., and F. H. G. CRICK: Nature (Lond.) **171**, 737, 964 (1953).
[*33*] WEISSMANN, C., P. BORST, R. H. BURDON, M. A. BILLETER, and S. OCHOA: Proc. nat. Acad. Sci. (Wash.) **51**, 890 (1964).

Diskussion

Vorsitz: *Beermann*

Bier: Da man von der Reparase nicht annehmen kann, daß sie gute und schlechte Mutationen von sich aus unterscheidet, würde das wohl bedeuten, daß Mutationen nur im Y (vgl. Abb. 2) auftreten.

Schaller: Man sollte vielleicht unterscheiden zwischen der Reparation von Mutationen, die durch äußere Einwirkungen (chemische Agentien, Strahlung) an vielen Stellen der DNS erfolgen können, und der Reparation von Replikationsfehlern. Während man sich im ersten Fall kaum vorstellen kann, daß eine Reparase zwischen dem „guten" und „schlechten" Teil eines Nucleotidpaares unterscheiden kann, ist dies bei der Replikation denkbar. Im Y könnte durchaus noch zwischen „gutem" Elternnucleotid und „schlechtem" neu synthetisiertem Nucleotid unterschieden werden, d. h. nur der neu synthetisierte Strang wird repariert. Voraussetzung dafür wäre natürlich, daß Reparation und Replikation gekoppelt sind.

Sachsenmaier: I have a question about the mechanism of unraveling the two parent strands. According to CAIRNS the energy for strand separation might be provided at the swivel and not at the Y. One would expect that in this case unraveling still proceeds after inhibiting DNA-replication with antimetabolites leading to the formation of large amounts of single stranded DNA.

Schaller: Nobody has yet been able to show this. On the other hand it is easy to imagine some kind of regulatory mechanism which would stop unwinding and DNA synthesis simultaneously. The new model of CAIRNS is based on experiments, where he induces breaks in the bacterial chromosome by ^{32}P decay and subsequently measures DNA synthesis. According to his model any break between swivel and Y would immediately stop DNA synthesis.

Klima: Wie soll man es verstehen, daß in beiden Richtungen synthetisiert wird? Ist die Richtung in diesen ringförmig geschlossenen Chromosomen durch die genetische Markierung vom Startpunkt aus festgelegt, oder bezieht man sich auf die Richtung (z. B. 5' — 3') innerhalb eines DNS-Stranges?

Schaller: On one hand we know that the two strands of a DNA double helix have opposite polarities, physical and biological data show on the other hand that both strands are growing (at the Y) in the same direction, that is chemically at different ends.

Klima: Yes. But you know that it sometimes proceeds on the circular chromosome counter clockwise and sometimes clock-wise?

Schaller: Concerning the direction of synthesis in a circular chromosome there is evidence from genetical work that in different strains of *E. coli* it may proceed in different directions. For *E. coli* the sequence of markers on the genetic map is known. By determining the sequence of multiplication of certain markers it is possible to state in what direction replication proceeds on the genetic map and thus along the DNA.

P. Sitte: Wenn im Doppelstrang die Information wesentlich sicherer gespeichert wird, warum wurden dann nicht die „Einstrangorganismen" in der Evolution ausgemerzt?

Schaller: In short chains of DNA the advantage of having the information in a single-stranded nucleic acid molecule (half the number of nucleotides for the same information, no strand separation during replication) outweighs the disadvantage of greater insecurity in replication. With greater amounts of information, double-strandedness becomes necessary, since the chance to loose information increases with the square of the chain length.

Taylor: With respect to the growth on the 5′ ends the evidence, I believe, is scant.

Beermann: How is your resolution on the radioautographic level, in respect of the number of nucleotides?

Schaller: 1% of the total DNA or so.

Beermann: Das Problem ist folgendes: Man beobachtet bei autoradiographischen Experimenten, daß das Wachstum in einer Richtung erfolgt, daß also das ganze Bakterienchromosom oder ein DNS-Molekül von einem Ende her repliziert wird. Biochemisch kennt man bis jetzt nur Enzyme, wie die E. coli-Polymerase, die vom 3′-Ende her synthetisieren können. Man müßte postulieren, daß gleichzeitig eine 5′-End-Vermehrung in der DNS stattfindet. Die findet man aber nicht.

Dabei muß man auch berücksichtigen, daß die Autoradiographie nur ein sehr stark vergröbertes Bild von dem gibt, was biochemisch passiert.

Schaller: Man könnte sich trotzdem ein Modell vorstellen, in dem die bekannte DNS-Polymerase für das Wachstum beider Stränge verantwortlich ist. Der eine Strang wird ohnehin von seinem 3′-Ende her repliziert, der Gegenstrang wächst in kurzen Abschnitten sozusagen rückläufig, ebenfalls vom 3′-Ende her. (Ähnlich arbeitet das Kornberg-Enzym in vitro, allerdings macht es dabei Fehler.) Diese Abschnitte würden in eine Größenordnung fallen, die autoradiographisch nicht mehr zu erfassen ist.

Beermann: Aside from the E. coli-enzyme, have any other DNA-polymerases been isolated which have been used in in vitro systems?

Schaller: The E. coli enzyme is the one most studied. Other polymerases have been isolated, e. g. T_2 polymerase from phage infected cells or the enzymes from calf thymus. All these enzymes elongate DNA at the 3′-end, all prefer single-stranded DNA as a template. Only the E. coli enzyme is able to separate the two strands of a native DNA during replication, but the resulting products differ from natural DNA.

Beermann: Obwohl also DNS-Replikation als Grundvorgang in allen Zellen sehr verbreitet ist, ist es bisher sehr schwierig, wenn nicht unmöglich gewesen, die Enzyme oder die Kombination der Enzyme zu finden, die das betreiben.

Wintersberger: The DNA-replication proceeds around the DNA-circle and ends when the new and the old ring are still sticking together at the starting point. CAIRNS believes that there is a protein which holds together these two rings, and that protein-synthesis is necessary to separate them.

Would it be possible that in order to start DNA-synthesis protein-synthesis is necessary primarily to separate these two rings which are held together at the starting point?

Schaller: This may be possible, but CAIRNS did not go into details about this. You can imagine bacteria in which replication has occured. If you would repress any protein-synthesis, you would end up in all bacteria with DNA which consists of the two completed rings. This has not been found.

Taylor: Do you know of experimental data that would indicate how much protein-synthesis is required to initiate a complete replication? Is it a very small amount or is it a relatively large amount of protein?

Schaller: It is a small amount of protein-synthesis.

Duspiva: Handelt es sich bei diesen Proteinen um basische Proteine, etwa um Histone, oder um Enzymproteine?

Schaller: Es ist nichts bekannt.

May I make some further comments on reparases. The fact that these enzymes can detect the mis-pairing of a single base has been shown by Dr. DÖRFLER (Stanford University). By hybridising the DNA of different strains of phage he obtained duplex DNA in one strand of which one base out of 10^5 was altered. If competent *E. coli* cells were infected with this DNA the information was expressed only according to expectation, if the cellular repair system was kept busy by UV irradiation prior to infection.

The Duplication of Chromosomes

By

J. Herbert Taylor*, Tallahassee

With 7 Figures

Introduction

Chromosome duplication first of all requires the replication of the DNA. It is now clear that the replication of most if not all of the chromosomal DNA follows the scheme based on precise base pairing as originally proposed by Watson and Crick [50]. The pattern of replication was first clearly indicated by experiments showing the semiconservative distribution of chromosomal DNA labeled with H^3-thymidine (Taylor et al. [48]) and later verified by density gradient centrifugation studies with *Escherichia coli* DNA labeled with the heavy isotope of nitrogen (Meselson and Stahl [24]). Similar results were obtained with phage T 7 and λ [23, 25] and with the alga, *Chlamydomonas* (Sueoka [38]). Subsequently, semiconservative distribution of DNA in mammalian cells was demonstrated by the use of bromouracil as a density label (Djordjevic and Szybalski [7]; Simon [36]; Chun and Littlefield [6]). Recently semiconservative replication of DNA has been demonstrated in higher plant cells with N^{15} as a density label (Filner [8]) and also with bromouracil as a density label (Haut and Taylor [13]).

The final proof for the Watson-Crick scheme of replication came from the nearest neighbor analysis of enzymatically synthesized DNA in cell-free systems (Josse et al. [17]). Although there have been doubts raised at various times concerning the complete separation of the two original polynucleotide chains of the DNA duplex during replication (e.g., Cavalieri and Rosenberg [5]), the data upon which these arguments were based probably have alternative explanations. The chain separation and reannealing experiments (Marmur et al. [22]) and the analyses of changes in density of hybrid and fully substituted bromouracil DNA with changes in pH (Baldwin and Shooter [1]) provide enough evidence for the existence of hybrid duplexes to warrant the assumption that complete separation of the two chains is a regular feature of replication. This feature, or indeed any exact copying of long helically coiled templates,

* This work was supported in part by a contract between the Division of Biology and Medicine, U.S. Atomic Energy Commission and the Florida State University. Data presented in Fig. 4—7 are from a doctoral thesis presented to Columbia University by William F. Haut, but the work was carried out under the above contract at the Florida State University. Further details of this work will be published elsewhere [13].

leads to problems in unwinding and sorting of strands which have not yet been solved, although there have been a variety of hypotheses put forward, particularly with respect to unwinding mechanisms. The problem of sorting and the prevention of entanglement will be considered in more detail later.

In addition to DNA, chromosomes contain a variety of proteins which must be synthesized at some stage in each cell cycle. The question which has never been satisfactorily answered is how many of these are essential to the structural integrity of chromosomes. One is tempted to generalize from the behaviour of lampbrush chromosome in amphibians where it has been clearly demonstrated that only by cutting the DNA chains is the continuity of the chromosome destroyed [4, 12, 26]. Treatments with deoxyribonuclease lead to rapid breakage of the loops as well as the chromosomal axis, while ribonuclease and various proteases erode away the matrix without affecting the linear integrity of the chromosomes.

However, at many stages the shape and compactness of chromosomes are probably maintained by proteins. Such structural elements are particularly evident in the prophase stages of meiosis, where a structure referred to as the synaptinemal complex (Moses [28, 29]) maintains a rather rigid axis which may be twisted into a helix. In addition the histones may be important in the supercoiling of the DNA. The recent experiments reported by Prensky and Smith [32] raise again the question of the permanence of the proteinaceous chromosomal components (see Prescott [33]). In Prensky and Smith's experiments arginine-H[3] incorporated into chromosomal proteins remained a part of the chromosome through the first division following incorporation, but almost completely disappeared from the chromosomes after one replication. Since most of the arginine was probably incorporated into histones, the question is raised concerning the disposition of these proteins during replication. Is the DNA stripped of its histone prior to or during replication? If histones fill one of the grooves in DNA, it is difficult to see how the duplex could unwind without displacement of the histone. If there is a displacement, how does the DNA form a complex with the same type of histone molecule following replication? Although there have been various suggestions, at the present time no experimental evidence is available to satisfactorily answer these questions.

The Sorting of DNA Strands during Chromosome Duplication

It is difficult if not impossible to visualize the organization and sorting of the long strands of DNA in chromosomes without some structural proteins to help maintain shape and orientation of the templates and products. Indeed all of our observations and experience indicate that structural components do exist. Light microscope pictures have for many years revealed a double fibrous component in the anaphase chromosomes of many plant and animal cells. This was once cited as evidence that chromosome reproduction occurred during or immediately following division stages. At other times it has been cited as evidence that chromosomes are two-stranded or multistranded. Recently Trosko and Wolff [49] treated isolated nuclei and chromosomes of *Vicia faba* with several enzymes in dilute phosphate buffer after a brief fixation in two percent formalin. None of the solutions produced any very striking changes except the one containing trypsin. It caused considerable swelling, and when the preparations were dried and stained by the Feulgen reaction the chromosomes were seen to have elongated and uncoiled to a considerable extent. Each chromatid consisted of two axes on which the DNA was precipitated. The DNA also formed fibrous extensions and connections between the axes which were separated by a space of 1—2 microns. In a few places the axes appeared double or bifurcated and on this basis the authors suggested that the chromosome is multistranded. Such a conclusion is obviously unjustified since many kinds of precipitation artifacts can be produced by the treatments. However, the regularity of the two axes emphasize the significance of the earlier observations made by many cytologists which indicated that chromatids are bipartite. The question to be answered concerns the structural significance of these axial elements. We have suggested several times in the past that they might be the basis for the segregation and sorting of the two polynucleotide chains of the DNA duplex during replication and chromosomal division (Taylor [43, 46]). The subunits (half-chromatids) are not seen in chromatids either in the living state or in well fixed preparations. Half-chromatids have been demonstrated in preparations which have been treated with fixatives at a low pH, or with low ionic strength buffers and distilled water, frequently combined with heat. The half-chromatids are usually relationally coiled several times and it is only in those cases where a swelling has been allowed before precipitation of proteins and nucleic acids that the subunits are well separated. It is clear that proteases are not a necessary treatment to demonstrate the separation. Several agents which cause swelling will allow their demonstration. We and various others have used distilled water to which is added a trace of potassium cyanide. One of the best agents we have tried was 25—30 percent formamide in water (unpublished observations), but none of these agents alone produced the amount of swelling which characterize the trypsin treated cells and the uncoiling of the half-chromatids did not proceed far enough to allow as good separation as that obtained by Trosko and Wolff [49].

In all such preparations the distance of separation of the axial units is clearly limited by a fibrous component which may be DNA. Certainly DNA strands extend between the units. The state of the DNA is uncertain in all such preparations for no attempt was made to control degradation by intracellular enzymes. In addition, trypsin is nearly always contaminated with nucleases; therefore, the state of preservation of the DNA in the preparations is unknown. Certainly the dilute buffers, or distilled water followed by drying or acid treatments all promote collapse of the secondary structure of DNA and some unwinding of the polynucleotide chains would be expected. Indeed, if the treatments which reveal half-chromatids have anything in common, it is most likely the denaturation of DNA along with the cleavage of an axial component. The structural properties of the axial component suggest that proteins make up an important part of it. Since little, if any, change occurs by the loss of histones, the axis is probably composed of non-histone proteins. The failure to demonstrate with the electron microscope any structural component in anaphase chromosomes equivalent to the half-chromatids is puzzling and probably indicates that the half-chromatids are in part an artifact as suggested above. However, the regularity with which they can be demonstrated indicates that the artifacts are significant.

Patel and Wang [30] have recently demonstrated a residue of calf thymus nuclei capable of incorporating tryptophan into non-histone protein. The incorporation is inhibited by DNase and stimulated by limited trypsin digestion. Electron microscopy of the isolated complex reveals spherical particles, 100—300 Å in diameter, attached to DNA strands. Perhaps these are the structural proteins or the organelles which are involved in their synthesis.

At least three possible interpretations of the half-chromatids will be mentioned. (1). Each half-chromatid represents a bundle of longitudinally oriented fibrils containing DNA. This is the interpretation favored by Trosko and Wolff [49] and is in line with the concept that a chromatid is multistranded as visualized by Ris [34], Steffensen [37], and Kaufmann et al. [18]. (2). Each half-chromatid is a folded or coiled DNA duplex. This concept is in line with suggestions made by LaCour and Pelc [21] and by Peacock [31]. (3). The half-chromatids indicate the plane of cleavage of a structural component which serves to untwist and sort the chains of a single DNA duplex which is folded and coiled into a chromatid. In some models the single duplex has linkers at intervals. Such a model has been described by Taylor [42, 46] and is in line with concepts of chromatid structure suggested by Schwartz [35], Freese [10], Gall [12] and Miller [26].

There is not enough evidence to decide unequivocally among these models, but the third possibility is favored by the evidence that semiconservative distribution of DNA occurs both at the level of the DNA duplex and the chromatid. The simplest assumption then is that the chromatid is a single duplex. The problems involved in the evolution of mechanisms of unwinding, sorting and rejoining of broken chromosomes appear simpler with this model than with a chromosome which has two or many DNA duplexes. The evidence that the two subunits of a chromatid have different polarities (Taylor [41, 42]) is most simply explained on the assumption that each chromatid consists of one duplex. Since a difference in polarity is a known characteristic of the two chains of a DNA duplex, this characteristic of a chromatid would naturally follow. Alternative explanations can be found for the above observations, but it is difficult to see how the kinetic studies of the breakage of lampbrush chromosomes can be explained with a multistranded model. Gall [12] found that loops, which are generally agreed to be segments of chromatids, break at the rate expected of a single DNA duplex. Perhaps of more significance the chromosomal axis which consists of two chromatids breaks at the rate predicted for four polynucleotide chains. MILLER's [26] electron micrographs of loops stripped of most of the matrix reveal a fiber which can consist of only one or two helices.

However, the hypothesis that each chromatid is composed of two DNA duplexes is interesting because it would explain a number of otherwise puzzling observations. The first is the occurrence of occasional mistakes in the regular semiconservative distribution of DNA. These are seen as regions with label in both chromatids at the second metaphase after labeling with H³-thymidine [41, 21, 31]. It might also explain the apparent doubleness of chromatids at the light microscope level if each duplex were folded and coiled separately so that the two could be separated by certain treatments during or preceding fixation. It might also explain the high spontaneous mutation rate observed in higher cells (SZYBALSKI [40]). If the two helices should carry different but homologous, closely linked mutant loci, perhaps in the same cistron, they could give rise to "mutants" by recombination between the two parallel duplexes.

Since the question of strandedness of chromosomes cannot be ascertained now, it would be useful to find a model of chromosome duplication which would be adaptable to either a single, double or multistranded chromatid. Such a model is possible. It is a modification of the models presented previously [46] but rather significant changes are made which are based on recent data with respect to DNA replication by density gradient centrifugation and the possibility of the occurrence of repeating sequence deoxyribonucleotide polymers. The model will be described in a way to account for the known properties of chromosomes and DNA, after which some additional supporting data will be cited.

A Model for DNA Replication in Chromosomes

The model will be based on a chromosome with a single duplex of DNA for simplicity, although, as pointed out above, the model can be adapted to either a two-stranded or a multistranded chromosome with slight modifications. Each arm of a chromosome at anaphase, telophase or the early stages of interphase will be assumed to consist of a single uninterrupted DNA duplex. As will be seen, no non-nucleotide linkers are required as assumed in most previous models, but the model could easily accommodate them should they be demonstrated. The point is that it does not require any if it is assumed that polynucleotides can be covalently linked when one chain is broken. All of the experiments which indicate repair by excision of nucleotides demand such a mechanism. A diagrammatic representation of a portion of such a duplex fully extended is shown in Fig. 1a. At intervals along the duplex are paired protein units coupled together in what will be the cleavage plane at the next replication. It will be noted that the polynucleotide chains pass through the protein units uninterrupted, but the chains are assumed to be firmly coupled, one to each of the paired protein units, which will be designated *replication guides* (RG). These pairs of RG's are assumed to be

stacked, at least during some stage of division, into two columns (Fig. 2)
with the DNA duplex forming loops on alternate sides along the axis.
A segment of the extended structure is redrawn in Fig. 1b after conver-

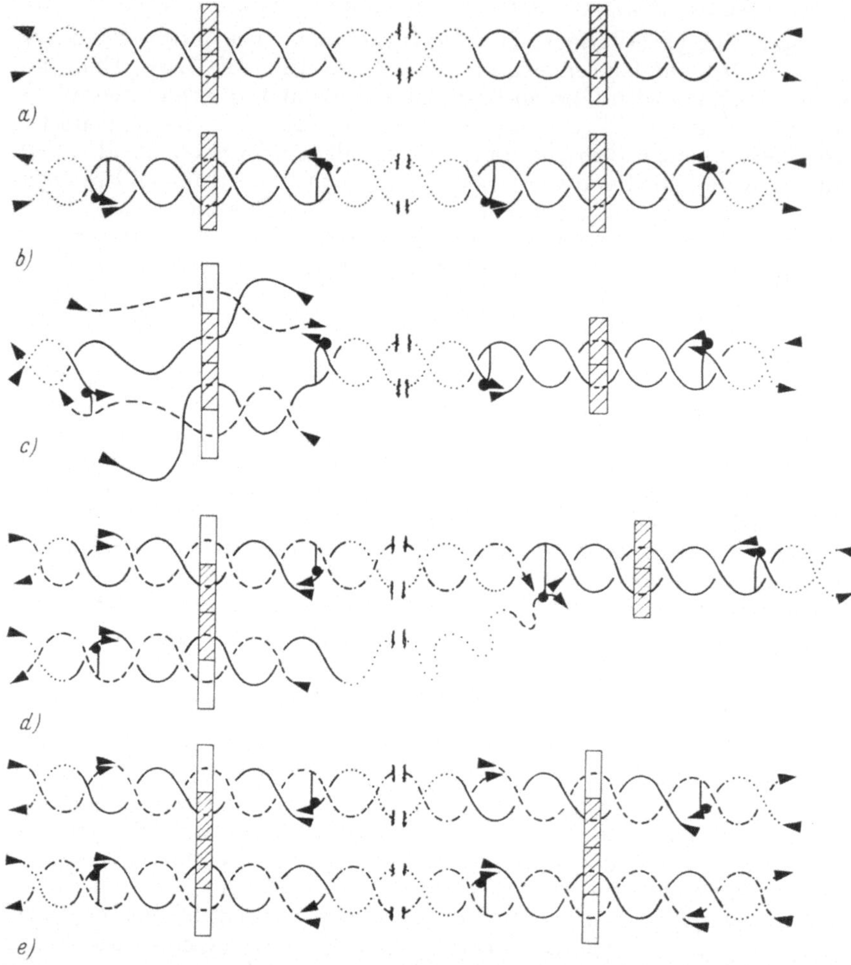

Fig. 1. Diagramatic representation of the mode of replication of chromosomal DNA. Only a
small segment of a chromosome is shown representing one replicon bounded on both ends
by a pair of replication guides (RG's). The solid lines (———) represent segments of the
original DNA chains composed of repeating sequence polymer (primer units). New primer
units are shown in dashed lines (--------). The original chains of informational DNA in each
replicon are shown as dotted lines (······) and their new complementary DNA chains are
shown with a dot-dash (— · — · — · — · —). Only a portion of each segment of informational
DNA is shown as indicated by the break in the middle of each replicon. The nucleases are
shown as spherical structures with a cross-linking arm. The arrows represent 3'OH ends of
DNA chains and therefore indicate polarity. a. Segment before change to the template state.
b. Segment after change to the template state. c. Attachment of the priming units. d. Replication is complete in the segment shown and the next step would be the separation of the
RG's and the rejoining of the interrupted chains

sion to the template state. This change, which is assumed to precede replication, is an enzymatic opening of one chain on each side of every RG. The rate and sequence of this event may, of course, vary with the type of cell under consideration. The proximal segments of polynucleotide chains on either side of the RG's will be assumed to consist largely of a repeating sequence polymer (shown by a solid line), although these regions may be assumed to code certain information for the control of replication and possibly transcription of RNA. The locus which includes

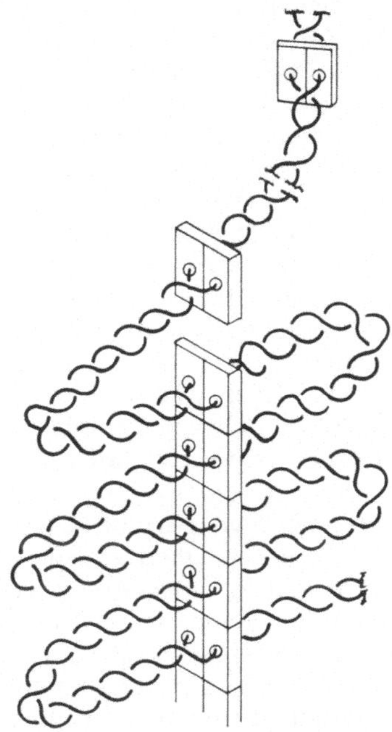

Fig. 2. A diagram of a segment of the chromosome model described in the text. The model consists of a single DNA duplex with pairs of proteinaceous replication guides (RG's) at intervals which separate the replicons. The RG's could be either permanently united end to end to form columns, or temporarily coupled. However, it should be noted that the RG's must have a polarity if they are not permanently stacked. In oder to maintain semiconservative segragation of strands, all of the RG's attached to a single strand would stack in the same column if they had interlocking polarized groups. The illustrated lengths of the replicons are not meant to indicate their actual relative lengths

a pair of RG's and the proximally located polymer on either side will be designated a *replicator locus*. A segment of DNA which is enclosed by two RG's will be designated a *replicon*, although it is realized that the term has been used to describe what may prove to be a different type of unit in *Escherichia coli* (JACOB et al. [16]). However, the term has also been previously used to descrive a unit of replication in higher cells (TAYLOR [47]). Each RG with its attached single chain of repeating sequence polymer will be designated a *primer unit*.

The events in replication will be divided into four stages: (1) change to a template state, (2) synthesis and attachment of the primer units,

(3) replication of the segments of DNA between primer units, and (4) cleavage and separation of the chromosomal components into chromatids. The order given need not necessarily indicate the sequence of events, but provides a framework for discussion.

The change to a template state is assumed to occur, as indicated above, by the action of specific nucleases. The nucleases might be assumed to recognize the RG's and to open a phosphodiester linkage a specified distance on each side of a pair of RG's. They would cleave one polynucleotide chain to reveal two ends, which might be a $5'PO_4$ end pointing away from the RG and a $3'OH$ end on the other side of the break. At the same time we will assume that the nuclease crosslinks the two strands distal to the break with respect to its RG (Fig. 1b). The crosslinking may be assumed to be a property of the nuclease molecule and no covalent linkage of the two polynucleotide chains need be involved, although it is by no means necessary to exclude covalent bonding. With completion of transition to a primer state, a chromosome would consist of a series of replicons separable by melting out the hydrogen bonds in the replicator loci and removal or separation of the RG's. It may be noted that these replicons can form rings by pairing of the repeating sequence polymers at their ends. They resemble phage λ rings (Hershey and Burgi [14]). Each would be free to rotate and unwind separately while still a part of a chromosome because of the single chain regions produced by opening opposite chains on each side of a pair of RG's.

The synthesis of *primer units* may occur at any time prior to replication. The details need not be specified except to point out that the unprimed synthesis of repeating sequence polymers by the *E. coli* polymerase might give a clue. Synthesis of *primer units* might occur as replication proceeds or in some cases a pool of primer units might be stored in the cell. Eggs and sperms of species in which rapid DNA replication follows fertilization might be expected to have a pool of the *primer units*. Attachment of the primer units would presumably be guided by base pairing of the repeating sequence polymer with the segments of the replicator loci in chromosomes in the template state. Alternatively, the primer units might be synthesized *in situ* by addition of monomers to the $3'OH$ ends of chromosomes in the template state. However, to be consistent for sorting in a semi-conservative fashion, such pieces would require excision at or near their starting points before replication could proceed.

The third stage of replication is initiated when a primer unit is in place and its terminus extends beyond the locus opened by the nuclease. At this point the assumption may be made that base pairing begins between the primer unit and the DNA chain with a $3'OH$ end (Fig. 1c). The crosslink is switched from the original strand to the new, growing strand attached to the new RG. Replication then proceeds to the next replicator locus (Fig. 1d). The complementary chain of DNA which is now being copied is temporarily unpaired. Rotation and unwinding of DNA is facilitated by the single chain regions as pointed out above.

The unpaired complementary chain would be copied from the opposite end in a similar manner. Eventually all replicons would be complete, either by sequential replication or by random or regulated replication of some type of chromosomal segment (Fig. 1e). The control that prevents replication two times in one cell cycle could be the RG units. When a new one was in place, no additional one could occupy the site until cleavage of the two original RG's had occurred.

The last stage would be the separation of the four RG's into pairs and the closure of the interrupted chains. Normally one new and one old RG would segregate together. This would involve unwindung the original DNA chains in the replicator loci as separation occurs if unwinding had not already occurred. Probably unwinding would occur when the new RG with its piece of repeating sequence polymer attached was added to the chromosome. Then the new and old chains in each replicator locus would coil into a duplex. No entanglement of the chains between RG would be possible if the proposed scheme operates. Any twists adjacent to RG's will be resolved by the separation and base pairing of a new and old chain of DNA in the replicator loci. The final act is the coupling of the broken chains. If segregation has separated a new and old RG, the 3'OH ends will rejoin with the same 5'PO$_4$ ends from which they were severed at the beginning of replication. However, should there be a misdivision, i. e., two new RG's segregate from two old ones, the result would be a sister chromatid exchange. If this misdivision of RG's should occur over a considerable length of a chromatid, the result would be isolabeling of the type reported in *Bellevalia* [41] and *Vicia* [31]. No peculiar segregation could be observed at the first division after labeling with H[3]-thymidine, but at the second division the label would be equally distributed to the two new chromatids because of the misdivision which occurred after the previous replication. The effect would be that of multiple sister chromatid exchanges — one at each replicator locus. The division into chromatids by the segregation and separation of the RG's would complete chromosome reproduction except for the mitotic distribution of chromatids to separate nuclei. The changes involved in this last stage might indeed be delayed until prophase of mitosis.

Although there may be several necessary steps which have been omitted in the model described above, it is sufficiently precise to test in various ways with experimental evidence. Furthermore, since the information available for formulating a model at this stage in our knowledge is limited, other modifications will likely be required. In any case, the principal features of the model are precise enough to allow it to be discussed in a critical fashion.

The Template State of DNA

Is a change required for converting DNA into a template state? Experimental evidence cannot yet provide the final answer, but the most highly purified polymerases do not utilize native DNA as a template or, if so, the initial reaction is quite slow. The calf thymus polymerase isolated by Bollum [2] is reported to be inactive when native DNA is used as primer. Heating converts a variety of DNA's into primer and template. Limited reaction with DNase also converts DNA to a primer, but less efficiently than heating. The one exceptional native DNA primer was salmon sperm DNA which primed before heating. Bollum attributed this to partial denaturation of a commercial product. However, the exception could be significant. It is quite possible that some sperm DNA's are largely or entirely in the template state. In addition, the finding of repeating sequence polymers (Sueoka and Cheng [39]) may indicate that the primer units are present as a pool in some types of cells. Of course, deproteinized DNA would have the RG's either lost or largely denatured and would not be expected to remain in the native template state. The other indication that sperm DNA may exist in a primer state is the formation of rings of various size in isolated boar sperm DNA (Hotta and Bassel [15]). Although these rings failed to open by the same treatment that opens phage λ DNA rings [14], it should not necessarily be assumed that they are covalently linked. If the overlap at the ends were rather long and composed of a high GC repeating sequence polymer, they would not be opened with heating even to higher temperatures than those employed. In summary, the evidence is meager and inconclusive, but the chance that the template state exists in some types of DNA makes the search for it worthwhile.

Replication of Chromosomal DNA by Segments

Autoradiographic studies indicate that large chromosomes have many growing points. For example, *Vicia* and *Bellevalia* chromosomes may be labeled along the whole length by the shortest pulse necessary to get enough label to detect (unpublished observation). Likewise, the X-chromosome of the hamster and human can be labeled along the whole length in a few minutes in spite of the fact that the entire replication requires several hours (Taylor [45]; Morishima et al. [27]). The clearest evidence for replication of short segments is the observation that single bands of dipteran salivary gland chromosomes may initiate replication out of phase with adjacent ones (Keyl and Pelling [19]; Gabrusewycz-Garcia [11]).

Evidence of another type has been obtained in studies of DNA replication by the use of the density label, bromouracil, in *Vicia faba* and Chinese hamster cells in our laboratory.

When cells are grown in a solution with the nucleoside, 5-bromo-deoxyuridine-C^{14} (BUdR) along with an inhibitor of thymidylate synthetase, 5-fluorodeoxyuridine (FUdR) or aminopterin, the heavy nucleoside is substituted for thymidine. Hybrid molecules of DNA, with one heavy strand and one light, unsubstituted strand, are formed as expected. Fig. 3 shows the profile obtained by centrifuging 20 μg of DNA isolated from Chinese hamster cells after five hours growth in BUdR-C^{14}. The DNA was centrifuged in a solution of cesium chloride with tris buffer, pH 7.8, and a starting density of 1.755 gm/ml for 48 hours at 37,000 rpm in the Spinco S 39 rotor. The solution was dripped from the bottom of the tubes and fractions collected with a decreasing density toward the top of the tube. Regular unsubstituted DNA banded with a peak corresponding to fraction 58 which had a density of about 1.710 gm/ml.

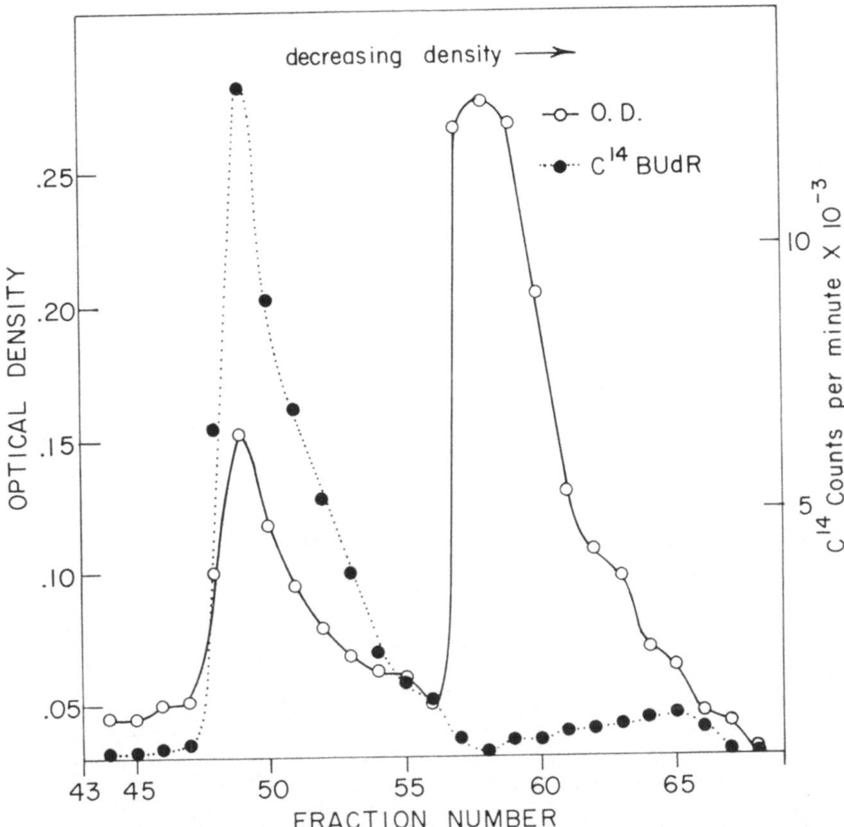

Fig. 3. Density profile in CsCl of Chinese hamster DNA isolated from cells grown 5 hours in BUdR-C^{14} (further details in text). Carbon-14 ($\cdots \bullet \cdots \bullet \cdots \bullet \cdots$) shows the position of the BUdR hybrid DNA and optical density (— o— o— o—) shows the position of the normal unsubstituted DNA as well as the smaller amount of hybrid

The hybrid DNA showed a peak at fraction 48 with a density of about 1.750. The carbon-14 counts, which extend into the band of unsubstituted DNA, are largely due to aggregation or the failure of all molecules to pass through the viscous barrier formed by the band of unsubstituted DNA. In any case the results are in accord with the semiconservative replication of DNA.

The profile of a similar sample of DNA isolated from *Vicia* root cells gave quite a different picture (Fig. 4). The roots were grown for

8 hours in thymidine-H³ to label a portion of the DNA and then after
3¹/₂ hours' growth in unlabeled thymidine (TdR), they were transferred
to a solution with BUdR-C¹⁴ (10⁻⁴ M) and aminopterin (10⁻⁶ M) for
10 hours. Hybrid DNA shows a peak in fraction 50 (density, 1.749),
while the normal DNA has a peak at fraction 63 (density, 1.702). There
is tritium in the hybrid DNA as would be expected if the H³-labeled

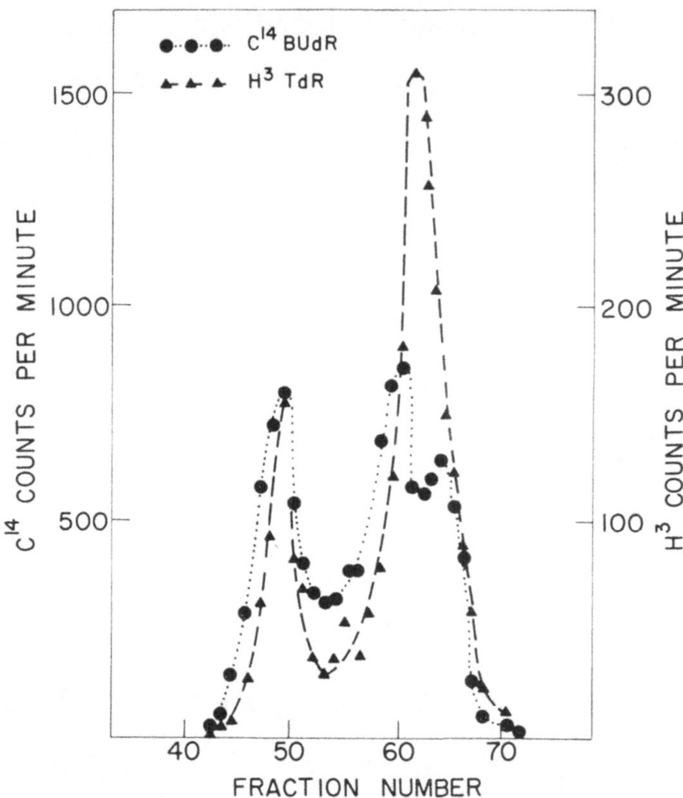

Fig. 4. Density profile in CsCl of *Vicia* DNA isolated after roots were grown for 8 hours in
H³-thymidine, 3¹/₂ hours in unlabeled thymidine and 10 hours in BUdR-C¹⁴. The H³
(- ▲ - - - ▲ - - - ▲ -) marks the position of the normal unsubstituted DNA. There is C¹⁴-
BUdR in a hybrid band with a peak at fraction 50. In addition, there are two bands of DNA
with a limited substitution of C¹⁴- BUdR, which are shifted slightly to the left under the
unsubstituted peak

DNA were replicating with substitution of BUdR in the new strands.
However, the unexpected result is the appearance of a large amount
of C¹⁴-labeled DNA in two bands, one of which is only slightly heavier
than normal DNA and the other banding on the light side of the major
band. When the polynucleotide chains were separated by centrifuging
another sample of the same DNA in cesium chloride at pH 12.0, the
density of all C¹⁴-BUdR labeled DNA increased more than that of the

normal tritium-labeled DNA (Fig. 5). Optical density measurements showed that the major part of the DNA was in the band labeled with tritium as expected. A density shift of the fully substituted DNA is expected since it separates from the light strands at the high pH. The

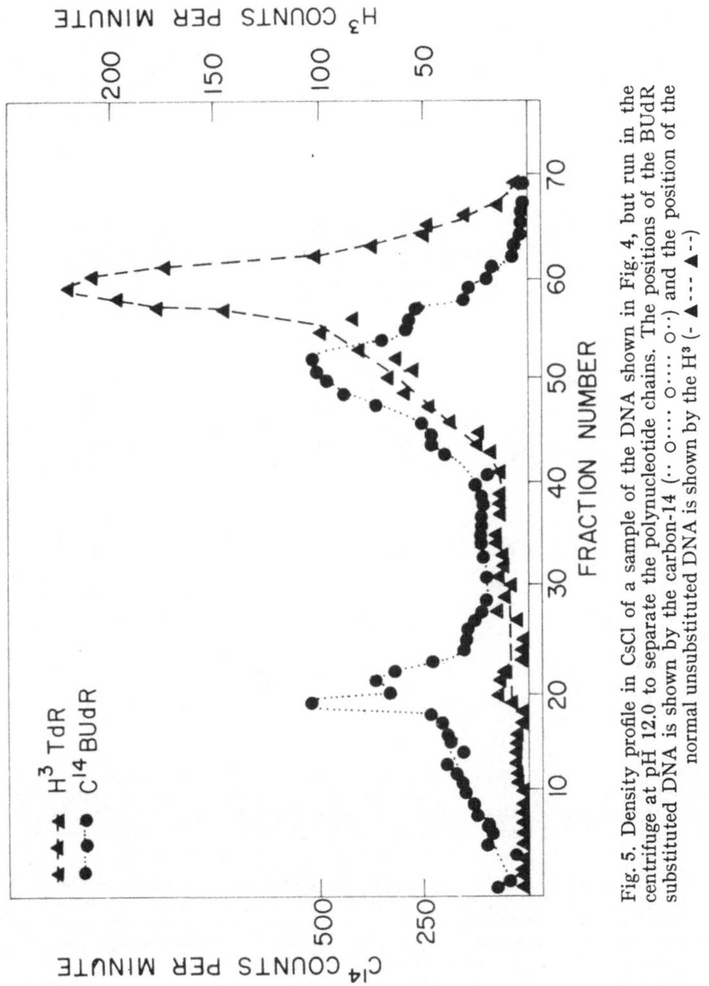

Fig. 5. Density profile in CsCl of a sample of the DNA shown in Fig. 4, but run in the centrifuge at pH 12.0 to separate the polynucleotide chains. The positions of the BUdR substituted DNA is shown by the carbon-14 (·· o···· o···· o··) and the position of the normal unsubstituted DNA is shown by the H³ (- ▲ --- ▲--)

shift in density of the slightly heavy DNA also indicates that these BUdR-labeled strands are separating from light strands. In other words, the BUdR in this DNA was probably in only one of the two chains and not randomly distributed through the molecules of DNA. Most of the tritium-labeled DNA in the hybrid shifted back to the unsubstituted band as expected, but a shoulder on the heavy side of the normal DNA, which is under the slightly substituted DNA, indicates that this DNA contains both tritium and BUdR-C^{14}. The amount of its density shift indicates that it is only partially substituted with BUdR, perhaps about 10 percent. This peculiar DNA always forms when BUdR is present, although it has not been possible to demonstrate two density species in the regular unsubstituted DNA. If roots are grown in BUdR without the

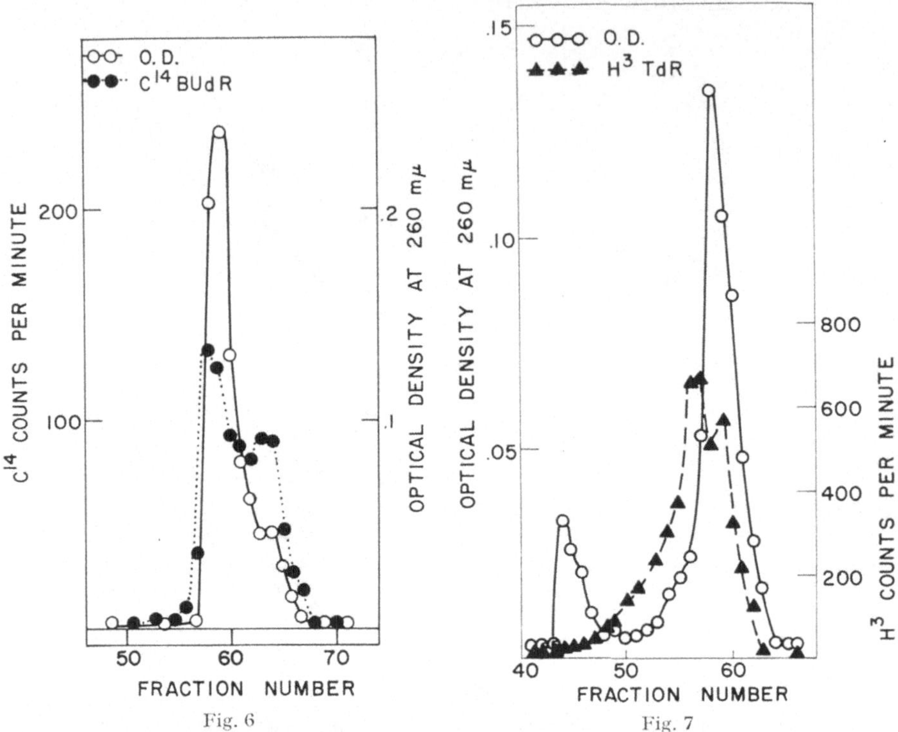

Fig. 6 Fig. 7

Fig. 6. Density profile in CsCl of the DNA obtained from *Vicia* roots grown for 5 hours in BUdR-C^{14} (10^{-4} M) and FUdR (10^{-7} M). Since this amount of FUdR is insufficient to completely block thymidylate synthetase, no fully substituted hybrid BUdR-DNA is formed, but the two slightly labeled bands are observed. BUdR appears to be more readily utilized in the synthesis of these two classes of DNA

Fig. 7. Density profile in CsCl of the DNA obtained from *Vicia* roots grown for 9 hours in 10^{-6} M FUdR and 10^{-5} M BUdR, followed by one hour in a solution without nucleosides, and then 6 hours in 10^{-5} M TdR-H^3 to label the newly synthesized DNA. Note that all of this new, labeled DNA contains a significant amount of BUdR since its density is greater than the majority of the DNA

inhibitor for thymidylate synthetase, no fully substituted hybrid is formed but two density species labeled with BUdR-C^{14} show up on the heavy and light side of the regular band of DNA (Fig. 6). Since these two bands now have only a partial substitution in any part of the molecule, a density shift is hardly if at all detectable.

Another characteristic of the slightly heavy DNA is shown in Fig. 7. If cells are grown for a number of hours in BUdR, with an inhibitor of thymidylate synthetase to force substitution, and then transferred to thymidine-H³ to label any new DNA formed after the transfer, all of this tritum-labeled DNA falls into the bands of slightly heavy DNA, which can be shown to contain BUdR when the density label also contains a radioactive isotope. Very little, if any, DNA of regular density is formed during the next 6 to 8 hours even though the roots continue to grow and synthesize DNA at a measurable rate.

The puzzling observations of the partially substituted DNA's have led to a consideration of possible repair mechanism and to hidden pools of precursor which are utilized for some hours after removal of the roots from BUdR. However, neither of these explanations account for all of the facts. A repair mechanism would not be likely to substitute the BUdR in a small fraction of the DNA which apparently does not include the hybrid DNA. If the hybrid DNA is repaired, it is changed from a hybrid molecule to one with about 10 percent substitution in one step, for there are few molecules of intermediate density compared to those slightly substituted ones.

The hypothesis that a precursor pool of nucleotides builds up does not appear to be an adequate explanation for the rapid appearance of the slightly substituted DNA at the same time that regular hybrid is forming unless a number of unlikely assumptions are made. However, the hypothesis of a hidden precursor pool makes sense if the precursor which builds up in the presence of BUdR is a polymer which is linked to the original (old) DNA chains. Upon incubation in thymidine, these segments of BUdR-containing polymer appear to be transferred to the newly labeled chains. It may be clear now that we wish to suggest that these segments of BUdR substituted DNA may be the hypothetical primer units in *Vicia*. One additional assumption is needed to explain the observations. The BUdR appears to be readily utilized by the cells to make the primer units, but is very slowly utilized in replication of the informational DNA, i. e., the pieces between replicator loci. This is in line with the observation that replication is very slow when only BUdR is available in place of thymidine. Normally, the cells do not appear to store a large pool of the primer units. Therefore, when BUdR is added to roots previously grown in thymidine-H³, rather little of the H³-labeled DNA is attached to the new BUdR DNA (Fig. 5) except by hybrid duplex formation. However, when BUdR has been substituted for thymidine for a number of hours, the pool of primer units increases so that only slightly heavy DNA is formed during the next 6—8 hours.

If the interpretation given above is correct, we can estimate the size of the primer units. Since the DNA is increased in density to about 10 percent of that of the fully substituted hybrid, the pieces should be about one-tenth the size of the DNA particles in the gradient. Although the average molecular weight of the DNA was not determined, it must have been as high as $4-6 \cdot 10^6$ to form narrow bands in cesium chloride. This means that the single chain segments of primer could have a molecular weight of the order of $2-3 \cdot 10^5$.

The sequence of events, if correctly interpreted, might indicate that primer units are produced by growth of free ends formed in the original duplexes by conversion to the template state. These would be excised later by cutting at or near the point of origin and then used as primer units for initiating the growth of new chains.

Replication of the DNA between Primer Units

There is little to add to this concept except to mention again that the evidence indicates that most replication is semi-conservative and a sequence complementary to the template is produced. The copying of one chain at a time is indicated by the specificity of the DNA polymerase, at least in *Escherichia coli*, where the most critical studies have been carried out (Kornberg [20]). The studies which indicate that both chains are copied in the same direction (Cairns [3]) cannot resolve what happens in the region near the growing Y. The autoradiographic evidence certainly does not exclude the alternate copying of short segments even in *E. coli*. However, the model suggested here was designed only for chromosomes of higher cells. There are some studies which indicate a denatured DNA exists during replication, but the evidence is not sufficient to decide whether one strand remains unpaired for a short time.

Cleavage and Separation of the Chromosomal Components into Chromatids

The evidence concerning the nature and timing of this event is meager. The chromatids are usually visibly separate when first observed in mitotic prophase. However, the sister chromatids are firmly attached along one edge until anaphase when they seem to rather quickly separate. Even when colchicine is used to block mitosis, the chromatids usually separate after a delay. Whether this represents the segregation of the RG's is not clear. Perhaps these units separate into pairs somewhat earlier and the open chains are repaired. Prophase condensation might then line up the RG's into columns if the assumption is made that they do not always remain united. One possible mechanism for this condensation would be the supercoiling of the loops. If the DNA helix should change from the B toward the A form [9, 50], enough torsion could probably be produced to cause the loops to twist and supercoil. This would bring sequentially arranged RG's along the duplex close together. If they fitted on contact, columns with alternating twisted loops projecting to the sides would be the equilibrium state. The coiling of the chromatids could result by shrinkage of the middle region of the chromatid axis in the stages of condensation. Such a differential contraction of a ribbon will make it fall into a helix (Taylor et al. [48], Taylor [44]).

References

[1] Baldwin, R. L., and E. M. Shooter: The alkine transition of BU-containing DNA and its bearing on the replication of DNA. J. molec. Biol. 7, 511—526 (1963).

[2] BOLLUM, F. J.: Intermediate states in enzymatic DNA synthesis. J. cell. comp. Physiol. **62**, Suppl. 1, 61—71 (1959).

[3] CAIRNS, J.: The bacterial chromosome and its manner of replication as seen by autoradiography. J. molec. Biol. **6**, 208—213 (1963).

[4] CALLAN, H. G., and H. C. MACGREGOR: Action of deoxyribonuclease on lampbrush chromosomes. Nature (Lond.) **181**, 1479—1480 (1958).

[5] CAVALIERI, L. F., and B. H. ROSENBERG: The replication of DNA. II. The number of polynucleotide strands in the conserved unit of DNA. Biophys. J. **1**, 323—337 (1961).

[6] CHUN, E. H. L., and J. W. LITTLEFIELD: The separation of the light and heavy strands of bromouracil-substituted mammalian DNA. J. molec. Biol. **3**, 668—673 (1961).

[7] DJORDJEVIC, B., and W. SZYBALSKI: Genetics of human cell lines. III. Incorporation of 5-bromo- and 5-iododeoxyuridine into deoxyribonucleic acid of human cells and its effect on radiation sensitivity. J. exp. Med. **112**, 509—532 (1960).

[8] FILNER, P.: Semiconservative replication of DNA in a higher plant cell. Exp. Cell Res. (1965, in press).

[9] FRANKLIN, R. E., and R. GOSLING: Molecular configuration in sodium thymonucleate. Nature (Lond.) **171**, 740—741 (1953).

[10] FREESE, E.: The arrangement of DNA in the chromosome. Cold Spr. Harb. Symp. quant. Biol. **23**, 13—18 (1958).

[11] GABRUSEWYCZ-GARCIA, N.: Cytological and autoradiographic studies in *Sciara coprophila* salivary gland chromosomes. Chromosoma (Berl.) **15**, 312—344 (1964).

[12] GALL, J. G.: Kinetics of deoxyribonuclease action on chromosomes. Nature (Lond.) **198**, 36—38 (1963).

[13] HAUT, W. F., and J. H. TAYLOR: Density gradient centrifugation studies of DNA replication in *Vicia* roots. (1965: In preparation.)

[14] HERSHEY, A. D., and E. BURGI: Complementary structure of interacting sites at the ends of lambda DNA molecules. Proc. nat. Acad. Sci. (Wash.) **53**, 325—328 (1965).

[15] HOTTA, Y., and A. BASSEL: Molecular size and circularity of DNA in cells of mammals and higher plants. Proc. nat. Acad. Sci. (Wash.) **53**, 356—362 (1965).

[16] JACOB, F., S. BRENNER, and F. CUZIN: On the regulation of DNA replication in bacteria. Cold Spr. Harb. Symp. quant. Biol. **28**, 329—348 (1963).

[17] JOSSE, J., A. D. KAISER, and A. KORNBERG: Enzymatic synthesis of deoxyribonucleic acid. VIII. Frequencies of nearest neighbor base sequences in DNA. J. biol. Chem. **236**, 864—875 (1961).

[18] KAUFMANN, B. P., H. GAY, and M. MCDONALD: Organizational patterns within chromosomes. Int. Rev. Cytol. **9**, 77—127 (1960).

[19] KEYL, H. G., and C. PELLING: Differentielle DNS-Replication in den Speicheldrüsen-Chromosomen von *Chironomus thummi*. Chromosoma (Basel) **14**, 347—359 (1963).

[20] KORNBERG, A.: Biologic synthesis of deoxyribonucleic acid. Science **131**, 1503—1508 (1960).

[21] LACOUR, L. F., and S. R. PELC: Effect of colchicine on the utilization of labelled thymidine during chromosomal reproduction. Nature (Lond.) **182**, 506—508 (1958).

[22] MARMUR, J., R. ROWND, and C. L. SCHILDKRAUT: Denaturation and renaturation of deoxyribonucleic acid. Progress in Nucleic Acid Research **1**, 231—300. New York: Academic Press 1963.

[23] MESELSON, M.: The deoxyribonucleic acid of coliphage T7 and its transfer from parental to progeny phages. *In* "The Cell Nucleus" (J. S. MITCHELL, ed.), pp. 240—245. New York: Academic Press 1960.

[24] —, and F. STAHL: The replication of DNA in *Escherichia coli*. Proc. nat. Acad. Sci. (Wash.) **44**, 671—682 (1958).

[25] —, and J. J. WEIGLE: Chromosome breakage accompanying genetic recombination in bacteriophage. Proc. nat. Acad. Sci. (Wash.) **47**, 857—868 (1961).

[26] Miller, O.: Fine structure of lampbrush chromosomes. National Cancer Institute Monograph (1964: in press).
[27] Morishima, A., M. M. Grumbach, and J. H. Taylor: Asynchronous duplication of human chromosomes and the origin of sex chromatin. Proc. nat. Acad. Sci. (Wash.) **48**, 756—763 (1963).
[28] Moses, M. J.: Chromosomal structures in crayfish spermatocytes. J. biophys. biochem. Cytol. **2**, 215—218 (1956).
[29] — The nucleus and chromosomes: a cytological perspective. *In* "Cytology and Cell Physiology" (G. Bourne and J. F. Danielli, eds.), pp. 423—558. New York: Academic Press 1964.
[30] Patel, G., and T. Y. Wang: Isolation of an active complex of DNA-RNA-protein from nuclear residual fraction. Life Sciences **4**, 1481—1486 (1965).
[31] Peacock, W. J.: Chromosome duplication and structure as determined by autoradiography. Proc. nat. Acad. Sci. (Wash.) **49**, 793—801 (1963).
[32] Prensky, W., and H. H. Smith: Incorporation of H^3-arginine in chromosomes of *Vicia faba*. Exp. Cell Res. **34**, 525—532 (1964).
[33] Prescott, D. M.: RNA and protein replacement in the nucleus during growth and division and the conservation of components in the chromosome. *In* Symp. Intern. Soc. Cell Biol. **2**, 111—149 (R. J. C. Harris, ed.) New York: Academic Press 1963.
[34] Ris, H.: Ultrastructure and molecular organization of genetic systems. Canad. J. Genet. Cytol. **3**, 95—120 (1961).
[35] Schwartz, D.: Deoxyribonucleic acid and chromosome structure. *In* "The Cell Nucleus" (J. S. Mitchell, ed.). New York: Academic Press 1960.
[36] Simon, E. H.: Transfer of DNA from parent to progeny in a tissue culture line of human carcinoma of the cervix (Strain HeLa). J. molec. Biol. **3**, 101—109 (1961).
[37] Steffensen, D.: A comparative view of the chromosome. Brookhaven Symp. Biol. **12**, 103—124 (1959).
[38] Sueoka, N.: Mitotic replication of deoxyribonucleic acid in *Chlamydomonas reinhardi*. Proc. nat. Acad. Sci. (Wash.) **46**, 83—91 (1960).
[39] —, and T. Y. Cheng: Fractionation of nucleic acids with metylated albumin column. J. molec. Biol. **4**, 161—172 (1962).
[40] Szybalski, W.: Chemical reactivity of chromosomal DNA as related to mutagenicity: studies with human cell lines. Cold Spr. Harb. Symp. quant. Biol. **29**, 151—159 (1964).
[41] Taylor, J. H.: Sister chromatid exchanges in tritium labeled chromosomes. Genetics **43**, 515—529 (1958).
[42] — The organization and duplication of genetic material. Proc. X Int. Gen. Cong., 1958, Montreal, Canada, Vol. 1, 63—78 (1959a).
[43] — Oklahoma Conf. Radioisotopes in Agr. 1959, U.S. Atomic Energy Comm. TID-7578, pp. 123—129 (1959b).
[44] — Autoradiographic studies of the organization and mode of duplication of chromosomes. *In* A Symposium on Molecular Biology, pp. 304—320 (R. E. Zirkle, ed.). Chicago: Univ. of Chicago Press 1959 (c).
[45] — Asynchronous duplication of chromosomes in cultural cells of Chinese hamster. J. biophys. biochem. Cytol. **7**, 455—464 (1960).
[46] — "The replication and organization of DNA in chromosomes" in Molecular Genetics (J. H. Taylor, ed.). Part I, pp. 65—111. New York and London: Acad. Press 1963.
[47] — DNA synthesis in relation to chromosome reproduction and reunion of breaks. J. cell. comp. Physiol. **62**, 73—86 (1964).
[48] — P. S. Woods, and W. L. Hughes: The organization and duplication of chromosomes as revealed by autoradiographic studies using tritium-labeled thymidine. Proc. nat. Acad. Sci. (Wash.) **43**, 122—128 (1957).
[49] Trosko, J. E., and S. Wolff: Strandedness of *Vicia faba* chromosomes as revealed by enzyme digestion studies. J. Cell Biol. **26**, 125—135 (1965).
[50] Watson, J. D., and F. H. C. Crick: The structure of DNA. Cold Spr. Harb. Symp. quant. Biol. **18**, 123—131 (1953).

Discussion

Chairman: *Beermann*

Tuppy: May it be possible that in the use of bromo-deoxyuridine as a label sometimes the repair mechanism might consist in a de-bromination of the used nucleoside, so that you get labeling but not as large an increase in density as one might expect.

Taylor: This certainly is a possibility. However, we would have the same difficulties in explaining our data on this basis as we do by irregular elimination of the analogue because we get these discrete density species. One would think that the de-bromination would occur at random.

Tuppy: My question would need the additional assumption that you get an induction of a de-brominating enzyme which, in the beginning, is not present yet, but, in the course of the experiment, becomes active, from a certain point on.

Taylor: This is the possibility that we cannot exclude because our label is in the ring (^{14}C-bromouracil). I have no data that would contribute to further understanding of this possibility.

Schaller: Have you any idea about the molecular weight of this slightly heavier DNA. And have you any observations which indicate a difference in the density shift depending on molecular weight?

Taylor: We have not determined molecular weights of this material in the density gradient. However, we can estimate from other experiments that the molecular weight would probably be between 6 and 10 millions.

The density-shift is enough to indicate that the material is substituted to the extent of 5 to 10% with bromouracil. With these two assumptions a molecular weight of let us say 10 million, and assuming that the bromouracil is all in one piece (or possibly in two pieces) its molecular weight would be of the order of 300 to 500 thousand. This would give us some idea perhaps of the size of the primer-pieces or of that unit with which we are dealing.

I meant to point out, that if these are primer-pieces it means that the unit of replication is very small in these organisms, of the order of a few millions molecular weight. If they were larger, we would have more pieces cut out without BU. But since we don't the interpretation is that which I have given. The replicating unit is quite small relative to the whole chromosome of *E. coli.*

Beermann: Would it be possible or is it hopeful to do some sonication of this particular material to try to break off the one end which is supposed to be heavier?

Taylor: Certainly the next experiment one would like to do is to sonicate this DNA which is slightly heavier and to see if it separates into higher and lower density species.

Beermann: Die neue Idee ist, daß diese besonderen replizierenden Segmente der DNS in den Chromosomen höherer Organismen, die hier gefunden wurden — oder die man gefunden zu haben glaubt — vielleicht so zu verstehen sind, daß sie nicht als Templates, sondern als Primer dienen, also als Starter der Replikation.

Eine Frage wäre natürlich, wo denn diese Primer herkommen, wo sie gemacht werden.

Taylor: In the bean root this material labeled by BU is attached to the other DNA apparently by covalent bounds. I think it is probably produced by a polymerization perhaps at the free ends where the strands open. However, to be consistent with a model of semi-conservative distribution it would have to be cut off at a later time at or near the point of its origin.

Beermann: Do you think that primers are probably AT-polymers?

Taylor: Certainly we considered this as a possibility since it would be the simplest situation that you have an AT-polymer which is one kind of primer and G-homopolymer and C-homopolymer which is the other kind of primer. Then it would be possible that this material is synthesized in quantity and perhaps stored in some special cells. Under special conditions this might be obtained; e.g. the crab *(Cancer borealis)* has a large amount of the AT-polymer present, but this seems to be present in many of the cells. Then we have in eggs large amounts of DNA-like polymers. These are usually cells that will replicate very rapidly.

Beermann: I just like to ask a question: How do you know that this is chromosomal DNA at all? And that it is not derived from cytoplasmic organelles, e.g. mitochondria?

Taylor: This is probably a question of some importance. We are separating the nuclei from the cytoplasm and plan to determine the cellular site of the slightly labeled DNA.

Schaller: Concerning the synthesis of the primer DNA, you are talking about dAT polymer and GC-poymer. But what is synthesized in your model is single-stranded DNA. However dAT- and dGC-polymer are always double-stranded DNA. It would not fit your model.

Taylor: That is right. It would from some points of view not be likely that such a primer produced in a quantity in a cell would not pair.

Schaller: The section of new DNA, is it in your model not primed by the parent strand? You would lose an information which is contained in the parent strand. You would not copy this when you elongate the DNA-chain.

Taylor: Well, one might visualize this as a region which codes a limited amount of information, perhaps for time of replication in the cell. The origin of the primer pieces could be such that they came from other places. But once this copying begins, it uses the informational strand as a template and does replicate, I assume, in a regular way. I suppose the breaks that I assume to occur for producing the replicative state are just a little distal to the point where the repeating sequence polymer occurs.

Duspiva: Ich möchte bezüglich der Herkunft der "primer" auf die kürzlich erschiene Arbeit von Stone, Miller und Prescott (J. Cell. Biol. **25**/2, Part 2, 171, 1965) über einen stabilen Pool von Desoxyribonucleotiden im Makronucleus von *Tetrahymena* hinweisen.

Schaller: It should be possible to find differences in molecular weight of DNA from resting cells and from replicating cells, if the transfer of resting DNA to the replicated state occurs by making chain-breaks. Is there any evidence in this direction?

Taylor: There is a considerable amount of data accumulated by Cavalieri. He made an analysis of a variety of DNA's that were in cells that were replicating and non-replicating.

However, depending on the amount of deproteinisation, if there is a cross-linking as suggested in the model, there might not be any demonstrable change. One might look for DNA that is in the replicative form, if it exists in quantity. In most cells, I would assume, that the amount present is small at any one time in synthesis. One possibility for finding larger amounts would be in mature sperms that are ready to replicate as soon as they enter the egg. There are some suggestions for this change of state: One is that generally sperm-DNA is a very good primer compared to other DNA's; the other is that recently Hotta and co-worker have found that DNA from sperm contains ring-structures. These are rather small rings. If DNA is in a replicative state according to this model it would tend to form rings by pairing of the complementary ends of replicons.

Beermann: Aus diesen Vorträgen ist klar geworden, wie sehr wichtig es ist und noch weiter werden wird, sich der komplexen Verhältnisse bei den höheren Organismen mehr anzunehmen. Man steckt da wirklich erst in den Anfängen. Vor allem wäre es lohnend, zunächst einmal die Grunddaten wie Molekulargewichte, Basenzusammensetzungen usw. zu sammeln.

Funktionelle und strukturelle Organisation der Lampenbürsten-Chromosomen

Von

OSWALD HESS, Tübingen

Mit 12 Abbildungen

A. Einleitung

Lampenbürstenchromosomen sind Chromosomen, die sich im Zustand starker Entspiralisierung befinden und an bestimmten Stellen seitlich herausragende schleifenartige Strukturen ausbilden. Der Lampenbürstenzustand ist phasenspezifisch und reversibel. Lampenbürstenchromosomen sind in den Kernen primärer Oocyten vieler Wirbeltier-Arten und einiger Wirbelloser gefunden worden [*10*]. Es ist möglich, daß bei allen Tierarten die Chromosomen in den Oocyten sich in einem lampenbürstenartigen Zustand befinden, ohne daß seitliche Schleifen in allen Fällen zu erkennen sind. Lampenbürstenchromosomen sind auch in den Spermatocytenkernen von mehreren Arten der Gattung *Drosophila* gefunden worden [*23, 24*]. In den Spermatocytenkernen verschiedener anderer Tierarten sind von mehreren Autoren ebenfalls seitliche Fortsätze an Chromosomen beobachtet und zum Teil als Lampenbürstenschleifen interpretiert worden [z. B. *26*].

B. Stukturelle Organisation

I. Amphibien

Besonders große Lampenbürstenchromosomen sind in den Oocytenkernen von urodelen Amphibien gefunden und untersucht worden (vgl. die Zusammenfassung von CALLAN [*11*]). Lampenbürstenchromosomen entstehen hier in jungen primären Oocyten und verbleiben während des ganzen Oocytenwachstums, das bei diesen Arten mehrere Wochen dauert, in diesem Zustand. Die Chromosomen befinden sich dabei im Stadium der meiotischen Prophase und zwar wahrscheinlich im Diplotän, jedenfalls in einem Stadium nach der Synapsis, denn homologe Chromosomen sind an einzelnen Stellen durch Chiasmata verbunden (Abb. 1). Als typische meiotische Prophasechromosomen zeigen die Lampenbürstenchromosomen eine lineare Gliederung in verdickte Chromomeren und Zwischenstücke. Jedes Chromomer bildet ein seitliches Schleifenpaar aus. Die Chromosomen besitzen zusammen also mehrere Hundert solcher Schleifenpaare und erhalten so ihr charakteristisches lampenbürstenartiges Aussehen (Abb. 1).

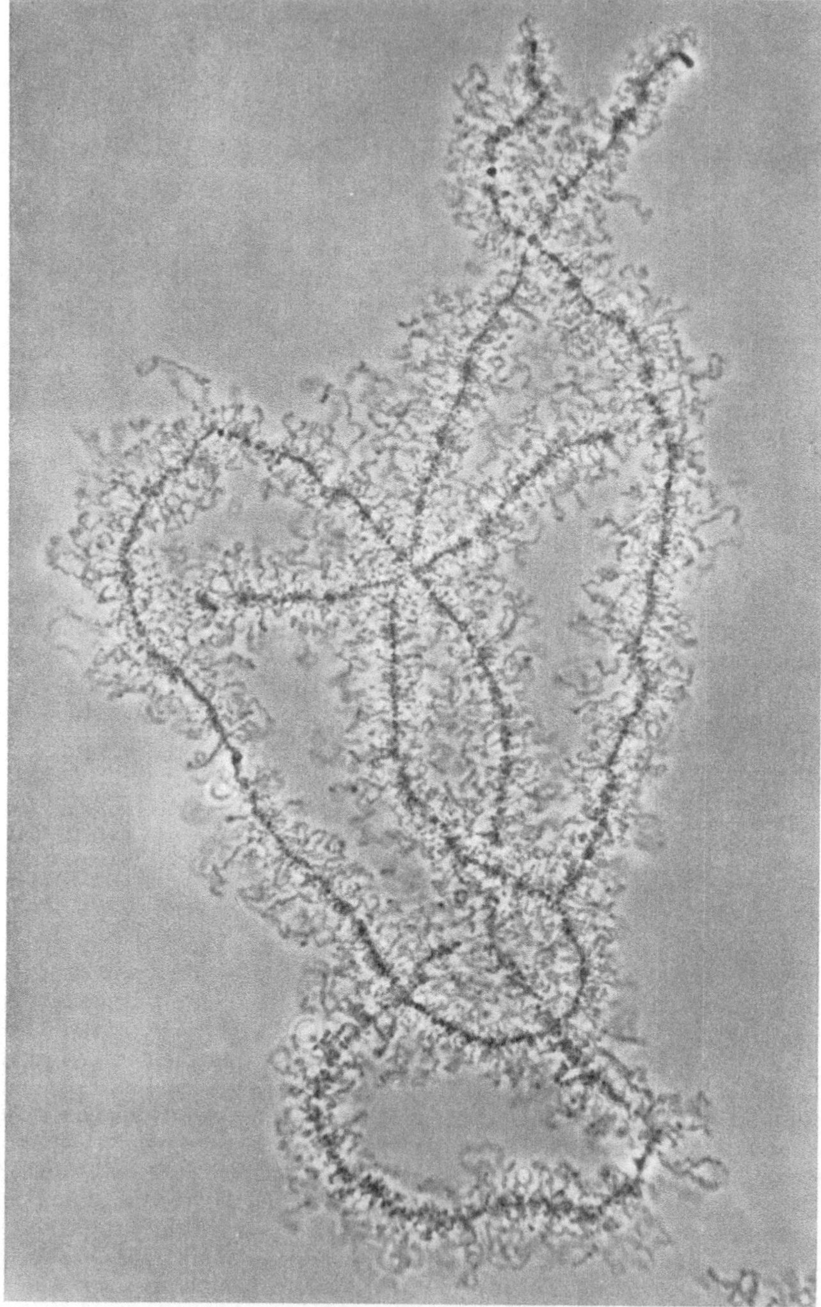

Abb. 1. Lampenbürstenchromosom (Bivalent) aus einem Oocytenkern von *Triturus spec.*, mit Chiasmata. — Phasenkontrastaufnahme eines Lebendpräparats, etwa 500 ×, von Prof. Dr. J. G. Gall, New Haven, Connecticut

Lampenbürstenchromosomen erreichen häufig eine Länge von 1 mm und übertreffen damit teilweise die polytänen „Riesen"-Chromosomen. Im Gegensatz zu den polytänen Chromosomen sind die Lampenbürstenchromosomen aber teilungsfähige Chromosomen, die sich am Ende der Wachstumsphase der Oocyten wieder kontrahieren und spiralisieren und, nachdem sie ihre normale Form zurückerlangt haben, die Reduktionsteilungen durchlaufen.

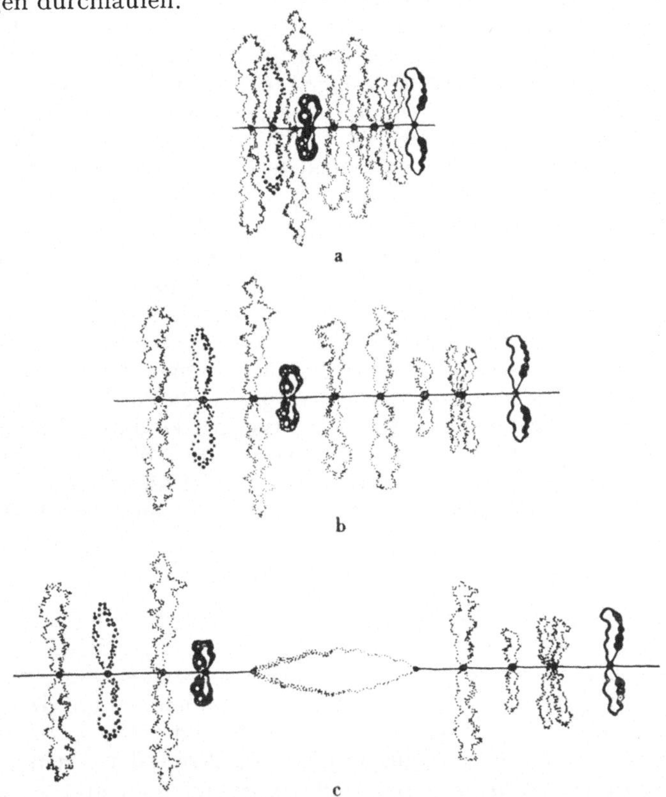

Abb. 2a—c. Ausschnitt eines Lampenbürstenchromosoms aus einem Oocytenkern von *Triturus spec.* — Aus CALLAN [11]. a normale Situation mit mehreren Schleifen verschiedener morphologischer Ausprägung; b nach Streckung innerhalb der Elastizitätsgrenze; c achsialer Bruch im Chromomer eines Schleifenpaars

Bei genauerer mikroskopischer Analyse zeigt sich, daß die Schleifen in morphologischen Eigenschaften wie Länge, Dicke oder Beschaffenheit der Matrix voneinander verschieden sein können. Bei manchen der größeren Schleifen besteht die Matrix aus radiär geordnetem Fibrillenmaterial, das einer deutlich erkennbaren Achse aufsitzt. Viele Schleifen sind polarisiert: das eine Schleifenende kann z. B. dünner sein als das andere, oft fehlt an diesem dünneren Ende das Fibrillenmaterial, nicht selten sind diese Fibrillen zuerst klein und werden im Verlauf einer Schleife allmählich größer. In der Regel ist eine derartige strukturelle

Asymmetrie bei den Schleifen aus kleineren Oocyten deutlicher zu sehen als bei Schleifen, die aus ausgewachsenen Oocytenkernen stammen. Während weitaus die meisten Schleifen dieses (normale) Aussehen haben, findet man auch einige, die von dieser Norm beträchtlich abweichen. So gibt es z. B. Schleifen, in deren Matrix viele Granula angehäuft sind. Häufig ist die Verteilung dieser Granula längs der Schleifen wiederum asymmetrisch. Bei anderen Schleifen besteht die Matrix aus größeren Mengen eines Materials, das die Tendenz zur Verschmelzung hat. Die ursprüngliche Schleifenform geht dabei nicht selten ganz verloren und es entstehen kugelige oder klumpige, oft vacuolisierte Gebilde, die Nucleolen ähneln. Alle diese morphologischen Besonderheiten sind locusspezifisch und konstant. Es ist deshalb in beschränktem Maß möglich, für den Chromosomensatz verschiedener Amphibien-Arten „Schleifenkarten" aufzustellen (Abb. 2a).

Die Lampenbürstenchromosomen lassen sich mechanisch sehr weit dehnen, wobei sich zunächst nur die Interchromomeren verlängern. Die Elastizitätsgrenze wird erst erreicht, wenn die Chromosomen etwa auf das $2^1/_2$fache ihrer ursprünglichen Länge ausgedehnt wurden (Abb. 2b). Danach zerreißen sie, und zwar in einer für Lampenbürstenchromosomen besonders charakteristischen Weise, nämlich immer in der Mitte eines Chromomers, so daß die Bruchstelle stets zwischen Vorder- und Hinterende eines Schleifenpaars liegt und die Bruchstücke durch die nunmehr geöffnete Doppelschleife miteinander verbunden bleiben (Abb. 2c). Diese Erscheinung des „achsialen Bruchs" [11] ist für die Interpretation der Organisation der Lampenbürstenchromosomen von Bedeutung (vgl. S. 36).

II. Drosophila

In den Kernen primärer Spermatocyten von *Drosophila*-Arten werden Strukturen gebildet, die, wie cytogenetische Untersuchungen gezeigt haben, den Schleifen der Lampenbürstenchromosomen von Amphibien nicht nur äußerlich ähneln, sondern ihnen auch homolog sind. Das mikroskopische Bild dieser Strukturen ist im einzelnen von Art zu Art sehr verschieden. Von den über 50 bisher geprüften Arten der Gattung haben *Drosophila hydei* und mehrere nahe verwandte Arten die größten und am weitesten differenzierten Schleifenstrukturen [24]. Bei *Drosophila hydei* sind die Verhältnisse näher untersucht worden. Die Spermatocytenkerne enthalten hier fünf Paare großer Schleifen, von denen jede ihren eignen charakteristischen Bau hat [34]. Wir unterscheiden ein Paar kompakter, stark lichtbrechender Fäden, ein Paar nucleolusähnlicher und ein Paar etwa keulenförmiger Körper, dazu eine Schleife, die aus verknäuelten Bändern besteht und im Lebendpräparat bei Beobachtung mit Phasenkontrast als ein Feld aus grauem Material erscheint, das einen großen Teil des Kernraums ausfüllt. Die fünfte Schleife hat die Form einer großen Schlinge aus dünnen, diffusen Fäden, die häufig mit den anderen Strukturen verschmelzen und deshalb sehr oft nicht zu sehen sind. Charakteristisch für diese Strukturen ist, daß sie außer der Paarigkeit eine Polarität erkennen lassen. So laufen z. B. die kompakten Fäden in einen diffusen Abschnitt aus, und die nucleolusähnliche Struktur (der sogenannte „Pseudonucleolus") hat immer zwei kompakte Zapfen an einer Seite (Abb. 3a).

XO-Männchen von *Drosophila hydei*, also Männchen, denen das Y-Chromosom fehlt, besitzen Spermatocytenkerne, die im Lichtmikroskop bis auf den Nucleolus und einige sehr kleine schleifenartige Strukturen leer erscheinen (Abb. 3b). Auf der anderen Seite enthalten Spermato-

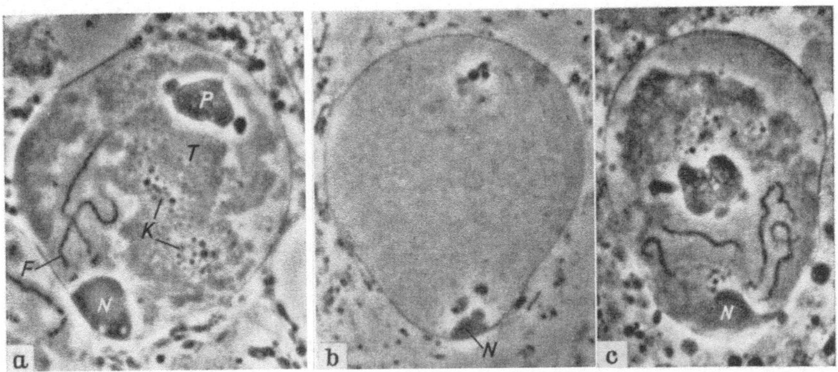

Abb. 3a—c. Lampenbürstenartige Strukturmodifikationen des Y-Chromosoms von *Drosophila hydei* in Kernen primärer Spermatocyten. — Phasenkontrastaufnahmen von Lebendpräparaten, etwa 1200 ×. — Aus HESS [20]. a Spermatocytenkern eines normalen Männchens (XY—♂); b eines Männchens ohne Y-Chromosom (X0—♂); c eines Männchens mit zwei Y-Chromosomen (XYY—♂), Verdopplung des Schleifensatzes. Bezeichnungen: *F* = fadenförmige, *K* = keulenförmige, *P* = nucleolusähnliche (sog. Pseudonucleolus), *T* = tubuläre Schleife; *N* = Nucleolus

cytenkerne von Männchen mit zwei Y-Chromosomen die doppelte Anzahl der Schleifenstrukturen, also von jeder einzelnen Schleife zwei Paare (Abb. 3c) [*23*].

Aus diesen Befunden ist zu folgern, daß es sich bei den beobachteten Strukturen um Bildungen des Y-Chromosoms handelt. In der Mitose ist am Y-Chromosom von *Drosophila hydei* eine Differenzierung in Chromomeren und Interchromomeren nicht zu erkennen. Und in den Kernen primärer Spermatocyten befinden sich alle Chromosomen, auch das Y-Chromosom, in einem stark entspiralisierten und diffusen Zustand. In diesem Stadium sind nur die schleifenartigen Strukturen zu identifizieren. Somit läßt sich durch direkte mikroskopische Beobachtung nicht feststellen, ob die Schleifen des Y-Chromosoms wie bei den Lampenbürstenchromosomen in den Oocytenkernen von Amphibien ebenfalls an besonderen Chromomeren entstehen und in welcher Reihenfolge sie gegebenenfalls auf dem Y-Chromosom angeordnet sind. Genauere Kenntnis über die Entstehung der Schleifen konnte aber mit Hilfe genetischer Methoden erhalten werden. Durch Röntgenbestrahlung von Weibchen, die ein markiertes Attached-X-Chromosom und ein zusätzliches Y-Chromosom besaßen, wurden Detachments induziert. Detachments sind in der Regel gleichzeitig Translokationen. Da in der meiotischen

Prophase X- und Y-Chromosomen teilweise gepaart sind, entstehen vorzugsweise X-Y-Translokationen. Diese Translokationen haben es ermöglicht, eine Reihe von verschiedenen Abschnitten des Y-Chromosoms getrennt auf ihre Bedeutung für die Bildung der Schleifen in den Spermatocytenkernen zu prüfen sowie ihre gegenseitige Beeinflussung zu studieren. Es ergab sich, daß jede einzelne Schleife tatsächlich ihren spezifischen, eng begrenzten Bildungsort auf dem Y-Chromosom hat. Diese Bildungsorte konnten auch lokalisiert werden und aus den Daten dieser Analyse eine Genkarte für das Y-Chromosom aufgestellt werden (Abb. 4) [22].

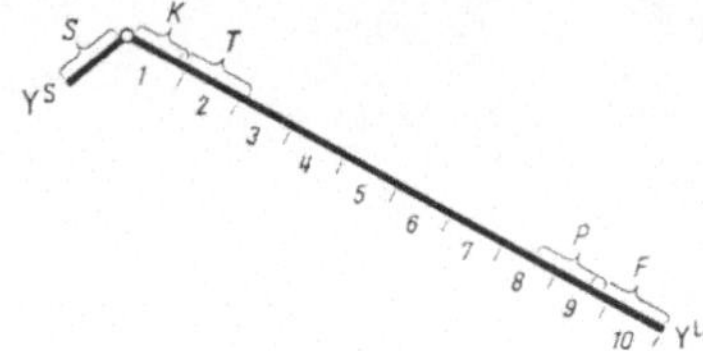

Abb. 4. Genetische Karte für die Schleifenbildungsorte auf dem Y-Chromosom von *Drosophila hydei*. Bezeichnungen: Y^L, Y^S, langer bzw. kurzer Arm des Y-Chromosoms; F, P, T, K, S, Bildungsort für die Fäden, Pseudonucleolus, Tubuli, Keulen und Schlinge

C. Funktionelle Organisation

Nach dieser knappen Darstellung der äußeren Morphologie der Lampenbürstenchromosomen soll nunmehr besprochen werden, wie sich das äußere Bild der Lampenbürstenchromosomen in unsere allgemeinen Vorstellungen über die Chromosomenstruktur einfügt und wie durch die weitere Analyse des Lampenbürstenzustandes Beiträge für die Vertiefung unserer Kenntnisse auf diesem Gebiet geliefert werden konnten.

Das grundlegende Problem, das sich bei dem Versuch, ein allgemein gültiges Modell für die Struktur der Chromosomen zu entwerfen, zu Anfang stellt, ist die Anordnung der Desoxyribonucleinsäure in den Chromosomen. Besonders einfache und deshalb schon sehr weitgehend aufgeklärte Verhältnisse liegen bei den Bakterien vor. Bei *Escherichia coli* z. B. besteht die DNA nur aus einem einzigen Molekül mit etwa 10^6-10^7 Nucleotidpaaren. Dies entspricht etwa 10^{-14} g DNA und ergibt nach dem Modell von Watson und Crick eine Doppelhelix von rund 1 mm Länge. Bei der Replikation verhält sich dieses Molekül als eine Einheit, als ein sogenanntes Replikon, bei dem die Replikation an einem bestimmten Punkt startet und von hier nach dem Reißverschlußprinzip allmählich über das Molekül fortschreitet. Dabei wickelt sich die Doppelhelix ab und jeder der beiden DNA-Einzelstränge dient als Matrix zum Aufbau eines neuen Einzelstranges von komplementärer Basensequenz. Die Replikation ist also semikonservativ und am Ende eines jeden Replikationsschrittes besteht die DNA von *Escherichia coli* aus je einem alten und einem neusynthetisierten Strang [33]. Es gibt verschiedene Stämme, die sich durch die Lage des Startpunktes bei der DNA-Replikation und durch die Richtung unterscheiden, mit der die Replikation allmählich über das Molekül fortschreitet. Bei Kreuzungsexperimenten ergab sich die zunächst überraschende Tatsache, daß die Genkarte von *Escherichia coli* circulär ist. Zirkularität der Genkarte kann zweierlei bedeuten: erstens kann das DNA-Molekül selbst Ringstruktur haben; zweitens kann das genetische Material am Anfang und Ende der Karte dupliziert sein (da der Abstand zwischen zwei genetischen Markierungen aus der Häufigkeit ihres Austausches erschlossen wird, müssen identische Verdopplungen am Anfang und Ende einer Kopplungsgruppe zu ringförmigen Genkarten führen). Die Entscheidung ist zugunsten der ersten Möglichkeit gefallen. Wenn man die DNA radioaktiv markiert und durch ein besonderes Verfahren isoliert und ausspreitet, kann man in autoradiographischen Präparaten direkt sehen, daß sie tatsächlich ringförmig ist (Abb. 5) [*8, 9*].

Bei den höheren Organismen sind die Verhältnisse wesentlich komplizierter. Schon rein quantitativ besteht ein Unterschied. In den Chromo-

a   b

Abb. 5. Schematische Darstellung der DNA von *Escherichia coli* in zwei Phasen der Replikation. — Nach Cairns [*8*], verändert

somen der höheren Organismen ist der DNA-Gehalt um zwei bis drei Größenordnungen höher als in Bakterien. Schwierig zu verstehen ist auch die Tatsache, daß nahe verwandte Arten mitunter nicht nur sehr verschieden große Chromosomen haben, sondern daß auch die DNA-Menge pro Chromosomensatz um das Zehn- bis Hundertfache verschieden sein kann. Man kann sich kaum vorstellen, daß diesem starken Unterschied im DNA-Gehalt ein gleich großer Unterschied an Information entspricht. Hinzu kommt, daß die arteigentümliche DNA-Menge keineswegs, wie man theoretisch erwarten könnte, der jeweiligen Organisationshöhe einer Art entspricht. So haben die Säugetiere und der Mensch bei weitem nicht den höchsten gemessenen DNA-Gehalt in ihrem Chromosomensatz.

Die einfachste Erklärung für diesen Sachverhalt wäre die Annahme, daß in den Chromosomen der höheren Organismen jeweils Vielfache der genetischen Information in Form parallel verlaufender homologer DNA-Stränge enthalten sind. Dabei könnte die Anzahl der vorhandenen Homologe von Fall zu Fall schwanken und eine charakteristische Eigenschaft jeder Art sein. Tatsächlich ist eine solche Theorie vertreten worden und wird z. T. auch heute noch vertreten [*40*].

Gegen diese Polynemiehypothese sprechen aber eine Reihe experimenteller Befunde. Ein zwar indirekter, aber doch sehr gewichtiger Widerspruch ergibt sich aus der semikonservativen Segregation der

3*

DNA in den Chromosomen bei mitotischen Teilungen, wie das zuerst von TAYLOR (vgl. [42] und den Beitrag von Herrn Dr. TAYLOR in diesem Symposium) nachgewiesen und inzwischen mehrfach bestätigt worden ist. Auf der Basis der Polynemiehypothese können die Ergebnisse dieser Experimente ebenso wie die Beobachtungen über die Induzierung von Mutationen nur unter Zuhilfenahme komplizierter Zusatzannahmen erklärt werden.

Einen weiteren Beitrag zu diesem Problem liefert die Analyse der Lampenbürstenchromosomen. Behandelt man aus Oocytenkernen isolierte, unfixierte Lampenbürstenchromosomen mit RNase oder mit verschiedenen proteolytischen Fermenten, so wird zwar die Schleifenmatrix oder Teile von ihr abgebaut, niemals aber werden die Schleifen fragmentiert. Bei einer Behandlung mit DNase dagegen werden nicht nur die Interchromomeren fragmentiert, sondern auch die Schleifen zerfallen in Stücke, die immer weiter zerkleinert werden [31]. Daraus ist zu folgern, daß die Schleifen eine aus DNA bestehende Achse haben.

Schon aus der Tatsache, daß die Schleifen Feulgen-negativ sind, ist zu erwarten, daß sie im Querschnitt nur wenige DNA-Elemente enthalten. Genauere Aufklärung haben Untersuchungen von GALL [16] gebracht. Es wurde die Kinetik des Abbaus der Schleifenachsen durch DNase genauer analysiert. Der Verlauf des Abbaus entspricht den theoretischen Erwartungen für den Fall, daß die Schleifenachse aus nur einer einzigen DNA-Doppelhelix besteht. Damit ist ein direkter experimenteller Beleg dafür erbracht, daß die Chromosomen höherer Organismen aus zwei Chromatiden mit nur je einem DNA-Strang bestehen.

Die direkte mikroskopische Beobachtung zeigt in vielen Fällen, daß die DNA nicht gleichmäßig über die ganze Länge der Chromosomen verteilt ist. Es sei in diesem Zusammenhang auf die Gliederung der Lampenbürstenchromosomen in Chromomeren und Interchromomeren und auf die Gliederung der polytänen Chromosomen in stark Feulgen-positive Querscheiben von unterschiedlicher Dicke und in schwach Feulgen-positive oder gar Feulgen-negative Interbanden hingewiesen. Für die Organisation der Lampenbürstenchromosomen ergibt sich daraus zusammen mit den Befunden über den Abbau der Schleifen durch DNase und die zuvor geschilderte Erscheinung des achsialen Bruchs nach mechanischer Dehnung die Vorstellung, daß die DNA in den Chromomeren besonders dicht aufspiralisiert ist und sich beim Übergang in den Lampenbürstenzustand nach außen ausstülpt und eine schleifenförmige Achse bildet. Die Schleifen werden im Lichtmikroskop dadurch sichtbar, daß Syntheseprodukte an der ausgefalteten DNA-Achse haften bleiben und eine Matrix bilden. Die bei manchen größeren Schleifen deutlich sichtbaren radialen Fibrillen sind wahrscheinlich solche Syntheseprodukte (Abb. 6).

Es gibt aber auch bereits experimentelle Hinweise dafür, daß die DNA in den Chromosomen der höheren Organismen untergliedert ist. Im Gegensatz zu den Verhältnissen bei *Escherichia coli*, wo, wie bereits geschildert, die DNA nur aus einem einzigen Replikon besteht, ergeben Untersuchungen über die DNA-Replikation in den Chromosomen höherer

Organismen eindeutig, daß während der Replikationsphase an sehr vielen Stellen, die über die ganze Länge der Chromosomen verteilt sind, Replikationsvorgänge gleichzeitig stattfinden (z. B. [*39, 30, 25*]). Die DNA ist also in den Chromosomen in viele unabhängige Replikationseinheiten aufgeteilt.

Es ist noch nicht bekannt, wie diese Unterteilung strukturell verwirklicht ist. Es sind dazu mehrere Modelle denkbar. Z. B. könnte die DNA Bruchstellen haben. Damit aber ihre Struktur trotzdem einen

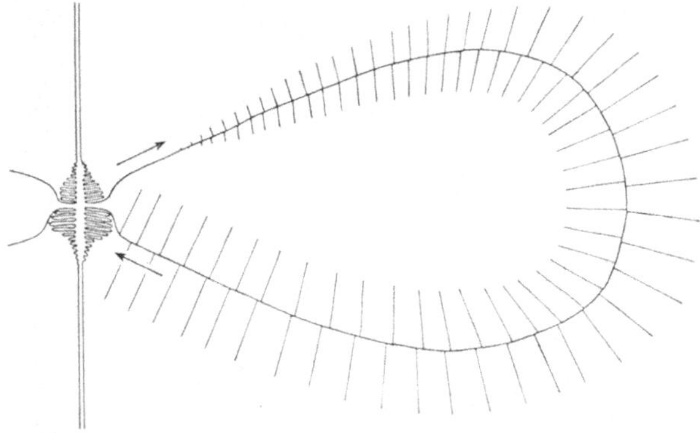

Abb. 6. Modell für ein Chromomer mit ausgefaltetem Schleifenpaar aus einem Lampenbürstenchromosom. Die beiden Pfeile bezeichnen die Richtung, in welcher sich der DNA-Achsenfaden vielleicht ab- bzw. wieder aufspult. Die radialen Fibrillen sind Stoffwechselprodukte. — Nach BEERMANN

Zusammenhalt hat, müßten diese Bruchstellen entweder alle in Abständen auf nur einem der beiden Einzelelemente der Doppelhelix liegen (Abb. 7d), oder aber, wenn Bruchstellen auf beiden Strängen vorkämen, müßten sie stets auf verschiedener Höhe liegen (Abb. 7e). Jede Bruchstelle könnte Startpunkt für eine DNA-Replikation sein. Die DNA könnte aber auch aus zahlreichen kürzeren Abschnitten bestehen, die als seitliche Fortsätze an einer doppelten Grundstruktur aus Protein sitzen. Jeder einzelne DNA-Seitenast wäre als ein selbständiges Replikon anzusehen (Abb. 7a). Ebenso könnten kürzere DNA-Untereinheiten in Längsrichtung hintereinander angeordnet und durch besondere Kupplungsstücke, wiederum am wahrscheinlichsten aus Protein bestehend, miteinander verbunden sein (Abb. 7b). Bei einer interessanten Variante dieses Modells (Abb. 7c) wird angenommen, daß sich die Proteinschaltstücke wechselweise aneinanderlagern oder auseinanderweichen können, also ihrerseits Stabilität und Zustandsform der Chromosomen weitgehend variieren können [*15, 43*].

Eine im Zusammenhang mit Fragen der DNA-Replikation interessante Beobachtung ist neuerdings bei den Lampenbürstenchromosomen einiger Molcharten gemacht worden. Von ganz bestimmten Regionen auf einem oder auf mehreren Lampenbürstenchromosomen werden Teile von Schleifen in Form von Ringen abgeschnürt (Abb. 8a). Die Ringe scheinen sehr stoffwechselaktiv zu sein, denn sie beladen sich mehr und mehr mit Material (Abb. 8b), bis sie das Aussehen von Nucleoli angenommen haben. Im Lebendpräparat findet man in einem Oocytenkern alle Zwischenstufen von einfachen Ringen bis zu kompakten Nucleoli. Durch Behandlung mit proteolytischen Fermenten wird Matrixmaterial von den Nucleoli abgebaut. Dadurch gewinnen sie ihre ursprüngliche

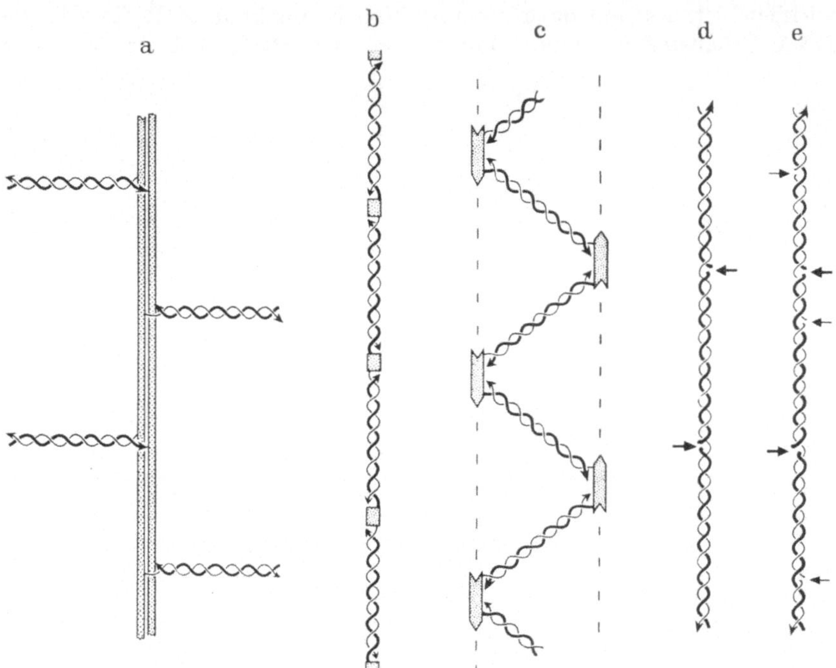

Abb. 7a—e. Modellvorstellungen über die Anordnung der DNA in unverdoppelten Chromosomen höherer Organismen. — Schematische Darstellung nach BEERMANN, verändert. a Modell nach TAYLOR, 1957, mit doppelter Proteinachse und seitlich angehefteten DNA-Molekülen. b Modell nach FREESE, 1958, mit Proteinkupplungen zwischen DNA-Molekülen. c Modell nach TAYLOR, 1960, mit zwei aus Proteinschaltstücken gebildeten Halbchromatiden entgegengesetzter Polarität und dazwischen ausgespannten DNA-Molekülen. d Durchlaufende Watson-Crick-Doppelschraube mit Bruchstellen (Startpunkte der Replikation) auf nur einem Strang. e Durchlaufende Watson-Crick-Doppelschraube mit Bruchstellen auf beiden Einzelsträngen.

Ringstruktur zurück. Werden die Ringe dagegen mit DNase behandelt, dann fragmentieren sie in ähnlicher Weise wie die Schleifen an den Chromosomen (KEZER, unveröffentlicht).

Nun enthalten die Oocytenkerne von Amphibien allgemein sehr viele kleine Nucleolen, die nicht, wie es der Definition eines Nucleolus entspricht, an spezifischen Loci (den Nucleolen-Bildungsorten) eines Chromosoms anzutreffen sind, sondern frei im Kernraum liegen. Diese Nucleolen sind nach Inkubation mit tritiiertem Uridin in autoradiographischen Präparaten markiert. Es muß also in ihnen eine RNA-Synthese stattfinden. Weiterhin ist die Tatsache bemerkenswert, daß die Oocytenkerne

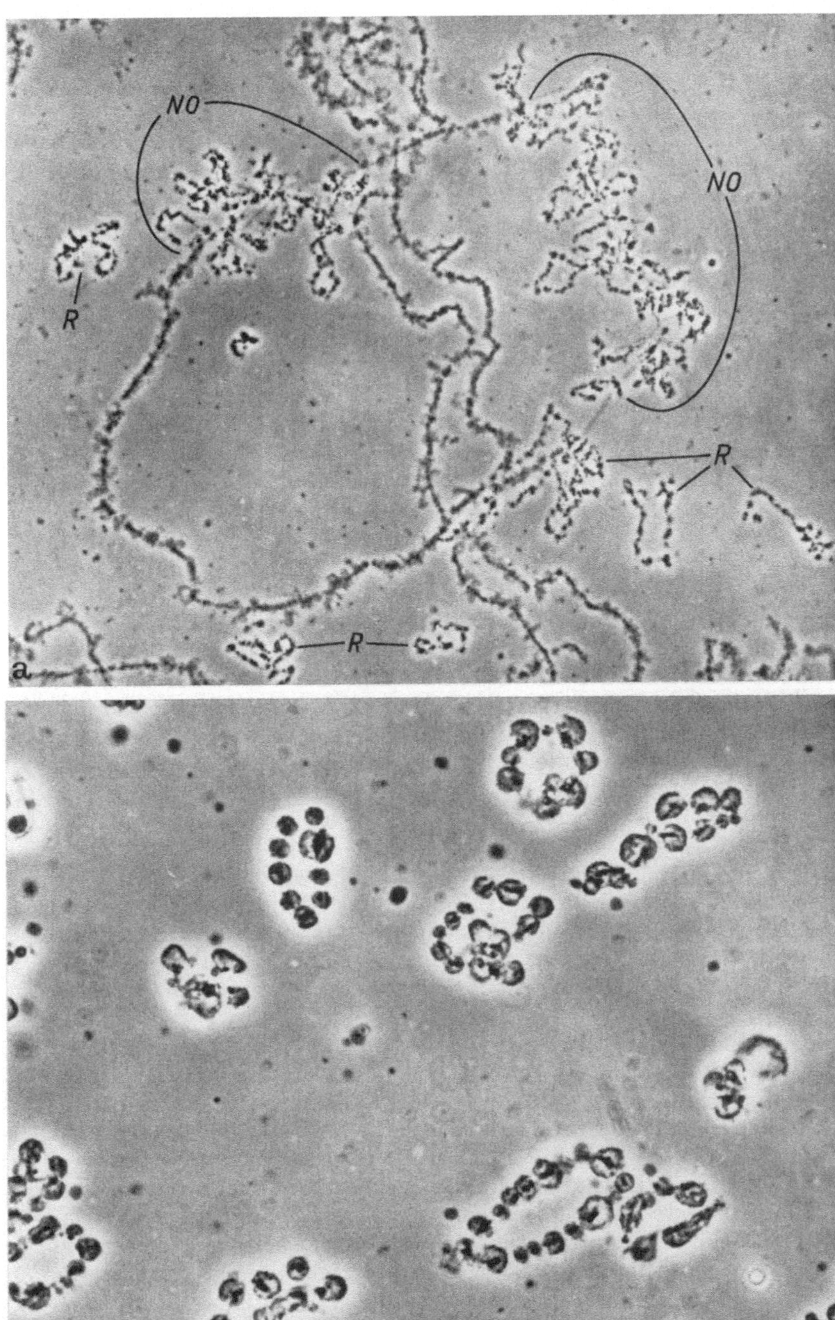

Abb. 8a—b. Lampenbürstenchromosom aus dem Kern einer Oocyte von *Plethodon cinereus*. — Phasenkontrastaufnahmen von Lebendpräparaten von Prof. Dr. J. KEZER, Eugene, Oregon. a Bivalent mit Nucleolenbildungsort (*NO*) und ringförmigen Bildungen (*R*) teils am Bildungsort, teils bereits abgeschnürt und vom Chromosom losgelöst. b Detailaufnahme mit einigen ringförmigen, freien Nucleoli

sehr viel mehr DNA enthalten als Kerne aus somatischen Geweben [28].
Es kann vermutet werden, daß an den oben beschriebenen ringab-
schnürenden Loci eine selektive DNA-Replikation stattfindet und daß
diese Loci die Nucleolenbildungsorte sind. Die unabhängig von der
übrigen chromosomalen DNA in großem Überschuß replizierte DNA
wird kontinuierlich von den Chromosomen abgegeben. Die Tatsache,
daß diese Abgabe anscheinend immer als Abschnürung von Ringen erfolgt, ist von besonderem Interesse und kann bei der weiteren Aufklärung der chromosomalen Organisation von Bedeutung sein. Die
freien Nucleolen sind durch den Besitz an DNA, und zwar sicherlich
einer DNA von ganz spezifischem Informationsgehalt, anscheinend in
der Lage, autonom DNA-abhängige RNA zu synthetisieren. Da nach
früheren Analysen ein sehr großer Bestandteil der RNA in den Oocytenkernen von Amphibien ribosomale RNA ist (nach den Untersuchungen
von Davidson, Allfrey und Mirsky [12] etwa 80%), liegt die Annahme
nahe, daß es sich dabei um ribosomale RNA handelt, was auch den derzeitigen Anschauungen über die Funktion der Nucleolen entspricht
(z. B. [32, 5, 14]).

Es bleibt abzuwarten, ob sich diese Annahmen durch Einbauversuche
experimentell bestätigen lassen. Entsprechende Experimente werden z. Z.
unternommen (Kezer und Wimber). Als Parallele hierzu sei erwähnt,
daß bei einigen Sciariden in polytänen Speicheldrüsenchromosomen zu
Puffs aufgelockerte Querscheiben gefunden worden sind, die ständig oder
doch zumindest während längerer Zeit markierte DNA-Vorstufen ein-
bauen. Hier wird also offenbar in eng begrenzten Loci während längerer
Perioden DNA synthetisiert. Vermutlich wird die lokal synthetisierte
DNA nicht als integrierender Bestandteil in die Chromosomenstruktur
eingebaut, sondern nach außen abgegeben. Über die Bedeutung der an
diesen Stellen, die als DNA-Puffs beschrieben worden sind [41], synthetisierten DNA wissen wir nichts. Beim Vergleich mit den ringabschnüren-
den Nucleolenbildungsorten in Lampenbürstenchromosomen ist es naheliegend, daran zu denken, daß auch in den DNA-Puffs eine selektive
Replikation einer DNA mit speziellen metabolischen Funktionen stattfindet.

Aus den zuletzt geschilderten Beobachtungen und Experimenten folgt,
daß die DNA in den Chromosomen bei der Replikation in Funktions-
einheiten gegliedert ist. Nun wird nach der Hypothese von der „differentiellen Gen-Aktivierung" von der in der Zelle zur Verfügung stehenden
genetischen Information immer nur ein kleiner Teil verwendet, ein Teil,
der für jedes Entwicklungsstadium und jeden Differenzierungszustand
charakteristisch und bestimmend ist. Diese Vorstellungen verlangen
also, daß die DNA in den Chromosomen auch bei der Transkription in
operative Einheiten untergliedert ist, die selbständig aktiviert werden
können. Es ist nun zu untersuchen, ob die Gliederung der DNA bei der
Transkription die gleiche ist wie bei der Replikation, mit anderen Worten, ob die Replikationseinheiten auch die operativen Einheiten bei der
Transkription sind.

Die ersten Ergebnisse, die bei höheren Organismen einen Beitrag zu diesem Problemkreis geliefert haben, wurden wiederum an polytänen Chromosomen erzielt. Es wurde gezeigt, daß eine als „Puffing" bezeichnete lokale Strukturauflockerung mit einer differentiellen Aktivierung des genetischen Materials in Beziehung steht. Es gibt gewebe- und stadienspezifische Puffs (BEERMANN [1]; weitere Literatur bei BEERMANN

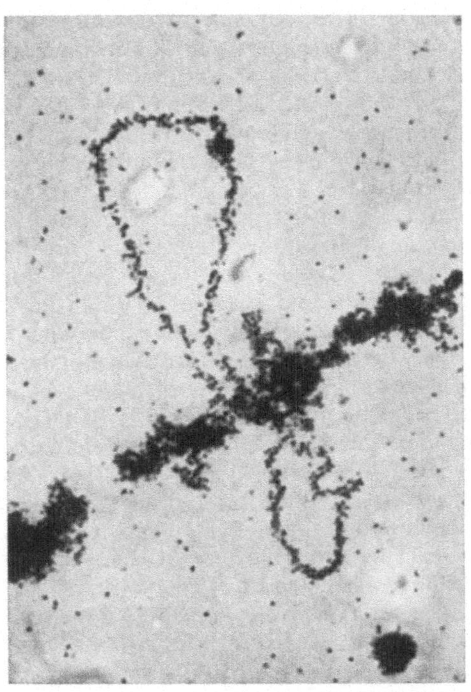

Abb. 9. Einbau von ³H-Uridin in ein Schleifenpaar eines Lampenbürstenchromosoms aus einem Oocytenkern von *Triturus spec*. Das Schleifenpaar ist in seinem ganzen Verlauf markiert. Die Chromosomenachse selbst ist wahrscheinlich unmarkiert; die Schwärzung über der Achse wird vermutlich durch markierte Schleifen verursacht, die bei der Fixierung verklumpt sind. — Photographie einer Radioautographie, etwa 1000 ×, von Prof. Dr. J. G. GALL, New Haven, Connecticut

[3]). Nach Inkubation mit tritierten RNA-Vorstufen sind die Loci der Puffs und (mit Ausnahme des Nucleolus) praktisch nur diese markiert [37]. Die Puffs sind demnach Stellen im Chromosom, an denen eine erhöhte RNA-Synthese stattfindet, in der Interpretation der Molekulargenetik also diejenigen Stellen, an denen Gene aktiv sind und messenger-RNA produzieren. Für unsere Betrachtungen über die Struktur der Chromosomen ist die Tatsache wesentlich, daß eben diese Struktur an Stellen mit aktivierten Genen in charakteristischer Weise verändert, und zwar, wie sich herausgestellt hat, reversibel verändert wird.

Es ist inzwischen erwiesen, daß auch die Ausfaltung der DNA zur Bildung von seitlichen Schleifen in den Lampenbürstenchromosomen eine solche reversible Modifikation der Chromosomenstruktur an Stellen ist, an denen sich aktivierte Gene befinden. Puffs und Lampenbürsten- schleifen kann man also gemeinsam als „chromosomale Funktions- strukturen" bezeichnen.

Die funktionelle Natur der Schleifenbildungen wurde zuerst auf auto- radiographischem Weg demonstriert [*17*]. Nach Inkubation von Ovarien von *Triturus spec.* mit tritiiertem Uridin waren die Schleifen in den Oocytenkernen markiert (Abb. 9), damit also angezeigt, daß hier eine RNA-Synthese stattfindet. Umgekehrt besteht auch eine Abhängigkeit zwischen RNA-Synthese und der strukturellen Stabilität der Schleifen. Wird nämlich die Synthese von DNA-abhängiger RNA in den Oocyten- kernen durch Actinomycin gehemmt, dann werden die Schleifenstruk- turen zurückgebildet. Der gleiche Effekt kann auch durch eine Behand- lung mit bestimmten Histon-Fraktionen erzielt werden [*27*]. Vergleich- bare Resultate wurden an den Lampenbürstenschleifen der Y-Chromo- somen von *Drosophila* in Spermatocytenkernen erzielt [*36*]. Wenn man männlichen Imagines oder Larven des letzten Stadiums von *Drosophila hydei* Actinomycin injiziert, und zwar nur 0,01 μg pro Tier, findet man bereits wenige Stunden nach der Injektion ebenfalls charakteristische Änderungen an den Schleifen. Die Veränderungen führen etwa 24 Std. nach der Injektion zu völliger Kollabierung der Y-Schleifen. Auch hier hat also offenbar die Hemmung der DNA-abhängigen RNA-Synthese eine Rückbildung der chromosomalen Funktionsstrukturen zur Folge. Ungefähr drei Tage nach der Injektion werden die Schleifen wieder neu gebildet und erlangen allmählich ihr normales Aussehen völlig zurück. Ähnlich wie die Schleifenbildungen der verschiedenen Arten von Lampen- bürstenchromosomen verhalten sich auch die Puffs in den polytänen Speicheldrüsenchromosomen von *Chironomus* [*2*].

Bei den Lampenbürstenchromosomen in Oocyten von Amphibien wurden die Vorgänge bei der RNA-Synthese in den Schleifen weiter untersucht [*17*]. Die größeren Schleifenpaare sind, wie bereits erwähnt, in der Regel deutlich polar gebaut. In diesen Schleifen ist nun nach einer verhältnismäßig kurzzeitigen Inkubation mit radioaktivem Uridin immer nur ein kurzer Abschnitt am dünneren Schleifenende markiert. Die markierte Zone wird größer, wenn die Inkubationszeit verlängert wird. Erst bei Inkubation von mehreren Tagen bis Wochen ist schließlich die ganze Schleife rundum markiert (Abb. 10).

GALL und CALLAN haben diese Beobachtung mit der Annahme inter- pretiert, daß die DNA in den Schleifen nur eine verhältnismäßig kurze Aktivitätsphase habe. Die DNA soll sich am dünneren (vorderen) Teil der Schleife aus der zugehörigen Chromomere entfalten und nur unmittel- bar nach der Entfaltung für kurze Zeit als Primer für eine Synthese von DNA-abhängiger RNA aktiv sein, vielleicht dadurch, daß die DNA am Vorderende der Schleife einen stationären Syntheseapparat durchläuft. Die Größe der Schleife soll dadurch konstant gehalten werden, daß in gleichem Maße, wie sich die DNA aus der vorderen Hälfte eines Halb- chromomers entfaltet, am Ende der Schleife Achsen-DNA wieder in die hintere Chromomerenhälfte aufgefaltet wird (vgl. Abb. 6). Auf diese

Weise soll bei beständiger äußerer Form eine kontinuierliche Wanderung der DNA-Achse zustande kommen. Die während der kurzen Aktivitätsphase zu Beginn der Entfaltung synthetisierten Produkte bleiben am Ort ihrer Bildung an der DNA-Achse haften und werden von ihr allmählich um die ganze Schleife mitgeführt. Damit wäre das allmähliche Herumwandern der Markierung erklärt.

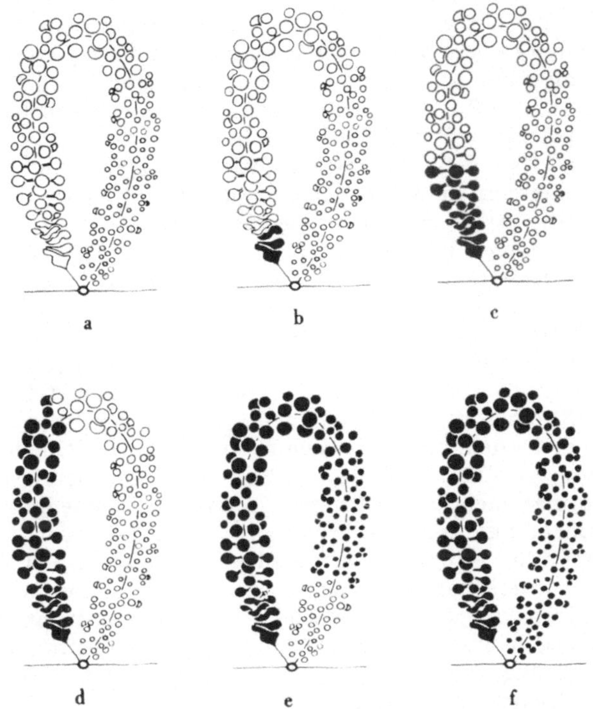

Abb. 10a—f. Fortschreitender Einbau von ³H-Uridin in ein großes Schleifenpaar von *Triturus cristatus*. Markierte Abschnitte schwarz. — Schematische Darstellung aus CALLAN [*11*]. a Vor der Injektion von ³H-Uridin, b einen Tag, c zwei Tage, d vier Tage, e sieben Tage, f vierzehn Tage nach Injektion

Wesentlich bei diesen Vorstellungen ist die Annahme, daß die an der DNA synthetisierte RNA nicht sofort nach der Bildung ins Cytoplasma abgegeben wird, wie das bei den Puffs der polytänen Chromosomen gefunden worden ist [*37*], sondern zumindest für eine längere Zeit in situ akkumuliert wird. Und nicht nur das, es findet offenbar am gleichen Ort auch eine Verbindung von RNA mit Proteinen statt, wobei vorläufig offen bleiben muß, ob diese Proteine ebenfalls an Ort und Stelle neusynthetisiert werden oder aber von anderen Stellen in der Zelle herangebracht werden. Aus diesem Modell ist außerdem die Folgerung zu ziehen, daß die DNA in einem Chromomer bzw. in einer Schleife aus einer ganzen Anzahl von hintereinander angeordneten Ableseeinheiten besteht, die zu verschiedenen Zeiten nacheinander aktiv sein können. Dies würde übrigens die beträchtliche Länge einzelner Schleifen, die unter Umständen bis zu 100 μ lang werden können, verständlicher werden lassen. Eine DNA dieser Länge muß ja ein Vielfaches an Information enthalten, als für ein Struktur- oder Enzymeiweiß notwendig ist.

Ein Teil der aus dem von GALL und CALLAN entwickelten Modell zu ziehenden Folgerungen kann an den Schleifen des Y-Chromosoms in den Spermatocytenkernen von *Drosophila* experimentell nachgeprüft werden. Das Y-Chromosom von *Drosophila hydei* bildet nur einige wenige Schleifen aus, die aber besonders groß und außerdem morphologisch sehr verschieden sind (Abb. 3a). Wir nehmen an, daß die starken Unterschiede in der Form jeder einzelnen Schleife durch den Besitz spezifischer Nucleoproteide geprägt werden. Diese Nucleoproteide scheinen nicht nur schleifenspezifisch, sondern auch artspezifisch zu sein. Homologe Schleifen von nahe verwandten Arten (z. B. *Drosophila neohydei, D. bifurca, D. eohydei, D. nigrohydei*) zeigen nämlich artspezifische morphologische Charakteristika [24]. Die fertilen Artbastarde von *Drosophila hydei* und *D. neohydei* haben in ihren Spermatocytenkernen niemals Mischformen, sondern stets Y-Schleifen des reinen Typs der Art, von der das Y-Chromosom abstammt. Auch wenn durch wiederholte Rückkreuzungen zur mütterlichen Art diejenigen Autosomen, die von der gleichen Art abstammen wie das Y-Chromosom, nach und nach eliminiert werden, bildet das Y-Chromosom in dem fremden Genom immer noch Schleifen von der für seine eigne Art charakteristischen Form [23]. Das deutet darauf hin, daß die Informationen für die Proteine, die jeder Schleife ihre spezifische Form verleihen, auf dem Y-Chromosom selbst lokalisiert sind. Noch weiteren Aufschluß erbrachten Untersuchungen von X-Y-Translokationen, bei denen verschiedene Fragmente des Y-Chromosoms an ein X-Chromosom transloziert waren. Noch verhältnismäßig kleine Fragmente des Y-Chromosoms können, wenn sie nur einen Schleifenbildungsort enthalten, selbständig Schleifen von ganz normaler Form bilden [22]. Die schleifenspezifischen Proteine müssen also von Genen codiert sein, die im Y-Chromosom unmittelbar bei den Bildungsorten liegen, wenn sie nicht überhaupt in den Schleifen selbst lokalisiert sind.

Durch Röntgenbestrahlung sind Mutationen induziert worden, durch die die Form der fadenförmigen Schleifen verändert wird. Bei der Mutante *tube-proximal* ist das aus stark lichtbrechenden kompakten Fäden bestehende Schleifenpaar in weite Schläuche umgewandelt (Abb. 11a). Durch eine zweite Mutation, *tube-distal*, werden die diffusen Fadenabschnitte in zwei Knäuel von dünnen Schläuchen umgewandelt (Abb. 11b). Bei Kreuzungsexperimenten sind beide Mutationen nicht vom Bildungsort der Fäden auf dem Y-Chromosom zu trennen. Lokalisation und Sequenz der Schleifenbildungsorte auf dem Y-Chromosom sind nicht verändert, und es sind auch keine cytologisch sichtbaren Veränderungen des Y-Chromosoms gefunden worden. Die Fertilität der Männchen wird von beiden Mutationen nicht beeinflußt. In Spermatocytenkernen von Männchen, die zwei genetisch verschiedene Y-Chromosomen haben, werden von jedem der beiden Y-Chromosomen Fäden von der ihrer genetischen Konstitution entsprechenden Form gebildet (Abb. 11c). Auch hier treten wie bei den Artbastarden niemals Mischformen auf. Die Faktoren, die die Form der Schleifen determinieren, sind also nur in Cis-Stellung wirksam [20, 21].

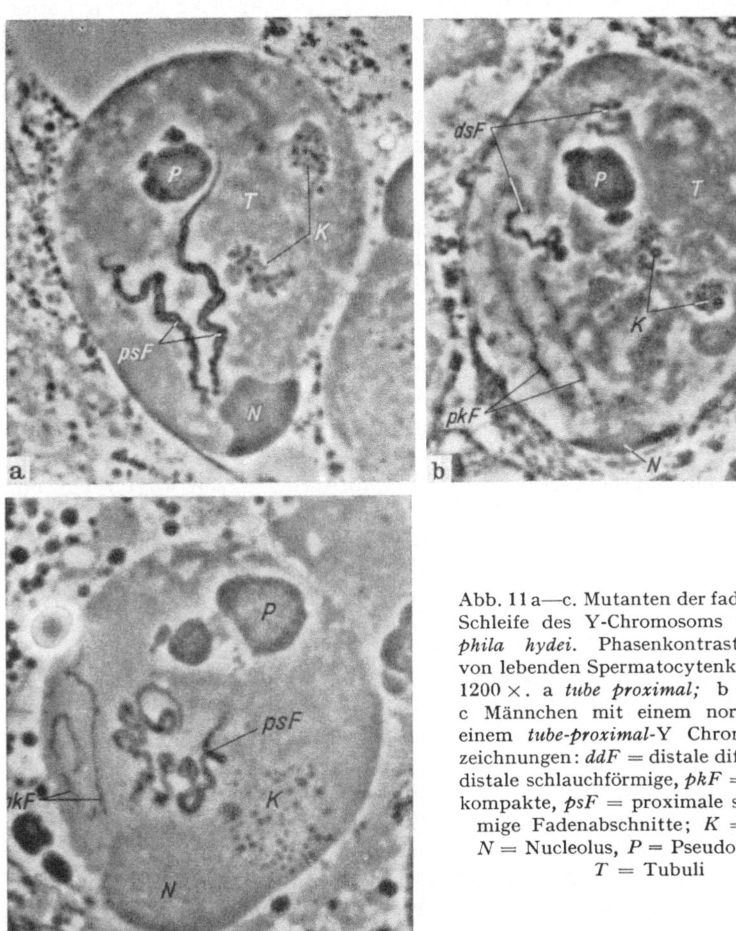

Abb. 11a—c. Mutanten der fadenförmigen Schleife des Y-Chromosoms von *Drosophila hydei*. Phasenkontrastaufnahmen von lebenden Spermatocytenkernen, etwa 1200 ×. a *tube proximal;* b *tube-distal;* c Männchen mit einem normalen und einem *tube-proximal*-Y Chromosom. Bezeichnungen: *ddF* = distale diffuse, *dsF* = distale schlauchförmige, *pkF* = proximale kompakte, *psF* = proximale schlauchförmige Fadenabschnitte; *K* = Keulen, *N* = Nucleolus, *P* = Pseudonucleolus, *T* = Tubuli

Die geschilderten Befunde passen zum Teil recht gut zu dem von GALL und CALLAN entworfenen Bild, zum Teil ergeben sich aber auch Widersprüche. Die Schleifen des Y-Chromosoms von *Drosophila hydei* sind 30—50 μ lang. Eine DNA dieser Länge reicht aus, um wenigstens ein Dutzend Proteine zu codieren. Das wäre in Einklang mit GALL und CALLAN, die ja annehmen, daß die DNA einer Schleife aus einer Anzahl von Informationseinheiten besteht, die hintereinander liegen und nacheinander als Primer für eine RNA-Synthese aktiviert werden. Es ist naheliegend anzunehmen, daß die von der DNA einer Schleife codierten Proteine dieselben sind, die sich an die Schleifenachse anlagern und ihr so eine bestimmte Form verleihen. Verschiedene Schleifen codieren jeweils andere Proteine und erhalten deshalb auch eine unterschiedliche Form. Da diese Form aber bei *Drosophila hydei* auf der ganzen Länge einer Schleife konstant ist, wäre zu schließen, daß die Untereinheiten der DNA einer Schleife identisch sein müssen. Dann aber ist nicht zu verstehen, wie durch einen verhältnismäßig einfachen Mutationsschritt alle diese Untereinheiten gleichzeitig und in gleichem Sinne verändert werden können; denn nur so wäre denkbar, daß die Form einer Schleife auf ihrer ganzen Länge verändert wird, wie das bei den oben beschriebenen Mutanten von *Drosophila hydei* tatsächlich gefunden worden ist. Andererseits kann man sich schwer vorstellen, wie eine (mutable) Spezifität zwischen DNA-Achse und bestimmten Proteinen beschaffen sein sollte, wenn man einmal annehmen will, daß das Gen für das jeweilige Schleifenprotein nicht in der Schleife selbst, sondern in der Nähe des Schleifenbildungsortes lokalisiert ist.

Die Ergebnisse aus unseren Untersuchungen an den Schleifenmutanten sind also ein Indiz dafür, daß die DNA einer Schleife nicht aus mehreren und identischen Informationseinheiten besteht. Das gibt Veranlassung, die Hypothese von Gall und Callan durch eine neue Arbeitshypothese zu ersetzen. Beermann [4] formulierte wie folgt: Nur ein kurzes Anfangssegment der DNA aus einem Chromosom kann als Primer für eine DNA-abhängige RNA-Synthese fungieren, z. B. für die Bildung von messenger-RNA für bestimmte Proteine. Die Wanderung der RNA-Markierung um die Schleife würde dadurch entstehen, daß sich die restliche DNA in der Schleife nach und nach mit der vom Anfangssegment synthetisierten RNA beläd. Die Bindung der RNA an die hinteren Abschnitte der DNA-Achse könnte durch spezifische Basensequenzen ermöglicht werden, die im Anfangssegment und in den zahlreichen Speichersegmenten identisch wären. Solche partiellen Sequenzhomologien könnten während der Evolution z. B. durch (partielle) Mehrfachreplikation des informativen Anfangssegments entstanden sein. (Vgl. hierzu die Ergebnisse von Keyl [29], nach denen homologe Querscheiben von Polytänchromosomen in mehreren nahe verwandten Arten von *Chironomus* DNA-Mengen enthalten, die dem 2-, 4-, 8fachen usw. eines Grundwertes entsprechen, was wahrscheinlich ebenfalls durch selektive Mehrfachreplikation einzelner Replikationseinheiten entstanden sein dürfte. Vgl. auch den Beitrag von Herrn Dr. Keyl in diesem Symposium). Bei dieser Situation ist also anzunehmen, daß jede Schleife eines Lampenbürstenchromosoms aus einem verhältnismäßig kleinen Abschnitt informativer DNA besteht, während die Hauptmenge der DNA die Funktion hat, als Substrat für eine spezifische Speicherung von messenger-RNA und deren Bindung mit besonderen Proteinen zu dienen. Dieses Modell würde auch noch eine bestehende theoretische Schwierigkeit einleuchtend erklären können. Man kann nämlich abschätzen, daß ein höherer Organismus die Information für etwa 10^4 verschiedene Enzym- und Strukturproteine benötigt. Auf der Grundlage eines Coding-Verhältnisses von drei Nucleotidpaaren für jede Aminosäure errechnet sich daraus eine DNA-Menge, die um Größenordnungen unter den tatsächlich gemessenen Werten liegt. Außerdem schwanken nicht selten die DNA-Werte pro Genom zwischen nahe verwandten Arten sehr stark. Die einfachste Erklärung nach dem oben entwickelten Bild wäre hierfür, daß eben nur ein Teil der im Genom enthaltenen DNA wirklich Informationen enthält, während ein von Art zu Art verschieden großer Anteil der DNA andere Hilfsfunktionen zu erfüllen hat.

D. Schlußbemerkungen

Am Schluß dieser zusammenfassenden Darstellung der strukturellen und funktionellen Organisation der Lampenbürstenchromosomen sind noch einige Betrachtungen über die Frage notwendig, welchen Sinn wohl die Ausbildung dieser doch recht komplizierten Maschinerie haben könnte. Beim heutigen Stand unseres Wissens darf man vermuten, daß in den Chromosomen der höheren Lebewesen die Aktivierung von Genen immer mit der Ausbildung spezieller Funktionsstrukturen einhergeht, daß diese aber im Normalfall unter der Sichtbarkeitsgrenze im Lichtmikroskop liegen. Was die Lampenbürstenchromosomen von diesem generellen Fall unterscheidet, dürfte die Akkumulation der primären Syntheseprodukte des genetischen Materials, also die Anhäufung von Ribonucleinsäuren sein. Weiterhin werden diese Ribonucleinsäuren in situ mit besonderen Proteinen verbunden, wodurch besonders große Strukturen von charakteristischer Form entstehen können. Die Bildung dieser komplizierten Strukturen mag wohl damit in Zusammenhang stehen, daß solche komplexer organisierten Lampenbürstenstrukturen, soweit wir wissen, stets in Keimbahnzellen unmittelbar vor der Reduktionsteilung entstehen. Hier hat aber der Organismus ein im Lebenszyklus einmaliges Problem zu lösen: Die Bereitstellung von genetischer Information außerhalb der Chromosomen. Bei *Drosophila* z. B. ist schon lange bekannt, daß die Spermiogenese genotypisch determiniert ist, entscheidend aber nicht der Genotyp der sich differenzierenden Spermatide, sondern der Genotyp der diploiden Spermatocyte ist. Obwohl auf dem Y-Chromosom wenigstens sieben Fertilitätsfaktoren lokalisiert sind, die alle vorhanden sein müssen, damit ein Männchen fertil ist und bewegliche Spermien bilden kann, differenzieren sich Spermatiden, die ein X-Chromosom und kein Y-Chromosom haben, zu funktionsfähigen Spermien aus. Andererseits können aber bei XO-Männchen die X-Spermatiden, obwohl sie genetisch mit den X-Spermatiden von normalen XY-Männchen völlig identisch sind, keine beweglichen Spermien bilden. Die auf dem Y-Chromosom liegende entscheidende Information muß also vor der Reduktionsteilung den Keimbahnzellen verwertbar mitgeteilt worden sein. Sie fehlt bei den XO-Tieren.

Entsprechende Beispiele lassen sich auch aus der weiblichen Keimbahn anführen. Bei manchen Schneckenarten gibt es z. B. Tiere mit links- und mit rechtsgewundener Schale. Die Windungsrichtung wird vererbt, und zwar wird sie durch ein einziges, mendelndes Genpaar determiniert.

Außer der Windungsrichtung der Schale werden durch dieses Genpaar auch die Windungsrichtungen der ersten Furchungsteilungen (die Mollusken besitzen eine sog. Spiralfurchung) bestimmt. Bei der Analyse des Erbganges wurde nun festgestellt, daß die Windungsrichtungen der Furchungsteilungen und damit der Schale nicht dem Genotyp des Embryos folgen, sondern dem der Mutter (Abb. 12). So können z. B. Tiere rechtsgewunden sein, wenn das so vom mütterlichen Genbestand determiniert worden ist, obwohl sie selbst für den Faktor „linksgewunden" homozygot sind. Ihr eigner Genotyp wird erst in der nächsten Generation manifest [38].

Dieses Phänomen, das als „Prädetermination" bezeichnet wird, bedeutet in die Sprache der Molekularbiologie übersetzt, daß die Information des mütterlichen Genotyps dem Ei im Ovar vor der Reifung mitgeteilt worden ist und daß während der ersten Entwicklungsstadien alle für das betreffende Merkmal entscheidenden Prozesse unter Verwendung dieser deponierten mütterlichen Information ablaufen, während das Genom des Embryos selbst in dieser Zeit inaktiv ist bzw. nicht abgelesen werden kann. Seit einiger Zeit gibt es direkte Beweise für diese Vorstellungen. Sowohl in Seeigel-Eiern als auch in Amphibien-Keimen findet während der ersten Entwicklungsstadien bis etwa zur Gastrulation hin im Keim selbst praktisch keine Synthese von DNA-abhängiger RNA statt. Alle Prozesse laufen mit Hilfe bereits vorgebildeter Messenger ab, die bei der Besamung aktiviert werden. Im reifen Ei müssen also große Mengen vorgebildeter messenger-RNA-Moleküle und andere Substanzen deponiert sein. Die ersten Entwicklungsvorgänge können deshalb völlig unabhängig vom Kern ablaufen (Seeigel: [18, 19]; Amphibien: [6, 7]).

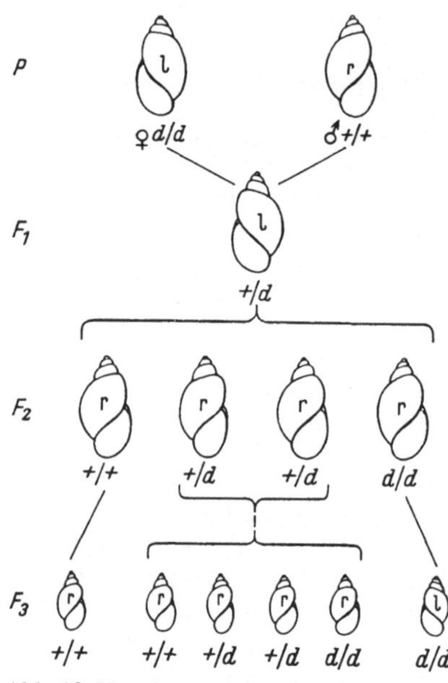

Abb. 12. Vererbung der Eigenschaft „Windungsrichtung" bei *Limnaea peregra*. — Nach Diver, Boycott u. a. [13], aus Kühn (Grundriß d. Vererbungslehre, Quelle und Meyer, 1961), etwas verändert. Bezeichnungen: *r* = rechts-, *l* = linksgewundene Schale; + = Gen für Rechtswindung, *d* = rezessives Allel für Linkswindung

Nun wissen wir aus zahlreichen Untersuchungen, daß messenger-RNA in der Regel sehr instabil ist. Man kann also voraussagen, daß es in Keimbahnzellen einen speziellen Mechanismus gibt, der die sonst instabile messenger-RNA stabilisiert. Es ist durchaus möglich, daß die Anhäufung von neusynthetisierter RNA in situ und deren Verbindung mit besonderen Proteinen in den Schleifen der Lampenbürstenchromosomen (vielleicht zu sog. „Informosomen") dieser Mechanismus ist.

Für den speziellen Fall der Schleifenstrukturen in Spermatocyten-kernen von *Drosophila* läßt sich noch folgendes anführen. Die Spermatozoen von *Drosophila* sind außergewöhnlich groß. Dabei besteht eine direkte Korrelation zwischen der artspezifischen Spermiengröße und dem Grad der Kompliziertheit der Schleifen des Y-Chromosoms in den Spermatocyten. *Drosophila hydei* hat nicht nur die größten und am meisten differenzierten Strukturen, sondern auch mit etwa 6,5 mm (!) die längsten Spermien unter allen bisher geprüften über 50 Arten der Gattung *Drosophila*. Es ist sicher, daß zur Differenzierung solcher großen und auch kompliziert gebauten Gebilde (vgl. [*35*]) in der Spermatide während der Spermiogenese ganz besondere Syntheseleistungen in relativ kurzer Zeit vollbracht werden müssen. Es könnte also sein, daß die Schleifen des Y-Chromosoms nicht nur die Aufgabe haben, messenger-Moleküle eine zeitlang zu stabilisieren, sondern auch durch Anhäufung der Ausgangsmaterialien die Möglichkeit zu schaffen, daß innerhalb kurzer Zeit in einer Zelle besonders große Mengen von einigen wenigen Substanzen synthetisiert werden können. Daß die Schleifenbildungen des Y-Chromosoms tatsächlich etwas mit der Erhöhung der Synthesekapazität für Prozesse während der Spermiogenese zu tun haben können, zeigt sich direkt in dem Befund, daß Männchen mit zwei Y-Chromosomen oder auch nur mit Verdopplungen eines Teils des Schleifensatzes wesentlich größere Spermatozoen haben als normale Männchen.

Falls sich diese Überlegungen als richtig erweisen sollten, wäre auch verständlich, warum die schleifenartigen chromosomalen Funktions-strukturen bei verschiedenen Arten so sehr verschieden sind, u. U. auch völlig fehlen können. Erstens sind die Spermatozoen bezüglich Größe und Morphologie von Art zu Art sehr verschieden, d. h. der Synthese-aufwand, der von der Spermatide während der Spermiogenese zu leisten ist, ist sehr unterschiedlich; zweitens mögen Lokalisation und damit Segregation der steuernden Informationen sowie deren Ablesung sehr verschieden sein.

Auch bei weiblichen Keimbahnzellen gibt es ähnliche Befunde. Anscheinend laufen nur bei sog. Regulationskeimen die Prozesse der frühen Embryonalentwicklung kernunabhängig, d. h. mit vorgebildeten und stabilisierten Informationen. Dagegen scheinen in sog. Mosaikkeimen, z. B. bei den Tunicaten, bereits die ersten Entwicklungsschritte von Informationen gesteuert zu werden, die im Kern der Blastomeren selbst neusynthetisiert worden sind [*44*]. Das mag eine Erklärungsmöglichkeit für die Tatsache sein, daß nicht in allen Tierarten lampenbürstenartige chromosomale Funktionsstrukturen in Oocytenkernen gefunden worden sind. Es bleibt aber abzuwarten, ob das hier entworfene Bild durch weitere Experimente bestätigt werden kann.

Verfasser dankt Herrn Prof. Dr. W. Beermann für anregende Diskussionen bei der Abfassung des Manuskripts, Herrn Prof. Dr. J. Kezer, Eugene, Oregon, für die Erlaubnis, seine noch nicht veröffentlichten Beobachtungen hier erwähnen zu dürfen und für die Überlassung von Fotos. Ebenso hat Herr Prof. Dr. J. G. Gall freundlicherweise Fotos zur Verfügung gestellt. Dank gebührt auch Herrn E. Freiberg für die Anfertigung einiger Zeichnungen.

Summary

Lampbrush chromosomes which are chromosomes in meiotic prophase with laterally projecting loops, have been observed in oocyte nuclei of several vertebrate, some invertebrate species, and in spermatocyte nuclei of *Drosophila* species. Especially large lampbrush chromosomes occur in oocytes of the urodele Amphibia and in spermatocyte nuclei of *Drosophila hydei* and some other species belonging to the *hydei*-subgroup.

In Amphibia the premeiotic chromosomes show a chromomeric organization and a pair of lateral loops projects from each chromomere. Often the loops show a thin axis and a matrix consisting of many radial projecting fibers attached to it. Moreover, the loops are often asymmetric. One of the insertions of the loops in the chromomere may have long fibers and as one follows around the loops from this end, the projecting fibers become progressively shorter and then disappear so that the second end of the loops is bare of fibers. Some of the large loops show morphological peculiarities in length or thickness, or in the nature of their matrix. These characters are constant and locus specific and have allowed to establish loop maps for the genome of several Amphibian species. Lampbrush chromosomes are highly extensible. Breakages after stretching beyond the elastic limits occur always transversely accross chromomeres in such a way that a pair of lateral loops span the break. The axis of the loops appears to contain DNA and there is evidence that the DNA consists of only one double helix in each loop.

In the primary spermatocytes of *Drosophila hydei* five pairs of loops are formed by organizers located on the Y chromosome. Each pair of these loop-like structures can be identified by its specific morphological appearance.

In oocyte nuclei of some newts, lampbrush chromosomes have been observed in which rings are continuously detached from the loops at specific loci. These rings are fragmented by DNase. Normally, they give rise to small free nucleoli which are usually found in abundance in oocyte nuclei of Amphibia. These observations indicate, that there is a selective replication of one or of a small number of particular loci which are presumably identical with nucleolus organizers. The DNA synthesized within these loci is detached from the chromosome, but it may still be metabolically active. This could explain three facts: the great number of free nucleoli in oocyte nuclei, the claim that in Amphibia the content of DNA is much higher in oocyte nuclei than in nuclei of any somatic tissue, and the high amount of ribosomal RNA in these nuclei. On the other hand, it demonstrates that chromomeres may act as independent units during replication.

The loops of the lampbrush chromosomes are reversible modifications of the chromosome structure at sites of active genes. After incubation with ³H uridine the loops are selectively labelled. Moreover, the loops collapse after inhibition of DNA-dependent RNA synthesis by actinomycin or histones.

Some giant loops of the lampbrush chromosomes of *Triturus* species label with ³H uridine sequentially along their lengths with passage of time. It was assumed that the loops have a restricted zone of RNA synthesis at the thinner insertions and that there is a movement around the loop, either of loop products alone, or of the loop axis and product together.

In *Drosophila hydei* mutations could be induced by X rays which cause alterations in the form of one of the loops. Investigations of these mutations and of species hybrids indicate that proteins which are attached specifically to the axis of the loops to give rise in this way to the specific form of each loop, are coded by genes which are localized on the X chromosome, either very near to the organizer of the corresponding loop, or on the loop itself. Moreover, these genes can act in cis relationship only. It therefore appears that the DNA of a chromomere and its respective pair of loops does not consist of a number of identical units of transcription, but of only one initial segment active in transcription which is followed by partially redundant posterior segments which may serve as substrate in packaging messengers produced by the initial segment. Also, with the aid of such a model the high content of DNA in the chromosomes of higher organisms as well as the variability of DNA content in related species may be explained without necessity of assuming differences in standedness of DNA. On the other hand, the complex machinery of lampbrush loops might act to stabilize messengers before the segregation of chromosomes during meiosis.

Literatur

[1] Beermann, W.: Chromosoma **5**, 139 (1952).
[2] — J. exp. Zool. **157**, 49 (1964).
[3] — Naturwissenschaften **52**, 365 (1965).
[4] — (im Druck). Differentiation at the level of the chromosomes.
[5] Brown, D. D., and J. B. Gurdon: Proc. nat. Acad. Sci. (Wash.) **51**, 139 (1964).
[6] —, and E. Littna: J. molec. Biol. **8**, 669 (1964).
[7] — — J. molec. Biol. **8**, 688 (1964).
[8] Cairns, J.: J. molec. Biol. **6**, 208 (1963).
[9] — Cold Spr. Harb. Symp. quant. Biol. **28**, 43 (1963).
[10] Callan, H. G.: Pubbl. Staz. Zool. Napoli **29**, 329 (1957).
[11] — Int. Rev. Cytol. **15**, 1 (1963).
[12] Davidson, E. H., V. G. Allfrey, and A. E. Mirsky: Proc. nat. Acad. Sci. (Wash.) **52**, 501 (1964).
[13] Diver, C., A. E. Boycott, and S. Garstang: J. Genet. **15**, 113 (1925).
[14] Edström, J. E.: Chromosomal RNA and other nuclear RNA fractions. In: "The role of chromosomes in development", 137—152. New York: Academic Press 1964.
[15] Freese, E.: Cold Spr. Harb. quant. Biol. Symp. **23**, 13 (1958).
[16] Gall, J. G.: Nature (Lond.) **198**, 36 (1963).
[17] —, and H. G. Callan: Proc. nat. Acad. Sci. (Wash.) **48**, 562 (1962).
[18] Gross, P. R., and G. H. Cousineau: Exp. Cell Res. **33**, 368 (1964).
[19] —, L. I. Malkin, and W. A. Moyer: Proc. nat. Acad. Sci. (Wash.) **51**, 407 (1964).
[20] Hess, O.: Verhandl. Dtsch. Zool. Ges., Zool. Anz. Suppl. **28**, 156 (1965).
[21] — Structural modifications of the Y chromosome in *Drosophila hydei* and their relations to gene activity. In: "Chromosomes today", C. D. Darlington and K. R. Lewis (eds.), Oliver and Boyd Ltd., Edinburgh (in press)

[22] — Chromosoma 16, 222 (1965).
[23] HESS, O., and G. F. MEYER: J. Cell Biol. 16, 527 (1963).
[24] —, — Port. Acta Biol. (A) 7, 29 (1963).
[25] HINEGARDNER, R. T., B. RAO, and D. E. FELDMAN: Exp. Cell Res. 36, 53 (1964).
[26] HSU, T. C.: J. Genet. 48, 311 (1948).
[27] IZAWA, M., V. G. ALLFREY, and A. E. MIRSKY: Proc. nat. Acad. Sci. (Wash.) 49, 544 (1963).
[28] — — — Proc. nat. Acad. Sci. (Wash.) 50, 811 (1963).
[29] KEYL, H. G.: Experientia (Basel) 21, 191 (1965).
[30] —, u. C. PELLING: Chromosoma 14, 347 (1963).
[31] MacGREGOR, H. C., and H. G. CALLAN: Quart. J. micr. Sci. 103, 173 (1962).
[32] McCONKEY, E. H., and J. W. HOPKINS: Proc. nat. Acad. Sci. (Wash.) 51, 1197 (1964).
[33] MESELSON, M., and F. W. STAHL: Cold Spr. Harb. Symp. quant. Biol. 23, 9 (1958).
[34] MEYER, G. F.: Chromosoma 14, 207 (1963).
[35] — Z. Zellforsch. 62, 762 (1964).
[36] —, u. O. HESS: Chromosoma 16, 249 (1965).
[37] PELLING, C.: Chromosoma 15, 71 (1964).
[38] PLAGGE, E.: Naturwissenschaften 26, 4 (1938).
[39] PLAUT, W.: J. molec. Biol. 7, 632 (1963).
[40] RIS, H.: Canad. J. Genet. Cytol. 3, 95 (1961).
[41] RUDKIN, G. T., and S. L. CORLETTE: Proc. nat. Acad. Sci. (Wash.) 43, 964 (1957).
[42] TAYLOR, J. H.: Int. Rev. Cytol. 13, 39 (1962).
[43] — The replication and organization of DNA in chromosomes. In: J. H. TAYLOR (Edit.), Molecular Genetics, Part 1, 65—111. New York: Academic Press 1963.
[44] URSPRUNG, H. (im Druck). Control mechanisms in cellular differentiation. In: Genetic control of differentiation. Brookhaven Symp. Biol. 18.

Diskussion

Vorsitz: *Beermann*

Beermann: I should just briefly point out which are the main points in this whole series of investigations. As shown by GALL and CALLAN, there should be a sequential incorporation of precursors of RNA-synthesis along the axes of the loop following the morphological polarity around the loop. This is so in all oocytes in the entire ovary of one animal regardless what phase of growth.

In GALL's model you have a fixed point of synthesis, sitting at the beginning of the loop, and all the DNA is walking through this point and become loaded with the RNA copies.

Now we propose an alternative model based on the fact that you can get mutations which concern the entire loop. This would say that in fact not all of the DNA around the loop is synthetically active, but only a small initial section. The rest of the DNA along the loop is only active in picking up the copies made by the initial master-segment and keeping them in position and perhaps stabilising them for later packaging as ribo-nucleo-protein molecules.

The findings of HESS could be easily explained if you assume that there is a mutation in the master-segment which changes the nature of the RNA-copies picked up by the rest of the loop. If you have a Y mutated and a Y non-mutated in the same nucleus you will find a pair of loops, one of which is mutated and the other is non-mutated. This fact is hard to explain by any model.

Karlson: Bei diesem neuen Modell sehe ich folgende Schwierigkeit: Wenn das entscheidende DNA-Segment, welches mutationsfähig ist, auf dem loop herumwandert, dann müßte es schließlich an dem Endpunkt der Schleife liegen. Eine jetzt vorgenommene Markierung mit Uridin dürfte sich nur hier bemerkbar machen.

Beermann: Unser Modell ließe sich so verändern, daß eine breitere Markierung in dem von Ihnen dargestellten Fall erklärlich wäre. Man muß dabei berücksichtigen, daß dieses Herumwandern ein wochenlanger Prozeß sein kann. Die Markierung der Schleifen gelingt übrigens nicht immer ohne weiteres, wie wir aus Versuchen mit *Drosophila hydei* wissen.

Karlson: Das Hauptargument für das neue Modell basiert auf dem Mutationsgeschehen. Sie nehmen gleichzeitig an, daß es mit dem alten Modell deshalb nicht vereinbar sei, weil es sich um eine Punktmutation handelt und weil die Struktur der Schleife bestimmt sein soll von allen Proteinen und allen Ribonucleinsäuren, die auf der Schleife sitzen.

Wenn man nun aber annimmt, daß für die morphologische Ausgestaltung der Schleife nur jenes Protein wichtig ist, das, sagen wir, am mutierten Anfang der Schleife gemacht wird, dann lassen sich doch beide Modelle vereinbaren.

Beermann: Man muß dann aber zusätzlich erklären, warum das innerhalb eines Kerns autonom ist, warum es in cis-Stellung geht und nicht in trans-, wieso es keine Mischformen gibt, wenn man zwei, ein mutiertes und ein nicht mutiertes Chromosom im gleichen Kern hat.

Taylor: How frequent are these changes?

Hess: The fact that we have found these mutations indicates already that the frequency must be rather high, because it is technically difficult to find them.

Taylor: Do they also affect the fertility?

Hess: No, as far as we see they don't effect the fertility.

Bajer: I understand from the discussion that the model you present here applies to lampbrush-chromosomes in the newt, but is based on observations in *Drosophila*. Is that correct?

Beermann: Yes, of course it is based on both. There is no restriction of this model to lampbrush-chromosomes. We think that the chromomeric organization is a general feature of all chromosomes.

Taylor: How many loops have you identified at this time?

Hess: Five loops each of which has its distinct organizer, and we know the loci of these five organizers. There may be some more.

Grundmann: Sie haben den Begriff der „DNS mit metabolischer Funktion" gebraucht. Gibt es irgendwelche Hinweise über die Zusammensetzung oder über die spezifischen Reaktionen dieser DNS?

Hess: Nein, über Zusammensetzung und Funktion sog. „metabolischer DNA" ist nichts bekannt. Bei den im Vortrag geschilderten Beobachtungen von Kezer ist es möglich, daß die Loci, die im Lampenbürstenzustand während des Oocytenwachstums diese DNA-haltigen Ringe abschnüren, mit Nucleolen-Bildungsorten identisch sind. Man kann also in diesem Fall annehmen, daß Nucleolen-DNA in großer Menge synthetisiert wird. Die Oocytenkerne enthalten, wie man gefunden hat, sehr viel ribosomale RNA. Andererseits nimmt man heute an, daß ribosomale RNA in den Nucleolen gebildet wird. Diese Befunde stimmen also alle gut mit der Vorstellung überein, daß es in diesem speziellen Fall die Funktion der außerhalb der normalen Replikationscyclen synthetisierten DNA ist, die Synthese großer Mengen von ribosomaler RNA zu dirigieren.

Beermann: We will have to go on speculating about the possible functions of the DNA in chromosomes of higher organisms. As Dr. Hess has reminded you, even very closely related species may differ in the DNA-content of their chromosomes by a factor of about eight or 10 or more. And this is not due to polyploidy of any kind.

So we have to think about some other functions for the DNA except of that of coding for specific protein structures. This may have something to do with the chromomeric organization which is also indicated by the fact that the number of chromomers in different organisms is very closely the same. Only the size of the chromomers varies from species to species.

Bier: Wahrscheinlich werden in der frühen Embryogenese nur einige wenige spezifische Proteine sofort gebraucht, etwa solche zum Aufbau der Spindeln usw. So ist es schwer zu verstehen, daß das ganze Genom in Form von Lampenbürstenchromosomen aktiviert ist, das heißt also, daß alle Gen-Orte spezifische Proteine bzw. spezifische RNS synthetisieren.

Könnte vielleicht RNS in Form von unspezifischen Polymerisaten gespeichert werden, um dem Ei einen Vorrat an Nucleotiden zu liefern?

Hess: Das weiß man nicht.

Bier: Wir haben in Münster angefangen, die Oocytenkerne von Insekten auf Lampenbürstenchromosomen zu untersuchen. Die Oocytenkerne panoistischer Ovarien enthalten sehr stark gestrecktes Chromosomenmaterial, an dem dünne Schleifen ansitzen. Gerade dieses stark entspiralisierte Chromosomenmaterial zeigt geringe RNS-Synthese, ähnlich wie die Loops der Spermatocyten von *Drosophila.* Das ist eine gewisse Schwierigkeit für die Deutung dieser Funktionsstrukturen.

Hess: In den Spermatocytenkernen von *Drosophila hydei* bilden neben dem Y-Chromosom auch die Autosomen (wesentlich kleinere) Schleifen aus; diese zeigen, soweit man das aus unseren ersten Einbauversuchen beurteilen kann, einen sehr hohen Umsatz (^3H-Uridin).

Sachsenmaier: Ist bei den Lampenbürstenchromosomen von *Drosophila* in dem Stadium, in dem die RNS-Synthese offenbar beendet ist, auch die Empfindlichkeit gegen Actinomycin geschwunden?

Hess: Während der ganzen Zeit, in der sich das Y-Chromosom von *Drosophila hydei* im Lampenbürstenzustand befindet, nämlich bis kurz vor Beginn der ersten Reduktionsteilung, kollabieren die Schleifen immer unter Actinomycin-Einwirkung.

Sachsenmaier: Wie erklären Sie das?

Beermann: Wenn man keinen Einbau findet, braucht das nicht unbedingt zu bedeuten, daß keine Synthese stattfindet. Es kann auch bedeuten, daß das UTP oder Uridin nicht in den Kern hineingeht. Diese Möglichkeit konnten wir noch nicht ausschließen.

Lokale DNS-Replikationen in Riesenchromosomen

Von

H.-G. Keyl, Tübingen

Mit 7 Abbildungen

Nach der Polytäniehypothese [*21, 1*] entstehen die Riesenchromosomen der Dipteren durch endomitotische Vermehrung der Chromatiden. Untersuchungen über den Verlauf des Chromosomenwachstums mit Hilfe von Kerngrößenmessungen [*12, 4*] und cytophotometrischen Bestimmungen des DNS-Gehalts [*22, 35, 36*] konnten zeigen, daß der Polytänisierungsprozeß rhythmisch und in Verdopplungsschritten abläuft. Darauf stützt sich die Annahme, daß der überwiegende Teil der chromosalen DNS bei der Polytänisierung in geometrischer Progression zunimmt. Zweifel daran, daß sämtliche Anteile der Chromosomen gleichmäßig an dem Wachstum beteiligt werden, bestehen seit Beginn der Riesenchromosomenforschung. Durch Untersuchungen an mitotischen und polytänen Chromosomen von *Drosophila*-Arten hat Heitz [*10, 11*] erkannt, daß sich Teile des Heterochromatins, das sog. α-Heterochromatin, in geringerem Ausmaß an dem Riesenwachstum beteiligen als die β-heterochromatischen und euchromatischen Abschnitte. Die Frage, ob in diesem Zusammenhang auch die DNS-Zunahme im α-Heterochromatin gehemmt wird, ist erst in jüngerer Zeit aktuell geworden. Mit Hilfe von Cytophotometrie [*31, 29*] und Autoradiographie [*5*] konnte inzwischen nachgewiesen werden, daß die DNS des Heterochromatins während des polytänen Wachstums der Chromosomen nicht in dem gleichen Ausmaß repliziert werden muß, wie die übrige chromosomale DNS. Damit ist grundsätzlich festgestellt, daß während des Riesenwachstums eine strenge Proportionalität der DNS-Zunahme längs der Chromosomen nicht obligatorisch ist.

Ähnlich dem Heterochromatinverhalten verursachen die lokalen DNS-Replikationen, über die hier referiert wird, in den Riesenchromosomen Veränderungen des DNS-Verteilungsverhältnisses, nur mit dem Unterschied, daß nicht durch lokale Ausfälle von Replikationsschritten, sondern durch lokalisierte zusätzliche Synthesen der DNS-Gehalt in bestimmten Querscheiben stärker anwächst als in den unmittelbar benachbarten Strukturen, die den normalen Replikationsrhythmus einhalten.

Die ersten Angaben über lokale disproportionale Vermehrung von DNS in Riesenchromosomen stammen von Breuer und Pavan [*6*], sie beziehen sich auf Untersuchungen an den Speicheldrüsen-Chromosomen von *Rhynchosciara angelae*. Bei Vorpuppen dieser Art tritt in drei bestimmten Querscheiben der Chromosomen B und C eine erhebliche Substanzzunahme auf, die als abnorme Steigerungen des DNS-Gehalts dieser Strukturen gedeutet wurde. An zwei Stellen in Chromosom B wird diese DNS-Zunahme erst nach Rückbildung von Puffs sichtbar. In Chromosom C ist sie schon vor der Puff-Bildung an der betreffenden Stelle zu erkennen.

Rudkin und Corlette [30] haben dieses Phänomen an verschiedenen Entwicklungsstadien des Chromosoms B durch Bestimmung der UV-Absorption quantitativ untersucht. Dabei zeigte sich, daß zwischen Prä- und Post-Puffstadien die Menge des absorbierenden Materials an dem distalen Puff-Locus um das 3,9fache und an dem proximal davon liegenden um das 2,5fache angestiegen war, während an anderen Stellen desselben Chromosoms nur das Doppelte an Zuwachs festgestellt werden konnte. Messungen nach RNase-Behandlung ergaben, daß die Substanzzunahme an den betreffenden Querscheiben mit großer Wahrscheinlichkeit durch abnorme Erhöhung des DNS-Gehalts hervorgerufen worden war. An dem Chromosom C wurden bisher keine quantitativen Untersuchungen vorgenommen.

Nach den Angaben von Swift [35] über totale und lokalisierte DNS-Zunahme in den Speicheldrüsen-Chromosomen von *Sciara coprophila*, dürften bei dieser Art ähnliche Verhältnisse vorliegen wie bei *Rhynchosciara*. Swift hat die durch E. Rasch ermittelten cytophotometrischen Daten interpretiert. Die durch Bestimmungen des Total-DNS-Gehalts von verschieden großen Kernen aus Speicheldrüsen des späten 4. Larvenstadiums erhaltenen Meßwerte konnten in eine Verdopplungsreihe eingeordnet werden. Die gleichen Messungen an Vorpuppen durchgeführt, ergaben noch eine weitere, höhere Verdopplungsklasse. Während dieses Entwicklungsstadiums treten an vielen Stellen der Speicheldrüsen-Chromosomen von *Sciara* Puffs auf, die als Orte lokaler DNS-Replikationen angesehen werden. Messungen im Vorpuppenstadium zeigten nur in den Puffs eine DNS-Zunahme und nicht in den pufflosen Vergleichsstücken; dadurch wird aber die Bedeutung der neuen Verdopplungsklasse in dieser Entwicklungsphase unklar. Die lokale DNS-Zunahme bewegt sich, wie den Daten zu entnehmen ist, nicht im Verdopplungsverhältnis. Diese quantitativen Angaben können noch kein sehr klares Bild über die lokalen Replikationen bei *Sciara* vermitteln. Da die Puffs vor Auflösung der Chromosomen normalerweise nicht zurückgebildet werden, war es unmöglich, vergleichende Messungen an demselben Locus vor und nach der Puff-Phase auszuführen; diese sind methodisch einfacher und dürften genauere Ergebnisse liefern.

Die photometrische Bestimmung des DNS-Gehalts großer Puffs ist wegen der niedrigen Absorptionen in diesen Strukturen mit einfachen Geräten nicht mit der gewünschten Präzision zu bewältigen. Auf einem methodischen Irrtum beruht offensichtlich auch der in der Literatur oft zitierte Fall von lokaler exzessiver DNS-Synthese in den Speicheldrüsen-Chromosomen von *Glyptotendipes barbipes*. Die Kinetochorregionen dieser Chromosomen enthalten heterochromatinähnliche Querscheiben [2] mit hoher DNS-Konzentration. Stich und Naylor [34] fanden, daß anstelle dieser blockartigen Strukturen Puffs treten können, wenn die Larven vorher einer Kältebehandlung unterworfen werden. Bei Erwärmung soll nach wenigen Tagen wieder die Rückbildung zum Blockstadium zu erreichen sein. Nach den cytophotometrischen Angaben, die durch STICH und NAYLOR gemacht worden sind, unterscheiden sich Puff- und Blockzustand der Kinetochorregionen um das 8—10fache ihres DNS-Gehalts. Dieses Ergebnis wurde als Beweis dafür angesehen, daß von den Puffs eine metabolische DNS abgegeben werden kann. Eine von uns vorgenommene Nachuntersuchung [15] hat gezeigt, daß diese quantitativen Angaben nicht zutreffen können. Messungen mit dem Zeiss-Mikrospektrophotometer an feulgengefärbten Speicheldrüsen-Chromosomen derselben Species haben keine Anhaltspunkte dafür ergeben, daß diese Chromosomenabschnitte ihren DNS-Gehalt ändern können. Abb. 1 zeigt die Kinetochorregionen der 3 mediokinetischen Speicheldrüsen-Chromosomen von *Glyptotendipes* in den beiden Strukturformen; auf Tab. 1 sind die Relationen zwischen den

Tabelle 1. *Relationen der Extinktions-Integralwerte der Kinetochorregionen aus den Speicheldrüsen-Chromosomen I, II und III von Glyptotendipes barbipes im Puff- und Blockstadium (s. Abb. 1) und einem Vergleichsstück aus Chromosom II (nicht abgebildet).* Messung mit dem Zeiss-Mikrospektrophotometer bei 546 nm an Feulgengefärbten Präparaten

Chromosom (Abb. 1)	Verhältnis Block/Vgl.	Chromosom (Abb. 1)	Verhältnis Puff/Vgl.
I	1,10	I a	1,08
II	1,23	II a	1,37
III	1,17	III a	1,05

Extinktionsintegralwerten der in Abb. 1 eingezeichneten Abschnitte und einem Pufflosen (nicht abgebildeten) Vergleichsstück aufgeführt. Aus den Abbildungen geht hervor, daß subjektiv der Eindruck einer DNS-Differenz zwar bestehen kann, durch die Messungen erweist er sich jedoch als unzutreffend. Bei unserem Material waren die Puffs auch nicht durch Kälte auszulösen, sie traten vielmehr spontan bei einigen aus Gelegen herangezogenen Larven auf. Es ist deshalb bisher nicht klar geworden, ob eine Umwandlung vom Puff- in das Blockstadium oder der umgekehrte Vorgang beliebig oft induziert werden kann.

Aufschlußreiche Informationen über lokale DNS-Replikationen sollten von der H³-Thymidin-Autoradiographie zu erwarten sein. Eine der ersten Untersuchungen mit dieser Methode an Riesenchromosomen wurden von FICQ und PAVAN [8] bei *Rhynchosciara* durchgeführt. Aus den dabei gewonnenen Ergebnissen und weiteren autoradiographischen Studien hat PAVAN [26, 27] ein recht eigenwilliges Bild von dem Verlauf der DNS-Synthese in den Riesenchromosomen und den möglichen Funktionen lokaler Replikationen entwickelt. Da nicht nur an den oben erwähnten „DNS-Puffs", sondern auch an vielen anderen Querscheiben lokalisierter Thymidin-Einbau gefunden werden konnte, wurde gefolgert, daß lokale DNS-Synthesen bei *Rhynchosciara* eine zwangsläufige Begleiterscheinung der Puff-Aktivität sei. Aus der Beobachtung, daß Querscheiben abgeschlossene DNS-Synthesen durchführen, schloß

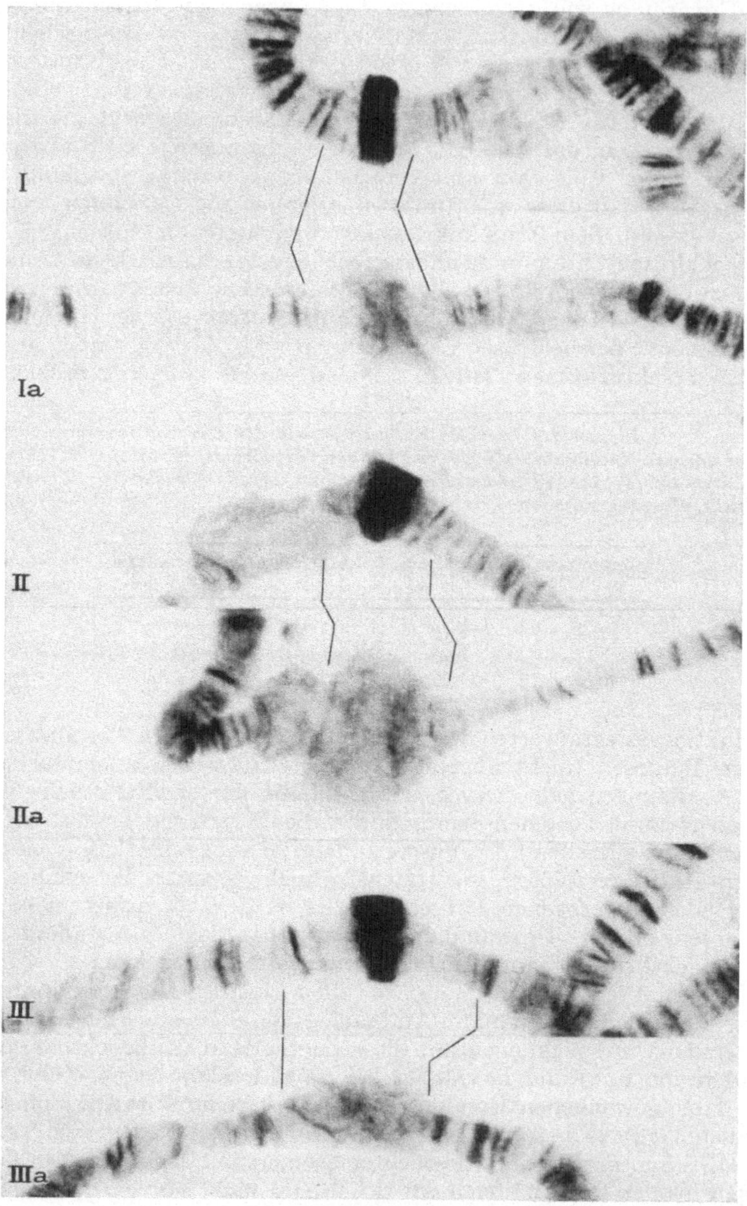

Abb. 1. Kinetochorbereiche der Speicheldrüsen-Chromosomen I, II und III von *Glyptotendipes barbipes* in Block- und Puff-Stadien. Die unter I, II, III und Ia, IIa, IIIa abgebildeten Chromosomen stammen jeweils aus einem Kern. Feulgen-Färbung. Angaben über den *DNS*-Gehalt der eingezeichneten Abschnitte s. Tab. 1

Pavan auf völlige Unabhängigkeit dieser Einzelreplikationen voneinander; daraus wurde die Annahme abgeleitet, daß nur wenige Abschnitte der Chromosomen im Verlauf der Polytänisierung ihren DNS-Gehalt in präzisen Verdopplungsschritten erhöhen. Diese Hypothesen

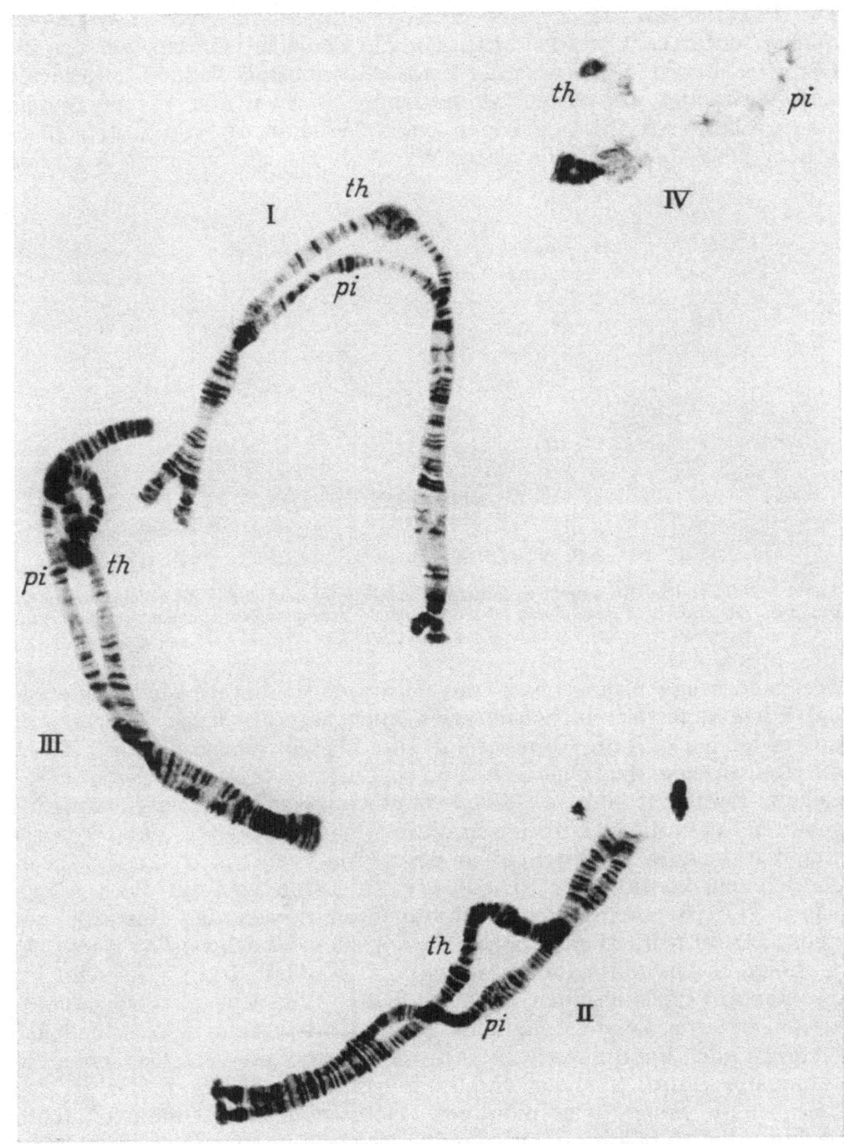

Abb. 2. Speicheldrüsen-Chromosomen des *Chironomus thummi*-Bastards. Die strukturdifferenten Abschnitte liegen innerhalb der ungepaarten Regionen. *th* Chromosomen von *Ch. th. thummi*, pi Chromosomen von *Ch. th. piger*. Feulgen-Färbung

wirken schon deshalb schwer verständlich, weil bisher keine genauen Angaben über das Puff-Aktivitätsmuster und das DNS-Synthesemuster von *Rhynchosciara* vorgelegt worden sind.

Das Problem der lokalen Markierungen bei H³-Thymidin-Autoradiographien von Riesenchromosomen konnte in Zusammenarbeit mit Dr. PELLING an den Speicheldrüsen-Chromosomen von *Chironomus thummi* untersucht werden [19]. Für die Versuche wurden von uns die Chromosomen des Bastards der beiden *Chironomus thummi*-Unterarten [20] verwendet, die gegenüber anderen Objekten den Vorteil bieten, eine größere Anzahl homologer Querscheibenpaare von unterschiedlichem DNS-Gehalt zu besitzen. Wie Abb. 2 zeigt, sind die 3 großen

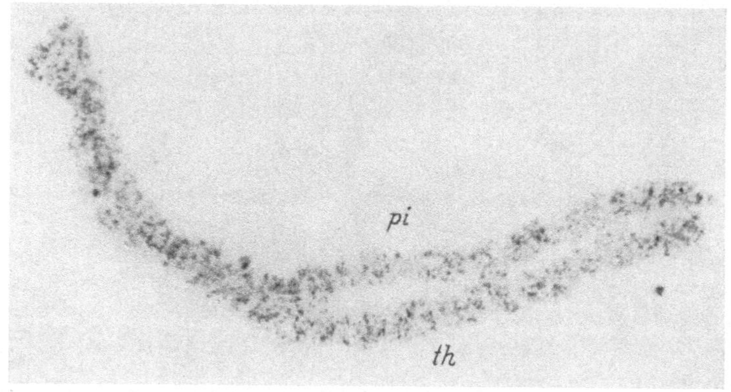

Abb. 3. H³-Thymidin-Einbau in Speicheldrüsen-Chromosom III des *Chironomus thummi*-Bastards. Gleichmäßige Beteiligung aller Querscheiben an der DNS-Synthese zu Beginn der Replikation. Nach KEYL und PELLING [19]

Speicheldrüsen-Chromosomen des Bastards in ihren Mittelregionen, das kurze 4. in der einen Endregion ungepaart. In diesen Abschnitten unterscheiden sich die Chromosomen der beiden Subspecies nicht durch die Reihenfolge der Querscheiben, sondern allein durch deren DNS-Gehalt. Zum Zeitpunkt unserer Untersuchungen wurde es für möglich gehalten, daß diese Differenzen durch eine festgelegte Folge lokaler Replikationen in frühen Stadien des Chromosomenwachstums hervorgerufen sein könnten [14]. Diese Ansicht stützte sich auf die cytologischen Verhältnisse an den Pachytänchromosomen des Bastards, wo weder Strukturdifferenzen noch Paarungslücken nachweisbar sind [20].

Lokalen Thymidineinbau fanden wir an allen dicken Querscheiben der Bastard-Chromosomen (Abb. 4). Durch Versuche mit Doppelmarkierungen unter Verwendung von C¹⁴- und H³-Thymidin in verschiedenen Zeitabständen konnte nachgewiesen werden, daß die verschiedenen Einbaumuster sämtlich einem Replikationsschritt angehören. Da in den *Chironomus thummi*-Chromosomen Heterochromatin offenbar fehlt, beginnt die Replikation an allen Querscheiben ungefähr gleichzeitig (Abb. 3), endet jedoch in den einzelnen Strukturen nach unterschiedlich langer Dauer. Entscheidenden Einfluß auf die Synthesedauer hat der

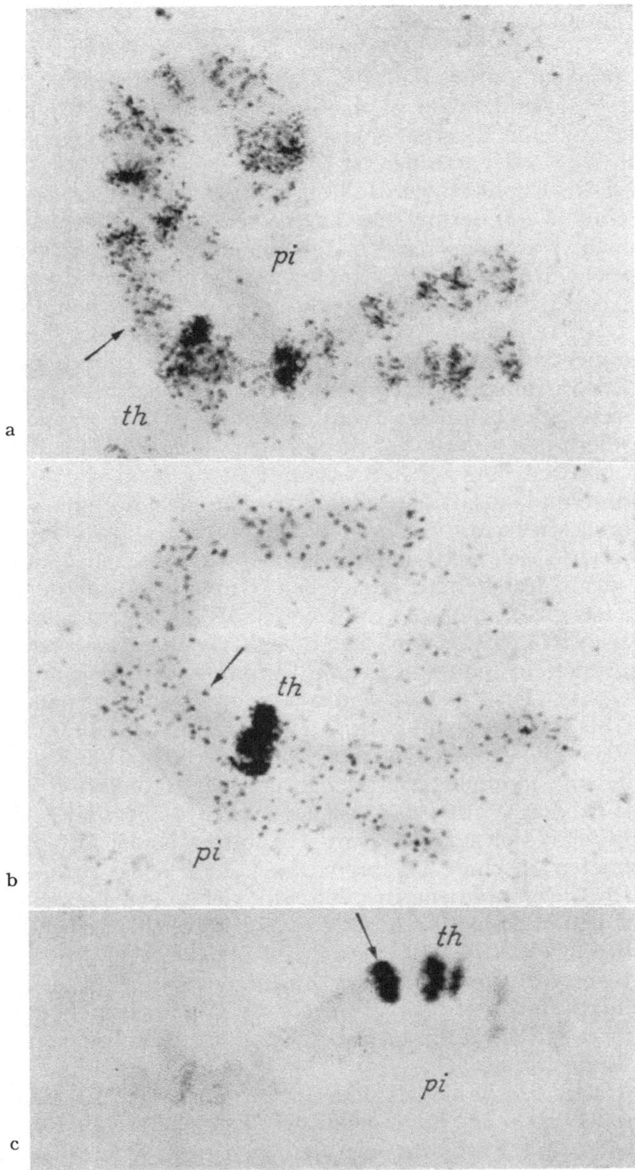

Abb. 4. H³-Thymidin-Einbau in Speicheldrüsen-Chromosom III des *Chironomus thummi*-Bastards während verschiedener Phasen der Replikation. *a* Mittel-, *b, c* Endphase. Der Unterschied zwischen der Replikationsdauer von Querscheiben mit hohem und niedrigem *DNS*-Gehalt zeigt sich besonders deutlich in den ungepaarten Abschnitten. Der Pfeil markiert die Lage der Querscheibe b3,11. Sie zeigt bei hohem *DNS*-Gehalt (*c*) länger andauernden Thymidin-Einbau als bei niedrigerem *DNS*-Gehalt (*a, b*). *a* und *c* nach KEYL und PELLING [*19*]

DNS-Gehalt der Querscheiben; dies beweisen die Markierungsmuster der strukturdifferenten Chromosomenabschnitte (Abb. 4). Das Auftreten lokaler DNS-Synthesen in bestimmten Phasen des Replikationsschritts ist ein Anzeichen dafür, daß die Querscheiben ihre DNS-Synthesen solitär durchführen können, d. h., daß sie jene Replikationseinheiten repräsentieren, deren Existenz schon aus den Untersuchungen an Mitosechromosomen gefordert worden ist [37].

Lokale DNS-Replikationen führen, wie eingangs dargelegt, zwangsläufig zu einer Veränderung des DNS-Verteilungsverhältnisses in den Chromosomen. Nach unseren Vorstellungen ist ein derartiger Zustand in den Speicheldrüsen-Chromosomen von *Chironomus thummi* auf die Dauer des Replikationsschritts beschränkt. In sämtlichen Phasen, die frei von DNS-Synthesen sind, müssen innerhalb der Chromosomen übereinstimmende Proportionen im DNS-Gehalt der Querscheiben bestehen. Zur Prüfung dieser Hypothese auf cytophotometrischem Wege erwiesen sich Speicheldrüsen von *Chironomus anthracinus*, die mit Mikrosporidiern infiziert waren, als besonders tauglich [16]. Diese Infektion, die schon seit längerer Zeit bekannt ist, bewirkt ein hypertrophes Zellwachstum und zusätzliche Replikationen der Riesenchromosomen [13]; dadurch wird eine größere Anzahl von Polytänieklassen, als sie normalerweise in den Chironomus-Speicheldrüsenkernen gefunden werden kann, der Untersuchung zugänglich. Unter der Annahme, daß sich dünne Querscheiben schneller replizieren als dicke, wurden mit dem *Zeiss*-Mikrospektrophotometer die Extinktionsintegrale verschiedener Scheibengruppen an feulgengefärbten Chromosomen bestimmt, die sich in ihrem Gesamt-DNS-Gehalt durch Verdopplung voneinander unterschieden (Abb. 5a und c), außerdem an Chromosomen, die in keine Verdopplungsklasse eingeordnet werden konnten (Abb. 5b). Von den letzteren wurde angenommen, daß sie während der Replikation fixiert worden waren. In den Chromosomenarmen von Abb. 5a und c, deren Gesamt-DNS-Gehalt sich im Verhältnis 1:4 unterscheidet, differieren auch alle bezeichneten Abschnitte in ihrem DNS-Gehalt in der gleichen Größenordnung. Im Chromosomenarm (Abb. 5b), der in seinem Gesamt-DNS-Gehalt von dem in Abb. 5a um das 1,5fache abweicht, haben die dünnen Querscheiben ihre DNS bereits verdoppelt, bei den Strukturen mit höherer DNS-Konzentration ist dieser Zustand noch nicht erreicht. Die heterochromatische Querscheibe, welche die Kinetochorregion markiert, liegt mit ihrer DNS-Synthese, wahrscheinlich infolge asynchronen Beginns, am weitesten zurück.

Lokale DNS-Synthesen als Phasen der normalen Replikation und die sich daraus ergebende Eigenschaft der Querscheiben als Replikationseinheiten, das sind Resultate, die auch den von anderer Seite durchgeführten autoradiographischen Studien an Riesenchromosomen [33, 28, 9] zu entnehmen sind. Eine ausführliche Untersuchung des DNS-Replikationsmusters der Speicheldrüsen-Chromosomen von *Sciara coprophila* wurde durch Gabrusewycz-Garcia ausgeführt [9]. Im Gegensatz zu dem von uns bearbeiteten Objekt *Chironomus thummi* verfügt *Sciara coprophila* an den Kinetochoren und den Telomeren über mehr Heterochromatin, das später mit der DNS-Synthese beginnt als die übrigen Strukturen. Der Verlauf der Replikation scheint den Verhältnissen bei *Chironomus thummi* weitgehend zu entsprechen. Die

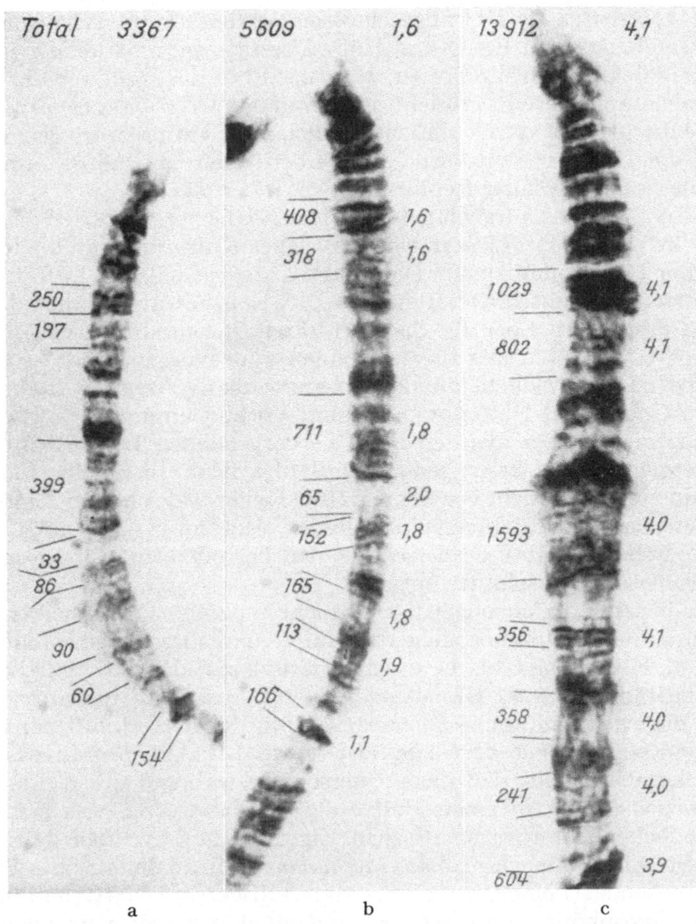

Abb. 5. Chromosomenarm D in verschiedenen Polytäniestufen aus einer Speicheldrüse von *Chironomus anthracinus*. *a* aus einer normalen Zelle, *b, c* aus Zellen mit Mikrosporidier-Infektion. Die links stehenden Zahlen geben die Höhen der Extinktionsintegralwerte der gesamten Chromosomenarme und der eingezeichneten Abschnitte an. Die in *b* und *c* rechts stehenden Werte bezeichnen jeweils das Verhältnis der zugehörigen Extinktionswerte zu denen des Chromosoms in *a*. *b* zeigt ein Chromosom während der Replikation. Nur die Abschnitte mit dünnen Querscheiben enthalten doppelt so viel *DNS* wie die entsprechenden von *a* (abgeschlossene Replikation), die übrigen Abschnitte haben die *DNS*-Synthesen noch nicht beendet. Feulgen-Färbung. Nach KEYL [*16*] verändert

Stellen, an denen im Vorpuppenstadium die Puffs auftreten, führen vorher ihre DNS-Synthese in dem gleichen Zeitabschnitt durch, wie zahlreiche andere dünne Querscheiben. Dies ändert sich, nachdem im Vorpuppenstadium die Puffs entwickelt worden sind; in dieser Phase tritt wahrscheinlich nur noch ein Replikationsschritt auf, ehe die Speicheldrüsen degenerieren.

Bei dieser Replikation konnte noch Thymidin-Einbau in den Puffs nachgewiesen werden, nachdem in den übrigen Chromosomenregionen die DNS-Synthesen bereits abgeschlossen waren. Damit scheinen sich die Meßergebnisse von E. Rasch [35] zu bestätigen, nach denen Replikationen in den Puffs weiterlaufen, wenn das Ende des regulären Chromosomenwachstums bereits erreicht ist. Gabrusewycz-Garcia [9] neigt zu der Ansicht, daß durch die späten lokalen Synthesen in den Puffs metabolische DNS erzeugt wird, wobei aber sicher ist, daß dies im Anschluß an einen normalen Replikationsschritt erfolgt.

Cannon [7] ist es kürzlich gelungen, nach Übertragung der Speicheldrüsen von *Sciara coprophila* in ein künstliches Kulturmedium die Rückbildung der Puffs zu indizieren. Dabei treten an die Stelle der Puffs blockartige Strukturen mit offenbar hoher DNS-Konzentration, ähnlich wie an den 3 erwähnten Loci der Speicheldrüsen-Chromosomen von *Rhynchosciara* [6]. Zur Deutung dieses Kondensationsvorgangs hat Cannon eine Hypothese entwickelt, die den Zusammenhang zwischen Inaktivierung von Genen und Heteropyknose zum Vorbild nimmt [23]. Die Art der Identifizierung der aus den Puffs hervorgehenden DNS-Strukturen als Heterochromatin kann jedoch vorläufig noch nicht überzeugen. Außerdem ist bisher nicht erwiesen, daß bei einer während der Embryonal- oder Larvalentwicklung ausgelösten Heteropyknose auch eine effektive Erhöhung des DNS-Gehalts der betreffenden Chromosomen oder Chromosomenabschnitte eintritt.

Nach unseren Erfahrungen ist es nicht möglich, Querscheiben der Chironomus-Riesenchromosomen nur auf Grund ihrer ungewöhnlichen Dicke und Färbbarkeit als heterochromatisch zu klassifizieren. In bestimmten Stämmen von *Chironomus thummi thummi* tritt in Chromosom III eine ungewöhnlich dicke Querscheibe auf, deren Herkunft zunächst auf lokale Replikationen zurückgeführt wurde [14] (Abb. 6c). Inzwischen hat sich herausgestellt, daß diese Querscheibe während des 4. Larvenstadiums vollständig in einen Puff aufgehen kann [18], was auch an ihren möglichen heterochromatischen Eigenschaften zweifeln läßt. Bei anderen Stämmen von *Chironomus th. thummi* kann an der gleichen Stelle des Chromosoms III eine Querscheibe von niedrigerem DNS-Gehalt liegen (Abb. 6a). Cytophotometrisch wurde ermittelt, daß sich die verschiedenen Strukturtypen dieses Locus nach ihrem DNS-Gehalt in eine Verdopplungsreihe einordnen lassen [17]. Isoliert betrachtet könnte dieses Ergebnis durchaus als Nachweis dafür angesehen werden, daß die Strukturdifferenzen und damit auch die ungewöhnliche Höhe des DNS-Gehalts bei dieser Scheibe durch zusätzliche lokale Replikationen während des Chromosomenwachstums entstehen. Gegen diese Vorstellung spricht, daß das an dieser Querscheibe verwirklichte Prinzip der Veränderung des DNS-Gehalts in Verdopplungsstufen die Grundlage der cytologischen Differenzierungen in den Speicheldrüsen-Chromosomen der beiden *Chironomus thummi*-Unterarten darstellt. An 21 Querscheiben von *Chironomus th. thummi* wurde cytophotometrisch nachgewiesen, daß ihr DNS-Gehalt von dem der homologen Strukturen bei *Chironomus th. piger* in Verdopplungsstufen abweicht [*18*]. Die aus den lokalen Erhöhungen des DNS-Gehalts in den Speicheldrüsen-Chromosomen von *Chironomus th. thummi* resultierenden Unterart-Differenzen im Gesamt-DNS-Gehalt der polytänen Kerne wurden in gleicher Höhe in den Spermatocyten gefunden. Damit konnte gezeigt werden, daß die lokalen Verdopplungen des DNS-Gehalts in den Speicheldrüsen-Chromosomen von

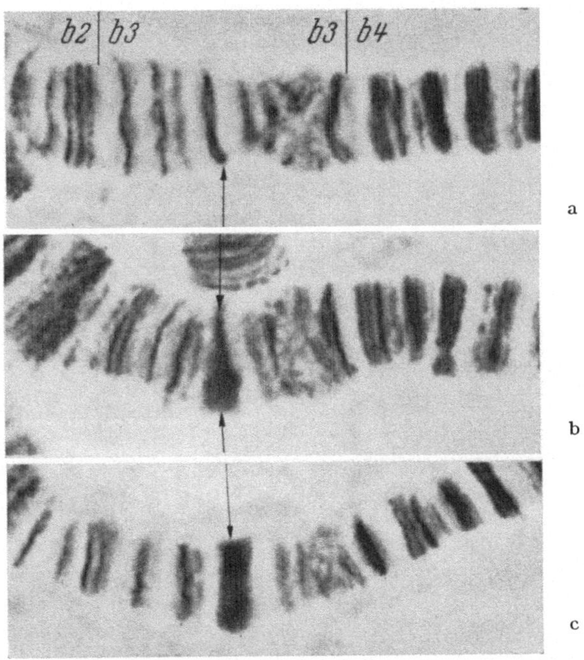

Abb. 6. Querscheibe b3,11 (Pfeil) aus Speicheldrüsen-Chromosom III von *Chironomus th. thummi*. *a, c* mit unterschiedlich hohem *DNS*-Gehalt. *b* Heterozygotie der Typen *a* und *c*. Orcein-Carmin-Essigsäure

Chironomus th. thummi auf Mutationsereignisse eines besonderen Typs zurückgehen müssen, die im Verlauf der Evolution spontan auftreten und als fixierte Strukturveränderungen weitervererbt werden. Aus den H³-Thymidin-Versuchen an dem *Chironomus*-Bastard [*19*] hatte sich bereits ergeben, daß bei der Erhöhung des DNS-Gehalts in den Querscheiben von *Chironomus th. thummi* keine neuen Replikationseinheiten gebildet worden sind, sondern, daß sich die bereits bestehenden vergrößert haben müssen.

Bei dem Versuch, einen Überblick über die erwähnten Erscheinungen von lokaler DNS-Replikation in Riesenchromosomen zu gewinnen, läßt sich nach Meinung des Referenten folgendes Bild entwerfen. Sämtliche DNS-Synthesen in den Riesenchromosomen laufen lokalisiert ab, d. h. innerhalb von Replikationseinheiten, die in Lage und Umfang mit den

Querscheiben übereinstimmen. Die lokalen DNS-Synthesen sind im allgemeinen koordiniert, so daß mit einer streng geometrischen Progression der gesamten DNS während der Polytänisierung gerechnet werden kann.

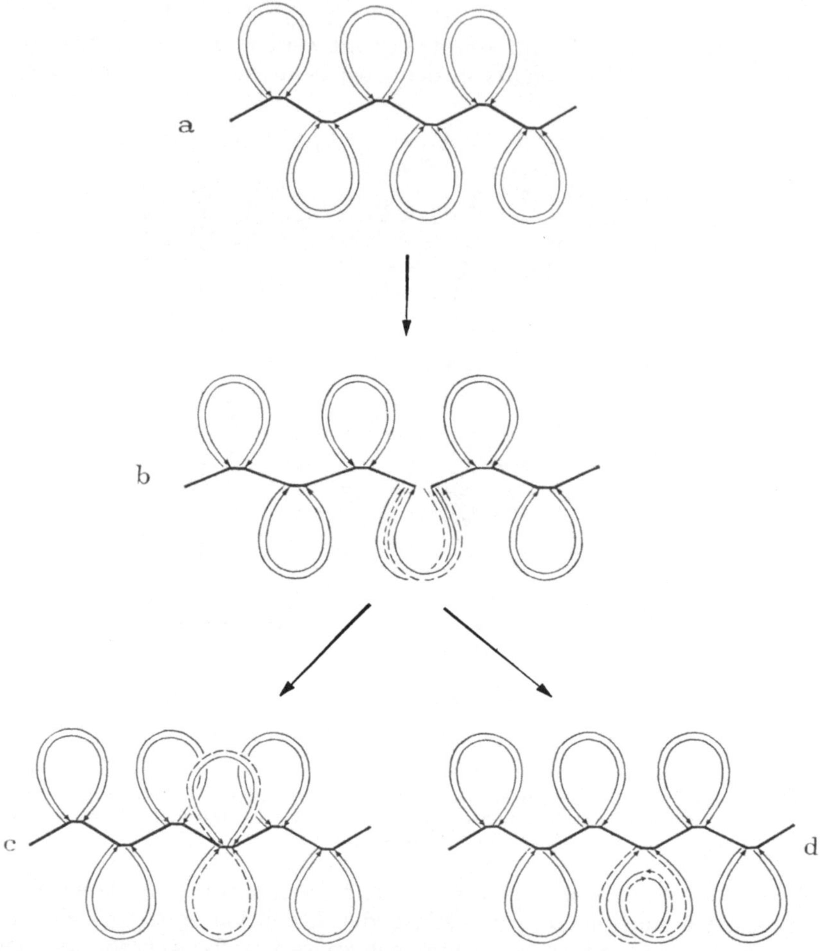

Abb. 7. Lokale *DNS*-Replikation an einem Chromatiden-Modell (s. KEYL [*17*]). Ringförmige *DNS*-Moleküle sind durch Verbindungsstücke unbekannter Zusammensetzung untereinander verbunden (*a*). Bei lokaler Replikation (*b*) muß der betroffene Ring geöffnet werden, um die Trennung der Schwesterstränge zu ermöglichen. Die entstandenen beiden gleichartigen Einheiten können nebeneinander am gleichen Chromatiden-Locus angebracht sein (*c*), dieser Zustand schließt die Möglichkeit der Löslösung eines Ringes ein. Beide Ringe können auch durch Hintereinanderschaltung eine doppelt so große Einheit bilden (*d*)

Die DNS des Heterochromatins macht in dieser Beziehung eine Ausnahme, indem sie sich nicht an sämtlichen Replikationsschritten beteiligen muß, die im Kern stattfinden. Dieses Verhalten ist nicht für jeden Heterochromatintyp verbindlich.

Lokale disproportionale Vermehrung von DNS tritt nach den bisherigen Befunden mit aller Wahrscheinlichkeit an einzelnen Querscheiben der Riesenchromosomen von *Rhynchosciara angelae* und *Sciara coprophila* auf. Über die damit verbundenen quantitativen Veränderungen des DNS-Gehalts an den Chromosomen besteht noch keine Klarheit, dasselbe gilt für die vermuteten Funktionen der lokalen Extrareplikationen bei der Genaktivierung und -inaktivierung. Zur Klärung des letztgenannten Problems müßte eigentlich zuerst untersucht werden, in welchem Ausmaß RNS-Synthesen an den DNS-erzeugenden Puffs stattfinden. Bei *Sciara* fehlen darüber noch jegliche Angaben.

Die lokalen DNS-Replikationen bilden auch einen Diskussionsgegenstand bezüglich der Modellvorstellungen des Chromatidenaufbaus. In Zusammenhang mit unseren Untersuchungen über die lokalen Duplikationen in den Replikationseinheiten der Chromosomen von *Chironomus thummi* wurde ein Modell entwickelt, das als Hilfsmittel für das Verstehen des Mechanismus der schrittweisen Verdopplungen gedacht ist [16, 13]. Das Prinzip der lokalen DNS-Zunahme ist relativ leicht verständlich, wenn angenommen wird, daß die Replikationseinheiten Ringform besitzen. Ähnlich wie bei den Modellen anderer Autoren [32, 25], wurden diese Ringe durch Zwischenstücke unbekannter Zusammensetzung untereinander verbunden. Obwohl es nicht ausgeschlossen ist, daß die Duplikationen im molekularen Bereich auch durch lokale Extra-Replikationen hervorgerufen werden könnten, wurde diese Möglichkeit zunächst umgangen und die Verlängerung der Replikationseinheiten auf das Doppelte durch einen besonderen Austauschmechanismus erklärt [17].

Für die echten lokalen Replikationen kann dieser Mechanismus nicht gültig sein, weil er nicht mit zusätzlicher DNS arbeitet, sondern in Zusammenhang mit einer totalen Replikation der Chromatide eine Synthese-Einheit von einem Strang auf den anderen verschiebt. Läßt man an diesem Chromatiden-Modell eine Replikation lokalisiert ablaufen (Abb. 7a, b), dann erscheint es sinnvoll, zwei Möglichkeiten zu berücksichtigen. Erzeugt die lokale Replikation integrierte Stücke der Chromosomenstruktur, dann kann der Einbau durch Hintereinanderschaltung der beiden identischen Einheiten erfolgen (Abb. 7d). Bleiben dagegen die beiden synthetisierten Einheiten isoliert am gleichen Locus der Chromatide angehängt (Abb. 7c), dann ist darin die Möglichkeit eingeschlossen, daß einer dieser Ringe abgetrennt werden kann, während der andere der Gesamtstruktur erhalten bleibt. Im Hinblick auf die Befunde über DNS-Ablösung von mitotischen Chromosomen [3], insbesondere die Angaben über ringförmige isolierte DNS in den Nucleolen der Oocyten von *Triturus* [24] erscheinen ähnliche Vorgänge auch in Polytänkernen nicht grundsätzlich ausgeschlossen.

Summary

This paper presents an investigation of the meaning of the local DNA synthesis in salivary gland chromosomes of *Sciara*, *Rhynchosciara*, and *Glyptotendipes*. The possibility of a local increase of the DNA content, other than during a replication phase is evidently proved for the chromosomes of *Sciara* and *Rhynchosciara*. However, this type of local increase in the DNA content is a less general event than may be concluded from the work of PAVAN. The occurrence of this type of local DNA increase (metabolic DNA) could not be proved for the chromosomes of *Glyptotendipes*.

At the end of each replication phase of the giant chromosomes a number of loci showing DNA synthesis can be observed. However, such local DNA synthesis is based on differential properties (DNA content) of the individual replication units.

A hypothesis for the construction of such replication units is proposed. This hypothesis presents a plausible basis for the occurrence and the origin of homologous replication units with different DNA content during evolution.

Literatur

[1] BAUER, H.: Z. Zellforsch. **23**, 280 (1935).
[2] — Zool. Jb., Abt. allg. Zool. u. Physiol. **56**, 239 (1936).
[3] BEERMANN, S.: Chromosoma (Berl.) **10**, 504 (1959).
[4] BESSERER, S.: Biol. Zbl. **75**, 205 (1956).
[5] BIER, K.: Verh. dtsch. zool. Ges. **25**, 102 (1961).
[6] BREUER, E. M., u. C. PAVAN: Chromosoma (Berl.) **7**, 371 (1955).
[7] CANNON, G. B.: J. cell. comp. Physiol. **65**, 163 (1965).
[8] FICQ, A., and C. PAVAN: Nature (Lond.) **180**, 983 (1957).
[9] GABRUSEWYCZ-GARCIA, N.: Chromosoma (Berl.) **15**, 312 (1964).
[10] HEITZ, E.: Z. Zellforsch. **19**, 720 (1933).
[11] — Biol. Zbl. **54**, 588 (1934).
[12] HERTWIG, G.: Z. indukt. Abstamm. u. Vererb.-L. **70**, 496 (1935).
[13] KEYL, H.-G.: Naturwissensch. **47**, 212 (1960).
[14] — Chromosoma (Berl.) **12**, 26 (1961).
[15] — Exp. Cell Res. **30**, 245 (1962).
[16] — Verh. dtsch. zool. Ges. §§, 78 (1963).
[17] — Experientia (Basel) **21**, 191 (1965).
[18] — Chromosoma (Berl.) **17**, (im Druck) (1965).
[19] —, u. C. PELLING: Chromosoma (Berl.) **14**, 347 (1963).
[20] —, u. K. STRENZKE: Z. Naturforsch. **11 b**, 727 (1956).
[21] KOLTZOFF, N. K.: Science **80**, 312 (1934).
[22] KURNICK, N. B., and J. H. HERSKOWITZ: J. cell. comp. Physiol. **39**, 281 (1952).
[23] LYON, M. F.: Nature (Lond.) **190**, 372 (1961).
[24] MILLER, O. L.: J. Cell Biol. **23**, 60 A (1964).
[25] MOSES, M. J.: In S. WOLFF, Radiation induced chromosome aberration, p. 155. New York: Columbia Univ. Press 1963.
[26] PAVAN, C.: Proc. X. Intern. Congr. Genetics 321 (1958).
[27] — Proc. XI. Intern. Congr. Genetics **2**, 337 (1963).
[28] PLAUT, W.: J. mol. u. Biol. **7**, 632 (1963).
[29] RUDKIN, G. T.: Proc. XI. Intern. Congr. Genetics **2**, 359 (1963).
[30] —, and S. L. CORLETTE: Proc. nat. Acad. Sci. (Wash.) **43**, 964 (1957).
[31] SCHULTZ, J.: Cold Spr. Harb. Symp. quant. Biol. **12**, 307 (1956).
[32] STAHL, F. W.: Proc. 11th annual reunion, Soc. chimie physique (1962).
[33] STEFFENSEN, D.: Genetics **48**, 1289 (1963).
[34] STICH, H., and J. M. NAYLOR: Exp. Cell Res. **14**, 442 (1958).
[35] SWIFT, H.: In J. M. ALLEN, The molecular control of cellular activity, p. 73. New York 1963.
[36] —, and E. M. RASCH: J. Histochem. Cytochem. **2**, 456 (1954).
[37] TAYLOR, J. H.: In J. H. TAYLOR, Molecular Genetics. New York: Acad. Press 1963.

Diskussion

Vorsitz: *Beermann*

Keyl: Um die Schwierigkeiten, dich sich aus verschiedenen Endopolyploidiegraden bei normalen somatischen Zellen ergeben würden, zu umgehen, haben wir an den praktisch tetraploiden Spermatocyten I gemessen. Spermien enthalten für die Cytophotometrie zu geringe DNS-Mengen.

Beermann: What is found here is a peculiar kind of mutational event which leads to a geometric increase in the DNA-content of single chromomers starting from a given base value in each chromomer. This would indicate that all chromomers are in some way multiple and gives rise to some evolutionary speculations about the development of the chromosomes.

Keyl: Ich möchte annehmen, daß zunächst DNS-Stücke mit identischen Informationen hintereinandergereiht werden. Im Laufe der Evolution können dann Teile mutieren, wobei aber die Grundinformation in den übrigen Teilen erhalten bleibt.

Karlson: Wieviel Proteine können auf einem solchen Verdoppelungsstück codiert sein? Läßt sich das aus der Länge ablesen?

Beermann: Der DNS-Gehalt — zurückgerechnet auf einzelne Chromatiden — ist bei den allerfeinsten Querscheiben, die man gerade noch im Lichtmikroskop sieht, ungefähr 50000 Nucleotidpaare; das hat RUDKIN für *Drosophila* angegeben. Das ist der Minimumwert; bei den Amphibien-Chromosomen dürfte dieser Wert mindestens zehnmal höher liegen.

These are really considerable amounts if you wish to speculate about genetic units in relation to chromomeres. They are approaching the entire DNA content of an E.-coli-chromosome.

Taylor: Were you referring to these giant chromosomes or to a similar situation?

Beermann: I was referring to accidental observations made by the amphibian lampbrush chromosome people. For instance in frogs it seems that the number of chromomeres is not very different from that in newts. What is different, is the size of the loops formed by the chromomeres.

Frösche und Molche unterscheiden sich ungefähr um einen Faktor 8—10 oder noch mehr im DNS-Gehalt ihrer Chromosomen. Das kann man auch für bestimmte Würmer — Turbellarien — feststellen. Bei Pflanzen (Hyazinthen oder manchen Alpenveilchen) gibt es wahrscheinlich noch extremere Fälle.

The pecularity in this particular case studied by KEYL is that it shows a general tendency of all chromosomes of this particular species, to increase the DNA-content in specific regions or by local duplication of the chromomeres. Why there exists such a general tendency to do this, is unclear.

Taylor: Do you mean a general tendency in the salivary gland chromosome only or are you referring to the evolution of normal mitotic chromosomes?

Beermann: We are referring to the evolution of mitotic chromosomes because I think here is good evidence that what one sees in the salivary chromosomes is only an expression of what has happened prior to this in the germ line.

Abel: Die geschilderten Befunde sprechen, wie mir scheint, doch sehr dafür, daß wir nur eine Duplex in einem Chromosom haben.

Beermann: Ja, das ist ein Argument in dieser Richtung.

Abel: Auch aus Experimenten zur Mutationslösung durch UV. (E. KNAPP, Z. Vererbungsl. **74,** 54, 1937) an *Sphaerocarpos* kann nach unseren heutigen Kenntnissen geschlossen werden, daß die Chromosomen der Spermatozoiden sehr wahrscheinlich nur *eine* Duplex enthalten.

Beermann: Die einfachste durch viele Befunde bekräftigte Annahme ist, daß alle Chromosomen höherer Organismen im Prinzip einer Kette von einfachen DNS-Doppel-Helices gleichkommen.

Allerdings müßte noch erklärt werden, wieso man bei einem Chromosom, das an sich unverdoppelt sein sollte, im Mikroskop wenigstens manchmal eine Doppelsträngigkeit sehen kann. Dr. TAYLOR hat den Versuch unternommen, dies zu erklären, vielleicht sollte man aber die Möglichkeit von 2 Doppelhelices doch nicht ganz ausschließen.

Beobachtungen zur Vermehrung und Funktion nucleolärer Strukturen*

Von

Norbert Weissenfels, Bonn

Mit 13 Abbildungen

A. Einleitung

Der Nucleolus wurde vor fast 200 Jahren erstmals beschrieben. Inzwischen ist eine nicht mehr überschaubare Anzahl von Arbeiten über Struktur, chemischen Aufbau und Funktion der Nucleolen erschienen. Bedingt durch die Verschiedenheit der Materialien und Methoden sind die Ergebnisse sehr uneinheitlich, z. T. sogar widersprechend. Dies drückt sich natürlich auch in einer Vielfalt von Deutungen aus, die u. a. Stockinger [31], Vincent [34], Hertl [21], Lettré und Siebs [26], Altmann, Stöcker und Thoenes [1], Weissenfels [36], Wessing [38] zusammenfassend behandelt haben.

In einer Reihe von Versuchen, bei denen es mir in erster Linie um eine lebensgetreue Fixation gezüchteter Hühnerherzmyoblasten ging, ergaben sich auch verbesserte Verfahren zur Darstellung der Kernbestandteile, insbesondere der Nucleolen, die in der lebenden Zelle fast homogen erscheinen (Abb. 1), in Wirklichkeit aber eine große Formen- und Strukturenmannigfaltigkeit aufweisen. Anzeichen für eine Nucleolus-Reduplikation im Hinblick auf die nächste Zellteilung und Hinweise bzgl. der Nucleolus-Funktion bildeten den Ausgangspunkt für die Untersuchungen, über die hier berichtet werden soll.

B. Die Zahlenkonstanz der Nucleoluseinheiten pro Zellkern

Hühnerherzmyoblasten besitzen in der Regel einen diploiden Kern mit zwei Nucleolen, und dieser Bestand ist durch die Mitose auch für die Tochterzellen sichergestellt. Bei Lebendbeobachtungen läßt sich dann und wann die Fusion zweier Nucleolen zu einem größeren, seltener auch der umgekehrte Vorgang verfolgen.

Bei der Durchsicht einer großen Anzahl von Kernen stellte sich eine Beziehung der Nucleolusanzahl zum Interphasealter der gezüchteten Myoblasten heraus, und zwar besitzen die jungen Zellkerne direkt nach der Mitose zwei Nucleolen, lassen aber mit zunehmender Größe im weiteren Verlauf der Interphase mehr und mehr nur noch einen, jedoch doppelt

---

* Herrn Prof. Dr. R. Danneel zur Vollendung seines 65. Lebensjahres in Dankbarkeit gewidmet.

so großen Nucleolus erkennen. In der folgenden Tabelle ist der Prozentsatz der diploiden Kerne mit zwei bzw. einem Nucleolus in drei Korngrößengruppen (I—III) dargestellt.

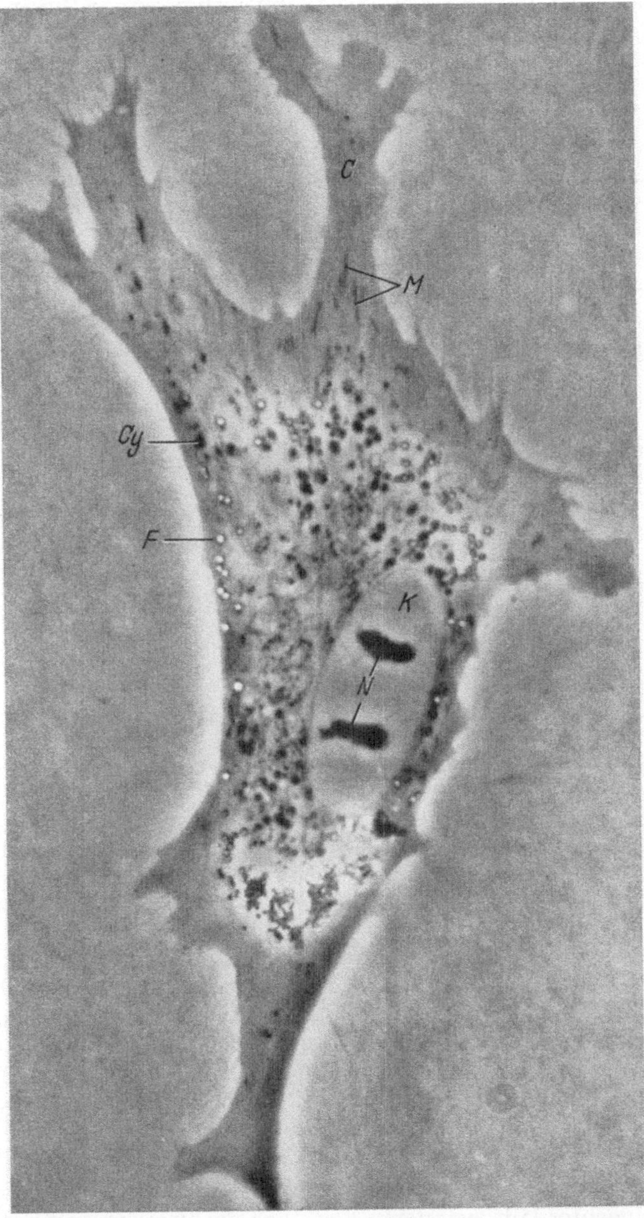

Abb. 1. Phasenkontrastmikroskopische Aufnahme eines lebenden Hühnerherzmyoblasten nach $2^{1}/_{2}$ Tagen Gewebekultur. K = Kern; N = Nucleolen; C = Cytoplasma; M = Mitochondrien; Cy = Cytosomen; F = Fetttropfen. Vergr.: 1500:1

Hieraus ist ersichtlich, daß 25% der kleinen (Gruppe I), 40% der mittelgroßen (Gruppe II) und 50% der großen Interphasekerne (Gruppe III) nur *einen* Nucleolus enthalten. Während der Interphase besteht also die Tendenz zur Nucleolenverschmelzung. Dieser Vorgang hat keinen Einfluß auf die Zahlenkonstanz der Nucleoluseinheiten pro Zellkern.

Tabelle

Kerngruppe	Kerngrößen in Flächeneinheiten	Prozentsatz der Kerne	
		mit 2 Nucleolen	mit 1 Nucleolus
I	24—77	75%	25%
II	78—131	60%	40%
III	132—185	50%	50%

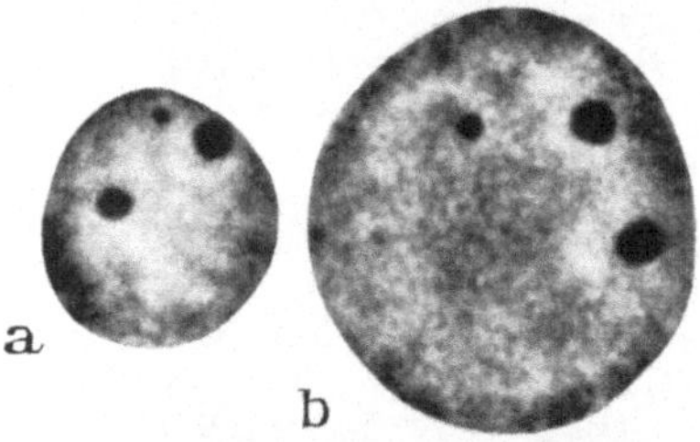

Abb. 2. Zwei Zellkerne aus einem Zwiebelhäutchen (*Allium cepa*). Der kleine Kern a enthält zwei große und einen kleinen Nucleolus, der große Kern b entsprechend größere Nucleolen. Vergr.: 1500:1

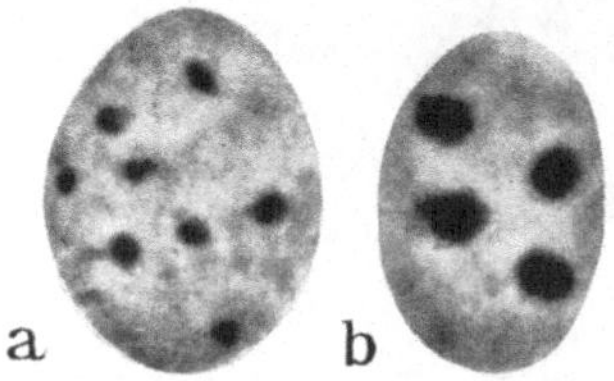

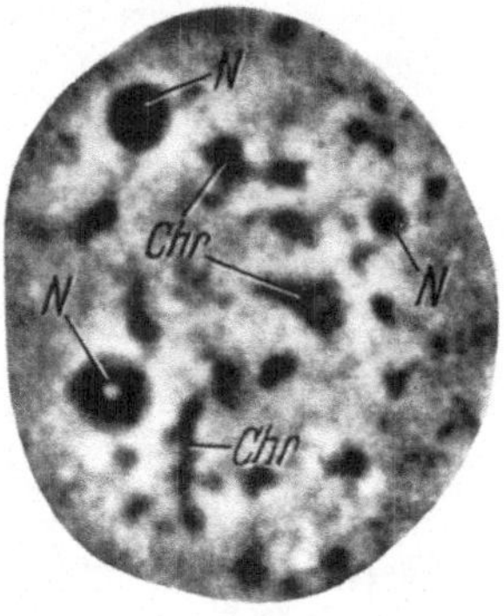

Abb. 3. Zwei Kerne von gezüchteten Myoblasten eines embryonalen Mäuseherzens. Kern a enthält 8 normale, Kern b dagegen 4 Doppelnucleolen. Vergr.: 1500:1

Abb. 4. Zellkern einer Teichmolchkultur. Der Kern enthält eine nicht genau bestimmbare Anzahl großer und kleiner Nucleolen (*N*) sowie mehrere Chromozentren (*Chr*). Vergr.: 1500:1

Eine konstante Nucleolenanzahl besitzen auch die Zellkerne im Zwiebelhäutchen *(Allium cepa)*, die überwiegend zwei gleich große, kugelförmige Nucleolen enthalten und dazu noch einen dritten, wesentlich kleineren. Beim Vergleich der verschieden großen Kerne a und b in Abb. 2 ist ferner eine Abhängigkeit der Nucleolen- von der Kerngröße ersichtlich [9], die auch bei Hühner- und Mäuseherzmyoblasten beobachtet werden kann und bei der Bestimmung fusionierter Nucleolen berücksichtigt werden muß.

Je größer die normale Anzahl der Nucleoluseinheiten im Zellkern ist, um so schlechter läßt sich deren Konstanz nachweisen, weil aus statistischen Gründen die Wahrscheinlichkeit der Nucleolenberührung und -ver-

schmelzung zunimmt. Diese Beobachtung wurde bei gezüchteten Mäuseherzmyoblasten gemacht, die im diploiden Zustand maximal 8 (Abb. 3a)
und nach Fusion entsprechend weniger, dafür aber größere Nucleolen
enthalten, wie z. B. der Kern b in Abb. 3 mit 4 Doppelnucleolen. Es ist
also verständlich, daß früher die Anzahl der Nucleolen „nicht als spezifisches und signifikantes Kennzeichen bestimmter Zell- und Gewebe-
typen" angesehen wurde [31].

Angaben über die Zahlenkonstanz der Nucleolen sind nahezu unmöglich (Abb. 4), wenn die Zellkerne nicht nur reguläre Nucleolen (N)
besitzen, sondern vorübergehend noch sog. Nebennucleolen (NN) ausbilden und außerdem auch eine wechselnde Anzahl nucleolusähnlicher
Chromozentren (Chr) aufweisen, bei denen es sich um mehr oder weniger
stark kondensierte Chromosomenbereiche handelt. Als Beispiel hierfür ist
ein diploider Kern aus einer Molchzellenkultur angeführt (Abb. 4).

Aus den Untersuchungen resultiert also, daß die diploiden Kerne der
Hühner- und Mäuseherzmyoblasten sowie die der Zellen von Zwiebelhäutchen jeweils eine konstante Anzahl von Nucleoluseinheiten besitzen,
die einzeln oder in Gruppen auftreten können. Die Gesetzmäßigkeit, mit
der die Nucleoluseinheiten an die Tochterkerne weitergegeben werden,
weist ferner darauf hin, daß zwischen den Nucleolen und dem Genom
enge Beziehungen bestehen müssen [33, 34].

C. Der Nachweis DNS-haltiger Nucleolusstrukturen

Die im lebenden Zustand fast homogen aussehenden Nucleolen der
gezüchteten Hühnerherzmyoblasten (Abb. 1) weisen nach geeigneter
Präparation charakteristische Innenstrukturen auf (Abb. 5a—c). Für die
Fixation bewährte sich das osmiumtetroxydhaltige Gemisch nach *Palade*
mit Zusatz von 3% Trauben- und 15% Rohrzucker [35]. Nach Aceton-
entwässerung über 7 Stufen wurden die Präparate in „Eukitt" eingeschlossen. Viele Nucleolen lassen dann im Phasenkontrastmikroskop
zwei gleich große Untereinheiten (rechter Kern in Abb. 5a) oder granulär-
fädige Strukturen erkennen, die von einer grauen, phasenoptisch homogen
aussehenden Substanz umgeben sind (linker Kern in Abb. 5a). Dieser
Befund bildete den Ausgangspunkt für meine Nucleolusuntersuchungen,
über die bereits zwei Veröffentlichungen vorliegen [36, 37].

Lettré und Siebs [24, 25] sowie Davis [11] bemerkten bereits
fädige Nucleolusinnenstrukturen und hielten sie aufgrund ihrer Anfärb-
barkeit mit Methylgrün und nach *Feulgen* für Chromosomenanteile. Diese
Auffassung steht in gutem Einklang mit dem Nachweis von DNS-
Synthese im Nucleolus [1, 18, 19, 28].

Daß bei der Nucleolusentstehung gewisse Chromosomenabschnitte
mit im Spiele sind, wurde auch von vielen anderen Autoren beschrieben
[1, 2, 3, 4, 5, 6, 7, 8, 10, 15, 16, 17, 20, 27, 29, 30]. Inzwischen liegen übereinstimmende Befunde bei pflanzlichen und tierischen Objekten vor, an
Protisten und Metazoen, an diploiden und höherploiden Zellen sowie an
Zellen mit polytänen Chromosomen. Die Chromosomenbereiche, an

denen die Nucleolenbildung erfolgt, werden allgemein „nucleolus organizers" genannt. Sie sind von Objekt zu Objekt verschieden gut nachweisbar und teils euchromatischer, teils heterochromatischer Natur.

Diese Auffassung steht im Gegensatz zur Ansicht von ESTABLE und SOTELO [12], die von fadenförmigen, *Feulgen*-negativen Strukturen („Nucleolonemata") im Nucleolusinnern berichten und diese für eine „neue Grundsubstanz der Zelle" halten.

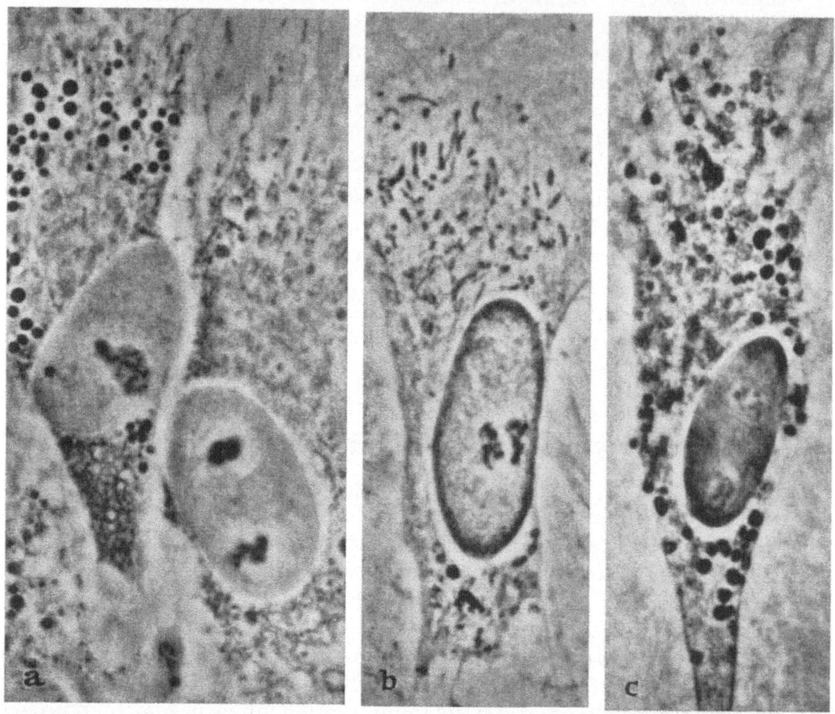

Abb. 5a—c. Darstellung DNS-haltiger Nucleolusstrukturen in gezüchteten Hühnerherz-myoblasten. Fix.: a und b modifiziertes *Palade*-Gemisch, b Nachbehandlung mit 1% Ribo-nuclease, d Kaliumpermanganat. Die DNS-haltigen Nucleolusanteile sind entweder kompakt (rechter Kern in a) oder fadenförmig (alle übrigen Kerne) und liegen beim linken Kern von a in einer grauen, RNS-haltigen Nucleolarsubstanz, die in b fermentativ herausgelöst und in c unkontrastiert geblieben ist. Vergr.: 1500:1

Da inzwischen durch neuartige Präparationsverfahren eine bessere Darstellung der Nucleolusstrukturen möglich ist, kann nun zu den strittigen Fragen erneut Stellung genommen werden.

Fixiert man gezüchtete Hühnerherzmyoblasten nur 2 min im modifizierten *Palade*-Gemisch und läßt nach guter Wässerung in 8,5%iger Rohrzuckerlösung 5 Std lang 0,1%ige Ribonuclease einwirken, so ist aus den Zellkernen die graue, RNS-haltige Nucleolarsubstanz weitgehend verschwunden, und die „demaskierten" Nucleolus-Fäden treten deutlich hervor (Abb. 5b). Ihr Gehalt an DNS wurde von mir [36] früher bereits durch die Feulgensche Nuclealreaktion und die Fast Green-Färbung nachgewiesen.

In Übereinstimmung mit den elektronenmikroskopischen Befunden sind die chromatischen Anteile des Zellkerns [13] und auch die Faden-strukturen innerhalb der Nucleolen nach Permanganat-Fixierung (3 min) sehr kontrastreich (Abb. 5c). Die RNS-haltige Nucleolarsubstanz bleibt erwartungsgemäß unkontrastiert und erscheint infolgedessen als helle Aussparung inmitten des dunklen Kernes.

Bei genauer Betrachtung der beiden Zellkerne in Abb. 5b und c sieht man deutlich, daß die Fadenstrukturen der Nucleolen paarig sind, und auch im rechten Zellkern der Abb. 5a erkennt man zwei, allerdings kompakte Untereinheiten. Der linke Kern in Abb. 5a besitzt in seinem durch Fusion entstandenen Doppelnucleolus 4 gekörnelte Fäden.

Faßt man die cytochemischen Versuchsergebnisse zusammen, so ergibt sich, daß die Nucleolen der Hühnerherzmyoblasten DNS-haltige Chromosomen bzw. Chromosomenstücke enthalten, die nach schonender Fixation und geeigneter Darstellung auch während der Interphase licht-mikroskopisch sichtbar bleiben.

D. Die Ermittlung des Interphasealters der Zellkerne

Beim Vergleich zahlreicher Mikroaufnahmen von Hühnerherzmyo-blasten erkennt man eine deutliche Abhängigkeit der Nucleolengröße und -differenzierung von der Kerngröße. Da alle Zellkerne während der Inter-phase bis zu ihrer nächsten Teilung eine Wachstumsperiode durchlaufen, versuchte ich das Interphasealter der Kerne aufgrund ihrer Größe zu bestimmen. Die weitere Aufgabe bestand darin, die biologische Wechsel-beziehung zwischen der Nucleolusdifferenzierung und dem jeweiligen Interphasestadium der Myoblasten festzustellen.

I. Die Bestimmung der Kerngrößen

Kerngrößenbestimmungen sind immer mehr oder weniger ungenau. Bei Gewebekulturen bleibt der Fehler aber in erträglichen Grenzen, weil sich die Kerne *in toto* vermessen lassen. Außerdem können die relativ großen Kernflächen als Maß für das jeweilige Kernvolumen gelten, weil die dritte Dimension bei den flachen Kernen der Gewebekulturen ver-nachlässigt werden kann.

Für die folgenden Untersuchungen wurden Myoblasten bei 385 facher Vergrößerung photographiert, auf 3000 : 1 nachvergrößert und die Kern-flächen planimetrisch bestimmt. Der Fehler, der durch die Vernach-lässigung der dritten Dimension entstehen könnte, läßt sich durch den Vergleich von Tochterkernpaaren abschätzen, deren Volumen im all-gemeinen ziemlich gleich ist.

Die prozentuale Flächenabweichung (FA) bei 37 Tochterkernpaaren wurde mit Hilfe des Quotienten aus der Flächendifferenz der Paarlinge $(K-k)$ und der mittleren Fläche des entsprechenden Kernpaares ermit-telt:

$$FA = \frac{K_n - k_n}{\frac{K_n + k_n}{2}} \cdot 100 \; (n = 1, \ldots, 37) \, .$$

Sie ist bei den Paaren kleiner und großer Kerne durchschnittlich gleich stark und beträgt im Mittel 12,8% oder 9,8 Planimeterflächeneinheiten. Eine Planimetereinheit entspricht dabei 1,21 μ^2.

II. Die Größenhäufigkeitskurve der Interphasekerne

In der Wachstumszone zweier, mit dem modifizierten *Palade*-Gemisch fixierter Schwesterkulturen wurden 344 Myoblasten ohne Auswahl photographiert, zur Ermittlung der Kerngrößen planimetriert und dann der Größe nach geordnet. Dabei stellte sich heraus, daß sehr kleine und sehr große Kerne nur in relativ geringer Anzahl vorkommen, während die überwiegende Menge mittelgroß ist. Die Extremwerte betragen 24 und 355 Flächeneinheiten.

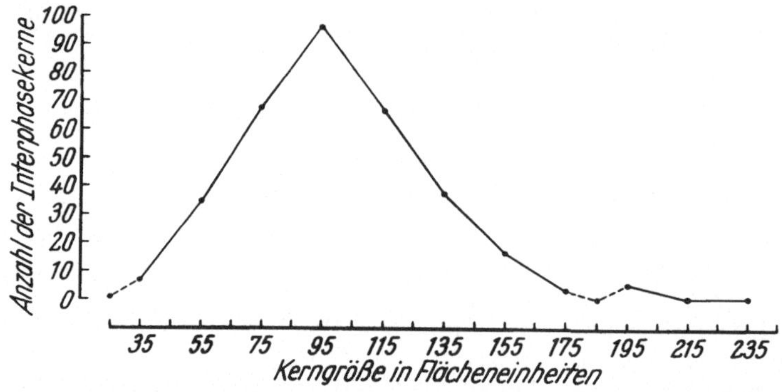

Abb. 6. Die Größenhäufigkeitskurve von 344 Interphasekernen. Nach der Aufteilung der Kerne in Größenklassen von je 20 Planimeter-Flächeneinheiten ergeben die Häufigkeitswerte die Kurvenpunkte. Die beiden Kurvenmaxima deuten auf das Vorhandensein von diploiden und tetraploiden Hühnerherzmyoblasten in den Gewebekulturen hin

Das Kernmaterial bis zur Größe von 245 Flächeneinheiten (die beiden größten Kerne mit 316 und 355 Flächeneinheiten blieben unberücksichtigt) wurde in 11 Größenklassen zu je 20 Flächeneinheiten aufgeteilt, deren Mittelwerte an der Abszisse von Abb. 6 vermerkt sind. Über den Klassenmittelwerten ist jeweils die Anzahl der Interphasekerne in das Koordinatensystem eingetragen. Ihre Verbindung ergibt die gesuchte Größenhäufigkeitskurve. Über den Abszissenwerten 25 und 185 sind noch zwei Hilfspunkte eingefügt, damit dort der nullnahe Bereich der Kurve (gestrichelt gezeichnet) den vorliegenden Werten entsprechend richtig zum Ausdruck kommt. Aus dem Verlauf der Kurve geht hervor, daß die Interphasekerne zwei Größenmaxima aufweisen, nämlich ein sehr ausgeprägtes bei 95 und ein schwaches, aber noch deutlich erkennbares bei 195 Flächeneinheiten. Die das zweite Maximum bestimmenden Kerne sind also doppelt so groß wie diejenigen des ersten. Alle übrigen Kerne verteilen sich normal um die beiden Maxima.

Dieser Befund muß erfahrungsgemäß so gedeutet werden, daß die Myoblasten in zwei verschiedenen Ploidiestufen vorliegen, und zwar überwiegend als diploide und zu einem kleinen Teil als tetraploide Zellen.

III. Die Bestimmung der Interphasedauer

Die Interphasedauer läßt sich durch direkte Beobachtung der Myoblasten unter dem Mikroskop nicht ohne weiteres ermitteln, da — wie schon frühere Versuche (nach *14*) gezeigt haben — der Zellteilungsrhythmus durch die für phasenkontrastmikroskopische Aufnahmen erforderliche helle Beleuchtung stark gestört wird.

Man kann die Interphasedauer jedoch berechnen [23]. Hierbei geht man aus von der Mitosedauer, die ja bedeutend kürzer ist als der Interphaseverlauf, und vom Mitoseindex, also dem Prozentsatz der in Mitose befindlichen Zellen. Als Durchschnittswert für die Mitosedauer ermittelte ich bei meinen Kulturen 48 min, als Mitoseindex 4,5%. Den gleichen Wert fanden auch HUPE und GROPP [22]. Die Auflösung der Gleichung:

$$\frac{\text{Mitosedauer}}{\text{Mitoseindex}} = \frac{\text{Interphasedauer}}{\text{Interphaseindex}} \quad \text{oder} \quad \frac{48\,\text{min}}{4,5\%} = \frac{x}{95,5\%}$$

ergibt als Interphasedauer 1020 min, mithin 17 Std.

Diese Methode ist allerdings nur bedingt zuverlässig, weil zwar der Mitoseindex anhand fixierter und gefärbter Präparate ziemlich exakt gewonnen werden kann, die Bestimmung der Mitosedauer dagegen nach subjektiven Gesichtspunkten erfolgt, da sich der Prophasebeginn und die Tochterzellentrennung beim lebenden Objekt nicht auf die Minute genau festlegen lassen.

IV. Die Wachstumskurve der Interphasekerne

Die Bestimmung des Interphasealters der Zellkerne hat für die Untersuchung der erwähnten Differenzierungsvorgänge in den Nucleolen eine entscheidende Bedeutung und kann am besten anhand einer Kernwachstumskurve erfolgen. Da die meisten diploiden Zellkerne einer Myoblastenkultur mittelgroß sind, während die kleinen und großen Kerne in relativ geringer Anzahl vorliegen (Abb. 6), durchlaufen sie das Anfangs- und Endstadium der Interphase schneller als den Mittelbereich, in dem sie also nur langsam an Größe zunehmen. Die Größenhäufigkeitskurve in Abb. 6 gibt demnach indirekt auch den Wachstumsverlauf der Interphasekerne wieder. Sie läßt sich in eine Wachstumskurve transformieren, in dem man an der Ordinate die Kerngröße in Flächeneinheiten aufträgt, dann die Ogivenkurve aller 334, der Größe nach geordneten, diploiden Interphasekerne zeichnet und schließlich an der Abszisse anstelle der vom 1. bis zum 334. Kern reichenden Strecke die zuvor ermittelte Interphasedauer als regelmäßige Zeitskala von der 1. bis zur 17. Std aufträgt (Abb. 7). Die Wachstumskurve weicht nur in ihrem Schlußabschnitt geringfügig von der Ogivenkurve ab. Diese Korrektur ist notwendig, weil die abgeflachten Kerne der gezüchteten Myoblasten gegen Ende der Interphase, also etwa 50 min vor Beginn der Prophase, sich zunehmend abkugeln und dann kleinflächiger erscheinen als zuvor. Für die Dauer der Mitose, deren Bereich in Abb. 7 durch zwei senkrechte Striche begrenzt ist, wurde der Erfahrungswert von 48 min zugrunde gelegt.

Beträgt die Interphasedauer der Myoblasten unter veränderten Kulturbedingungen mehr oder weniger als 17 Std, so läßt sich das Interphasealter der Zellkerne anhand der gleichen Wachstumskurve auch in relativen Werten angeben.

Abschließend muß noch daran erinnert werden, daß die Ermittlung des Interphasealters der Kerne anhand ihrer Wachstumskurve um so ungenauer wird, je fehlerhafter die Bestimmung der Kerngrößen ist. In der vorliegenden Arbeit beträgt der Fehlerbereich, wie eingangs erwähnt, durchschnittlich 9,8 Flächeneinheiten oder 12,8%.

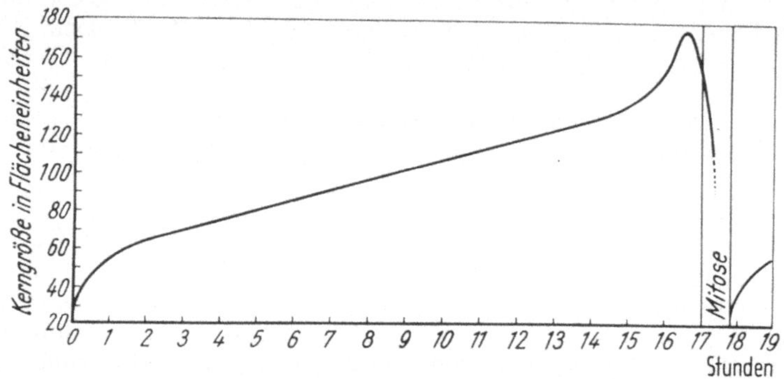

Abb. 7. Die Wachstumskurve der diploiden Kerne von gezüchteten Hühnerherzmyoblasten im Verlauf der Interphase

E. Die Reduplikation der Nucleolen

Wenn die Chromosomen unmittelbar nach der Zellteilung noch relativ kondensiert erscheinen, die während der Mitose verschwundenen Nucleolen aber schon wieder sichtbar geworden sind, tritt ihre Zugehörigkeit zu zwei Chromosomen klar zutage. Mit fortschreitender Interphase dekondensieren die extranucleolären Chromosomenanteile und entziehen sich damit der mikroskopischen Betrachtung.

Die weitere Entwicklung der Nucleolen verfolgte ich an den schon erwähnten 344 Kernen aus zwei Myoblastenkulturen unter Einbeziehung einiger höherploider Kerne von gezüchteten Zellen der Leber und des Darmes 9 Tage alter Hühnerembryonen. Alle Kulturen wurden mit dem modifizierten *Palade*-Gemisch fixiert.

Die Nucleolen der jungen Interphasekerne erscheinen im osmierten Zustand kompakt und abgerundet. Frühestens 1 Std nach Beginn der Interphase kann man in ihnen gelegentlich zwei gleich große Untereinheiten erkennen. Mit zunehmendem Interphasealter vergrößert sich die osmiophile Masse der Nucleolen etwas, wobei ihre beiden Untereinheiten weiter auseinanderrücken können (Abb. 8a). (Durch Angabe der Kerngrößen unter den Abb. 8 und 10 kann in Verbindung mit der Wachstumskurve in Abb. 6 das Interphasealter der abgebildeten Kerne bestimmt werden.)

Die Nucleolen bereiten offensichtlich schon in einem relativ frühen Stadium der Interphase ihre Verdoppelung im Hinblick auf die nächste Mitose vor; in diesem Sinne müssen auch die Ergebnisse von HARRIS [*18*] bezüglich der DNS-Synthese in den Nucleolen gedeutet werden.

Etwa 2 Std nach Beginn der Interphase strecken sich die Nucleolen in die Länge, wirken in diesem Stadium zunächst noch relativ kompakt,

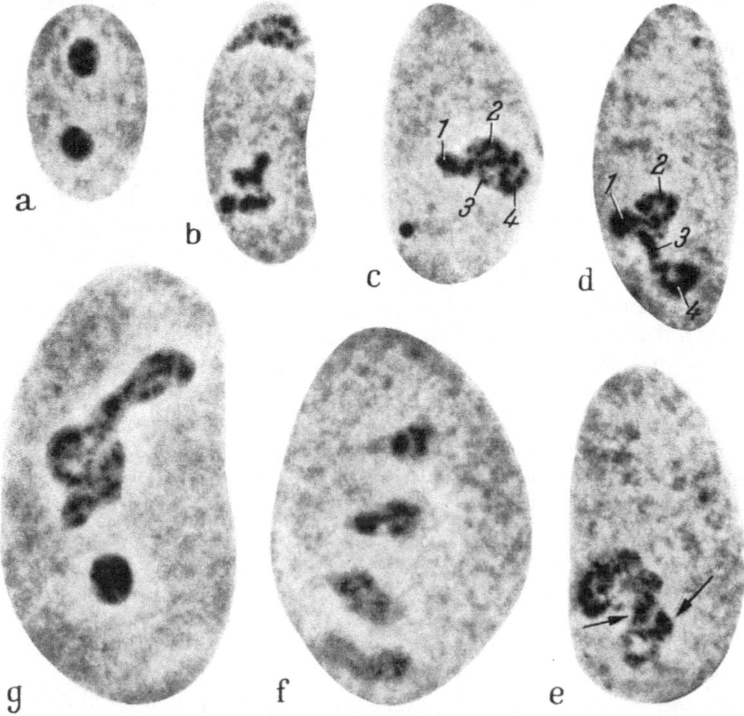

Abb. 8a—g. Phasenkontrastmikroskopische Aufnahmen von sieben der Größe nach geordneten Kernen aus gezüchteten Zellen eines Hühnerembryos. Kerne a—e von Herzmyoblasten, Kern f von einer Leberzelle, Kern g von einer Darmzelle. Fix.: Modifiziertes *Palade*-Gemisch. Die Kerne a—d sind diploid und enthalten unabhängig von der Nucleolusanzahl 4 Nucleolusuntereinheiten (Kern a) bzw. 4 kontrastreiche Fadenstrukturen. Die Kerne e—g sind höherploid. Hierfür spricht ihre Größe und auch die vermehrte Anzahl der chromatischen Nucleolusstrukturen. Planimeterwerte der Kerne: a 74 FE, b 88 FE, c 129 FE, d 141 FE, e 189 FE, f 281 FE, g 316 FE. Vergr.: 2000:1

lassen aber bei genauer Betrachtung in ihrem Innern schon zwei nebeneinanderliegende Granulareihen erkennen. In vielen Fällen rücken die beiden Körnerfäden etwas auseinander. Dabei wird deutlich sichtbar, daß es sich hier um fädige Strukturen mit granulären Verdickungen handelt. Wenn die Fadenpaare annähernd in einer optischen Ebene liegen, kann man sogar die Verdickungen abzählen und aufgrund ihrer Größe und Form paarweise homologisieren (Abb. 8b). Der Kondensationsgrad der Nucleolusfäden ist von Kern zu Kern unterschiedlich, mitunter auch schon bei 2 Nucleolen desselben Kernes (Abb. 8b).

Liegt im diploiden Zellkern nur ein Nucleolus vor (Abb. 8c), oder sind die beiden Nucleolen so nahe zusammengerückt, daß sie fast eine Einheit darstellen (Abb. 8d), so sieht man unter günstigen optischen Bedingungen 4 Nucleolusfäden. Im Kern c der Abb. 8 haben sie ihre paarige Anordnung aufgegeben und sich im Kern d der Abb. 8 sogar hintereinander gelegt. Im Mikroskop sind die 4 Fäden klar zu unterscheiden; da sie sich in der Photographie aber teilweise überlagern, wurden sie fortlaufend numeriert.

Die sehr kontrastreichen, DNS-haltigen Fadenstrukturen des Doppelnucleolus in Abb. 8c sind von RNS-haltiger Nucleolarsubstanz umgeben (vgl. Abb. 5), die phasenoptisch unstrukturiert dunkelgrau erscheint und sich deutlich vom übrigen Kerninhalt abhebt. Hiervon wird im nächsten Kapitel noch die Rede sein.

Im Gegensatz zu den diploiden Zellkernen (a—d) der Abb. 8 sind die folgenden Kerne (e—g) alle über 185 Flächeneinheiten groß, also mindestens tetraploid. Der Kern e hatte zur Zeit der Fixierung die Größe von 189 Flächeneinheiten erreicht und somit die kritische Grenze vom diploiden zum tetraploiden Zustand gerade überschritten. Die bis zum Ende der diploiden Phase immer nur in der Vierzahl vorliegenden Nucleolusfäden haben sich hier bereits verdoppelt. Aus optischen Gründen war es nicht möglich, alle Nucleolusbestandteile des Kernes gleichmäßig scharf abzubilden; daher sind nur die Spalthälften der beiden unteren Nucleolusfäden gut focussiert (Pfeilmarkierung der Abb. 8e).

Dieser Befund eröffnet eine weitere Möglichkeit zur Unterscheidung der diploiden von den tetraploiden Kernen. Erstere enthalten, wie gesagt, 2 Nucleoluseinheiten mit je zwei identischen, insgesamt also 4 Fadenstrukturen. Durch die Verdoppelung entstehen im tetraploiden Kern (Abb. 8f) 4 Nucleoluseinheiten mit wiederum je 2 Nucleolusfäden.

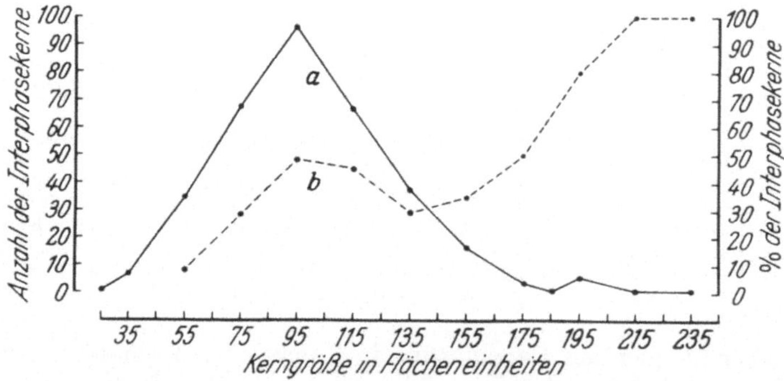

Abb. 9. Im Vergleich zur Größenhäufigkeitskurve a (linke Ordinate) stellt die gestrichelt gezeichnete Kurve b den prozentualen Anteil der diploiden und tetraploiden Interphasekerne mit zwei Untereinheiten bzw. zwei Fadenstrukturen pro Nucleoluseinheit dar (rechte Ordinate)

Bei dem höherploiden Kern in Abb. 8g sind die optischen Verhältnisse noch ungünstiger, weil die Nucleoluseinheiten mit wachsender Anzahl stärker zur Verschmelzung neigen und ein nicht mehr analysierbares Fadengewirr bilden.

Alle diese Beobachtungen lassen sich auch statistisch erfassen. Die gestrichelt gezeichnete Kurve *b* in Abb. 9 gibt im Vergleich zu der bereits bekannten Größenhäufigkeitskurve *a* der Interphasekerne den prozentualen Anteil der diploiden und tetraploiden Kerne wieder, deren Nucleoluseinheiten zur Zeit der Fixierung zwei Untereinheiten (vgl. Abb. 8a) bzw. zwei Fadenstrukturen (vgl. Abb. 8b) erkennen ließen.

Obwohl mit steigender Kerngröße der morphologische Hinweis für eine Nucleolusverdoppelung immer deutlicher wird, fällt das Kurvenminimum bei den Kernen von etwa 135 Flächeneinheiten besonders auf. Die Erklärung hierfür ergibt sich aus dem funktionellen Verhalten der Nucleolen, das im nächsten Kapitel beschrieben werden soll.

Zuvor muß noch betont werden, daß bei insgesamt 37 % aller fixierten Interphasekerne deutliche Merkmale einer Nucleolusverdoppelung erkennbar sind. Dieser Prozentsatz ist sehr hoch, wenn man bedenkt, daß für die statistische Auswertung jeweils nur ein Momentbild des Interphaseverlaufes zur Verfügung stand. Die gesetzmäßige Verdoppelung der Nucleoluseinheiten und deren Beziehung zum Genom steht somit außer Zweifel.

F. Zur Funktion der Nucleolen

Die Nucleolen der Hühnerherzmyoblasten sind an keinen bestimmten Platz im Zellkern gebunden, befinden sich aber meistens im Innenbereich, und fallen deshalb besonders auf, wenn sie an der Kernwand liegen (rechter Nucleolus in Abb. 10a).

Durch Lebendbeobachtungen konnte sichergestellt werden, daß die Nucleolen im Verlauf der Interphase mit Nucleolarsubstanz beladen an die Kernwand herantreten und später nur mit einer kleinen Restmenge oder ganz ohne diese wieder ins Kerninnere zurückkehren. Der chromosomale Nucleolusanteil ist danach im fixierten Präparat sehr deutlich sichtbar (Abb. 10b).

Die gleichen Rückschlüsse auf die Funktion lassen sich aus dem Nucleolusverhalten der tetraploiden Schwesterkerne *c* und *d* in Abb. 10 gewinnen. Die oberen Nucleolen der beiden Kerne gleichen sich sehr, denn beide liegen in einem Zustand vor, der für die Wandständigkeit charakteristisch ist. Der Nucleolus im rechten Schwesterkern (*d*) hat die Kernmembran zwar schon erreicht, derjenige im linken Kern (*c*) aber noch nicht.

Im Gegensatz zu den oberen sind die unteren Nucleolen des gleichen Kernpaares stark aufgelockert und enthalten nur wenig Nucleolarsubstanz. Ihre chromosomalen Strukturen treten daher deutlich in Erscheinung. Beide verhalten sich insofern typisch, als sie keine Verbindung mit der Kernmembran besitzen.

Dieser Befund spricht dafür, daß die kernwandständigen Nucleolen ein inzwischen cytochemisch (vgl. Abb. 5b) und autoradiographisch als RNS-haltig erkanntes Produkt an das Cytoplasma abgeben [*32*]. Obwohl hierfür bei den gezüchteten Hühnerherzmyoblasten im Stadium der Interphase noch kein direkter morphologischer Beweis erbracht werden konnte, darf man mit einem langsamen „Abfließen" RNS-haltiger Partikel in das Cytoplasma rechnen, zumal die Kernmembran submikroskopisch kleine Poren in genügend großer Anzahl besitzt.

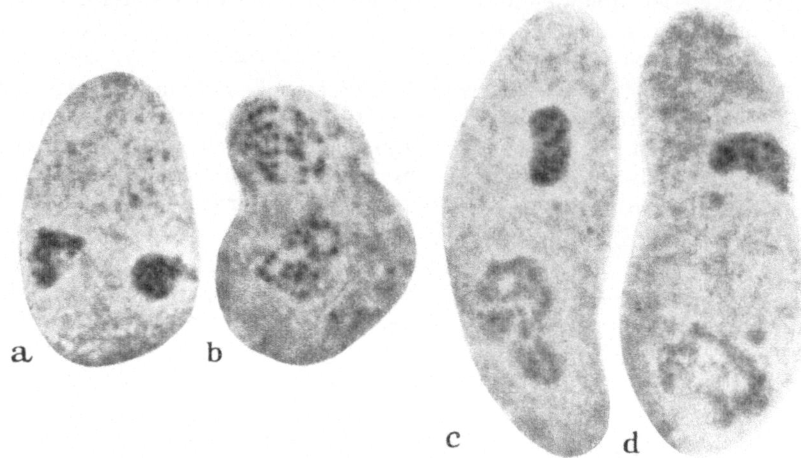

Abb. 10. Vier der Größe nach geordnete Kerne von gezüchteten Hühnerherzmyoblasten im Phasenkontrast. Fix.: Modifiziertes *Palade*-Gemisch. Die Kerne a und b sind diploid, das Schwesterkernpaar f und g ist tetraploid. Der rechte Nucleolus im Kern a und der obere in Kern d stehen in Verbindung mit der Kernwand. Der Gehalt an grauer Nucleolarsubstanz (RNS) ist bei den einzelnen Nucleolen sehr verschieden und erlaubt einen Rückschluß auf die Nucleolusfunktion. Planimeterwerte der Kerne: a und b 150 FE, c 230 FE, d 250 FE. Vergr.: 2000:1

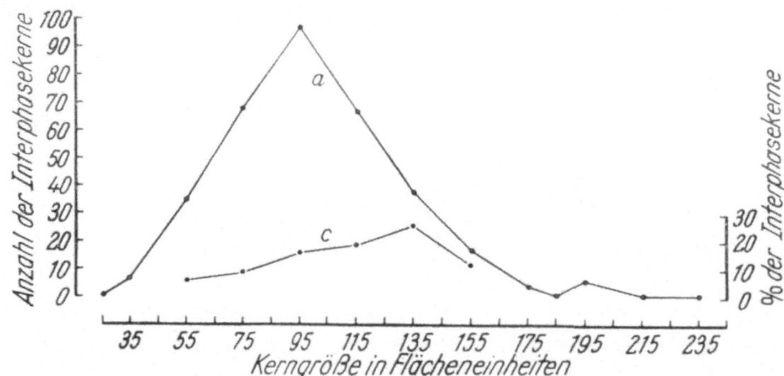

Abb. 11. Im Vergleich zur Größenhäufigkeitskurve a (linke Ordinate) stellt die Kurve c den prozentualen Anteil der diploiden Interphasekerne mit kernwandständigen Nucleolen dar (rechte Ordinate)

Auch das Vorkommen der kernwandständigen Nucleolen läßt sich statistisch erfassen. Als Bezugssystem diente wieder die Größenhäufigkeitskurve der Interphasekerne in Abb. 6. Dabei konnte ich mich auf die 334 diploiden Kerne beschränken, weil die 10 höherploiden Kerne keine wandständigen Nucleolen besaßen.

Der prozentuale Anteil der Kerne mit wenigstens einem wandständigen Nucleolus wurde für jede Kerngrößenklasse berechnet und in Abb. 11

eingetragen. So entstand die Kurve *c*. Wie aus einem Vergleich mit der Kernwachstumskurve hervorgeht, besitzen die Myoblastenkerne nur während der ersten und letzten Interphasestunde keine wandständigen Nucleolen. Zum Zeitpunkt der Fixierung hatten Nucleolen von 15% aller diploiden Interphasekerne Kontakt mit der Kernwand. Dies ist vor allem bei den etwa 135 Flächeneinheiten großen Kernen der Fall.

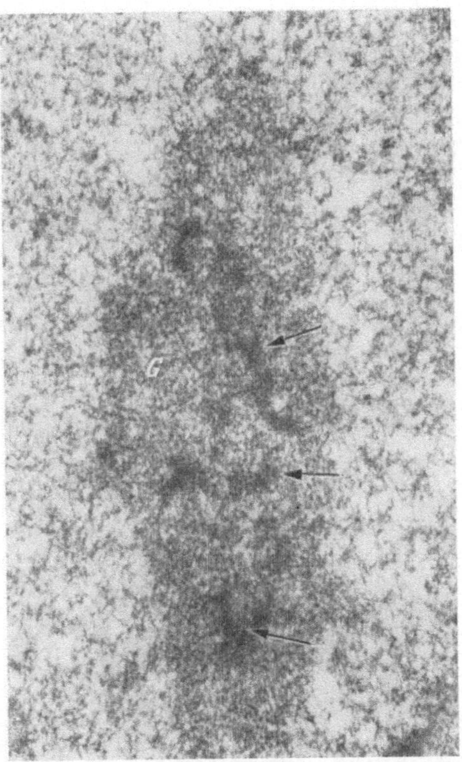

Abb. 12. Feinstruktur eines Kernausschnittes mit einem Nucleolus. Objekt: Hühnerherz-myoblast nach 2¹/₂ Tagen Gewebekultur. Fix.: Modifiziertes *Palade*-Gemisch. Pfeilmarkie-rungen = dunkle Nucleolusstränge (Chromatiden); *G* = granuläre Nucleoluskomponente (Nucleolus-Ribosomen). Vergr.: 23000:1

Da die Nucleolen im Stadium der Kernwandständigkeit stark kontra-hiert und wegen ihres hohen Gehaltes an RNS kaum einen Einblick in ihren Aufbau zulassen, wird jetzt auch verständlich, wie in Abb. 9 das Minimum der Kurve *b* zustande kommt, die ja den prozentualen Anteil der Kerne mit zwei deutlich sichtbaren Untereinheiten bzw. Fadenstruk-turen je Nucleoluseinheit wiedergibt.

Die lichtmikroskopischen Befunde konnten durch elektronenmikro-skopische Untersuchungen ergänzt werden. Wie schon von anderen Objekten hinlänglich bekannt ist [*38*], besteht auch die phasenoptisch homogen aussehende, RNS-haltige Nucleolarsubstanz der Hühner-herzmyoblasten vorwiegend aus einer Anhäufung kontrastreicher Granula

(*G* in Abb. 12) von der Größenordnung der Ribosomen des Cytoplasmas. Dazwischen liegen dunklere Stränge, die wegen ihres gewundenen Verlaufs hier nur im Anschnitt zu sehen sind und offensichtlich den chromosomalen Fadenstrukturen entsprechen.

Für eine direkte Beziehung der Nucleolen zu den Ribosomen des Cytoplasmas spricht schließlich noch ihr Verhalten in der Mitose. Lebend-

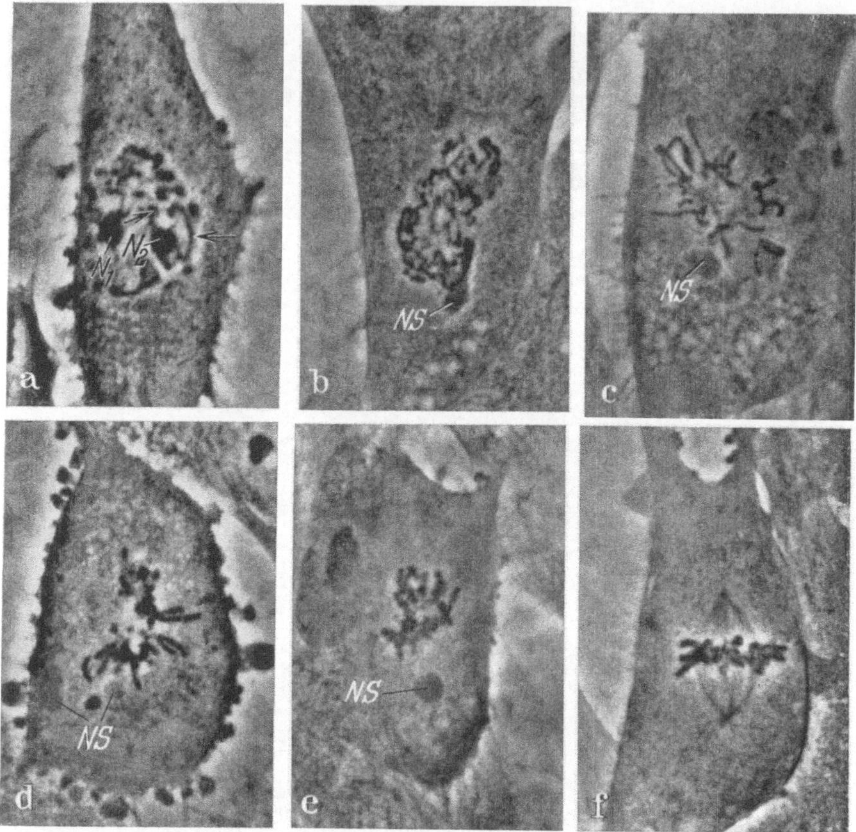

Abb. 13a—f. Sechs verschiedene Mitosestadien von diploiden Hühnerherzmyoblasten im Phasenkontrast. Vorfix.: modifiziertes *Palade*-Gemisch (3 min), Nachfix.: Alkohol-Eisessig (30 min). In dem Prophasekern a ist die Zugehörigkeit des rechten Nucleolus (N_2) zu einem kondensierten Chromosom (Pfeilmarkierungen) deutlich sichtbar. Wenn die Kernmembran aufgelöst ist (*b*) tritt die Nucleolarsubstanz (*NS*) in das Cytoplasma aus. Dort bleibt sie während der Chromosomeneinordnung in die Metaphaseplatte noch sichtbar (*NS* in c, d und e). Bei Beginn der Anaphasebewegung (f) ist die Nucleolarsubstanz nicht mehr nachweisbar. Vergr.: 1500:1

beobachtungen sind für solche Untersuchungen leider ungeeignet, weil sich die Kerne gezüchteter Zellen zu Beginn der Mitose abkugeln und phasenoptisch nicht genügend Einzelheiten erkennen lassen. Eine gute Darstellung der Nucleolen und der Prophasechromosomen wird jedoch durch kurzfristige Osmierung (modifiziertes *Palade*-Gemisch, 3 min) und nachfolgende Fixierung in Alkohol-Eisessig (Verhältnis 3:1, 30 min) gewährleistet.

Die Nucleolen der gezüchteten Hühnerherzmyoblasten gehen in der Regel mit einem Rest an RNS-haltiger Nucleolarsubstanz in die Mitose. Wenn in der Prophase die Chromosomen genügend spiralisiert sind, tritt die Zugehörigkeit der Nucleoluseinheiten zu bestimmten Chromosomen wieder deutlich zutage, und zwar noch etwas besser als in der Rekonstruktionsphase. Die Nucleolen erscheinen dann doppelt geschwänzt, da jeweils ein langes und ein kurzes Chromosomenstück in sie einmündet (Pfeilmarkierung in Abb. 13a).

Sobald die Kernmembran aufgelöst ist, trennt sich die RNS-haltige Nucleolarsubstanz (NS) vom Chromosom, verläßt den Kernbereich in Gestalt eines Tropfens und tritt in das Cytoplasma ein (Abb. 13b), wo sie während der Chromosomeneinordnung in die Metaphaseplatte noch eine Zeitlang sichtbar bleibt (Abb. 13c−e), bei Beginn der Prophasebewegungen aber verschwindet (Abb. 13f). Der Nucleolusbereich der Nucleolus-Chromosomen ist dann nicht mehr zu erkennen.

Somit ist also auch bei Myoblasten der Beweis für die Abgabe von Nucleolus-Ribosomen an das Cytoplasma erbracht. Da die recht kompakte Nucleolarsubstanz zwischen Meta- und Anaphase innerhalb weniger Minuten vom Cytoplasma vollständig „resorbiert" wird, ist es nun verständlich, daß sich in der Interphase das langsame „Abfließen" des gleichen Nucleolusproduktes ins Cytoplasma der mikroskopischen Beobachtung entzieht.

G. Schlußwort

Hühner- und Mäuseherzmyoblasten sowie Zellen von Zwiebelhäutchen enthalten in jedem Ploidiegrad eine statistisch verteilte Anzahl von Nucleolen. Eingehendere Untersuchungen haben gezeigt, daß die Nucleolusanzahl zwar in jungen Tochterzellen weitgehend konstant ist, im Verlauf der Interphase aber durch Fusion von zwei oder mehr Nucleolen zunehmend verringert wird. Da sich die Anzahl der „Nucleoluseinheiten" mit Hilfe neuer Präparationsmethoden auch in Sammel-Nucleolen bestimmen läßt, kann die Zahlenkonstanz der Nucleoluseinheiten pro Zellkern gleichen Ploidiegrades als gesichert gelten. Sie bleibt nach der Mitose auch für die Tochterzellen gewährleistet.

Wenn die Chromosomen unmittelbar vor und nach der Mitose infolge ihrer starken Kondensierung sichtbar und auch die Nucleolen eben noch bzw. schon wieder erkennbar sind, zeigt sich, daß sie integrierende Bestandteile bestimmter Chromosomen sind. Mit fortschreitender Interphase dekondensieren die extranucleolären Chromosomenanteile und entziehen sich im Gegensatz zu den Nucleolen der mikroskopischen Betrachtung.

Bei gezüchteten Hühnerherzmyoblasten macht sich bald darauf die Nucleolus-Reduplikation bemerkbar, die im Hinblick auf die nächste Zellteilung erfolgt. Nach der ersten Interphasestunde weisen nämlich zunächst vereinzelte und dann immer mehr Nucleoluseinheiten zwei DNS-haltige Fadenstrukturen auf, bei denen es sich jeweils um einen kondensiert gebliebenen Chromosomenbereich handelt, der seine Verdoppelung durch zwei identische Chromatidenstücke zu erkennen gibt. Diese liegen mehr oder weniger dicht beieinander, manchmal auch hintereinander und sind fast für die ganze Dauer der Interphase von RNS-haltiger Nucleolarsubstanz umgeben.

Aus zwei (diploide) oder mehr Nucleoluseinheiten bestehende „Sammelnucleolen" (höherploide Hühnerherzmyoblasten) enthalten die doppelte bzw. vielfache Anzahl Chromatidenpaarstücke; deren genaue Identifizierung ist allerdings schwierig.

Das Verhalten der Nucleolen gibt auch Hinweise auf ihre Funktion. Sie treten nämlich im Verlauf der Interphase mit Nucleolussubstanz beladen an die Kernmembran heran und geben diese in submikroskopisch kleinen Mengen an das Cytoplasma ab. Das RNS-haltige Nucleolus-Produkt dürfte eine Bedeutung für die Eiweißsynthese der Zellen haben, die bei der raschen Teilungsfolge in der Gewebekultur natürlich sehr rege ist.

Nach der Aktivitätsphase löst sich der chromosomale Anteil der Nucleolen mit einem Rest Nucleolarsubstanz wieder von der Kernwand ab, wandert zum Kerninnern zurück und läßt dann einen genaueren Einblick in seine chromatischen Strukturen zu. Der den Nucleolen verbliebene Substanzrest wird noch vor der Zellteilung, nämlich nach der Auflösung der Kernmembran während der Prophase, in mikroskopisch sichtbarer Form dem Cytoplasma zugeführt.

Gelegentlich erfolgen während der Interphase Nucleolusextrusionen. Hierbei können außer der RNS-haltigen Substanz auch chromosomale Nucleolusanteile als Knospen in das Cytoplasma ausgeschleust werden. Dieser Vorgang ist zwar sehr augenfällig, kann aber bei Hühnerherzmyoblasten aus statistischen Gründen kaum eine besondere Bedeutung haben, weil er keine regelmäßige Versorgung des Cytoplasmas mit RNS-haltigen Substanzen gewährleistet.

Für das Vorhandensein eines DNS-freien „Nucleolonemas" im Sinne von Estable und Sotelo liegen bei den untersuchten Zellen keine Anhaltspunkte vor.

Summary

Chicken and mouse heart myoblasts as well as onion epidermis cells contain a number of nucleoli which are distributed statistically at every ploidy level. More detailed studies have shown that the number of nucleoli, being fairly constant in young daughter cells, is reduced progressively during interphase by fusion of two or more nucleoli. With the aid of new preparation methods, which allow to determine the number of nucleolar units even in aggregate nucleoli, it could be demonstrated that the number of nucleolar units per nucleolus is constant at a given ploidy level and remains also constant in doughter cells after mitosis.

Immediately before and after mitosis, when the chromosomes, due to their condensation, are visible at the same time, the nucleoli are seen to be integral parts of certain chromosomes. Later in interphase the extranucleolar chromosome parts become less dense and are no longer observable in the microscope in contrast to the intranucleolar chromosome parts and the nucleoli, which remain visible.

The duplication of the nucleolus, preceeding the next cell division, can be observed in chicken heart myoblasts. After the first hour of interphase an increasing number of nucleolus units show two DNA-containing filaments. These are the condensed chromosome parts, which have doubled to two identical chromatids. They lie together rather closely, sometimes one after another, and are surrounded by RNA-containing nucleolar substance during most time of the interphase.

Aggregate nucleoli composed of two (diploid) or more nucleolus units contain the double, respectively multiple number of parts of chromatid pairs, the exact identification of which is more difficult at higher ploidy levels.

Charged with nucleolar substance, the nucleoli approach the nuclear membrane during the interphase. These substances are transferred to the cytoplasm in submicroscopic quantities. The RNA-containing nucleolar product may play a role in the protein synthesis of the cell.

After this active phase, the chromosomal part of the nucleolus, together with a remaining nucleolar substance, retreat from the nuclear membrane. Thereafter the chromatin structures of the nucleolus become more clearly visible. Before the next cell division (after the disappearance of the nuclear membrane during prophase), the remaining nucleolar substance invades the cytoplasm in form of droplets, which are visible in the light microscope.

Occasionally nucleolar extrusions occur during interphase. During this process not only RNA-containing substances, but also chromosomal nucleolar parts may be extruded into the cytoplasm in form of buds. Due to statistical reasons, this process, although conspicuous, cannot have a special importance in chicken heart myoblasts, because it does not support a constant supply of RNA-containing substances to the cytoplasm.

In the cells examined, there are no indications for the presence of a DNA-free "nucleolonema" as described by ESTABLE and SOTELO.

Literatur

[1] ALTMANN, H. W., E. STÖCKER, u. W. THOENES: Z. Zellforsch. **59**, 116 (1963).
[2] BALBIANI, E. G.: Zool. Anz. 4, 637 (1881).
[3] BARIGOZZI, C., and P. FELETIG: Nature (Lond.) **166**, 36 (1950).
[4] BAUER, H.: Z. Zellforsch. 23, 280 (1935).
[5] BEERMANN, W.: Chromosoma (Berl.) **11**, 262 (1960).
[6] — Protoplasmatologia, Bd. VI D. Wien: Springer 1962.
[7] BINDER, S.: Z. Zellforsch. 5, 293 (1927).
[8] BOLOGNARI, A.: Acta histochem. (Jena) 2, 229 (1956).
[9] BUCHER, O., u. R. GATTIKER: Z. mikr.-anat. Forsch. **60**, 467 (1954).
[10] CHEN, T. T.: J. Morph. 83, 281 (1948).
[11] DAVIS, J. M. G.: In: The Cell Nucleolus, S. 3. London: Butterworths 1960
[12] ESTABLE, C., y J. R. SOTELO: Publ. Inst. Ciencias Biol. 1, 105 (1951).
[13] FALK, H.: Protoplasma (Wien) **54**, 432 (1962).
[14] FISCHER, J.: Grundriß der Gewebezüchtung. Jena: Gustav Fischer 1942.
[15] GEITLER, L.: Planta (Berl.) **17**, 801 (1932).
[16] — Z. Zellforsch. **26**, 641 (1937).
[17] GODWARD, M. B. E.: Ann. of Bot. **14**, 39 (1950).
[18] HARRIS, H.: Biochem. J. **72**, 54 (1959).
[19] HAY, E. D., and J. P. REVEL: 5. Intern. Congr. of Electron Microscopy, Bd. 2, S. 1. Philadelphia-New York-London: Academic Press 1962.
[20] HEITZ, E.: Planta (Berl.) **12**, 775 (1931).
[21] HERTL, M.: Z. Zellforsch. **46**, 18 (1957).
[22] HUPE, K., u. A. GROPP: Z. Zellforsch. **46**, 67 (1957).
[23] KNOWLTON, N. P., and W. R. WIDNER: Cancer Res. **10**, 59 (1950).
[24] LETTRÉ, R., u. W. SIEBS: Z. Krebsforsch. **60**, 19 (1954).
[25] — — Z. Krebsforsch. **60**, 564 (1955).
[26] — — Proc. intern. Union Physiol. Sci., XXII. Intern. Congr., Leiden 1962, S. 837.
[27] McCLINTOCK, B.: Z. Zellforsch. **21**, 294 (1934).

[28] Revel, J. P., and E. D. Hay: Exp. Cell Res. **25**, 474 (1961).
[29] Ries, E.: Grundriß der Histophysiologie. Leipzig: Akad. Verl. Ges. 1938.
[30] Schultz, J., and P. St. Lawrence: J. Hered. **60**, 31 (1949).
[31] Stockinger, L.: Protoplasma (Wien) **42**, 365 (1953).
[32] Stöcker, E.: Z. Zellforsch. **57**, 145 (1962).
[33] Uehlinger, V.: Rev. suisse Zool., **70**, 331 (1963).
[34] Vincent, W. S.: Int. Rev. Cytol. **4**, 269 (1955).
[35] Weissenfels, N.: Vierter intern. Kongr. f. Elektronenmikroskopie, S. 60. Berlin-Göttingen-Heidelberg: Springer 1960.
[36] — Z. Zellforsch. **62**, 667 (1964).
[37] — Verh. dtsch. Zool. Ges. ,Kiel 1964, Zool. Anz. Suppl. **28**, 144 (1965).
[38] Wessing, A.: Z. Zellforsch. **65**, 445 (1965).

Diskussion

Vorsitz: *Lettré*

Lettré: I like to say that I completely agree with the interpretation of these beautiful findings which in some way are complementary to what we found.

Bajer: To what extent does the size of the nucleus really diminish before the mitosis starts?

Lettré: In the case of tissue culture one must consider that we have very flattened nuclei in the cells. At the beginning of mitosis the cytoplasm retracts its processes, the cell rounds up and by rounding up the diameter of the nucleus becomes smaller. That is the reason for the apparent diminution in size.

Weissenfels: Die Volumenbestimmung der Kerne aus ihrer Fläche unter Vernachlässigung der dritten Dimension ist nur möglich, wenn die Zellen während der Interphase flach ausgewachsen sind und sich der Kern dem flachen Zellkuchen weitgehend angepaßt hat. Kugelt sich aber die Zelle und mit ihr der Kern in Vorbereitung der Mitose mehr und mehr ab, dann kann aus der Projektionsfläche nicht ohne weiteres auf das Kernvolumen geschlossen werden. Ich habe deshalb die statistisch ermittelte Wachstumskurve (Ogivenkurve) der Kerne in ihrem Endbereich durch die Meßwerte aus Lebendbeobachtungen korrigiert.

Bajer: In time-lapse studies of some living cells it is very clearly visible that the nucleolus is composed of two components which have different refractive index. One of these seems to be more liquified than the other.

Is it possible that you have seen such phenomena?

Lettré: I don't think so. It may be explained, in my opinion, in the following way: we were able to get a positive lipid reaction of the filamentous structures or at least a little stronger reaction of these structures than of the ground material of the nucleolus. This speaks in favour of a coating of the chromosomal part with some lipids. I think that Bungenberg de Yong was quite right in speaking of a coacervate-like substance, and that in some instances we have a disproportion of the coacervate so that you have vacuoles inside of the nucleoli. But this is a disintegration of the nucleolar material.

Weissenfels: Die Herzmyoplasten haben extrem einfach gebaute Nucleolen, in denen sich fadenförmige Strukturen befinden, die eindeutig DNS-haltig sind. Wir haben hier aber keine Nucleolonemata, also keine besondere, strangförmige Anordnung der RNS-haltigen Substanz.

Grundmann: Kann man aus der Zweigipfligkeit der Volumen-, genau genommen Flächenkurve Polyploide sicher erschließen? Man bekommt die gleiche Kurve bei Kulturen, die mit Sicherheit keine höhere Ploidie aufweisen, einfach auf Grund der G-2-Phase.

Weissenfels: Natürlich könnte es sein, daß eine zweigipfelige Kurve auch im diploiden Bereich aufträte. Aber gerade zum Zeitpunkt des Kurvenminimums sehen wir nicht mehr zwei Fadenstrukturen, sondern vier im Nucleolus. Damit läßt sich rein morphologisch schon feststellen, wann der Kern tetraploid zu werden beginnt.

Grundmann: Zur Frage der Nucleolenbewegungen an die Kernmembran im Zusammenhang mit der Nucleolenfunktion: Sie haben damit eine ganze Reihe von Befunden bestätigt, übrigens auch die unserer Chair-Lady.

Trotzdem sollte man in der Deutung vorsichtig sein. Diese Randlagerung hängt nicht unbedingt mit der Funktion des Nucleolus zusammen. Zum zweiten sollte man nicht etwa folgern, daß die Nucleolusfunktion immer mit einer Randverlagerung einhergehen muß. In den Ganglienzellen von Säugern verlagert sich der Nucleolus nie an den Rand — es sei denn, die Zelle ist krank —, und trotzdem ist er nach autoradiographischen Untersuchungen sehr intensiv am Ribonucleinsäurestoffwechsel beteiligt. Wenn in der Rattenleber Krebs entsteht, steigert sich der RNS-Stoffwechsel in dem Augenblick, in dem die Leberzelle in die Tumorzelle umschlägt, ungefähr um das Achtfache, der RNS-Gehalt im Cytoplasma um das Zwei- bis Dreifache. Trotzdem verlagern sich gerade in diesen primären Leberkrebszellen die Nucleolen niemals an den Rand.

Weissenfels: Ich weiß nicht, ob sich das ohne weiteres auf die gezüchteten Hühnerherzmyoblasten übertragen läßt. Jede Funktion ist bei diesen in Kultur stark modifizierten Zellen praktisch ganz auf die Zellvermehrung hin ausgerichtet.

Wenn man bei der Vitalbeobachtung von Nucleolen, die gerade wandständig werden oder geworden sind, und von solchen, die die Wand wieder verlassen haben, feststellt, daß in der Zwischenzeit jene Substanz verloren gegangen ist, die man zuvor als RNS-haltig identifiziert hat, dann erscheint der von mir vorgetragene Schluß doch wohl berechtigt.

Das morphologische Verhalten der Nucleolen mag natürlich bei den einzelnen Zellarten verschieden sein. Grundsätzliche Unterschiede in der Nucleolusfunktion brauchen deshalb aber nicht gefordert zu werden. Es könnte z. B. so sein, daß bei den von Herrn GRUNDMANN genannten Zellen sich die Nucleolen — in Abhängigkeit von der Lage der Nucleoluschromosomen — permanent im Kerninnern befinden und die Nucleolus-RNS entweder in größeren Portionen tropfenartig oder in einer Serie von kleinen Partikeln durch die Kernmembran hindurch an das Cytoplasma abgegeben wird.

Lettré: Die Befunde über die Verdoppelung der fädigen Strukturen stellen eine schöne Ergänzung zu der Arbeit von HARRIS dar, der als erster den Einbau von ^{3}H-Thymidin in die Fadenstrukturen der Nucleolen nachgewiesen hat. Er kam auf Grund von pulse-label-Experimenten zu der Überzeugung, daß diese Strukturen als erste in der Zelle verdoppelt werden.

Wir haben übrigens bei unseren Untersuchungen an lebenden Kulturen den Eindruck gewonnen — damit möchte ich an den Vortrag über die Lampenbürsten-Chromosomen anknüpfen —, daß Loop-Strukturen in den Nucleolen vorliegen, die man besonders nach Verflüssigung der Nucleolarsubstanz mit Adenosin nachweisen (sozusagen bloßlegen) kann.

P. Sitte: Sie meinten in Ihrem Vortrag, daß die Strukturen, die Sie im Elektronenmikroskop im Nucleolus sehen, zu dünn seien, als daß man sie im Lichtmikroskop sehen könnte. Man muß aber zwischen optischer *Auflösung* und *Sichtbarkeit* unterscheiden. Man sieht viele Strukturen ohne weiteres, die man keineswegs optisch auflösen kann. — Ist es möglich, die Schädigung der Zellen infolge langdauernder Lichtbelastung durch Grünfilter oder Blitzlicht zu mindern, wie es Herr BAJER macht?

Weissenfels: Von diesen Möglichkeiten machen wir weitgehend Gebrauch. Da wir aber im Gegensatz zu Herrn BAJER bei sehr starken Vergrößerungen arbeiten müssen, benötigen wir vergleichsweise immer recht hohe Lichtintensitäten.

Morphological Aspects of normal
and abnormal Mitosis

By

ANDREW BAJER, Eugene

Department of Biology, University of Oregon, Eugene, Oregon, U.S.A.

With 9 Figures

Introduction

More than one-hundred years of studies on mitosis have resulted in
the accumulation of an enormous amount of facts. The early period of
study has been reviewed by WILSON [55] and some general ideas concerning attempts of interpretation of previously observed phenomena have
been cited by SCHRADER [51]. MAZIA [38] presents a synthesis of our
current knowledge.

First, the approach used in this paper regarding mitosis will be briefly
outlined. It is not the purpose of this article to review the literature. An
effort, however, is made to support the statements through experiments with one material and one type of division (plant endosperm
division, Fig. 1), that the essential pattern and mechanisms of movement
during mitosis are the same in most if not in all organisms. From
hundreds of facts only a few are selected which are pertinent to an understanding of the mechanism of chromosome movement. The importance
of some facts to be reported in this paper is not generally recognized.
However, in any attempt to explain some aspects of the mechanism of
mitosis, these facts need to be considered. We are fully justified to speak
about several different mechanisms and processes taking part simultaneously during mitosis. Only one mechanism, that of chromosome
movements, will be considered here. A very general outline of the problem
of movement will be given. Division of the cell body, cytokinesis, will be
considered only fragmentarily.

From a morphological point of view there are two distinct cycles in
mitosis: the spindle cycle; and the chromosome cycle. These two cycles
are partially independent. In normal mitosis the cycles are correlated
precisely and the basic autonomy is difficult to visualize. Further, an
understanding of spindle function is needed before any detailed application can be made to chromosome movements and spindle structure. These
basic points will be discussed before any interpretations are made.

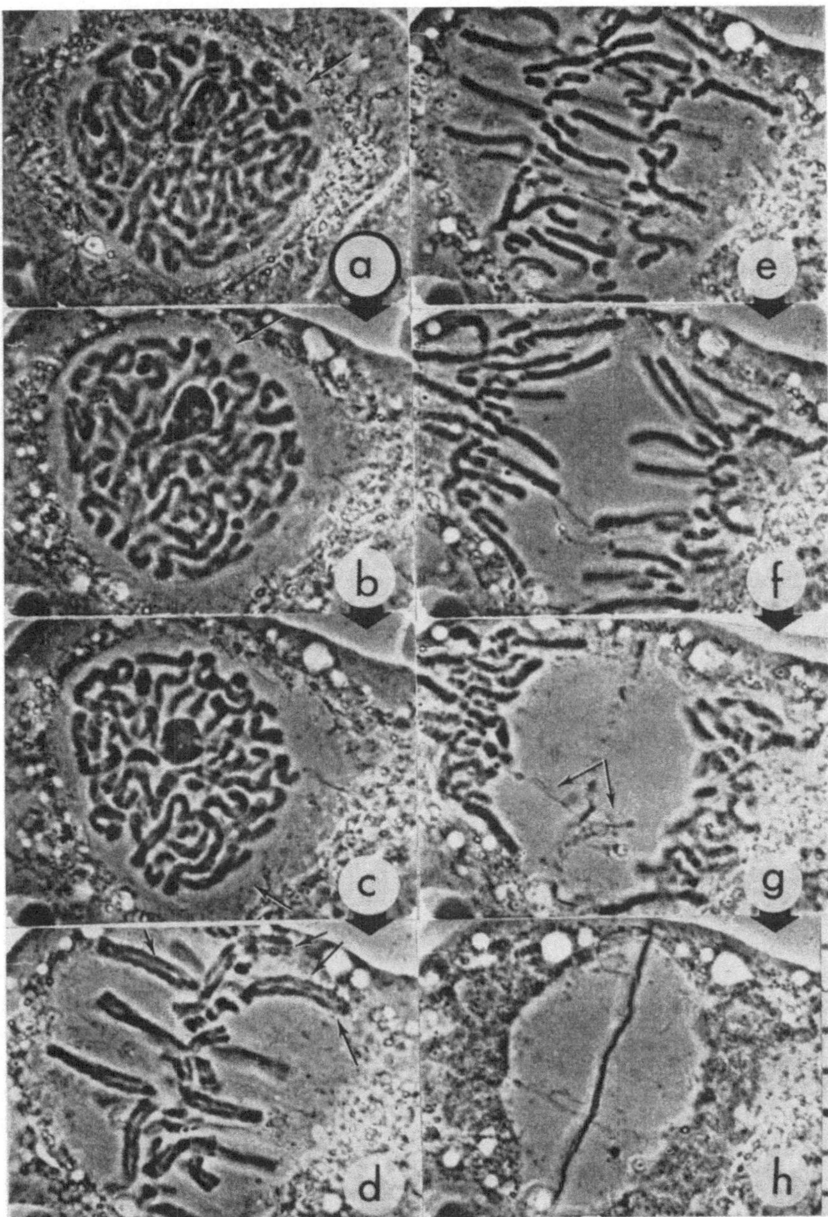

Fig. 1. Mitosis in endosperm of *Haemanthus katherinae* — prints from 16 mm film. (*a*), First stages of clear zone formation and its growth (*b*) around the nucleus. In this cell the development of the clear zone is not uniform around all sides of the nucleus and on the right side it is half-moon shaped, forming what has been described as a polar cap. In (*a*) and (*b*), nuclear membrane is well visible (arrows); (*c*), just after its disappearance (arrow, nuclear membrane indistinct but still visible). (*d*), Metaphase plate just before start of anaphase. Double structure of each of the sister chromosomes (arrows) is not distinct but discernible. (*f*), Late anaphase with first stages of phragmoplast formation; very indistinct fibrils appear in the lower part of the phragmoplast. These fibrils fuse (*g*) and finally middle lamella is formed. Long threads in the phragmoplast are mitochondria (arrows in (*g*)). Times after (*a*): (*b*), 1 hr; (*c*), 1 hr 24 min; (*d*), 2 hr 43 min; (*e*), 3 hr 7 min; (*f*), 3 hr 26 min; (*g*), 3 hr 36 min; (*h*), 3 hr 56 min. Scale in (h) — 10 micron intervals. (Fig. 8. BAJER and MOLÈ-BAJER 1963 [*11*])

Development and structure of the spindle and activity of kinetochores

In most plant and animal cells the spindle begins its formation outside the nucleus. This is true especially in higher organisms. The problem arises as to whether a border between the cytoplasm and the spindle exists. Occasionally in vacuolized plant cells, vacuoles may be situated close to the spindle giving the impression that the spindle is enveloped by the membrane. The border between the cytoplasm and the spindle is often distinct because of the great difference in structural organization of these two components. This difference is exaggerated after fixation and sometimes is seen *in vitro* [cf. micrographs of living cells (*2a* Fig. 2)]. This is however only the phase border and is not correlated with the nuclear membrane. No evidence has ever been furnished that it arises from the nuclear membrane. A phase border is seen often when the nuclear membrane is still present. However, the vacuolar membrane (tonoplast) or in other cases phase border has not been distinguished by some authors from the nuclear membrane and is interpreted as evidence of persistence of the nuclear membrane throughout the mitotic cycle both in plants [53, 54] and in animals [54a]. Evidence from observations of cells *in vitro*, using the phase contrast time lapse technique, leaves no doubt that the nuclear membrane disappears. This fact is seen clearly in material with large nuclei [2a]. The disappearance of the nuclear membrane is established within precisely one to two minutes. Changes connected with the disappearance of the nuclear membrane are sometimes very drastic: the whole chromosome group is compressed into a tighter mass (contraction stage [10]). Such changes connected with the abrupt mixing of the substances of the nucleus and the clear zone can be explained as a sudden disappearance of a physical border (nuclear membrane). The further role and fate of the nuclear membrane is not known. It is not proved at present whether the nuclear membrane breaks into small pieces not further detectable by the light microscope or dissolves completely. Porter and Machado [49] report migration of fragments of the nuclear membrane inside the spindle in plant cells. Daniels and Roth [20] find that small pieces of the nuclear membrane are seen in late anaphase and suggest that these pieces are used in the formation of the new membrane. On the other hand there is no evidence of the existence of such a membrane inside the spindle in the light microscope investigations and on the basis of observations of movements, it is difficult to imagine how the membranes could remain inside the spindle without being removed (cf. pg. 102, 107).

Description of the formation of the spindle will be restricted to higher plants. Though the observations are based on studies of endosperm, the course of this process is, from our present knowledge, the same in all higher plant cells. Formation of the spindle is noted by an appearance of the clear zone around the nucleus. In elongated cells (Fig. 1) the clear zone is often shaped in two new moon-like areas facing the long axis of the cell. Such structures have been described in the earlier literature as polar caps. These polar caps are the special cases of the formation of the clear zone and depend entirely on the mechanical conditions within the cell. At the beginning the clear zone forms slowly but later formation is rapid. Clear zone formation in *Haemanthus* endosperm takes from two to six hours (usually not longer than three). The spindle, from its beginning, stains with nuclear vital stains and this fact leads to the conclusion of some earlier workers (review of literature concerning this problem given by BECKER [*13*]) that the spindle is entirely of nuclear origin. This view can be retained no longer. The spindle forms outside the nuclear membrane and the nucleus contributes quantitatively to its formation to a very small extent. The volume of the nucleus decreases not more that five to ten percent [*2*]. Though the amount of the material does not contribute quantitatively, it seems to stimulate formation of the clear zone. Thus it can be stated that the spindle forms mostly from transformed cytoplasm. Formation of the clear zone is probably connected with the organized orientation of its elements. The electron microscope observations have not furnished data up to now concerning the ultrastructure of these elements. However, the clear zone shows uniform birefringence (Fig. 2 [*32*]).

The nuclear membrane disappears at the moment the clear zone reaches its maximal dimension. Immediately after disappearance, cooperation of kinetochores with the spindle begins, or stated in another way "the spindle fibers attach to kinetochores until the beginning of telophase. Cooperation does not begin at precisely the same moment for all kinetochores. Kinetochores are stretched often visibly in the direction they move [*2a*]. Usually however the kinetochores start to move without any visible stretching.

Kinetochores are not the only portion of the chromosome where spindle fiber attachment is possible. They can attach temporarily to any part of the chromosome (like chromosome ends) for a time usually not exceeding ten minutes (which is a small fraction of the total time of cooperation of kinetochores with spindle fibers). After attachment chromosome arms are "pulled" toward the plate or toward the pole. Often this "pulling" appears as first a sudden bending of the chromosome arm toward the plate, and second a restraightening. This phenomenon is termed "neocentric activity" [*8*]. Further a small outpushing of chromosomal material is often seen in places where fibers attach. These movements were described before their motion was understood [*2a*] and were analyzed in detail [*8, 47*]. The role that neocentric activity plays in mitosis is not well understood. There seems to be no doubt that fibrillar material is responsible for such movement. From studies with the electron microscope it is evident that in *Haemanthus* microtubular material penetrates the chromosomes [*28*]. Evidence from other material [*20, 50*] shows the same penetration. Unfortunately no detailed analysis of the behavior of chromosomes in the living stage has been applied to any other material except *Haemanthus* endosperm subsequent to the electron microscope studies.

Kinetochore and spindle cooperation are connected with changes in the spindle organization. Uniform, birefringent, clear zone differentiate into zones of stronger and weaker birefringence. The zones of stronger birefringence connect with kinetochores and are what is being described as chromosomal fibers (Fig. 2). Chromosomal fibers in higher organisms usually are not seen with the standard light microscope. Few reports of

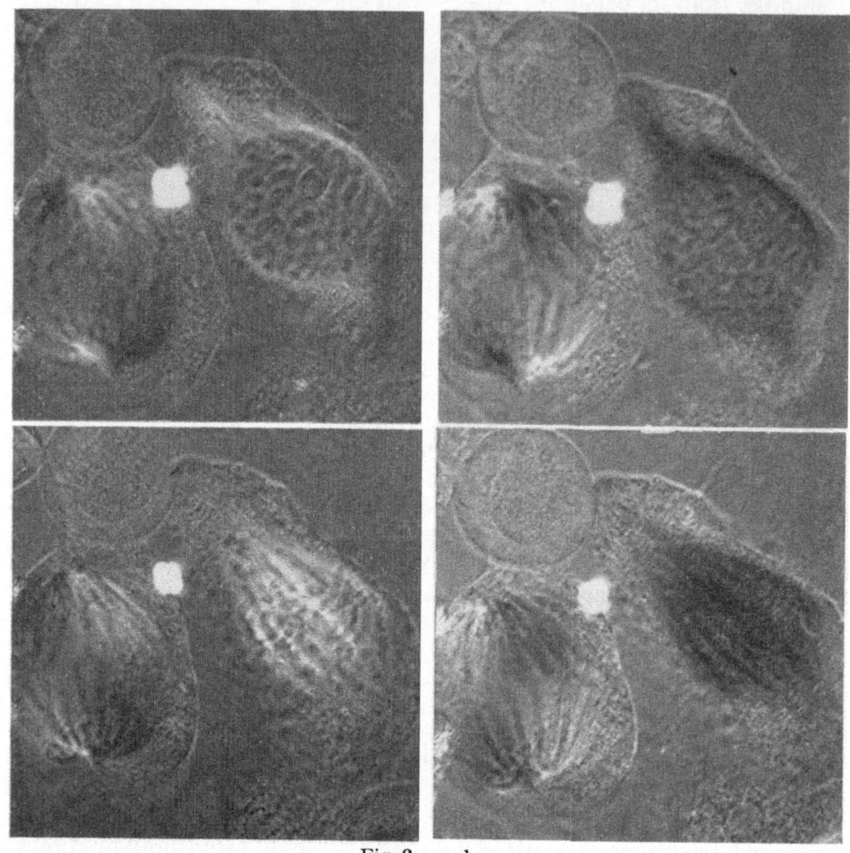

Fig. 2 a u. b

Fig. 2. Development in endosperm of *Haemanthus katherinae* and differentiation of the spindle in polarized light. Black and white compensation. (*a*) and (*a'*), formation of the clear zone. It shows uniform and not differentiated birefringent structure. Nuclear membrane is clearly visible. Starch grain in the center. (*b*) and (*b'*), early prometaphase, clear zone begins differentiation into more or less birefringent zones i. e. spindle fibers and interfibrillar material. (*c*) and (*c'*), long thin fibers — some of them are chromosomal fibers, others look like continuous fibers. (*d*) and (*d'*), metaphase fibers are broader but shorten and are more diffuse at the polar region. (Part of Fig. 7. Bajer and Molè-Bajer, 1963 [*11*])

their visibility have been made so far. In animal material Cooper [*19*] observed chromosomal fibers in blastomeres of the mite, and in plant material chromosomal fibers were observed in a few percent of cells with the use of phase contrast in endosperm [*6*]. The Nomarski interference contrast system (further abbreviated as Nomarski system) used successfully for the first time in studies of the large mitotic spindle *in vitro* [*12*], shows many details of structure (Figs. 3, 4). Comparison of the phase contrast (Fig. 3a) with the Nomarski system (Fig. 3b) shows the superiority of the later in detecting detailed structure. The small depth of focus (about 0.2 microns) of the Nomarski system makes possible optical

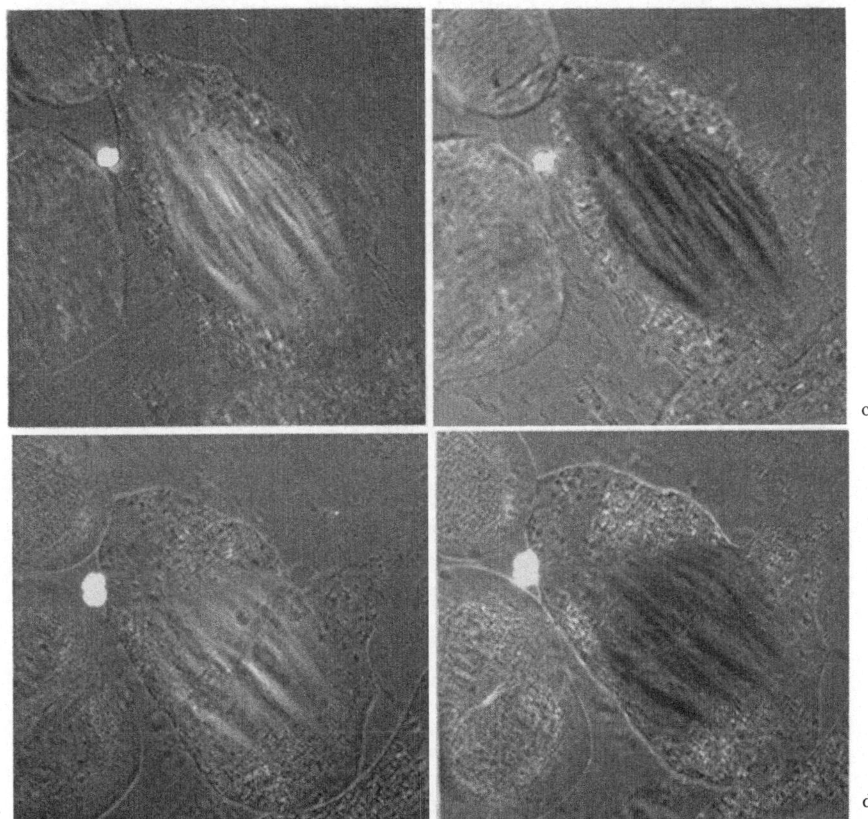

Fig. 2 c u. d

sectioning through the cell. Fibers appear to be connected with the kinetochores. A bundle of fibers connected to a single kinetochore diverges toward the pole while adjacent chromosomal bundles intermingle with each other at the polar regions or between the polar regions and kinetochores. Thus the spindle pole is composed of loosely arranged fibers from different bundles. Figurs 3b and 4 shows very clearly the intermingling of fibers from different kinetochores. The difference of the Nomarski system picture with the picture obtained using polarized light is conceptually important. Birefringence in the polar region is too weak to be detected even by the most sensitive polarizing microscope. Therefore bundles connected with single kinetochores appear to converge toward the pole and bundles of adjacent chromosomes appear to have no connections

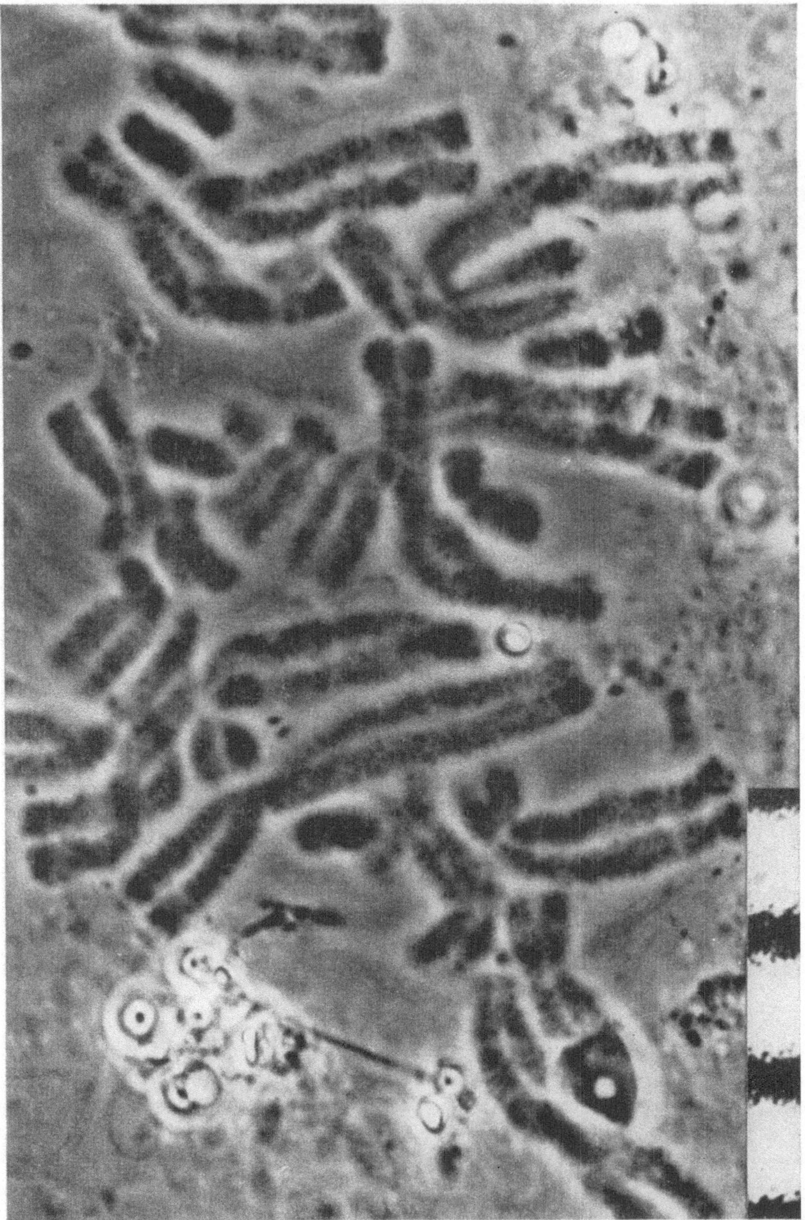

Fig. 3a

Fig. 3. (a), metaphase in very flattened cell of *Haemanthus katherinae*. With dark phase contrast the chromosomal fibers are seen in several places (cf. Fig. 3b). The cell is very thin and as a result the metaphase plate is not very regular. Scale: 10 micron intervals. (b), the same cell as Fig. 3a in Nomarski interference phase contrast system. Chromosomal fibers connected with several kinetochores are well seen. The beaded appearance of fibers is well discernible. Fibers from neighboring kinetochores join close to the poles. Scale: 10 micron intervals

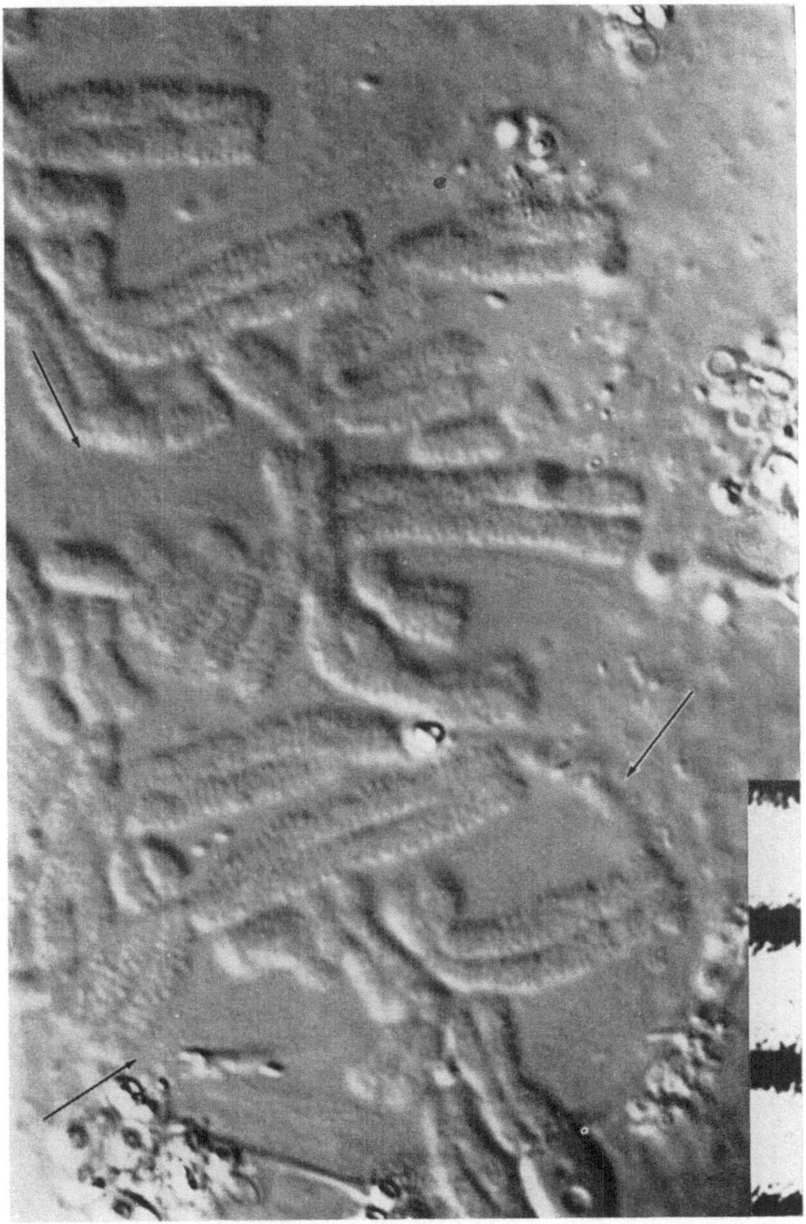

Fig. 3 b

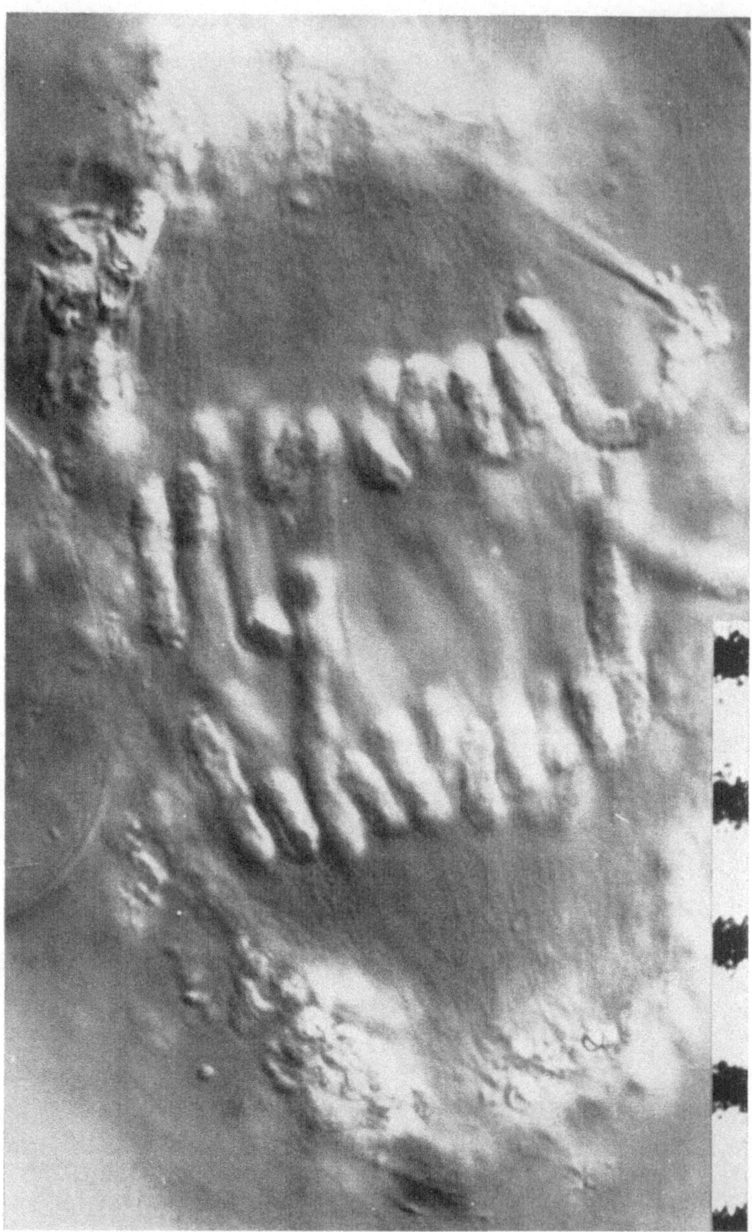

Fig. 4. (a—g), organization of the spindle in endosperm of *Haemanthus katherinae* with the Nomarski interference contrast system. The cells are unflattened and chromosomal fibers connected with kinetochores are seen in several places. (a), the fibers of neighboring chromosomes intermingle at the poles. No distinct fibers are in the interzonal region. The fibers are still clear in (b) (arrow) and are very unclear in (c). No traces of fibers can be detected in (d). Fibers in the interzonal region appear in mid-anaphase (b). Their number increase as the phragmoplast develops (d—g) and the cell plate forms (e—g). Some particles are eliminated from the phragmoplast toward the pole (short arrow d—e) while others move temporarily toward the plate (long arrow e—f). In later stages granules accumulate in the polar region of the phragmoplast close to the sister nuclei (g). Scale: 10 micron intervals

(Fig. 2d). On the basis of observations in polarizing light it can be concluded that the polar region is very loosely organized and that chromosomal fibers do not intermingle at the poles. Comparing the

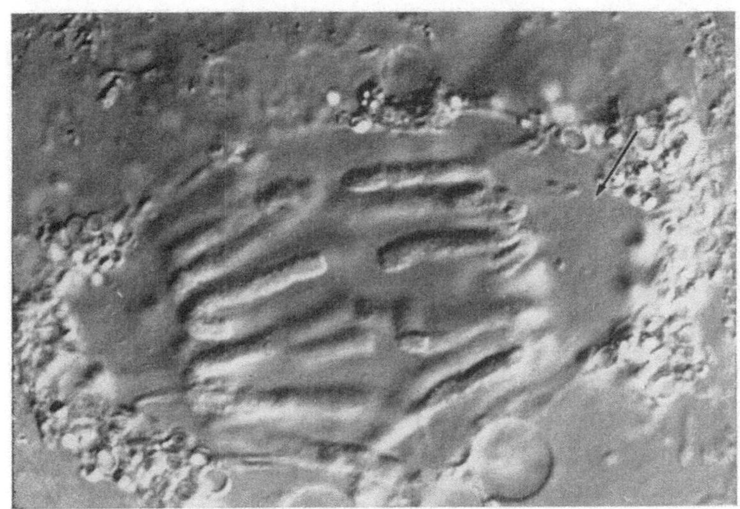

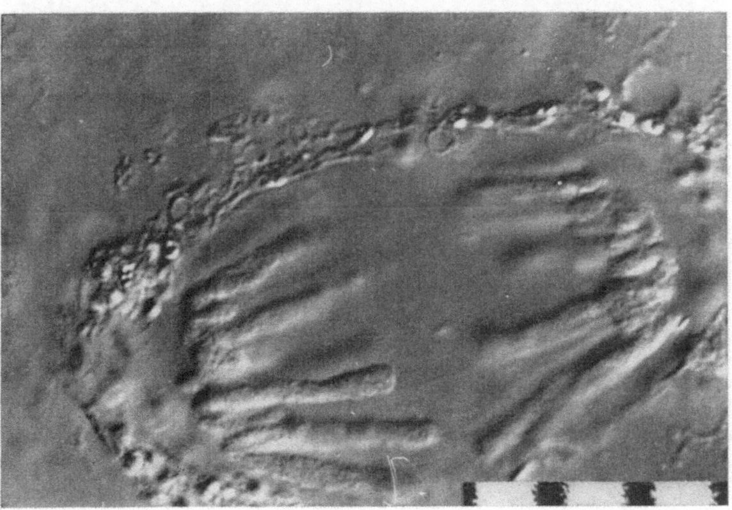

Fig. b, c

arrangement of different stages will be very important in understanding the processes occurring inside the spindle during chromosome movement. Data available at present are however very fragmentary and do not permit the drawing of the foregoing conclusions.

7*

During anaphase the fibers connected with kinetochores shorten and are no longer detectable at the end of this division stage. The fibers never thicken during their movements. Electron microscope studies using

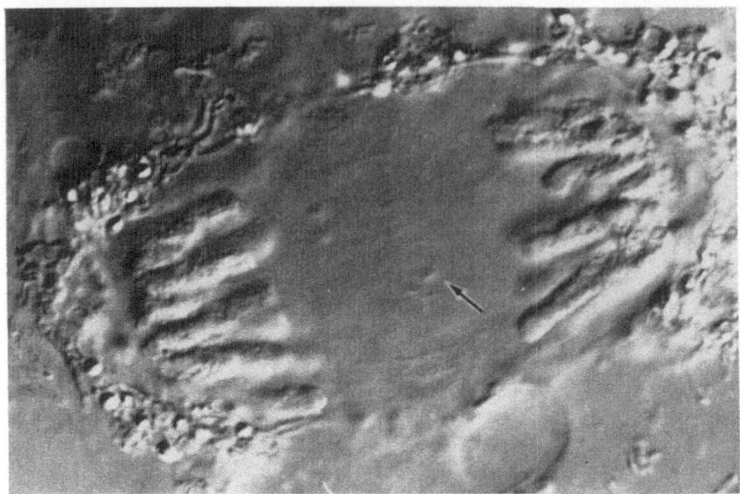

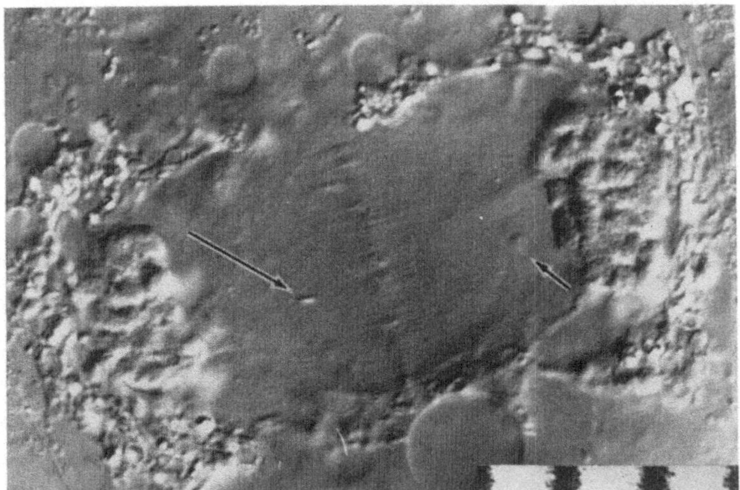

Fig. d, e

different materials [*15, 20, 26, 27, 36*] show that fibers are in fact microtubules about 130 to 270 Å in thickness. A comparison of pictures from different material indicates that the number of microtubules varies depending possibly on the size of the chromosome. There is no doubt that these microtubules play a role in movement and some suggestions will be discussed later (pg. 112).

There is considerable controversy concerning the fiber type that functions in arrangement of the spindle. The classical picture obtained using fixed material in the light microscope distinguishes two types of

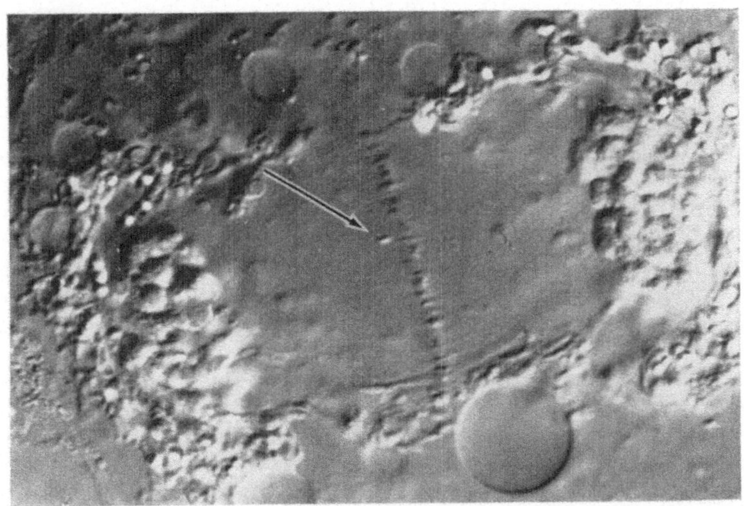

f

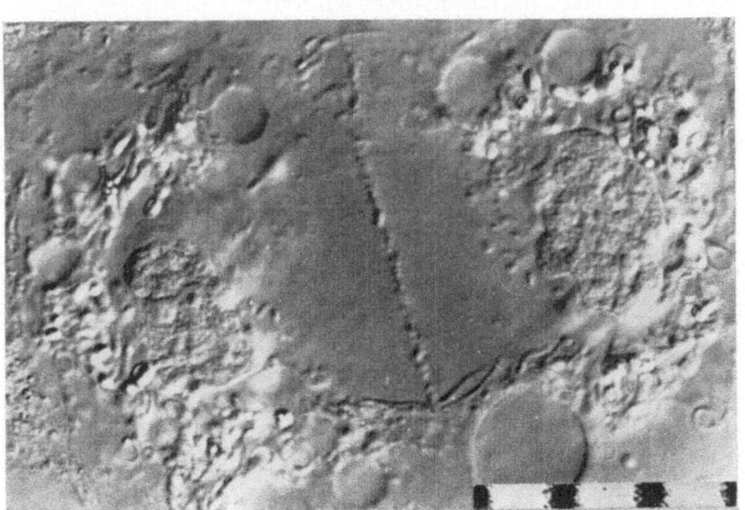

g

Fig. f, g

fibers (cf. SCHRADER [51]): chromosomal fibers connecting the kineto-chores; and continuous fibers connecting the two poles. During anaphase, usually middle anaphase, fibers between the two chromosomal groups are visible. It is not clear whether these are continuous fibers described in prometaphase-metaphase or are formed *de novo*. Continuous fibers are seen in early prometaphase but are difficult to detect in metaphase. Gimenez-Martin and Lopez-Saez [24] were not able to detect them after using piperazine hydrate, which exaggerates visibility of chromosomal fibers. Thus the origin of fibers in the interzonal region is unclear. In the early stages of their appearance, the fibers possibly may be continuous (if they exist) or they may form *de novo*. In later stages there is no doubt that they form *de novo* as their number increases appreciably during the progression of anaphase. In plant material the interzonal region is transformed into the phragmoplast which is both birefringent and fibrillar. The relation of the early fibers found in the interzonal region to those found in the phragmoplast is uncertain. Again, the microfibrills are in fact microtubules of the same size as those found in the spindle.

The above remarks concern the higher plant mitotic spindle. Due to the existence of centrioles and asters in animal cells the spindle shows some differences in its formation and structure e.g. the polar region is more condensed, cytokinesis is preformed differently in comparison to plants, the cleavage furrow replaces the phragmoplast and cell plate etc. However these differences are not essential because of the many similarities. It may be argued that the spindle with centrioles is considerably different. Dietz [21] nevertheless observed division both with and without centrioles in the same material. In both cases the distribution was undisturbed. This indicates that chromosome distribution can be performed in a perfectly normal manner without the centrioles.

Movement of chromosomes and akinetic bodies during mitosis

A general outline of movements in plant cells based on analysis of films will be presented. (For a previous detailed analysis see [11].) The following description will be applied mainly to the normal course of mitosis with some reference to the abnormal course of mitosis.

Chromosome movement starts after the establishment of cooperation of kinetochores with the spindle i.e. immediately after the disappearance of the nuclear membrane on the onset of prometaphase. The most characteristic feature of prometaphase is the simultaneous movement in two opposite directions of different bodies inside the spindle: kinetochores show a strong tendency to move toward the equator while all bodies that were not attached to the spindle move toward the spindle poles (Figs. 5—6). Kinetochores do not always take the shortest and simplest path to their final destination on the equatorial plate. They often move first toward the pole and then toward the plate. Also some kinetochores move across the spindle or several times across the equator before reaching the fixed position. Kinetochores in prometaphase retain considerable individuality of movement. Often neighboring kinetochores move simultaneously in opposing directions. Most movements are parallel to the long axis of the spindle. All bodies which are unattached to the spindle, like persistent nucleoli, small granules, akinetic fragments formed after irradiation, etc., move toward the poles with the same speed independent of their size (Figs. 6—7). Chromosome arms are subjected to the same forces and as a result tend to arrange themselves parallel to the long axis of the spindle

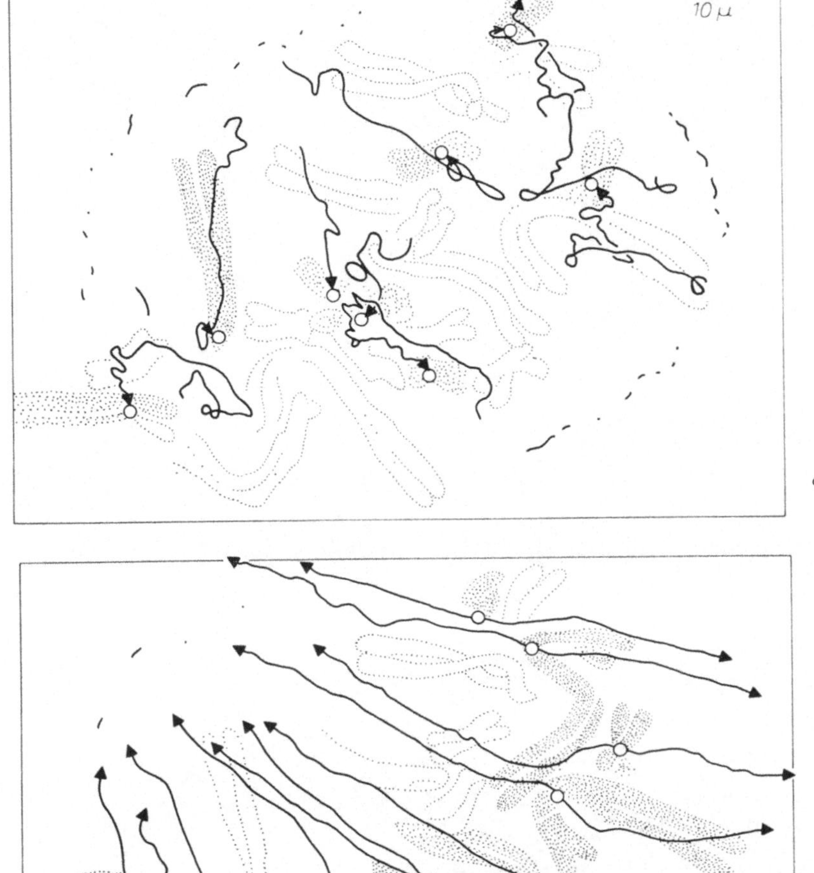

Fig. 5. Path of kinetochores during the formation of the metaphase plate (*a*) and anaphase in endosperm of *Haemanthus katherinae*. Circles show the position of kinetochores (*b*) at the start of anaphase. Chromosomes for which the paths are drawn are dotted. Paths of kinetochores during prometaphase are rather complicated (*a*). During anaphase the kinetochores move toward the pole but the sharply defined polar center toward which they move is not present

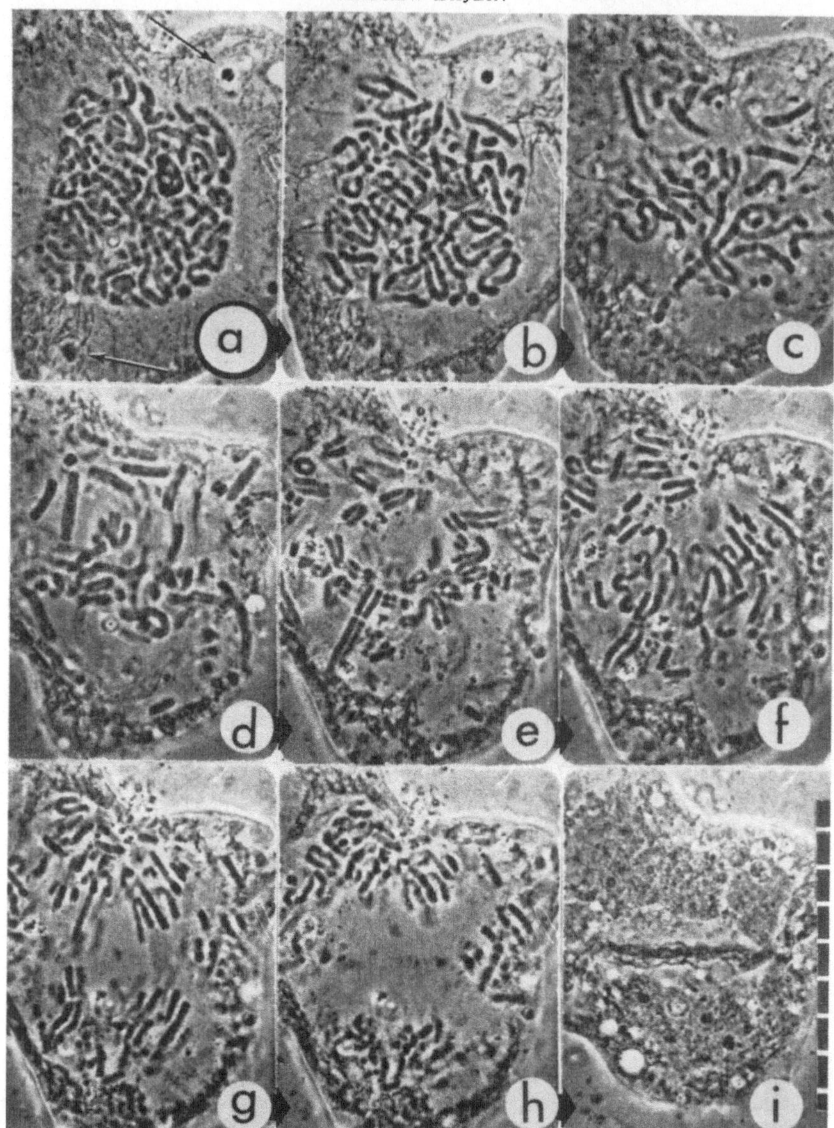

Fig. 6. Poleward elimination of akinetic fragments from the prometaphase spindle. β-irradiated cell. At least second generation of cells after irradiation. Material is from endosperm of *Haemanthus katherinae*. This is indicated by amorphous globule (right upper corner and left lower corner: arrows in (*a*) which probably are composed partly from degenerated fragments. (*a*), prophase with clear zone normally formed. (*b*) and (*c*), prometaphase; akinetic fragments are eliminated toward the poles. (*d*), metaphase; akinetic fragments accumulate at both poles. (*e*), start of anaphase: akinetic fragments stay at the poles and do not change their position until late anaphase. In later stages of anaphase (*g*) some akinetic fragments move toward the equator (cf. Fig. 7). These fragments are eliminated from the phragmoplast in later stages of the phragmoplast development and the results of elimination are seen in (*i*) where not many fragments are found between sister nuclei. Time after (*a*): (*b*), 1 hr 6 min; (*c*), 1 hr 29 min; (*d*), 2 hr 7 min; (*e*), 3 hr 5 min; (*f*), 3 hr 20 min; (*g*), 3 hr 46 min; (*h*), 3 hr 50 min; (*i*), 7 hr 12 min. Scale: 10 micron intervals (Fig. 11 Bajer and Molè-Bajer, 1963 [*11*])

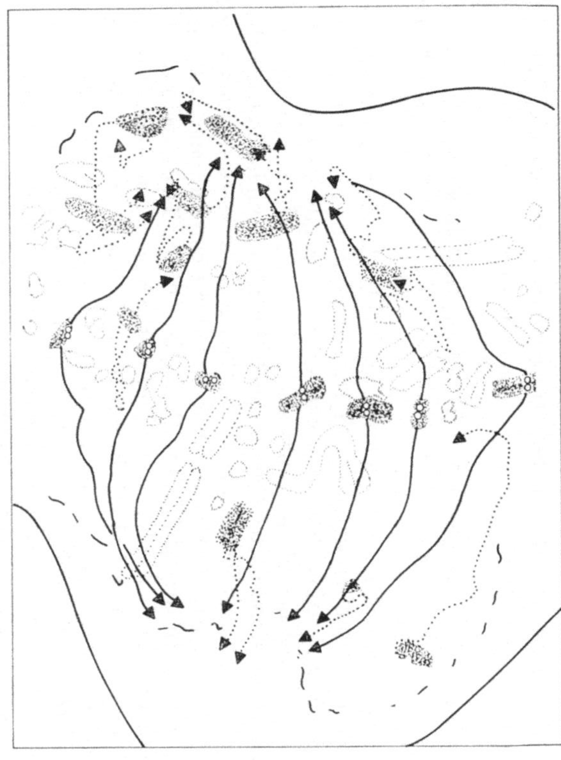

Fig. 7. Paths of akinetic fragments (dotted) during prometaphase and anaphase and paths of kinetochores (continuous lines) during anaphase. The position of chromosomes in the metaphase position just before the start of anaphase is drawn. Chromosomes — paths of movement are drawn dotted. Kinetochores — white circles. During prometaphase (a) akinetic fragments are transported toward the poles. Some execute complicated movements, move across the spindle (lower spindle pole) etc. In later stages (metaphase) many fragments are transported toward the plate due to shortening of the length of the spindle(cf.pg.113). During anaphase (b) kinetochores move toward the poles and as they move they push most of the akinetic fragments in front of them. However, some akinetic fragments move toward the plate and then move toward the poles again (explanation in the text). Micrograph of this cell in Fig. 6

or more precisely, parallel to the ultrastructure of the spindle. Thus the essential feature of prometaphase is the simultaneous movement toward the pole and the plate. During a typical metaphase the kinetochores are arranged precisely on the equatorial plate, the chromosome arms are situated parallel to the long axis of the spindle, and the akinetic bodies accumulate in the polar regions.

Though detailed analysis on other material is lacking, as judged from the literature, similar events with some modifications occur in other material. Modification during prometaphase depends on many factors such as the presence of centrioles and asters, the size of chromosomes, the size of the spindle, etc. Even in such diverse examples as the hollow spindle found in cold-blooded animals, the basic processes are unchanged. The spindle is too small to accumulate all of the chromosomes and elimination both from the spindle and the asters occurs. These two processes counteract each other and as a final result the long chromosome arms are in a star-like arrangement on the equatorial position around the spindle [48]. Also "up and down" movement of kinetochores (chromosome dance) seen in most animal cells and reported by LEWIS and LEWIS [37] is in the author's opinion not essentially different in plant and animal mitosis.

During anaphase direct observations indicate that two mechanisms are responsible for chromosome movement: shortening of the chromosomal fibers; and movement apart of the whole half-spindles. The increase of the spindle in length is due to elongation of the interzonal region. BĚLAŘ [14] proposed the "pushing body" hypothesis which assumes that the whole half-spindles move apart due to expansion of the interzonal region (pushing body — Stemmkörper). Elongation of the whole spindle with simultaneous shifting of chromosome fibers within the spindle has been suggested [4]. According to this suggestion the interzonal region would be stretched due to the moving apart of the half-spindles. This process does not prevent any transport phenomena within the interzonal region. Other explanations of elongation of the spindle are also possible. We must also consider that the interzonal region in plants is transformed into the phragmoplast with typical transport properties (to be discussed later).

There are a few other phenomena that occur during anaphase and that are important. One well recognized event is that all kinetochores of the chromosome set enter anaphase at precisely the same moment. If this does not occur, as is sometimes observed if the metaphase plate is very wide, the movement starts in one place and spreads to adjacent chromosomes. This suggests that the movement of chromosomes is influenced by one another. However the movement in the normal course of anaphase is synchronized precisely and the individuality of the single kinetochore is not exhibited. Therefore the questions arise as to what extent the individuality of this movement is retained and to whether there is any critical influence of one kinetochore on another. Observations of movement of kinetochores connected by sister-reunion chromatid bridges [6] shows that one possibility of behavior of such bridges is as follows: kinetochores connected by bridges are retarded and speed up their movement when the bridge breaks. Kinetochores farther from the bridge show no effect, and adjacent kinetochores are retarded in spite of the fact that they have no bridges. Thus there is an influence of one chromosome on the movement of another one in anaphase.

The interzonal region transforms into the phragmoplast which is responsible for the formation of the cell plate that later transforms into new middle lamella. The role of the phragmoplast in the formation of the new cell wall is well recognized and this problem will not be discussed here. However, the phragmoplast plays an important role in the distribution of chromosomes. There is some evidence that the interzonal region in animal cells plays an important role also but unfortunately reported literature is scarce. STICH [52] described similar transport phenomena in the interzonal region during anaphase in *Cyclops*. In plant mitosis transport properties of the early phragmoplast are similar to the prometaphase spindle: all bodies like small granules (Fig. 4), fragments after irradiation, laggard chromosomes, etc., are transported toward the poles. The phragmoplast is responsible partly for stretching bridges which tend to be eliminated toward the poles. As this takes place in early telophase after the anaphase movements have ceased, the impression is that a "second anaphase" occurs in telophase. Such "second anaphases" have been observed previously in detail [2b].

During the organization of the cell plate the elements fuse and this process is connected with a lateral (perpendicular to the long axis of the phragmoplast) transport of material [9]. Due to this property e.g. fragments are transported toward the side and arms swing to the side and often in telophase diverge away from the central part of the phragmoplast (Figs. 1f—g). Such behaviour was often attributed to the action of the "pushing body" [28].

Abnormal course of mitosis

The literature concerning abnormalities observed during mitosis is probably more extensive than that concerning normalities [16, 22, 40]. Studies of abnormalities often throw important light on the course of normal mitosis for some factors are exaggerated. There are various aspects of disturbances and it is possible to classify them in different ways. A simple classification will be applied such as disturbances depending on 1) chromosomes, 2) spindle, 3) both spindle and chromosomes. Most disturbances will fit one of these classes and only the first two classes will be considered shortly in this paper. Also only one aspect of the disturbances will be discussed — namely the disturbances connected with chromosome movements and their importance for an understanding of the spindle mechanism. A very few selected facts will be reported and some general conclusions will be made.

It is comparatively easy to produce both chromosome and spindle aberrations. The difficulty arises because most effective agents prevent spindle formation or completely stop its activity. It is difficult to produce partial disorganization of the spindle which would further produce disturbances in chromosome movement.

Disturbances observed are either spontaneous or experimentally produced. Serious spontaneous chromosome aberrations are rarely observed except in some cases as described by HENEEN [30]. He found a vegetatively propagating plant with very strong radiomimetic effects. Fragments were found in this plant in practically all cells. Spontaneous disturbances in the spindle are even more seldom reported.

Probably most experimental treatments of mitosis delay the start of mitosis or even do not permit entry into mitosis. Most chemical agents will produce this effect at certain concentrations. These chemicals generally will kill the cell before producing the desired effect and the same chemical agent produces as a rule different effects at different concentrations. Thus in colchicine agent studies [22] only an irregular anaphase is caused at very low concentrations and disorganization of the spindle at higher concentrations.

Chromosome disturbances

The breakage and reunion cycle is observed most commonly as a result of ionizing radiation or of treatment with several chemicals (review by KIHLMAN [34]). As a result chromosome fragments and bridges form. The spindle, in spite of the treatment, functions as a rule in a normal way. Behaviour of akinetic fragments has been described above (Figs. 6—7). Bridges behave differently depending on their number per cell and their length. A detailed analysis of some aspects of bridge behavior in living cells was reported previously [5, 6]. If many bridges are present in anaphase, the spindle fibers are not strong enough to tear them apart or stretch them, and as a result one restitution nucleus is formed. Retardation of neighboring kinetochores not connected by bridges plays an important role here. If a few bridges are present, they are torn apart or stretched and their behavior is length dependent. In anaphase, the elastic and plastic properties of chromosomes change rapidly. During early anaphase the chromosomes break easily, while at later stages they can stretch indefinitely. Thus during anaphase short bridges will stretch to different degree and break while long ones only stretch. Very long bridges may be long enough not be stretched at all and as a result connect the two chromosome groups in late anaphase. Such bridges generally are stretched and are cut finally due to the transport activity of the phragmoplast.

Broken chromosome ends show considerable activity during prometaphase and sometimes fuse at the start of anaphase. Fusing of chromosomes is observed especially in the case of interlocked dicentrics or interlocked rings [5]. In this way dicentric chromosomes pass from one generation to another. Breakage and fusion of chromosomes in the living stage explains why some observations on fixed material do not agree always with the theoretical expectations (e.g. CONGER [18]).

Occassionally fragments which are close to the poles or among chromosome groups move rapidly toward the equatorial plate while others move toward the pole (Fig. 7b). Neocentric activity is the phenomenon responsible for such movements. Small particles sometimes move toward the equatorial position (Fig. 4c—f) and the mechanism of their movement is not clear. Fragments which are on the sides of the spindle also occassionally move toward the plate and the mechanism of this movement is different. In cells with small amounts of cytoplasm each group of chromosomes act as a piston in a very loose cylinder. As a result material accumulated at the polar areas is squeezed in the direction of the equatorial plate (Fig. 7).

Spindle disturbances

Kinetochores are probably responsible for the final organization of the spindle as described above (pg. 95). Consequently the mechanical conditions in the cell, like the relation of the size of the spindle to the cell and the number of chromosomes, can produce multipolar divisions. In endosperm this division is often observed in comparatively small cells with a large number of chromosomes where there is no space to properly arrange all kinetochores on the equatorial plane. The behavior of chromosomes in such spindles is exactly as predicted from studies of dipolar mitosis. Often bridges are formed.

Complete inhibition of spindle development and complete disorganization of the ultrastructure of the existing spindle are probably the effects most extensively studied. Colchicine, in proper concentrations, produces this effect. The spindle is abruptly disorganized (in about 30 minutes at 40 p.p.m. in *Haemanthus* endosperm). Chromosome movement in prometaphase or anaphase is stopped completely and one restitution nucleus is formed. The cycle of the chromosomes is typically not disturbed except for supercontraction. With prolonged treatment the process is more complex. The process was described in detail in living cells in endosperm [41]. During c-mitosis chromosomes change arrangement due to their shortening and repulsing forces. The question of when the chromosomes repulse each other is often raised especially in the older literature. C-mitosis shows this effect very clearly: chromosomes tend not to touch each other and arrange themselves either parallel or perpendicular to each other which indicates repulsion. The organized ultrastructure of the spindle is completely disturbed and birefringence disappears [31].

According to ÖSTERGREN (personal communication) many chemicals at certain concentrations produce the c-mitotic effect. However most of them are too toxic with many side effects. Usually the cells are poisoned and all processes gradually stop before the concentrations necessary to disorganize the spindle are reached. The c-mitotic effect of chloral-hydrate is also well known after the experiments of NEMEC [43] were reported. C-mitotic effects from methanol have also been reported [42]. The present author observed c-mitosis after treatment of endosperm cells with ribonuclease (BAJER unpublished).

The fact that similar results can be obtained from the action of different chemicals has important theoretical implications. The ultrastructure of the spindle is very labile and it is obvious that once it is disturbed strongly without effecting the chromosomes, only the chromosomal cycle will continue and c-mitotic effects will be observed. In the opinion of the present author, several factors are responsible for the organization of the spindle (hydration, surface properties of the membranes, etc., etc.) and therefore distortions of any kind will prevent completely normal functions. In plant material it is comparatively difficult to produce intermediate effects such as to produce partial distortions of the function of the spindle. However once they are produced several important observations are furnished concerning chromosome movement. MOLÈ-BAJER (unpublished) has studied in detail the effect of chloral-hydrate on living cells (Fig. 8). At low concentrations (about 0.01 %) regular prometaphase or metaphase disorganizes and resembles c-mitosis. After a lengthy time in c-metaphase, very irregular movements start and kinetochores move independently from each other in all directions with no definite polar orientation. This is convincing proof that kinetochores with the structures responsible for their movement are to a great extent separate independent units capable of distribution of chromosomes. The normal spindle is formed from several such units. Chloral-hydrate discorrelates their mutual arrangement without effecting the single units (bundle of chromosomal fibers). Similar effects were observed by the author after ribonuclease treatment. After ribonuclease all possible transitions between normal and c-anaphase were observed. Also synchrony of anaphase movements and the start of anaphase was greatly disturbed (BAJER unpublished).

As it was stated before, kinetochores play an important role in the organization of the spindle. It has been found that kinetochores can organize fibers from the phragmoplast [42]. In cells treated with methanol, in a few examples, the cycle of chromosomes proceeds without any evident development of the spindle and the process resembles c-mitosis. However some chromosomes show kinetochore activity during development of the phragmoplast. The stage of the chromosomes was determined on the basis of measurements of chromosomal length change during normal and c-mitosis. From the comparison of how the chromosomes change in length in normal and methanol treated mitosis, it is possible to determine that when the chromosomes move they are in the telophase chromosome cycle of normal mitosis.

As already mentioned previously (pg. 107), the phragmoplast and interzonal region play an important role in chromosome distribution. Though detailed studies have not been made, it is very likely that the interzonal region in animal cells shows similar properties. Transport properties of the phragmoplast prevent the approaching of the two sister nuclei in telophase to each other. If the phragmoplast activity is stopped the nuclei may fuse or binucleate cells may be formed. Formation of binucleate cells has been frequently observed [25]. Tearing apart of bridges in telophase is another result of phragmoplast activity.

Sometimes additional phragmoplasts are formed in different positions in relation to the primary one and often form in sister nuclei perpendicular to the primary one. As a result the nuclei or nucleus are cut into two and after a regular bipolar anaphase, three or four nuclei are formed. Occassionally this formation is not correlated with the appearance of an extra cell plate. This process can take place in late stages between nuclei which resemble interphase nuclei. Such cases appear similar to amitosis [42].

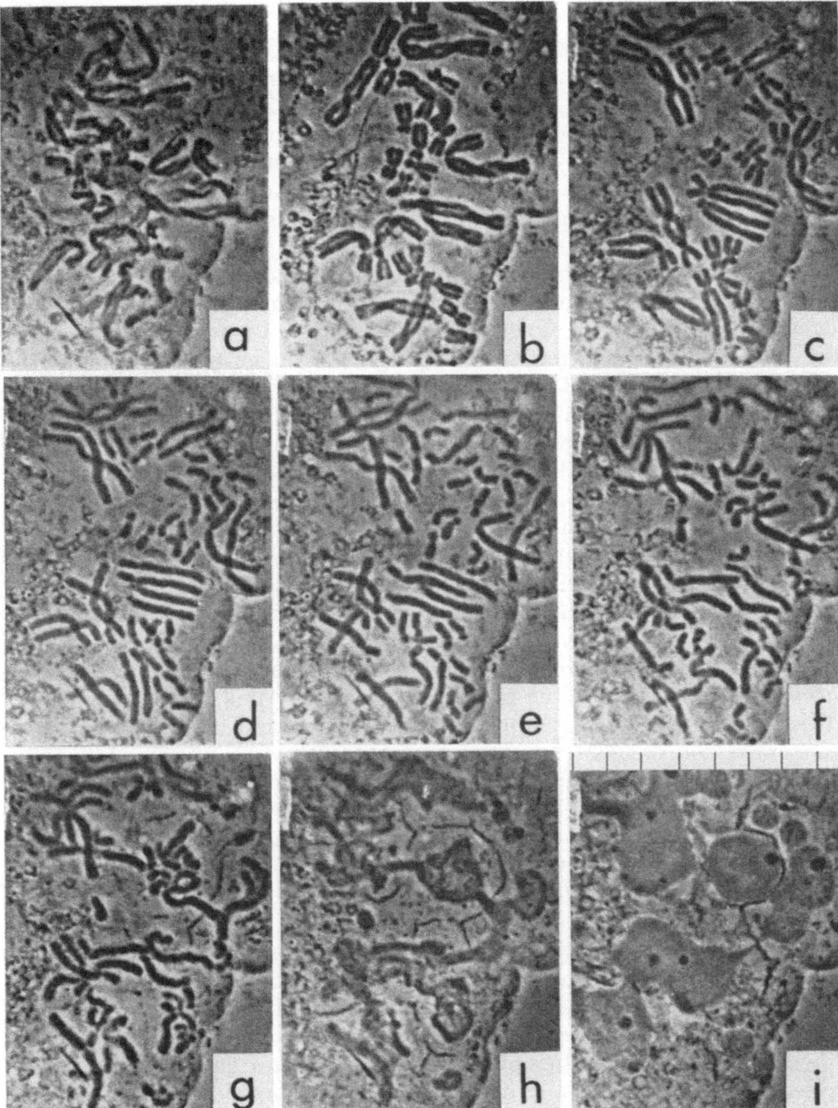

Fig. 8. Mitosis under the influence of chloral-hydrate in endosperm of *Haemanthus katherinae*. Time after (*a*): (*b*), 1 hr 3 min; metaphase with more or less regular metaphase plate (*c*), 3 hr 51 min; late metaphase after reorganization of metaphase plate (*d*), 4 hr 1 min; (*e*), 4 hr 26 min; (*f*), 4 hr 39 min; anaphase. The shape of chromosomes during anaphase (they are bent in kinetochore region) indicates that an active role of kinetochores takes place during this movement. (*g*), 4 hr 59 min; telophase and formation of multipolar phragmoplast. (*h*), 5 hr 20 min; (*i*), 6 hr 46 min; formation of cell plates and daughter nuclei. Scale: 10 micron intervals in (*i*). (Unpublished micrographs of Dr. J. MOLÈ-BAJER)

Conclusions

Any acceptable hypothesis of chromosome movements has to explain the above summarized movements and take into consideration facts concerning the spindle reported before. In general outline, the hypothesis of ÖSTERGREN et al. [48] fulfills these requirements and is the only hypothesis so far that explains both movement of kinetochores and akinetic bodies. At the time the hypothesis was reported many facts were unknown and some modifications now might be proposed. Some points of this hypothesis, partly modified, are presented below. The key for understanding chromosomal movement is the spindle which is composed basically from microtubules and intertubular material. Microtubules attached to kinetochores form passive support for chromosome movements and are shifted as a rule parallel to the long axis of the spindle due to the activity of the intertubular material. As the kinetochores are attached to the microtubules, shifting of bundles of microtubules results in a change of a position of the kinetochore. Thus kinetochores accumulate microtubules into denser bundles closer to the kinetochores, and at the same time microtubules tend to be transported toward the poles. The intensity of these two factors is different in the various stages of mitosis. The microtubules are supported or are kept in tension by the activity of the intertubular material. Activity of intertubular material eliminates all bodies, granules, fragments, etc. along regulary arranged microtubules toward the poles. A tendency to extrude the microtubules out of the spindle results in keeping microtubules in tension. Thus microtubules do not need to be anchored at the poles and are able to pull the chromosomes. This takes place during anaphase when the mechanical bond at the kinetochore region breaks triggering anaphase. The whole bundle of chromosomal fibers (microtubules) is shifted (or slides) parallel to the long axis of the spindle and the regular orientation is disarranged at the poles. The evidence for shifting is found in the fact that the chromosomes in anaphase move faster than the chromosomal fibers shorten after the component of movement due to the elongation of the spindle is subtracted [4]. The shifting or sliding of microtubules probably takes place also during prometaphase and metaphase. FORER [23] has found that the disorganized part of chromosomal fibers is transported toward the pole. Once microtubules reach the poles their regular arrangement is disorganized to bundles of microtubules and kinetochores. A bundle of chromosomal fibers and part of the surrounding material form a unit which is partly independent. This is well demonstrated by some disturbances [41]. Independence is greater in prometaphase than anaphase and this should be reflected by the ultrastructure of the spindle (no data are available as yet). Microtubules intermingle at the poles and the surrounding material keeps the microtubules of different kinetochores in tension. This explains the phenomena during the stage of anaphase and retardation of kinetochores due to bridges reported above.

During early prometaphase microtubules are regularly and uniformly arranged. Kinetochores are responsible for the secondary organization of the spindle such as arrangement of microtubules into areas where there are densely packed chromosomal fibers. This tendency of denser packing of organized material close to kinetochores is well seen in the shape of chromosomal fibers during the transition between prometaphase and metaphase (cf. Figs. 3—4). During prometaphase the bundles of chromosomal fibers are long and thin and in metaphase they are wider and thicker. As a result the spindle is shorter in metaphase than in prometaphase. Thus the important property of kinetochores is in arranging microtubules into denser bundles of chromosomal fibers from any cell component where microtubules are found abundantly such as in the prometaphase spindle and occassionally the phragmoplast [41].

It is too early to state what chemical changes are connected with chromosome movement and what is the source of energy, etc. It is very likely that many processes take place here. Hydration is a likely change as suggested by INOUÉ [32]. The importance of hydration and dehydration was realized a long time ago by Japanese workers and summarized by KUWADA [35]. Hydration does not exclude other processes like electrophoresis and membrane potentials [1].

The few examples mentioned above indicate that various disturbances are easily understood on the basis of changes of the ultrastructure of the spindle, discorrelation of basic units (bundles of chromosomal fibers), disorganization of the spindle structure, and chromosome disturbances. Examples given here were chosen when only one phenomenon was shown clearly. More often several types of aberrations occur simultaneously and the picture becomes complicated.

Several deviant types of mitosis and especially meiosis are reported in the literature. SCHRADER [51] reports some classical examples. In the opinion of the present author all such deviations can be explained on the basis of the ultrastructure outlined here and on the assumption that in such deviant divisions some factors play a greater role and obscure the activity of others. As an axample at certain stages the metaphase elimination activity can decrease or increase, the activity of some kinetochores can be delayed, etc. If kinetochores attach to the spindle with delay the chromosomes will behave like akinetic fragments during prometaphase or metaphase. They may accumulate at the poles and when their activity starts, they move toward the plate in mid-anaphase and telophase. The later is a classic picture found in meiosis of plant hybrids [17]. Monopolar mitosis in *Sciara* [39] can be explained also in similar terms. Obviously when there are more than one kinetochore in a chromosome, the problem of coorientation arises [42] but this problem can be explained in the above mentioned terms and is beyond the scope of the present article.

Summary of chromosome movements

There are several factors which influence chromosome movement. They have been enumerated by ÖSTERGREN [44, 45]. The summary will be restricted however to factors reported here. Schematic drawings represent the direction of movements in plant cell mitosis. Very few

Fig. 9. Schematic representation of movements and direction of acting forces (arrows) within the spindle during plant mitosis. Kinetochores — white circles, n — persistant nucleolus, f — akinetic fragment, p — small particle, nc — neocentric activity. Mitochondria and small particles accumulate on the border between the cytoplasm and the spindle. (*a*), prometaphase: transport of akinetic bodies toward the poles; some chromosome arms execute neocentric movements. (*b*), metaphase: no or very few akinetic bodies inside the spindle; chromosome arms arranged parallel to the long axis of the spindle. (*c*), anaphase: sister kinetochores move toward the poles; some akinetic fragments move toward the plate; the spindle elongates; the action of pushing body (?) is not certain as far as existence is concerned; poleward transport from the interzonal region begins. (*d*), telophase and phragmoplast activity: kinetochores and akinetic fragments are on the poles. Fragments which moved toward the plate move to the pole and those which were caught by the cell plate are torn apart. Paths of chromosome ends (dotted) do not follow the paths of kinetochores (dashed). General poleward movement and cell plate area transverse transport tendency is clearly exhibited. Further explanations are in the text

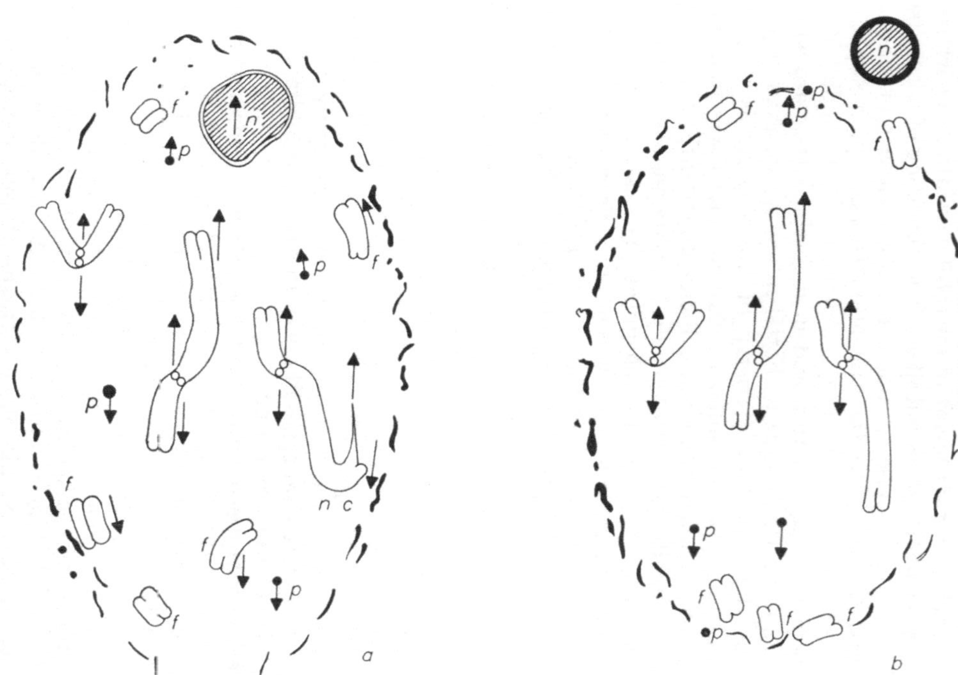

Fig. 9 a, b

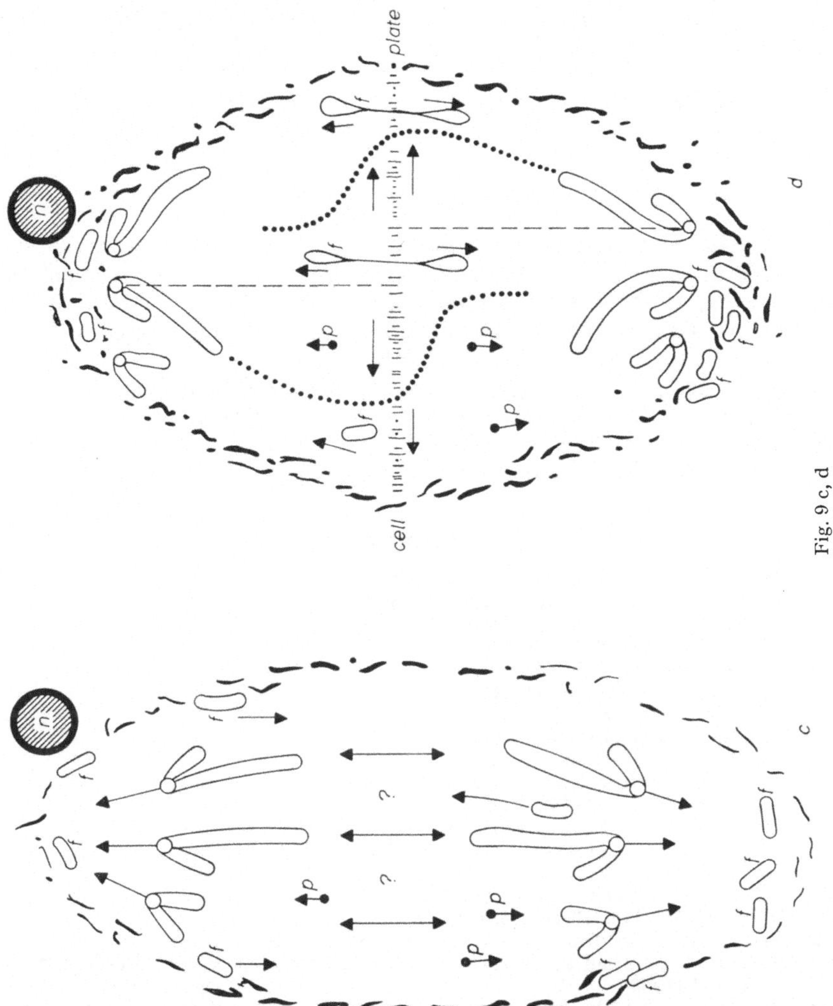

modifications are needed for animal cell mitosis. Typically, during prometaphase chromosomes move in an opposite direction to all other
bodies within the spindle (Fig. 9a). During metaphase (Fig. 9b)
kinetochores occupy the equatorial position and chromosome arms are
arranged parallel to the long axis of the spindle. Eliminated bodies are
accumulated at the poles of the spindle. Some chromosome arms bend
rapidly due to a short lasting attachment of chromosomal fibers such as
neocentric activity both in prometaphase and metaphase. During anaphase the spindle elongates, chromosomal fibers shorten, and some fragments (usually those which are in the middle of the spindle) may move
toward the equator due to neocentric activity (Fig. 9c), or to piston

8*

action of the half-spindle (usually those which are on the periphery of the spindle). The events of telophase (Fig. 9d) are: the shortening of chromosomes; the transportation of cell bodies from the phragmoplast to the poles; and the tearing of bridges. Due to lateral movements during telophase the long trailing chromosome arm does not follow the path of the kinetochore (Fig. 9d).

Acknowledgments

The work was supported by grants: NSF GB 3335 to the author; and NIH GM 08 991 to R. D. ALLEN (Princeton University — investigation with the Nomarski interference contrast system). The author is grateful to Prof. G. ÖSTERGREN for permission to publish some of his personal communications, to Prof. S. INOUÉ for the facilities he offered in 1958 which resulted in the taking of pictures in polarized light (Fig. 2) and to my wife Dr. J. MOLÈ-BAJER for permission to include some of her unpublished material. Finally I would like to thank Mrs. MARTHA ANN KAPLAN for her considerable and patient help in the preparation of the manuscript.

Literature

[1] AMBROSE, E. J.: Endeavour 44, 27 (1965).
[2] BAJER, A.: Exp. Cell Res. 14, 245 (1958).
[2a] — Exp. Cell Res. 14, 245 (1958).
[2b] — Chromosoma 9, 319 (1958).
[3] — Hereditas 45, 579 (1959).
[4] — Chromosoma 12, 64 (1961).
[5] — Chromosoma 14, 18 (1963).
[6] — Chromosoma 15, 630 (1964).
[7] — Chromosoma 16, 381 (1965).
[8] —, and G. ÖSTERGREN: Hereditas 47, 563 (1961).
[9] — — Hereditas 50, 179 (1963).
[10] —, and J. MOLÈ-BAJER: Chromosoma 7, 558 (1956).
[11] — — In: Cinematography in Cell Biology, S. 357. New York: Academic Press 1963.
[12] —, and R. D. ALLEN: (in press 1965).
[13] BECKER, W.: Bot. Review 4, 446 (1938).
[14] BĚLAŘ, K.: Arch. Entwickl.-Mech. Org. 118, 359 (1929).
[15] BERNHARD, W., and E. DE HARVEN: Proc. IV Internat. Congr. Electron Micr. 2, 217 (1958).
[16] BIESELE, J. J.: In: Mitotic Poisons and the Cancer Problem. New York: Elsevier 1958.
[17] BLEIER, H.: Arch. exp. Zellforsch. 22, 257 (1939).
[18] CONGER, A. D.: Radiation Botany 5, 81 (1965).
[19] COOPER, K. W.: Proc. nat. Acad. Sci. (Wash.) 27, 480 (1941).
[20] DANIELS, E. W., and L. E. ROTH: J. Cell Biol. 20, 75 (1964).
[21] DIETZ, R.: Z. Naturforsch. 14b, 749 (1959).
[22] EIGSTI, D. H., and P. DUSTIN JR.: In: Colchicine in Agriculture, Medicine, Biology and Chemistry. Ames, Iowa: Iowa State College Press 1955.
[23] FORER, A.: J. Cell Biol. 25, 95 (1965).
[24] GIMÉNEZ-MARTIN, G., and J. F. LÓPEZ-SÁEZ: Fyton 20, 183 (1963).
[25] —, A. GONZÁLEZ-FERNÁNDEZ, and J. F. LÓPEZ-SAEZ: Fyton 21, 77 (1964).
[26] HARRIS, P.: J. biophys. biochem. Cytol. 11, 419 (1961).
[27] — J. Cell Biol. 14, 475 (1962).
[28] —, and A. BAJER: Chromosoma 16, 624 (1965).
[29] HEITZ, E.: Ber. d. Bot. Gesell. 60, 28 (1942).

[30] HENEEN, W. K.: Hereditas **49**, 1 (1963).
[31] INOUÉ, S.: Chromosoma **5**, 487 (1953).
[32] — In: Primitive Motile Systems in Cell Biology, S. 549. New York: Academic Press 1964.
[33] —, and A. BAJER: Chromosoma **12**, 48 (1961).
[34] KIHLMAN, B. A.: In: Radiation-induced Chromosome Aberrations, S. 100. New York: Columbia Univ. Press 1963.
[35] KUWADA, Y.: Cytologia **10**, 213 (1939).
[36] LEDBETTER, M. C., and K. R. PORTER: J. Cell Biol. **19**, 239 (1963).
[37] LEWIS, W. H., and M. R. LEWIS: In: General Cytology, S. 383. Chicago: Univ. Chicago Press 1924.
[38] MAZIA, D.: In: The Cell, Vol. III, S. 77. New York: Academic Press 1961.
[39] METZ, C. W.: Biol. Bull. **64**, 333 (1933).
[40] MILOVIDOV, P.: In: Protoplasma Monographien, Vol. 20, Pt. 2. Berlin: Borndräger 1954.
[41] MOLÈ-BAJER, J.: Chromosoma **9**, 332 (1958).
[42] — J. Cell Biol. **25**, 79 (1965).
[43] NEMEC, B.: Jb. wiss. Bot. **39**, 645 (1903).
[44] ÖSTERGREN, G.: Hereditas **35**, 525 (1949).
[45] — Hereditas **36**, 1 (1950).
[46] — Hereditas **37**, 85 (1951).
[47] —, and A. BAJER: Coll. Int. Centr. Natl. Rech. Scien. **88**, 199 (1960).
[48] —, J. MOLÈ-BAJER, and A. BAJER: Ann. N. Y. Acad. Sci. **90**, 381 (1960).
[49] PORTER, K. R., and R. D. MACHADO: J. biophys. biochem. Cytol. **7**, 167 (1960).
[50] ROBBINS, E., and N. K. GONATAS: J. Cell Biol. **21**, 429 (1964).
[51] SCHRADER, F.: In: Mitosis, 2nd ed. New York: Columbia Univ. Press 1953.
[52] STICH, H.: Chromosoma **6**, 199 (1954).
[53] WADA, B.: Cytologia **11**, 93 (1941).
[54] —, and F. KUSUNOKI: Cytologia **29**, 109 (1964).
[54a] — — Cytologia **29**, 118 (1964).
[55] WILSON, E. B.: In: The Cell in Development and Heredity, 3rd ed. New York: Macmillan 1925.

Discussion

Chairman: *Taylor*

Karlson: With regard to the forces, which could pull the chromosomes, I have two proposals: The first possibility would be, that the spindle fibres might be myosine or an equivalent to it. The kinetochore would produce a short thread of actin; this would act similar to muscle contraction (sliding filament theory).

There are some papers about the forces in mitosis by H. H. WEBER's school (especially by HOFFMANN-BERLING), which show, that the spindle-fibres have some properties in common with myofilaments, at least that they have ATPase-activity and that the contractile filaments might be similar.

Bajer: As far as I know from the literature this is likely but I myself did not do any experiments in relation to this problem.

Karlson: Another model to explain the chromosome movement is based on peristaltic movements, pulsations of something like a bacterial flagellum, which is attached to the kinetochore.

Bajer: There is some kind of peristaltic movement seen very clearly in the phragmoplast, and there is probably no essential difference in basic structure of the phragmoplast and the prometaphase of a spindle. I believe there is also some kind of peristaltic movement of the spindle. (After this discussion Dr. R. ALLEN and I were able to take a time lapse movie of movements within prometaphase spindle which resemble peristaltic movements).

Bier: Is the shortening of the chromosomal fibres during prometaphase caused by contraction or by decomposition?

Bajer: I don't really think, that movement in prometaphase is caused by shortening of chromosomal fibres. Chromosomal fibres shorten very slowly during prometaphase, which in this plant lasts about two hours. Some comparatively quick movements of the chromosomes, which you can see, last for a very short time only (5—10 min). So we can't say that shortening of the chromosomal fibres is responsible for movement. My feeling is that both chromosomes and fibres connected to chromosomes move together as a whole system, i.e., they are shifted or slide up and down parallel to the long axis of the spindle.

Grundmann: What do you think about the development of spindle fibres? I think, there are three possibilities. Firstly fibres pre-existing in the mixoplasm are attached to the kinetochores. Secondly they are growing out from the kinetochore as an orientation of macromolecules. The third possibility, which is often to be found in egg cells, is that the orientation of the macromolecules occurs from the centrosome, and that thereafter the attachment to the chromosomes takes place.

Bajer: There is no clear evidence that centrosomes are present in plants. It is my feeling that if centrosomes are present in animal cell they are a kind of additional structure which helps to organize spindle in a quick and precisely oriented way. The spindle can develop without centrosomes. I also think that at present we do not have enough data to state how micro-tubulus are formed.

Stoeckenius: In the case of plant cells, PORTER has speculated that they are stored under the plasmalemma and may be re-used for spindle formation during division.

Grundmann: In animals, there are lots of cases where you have a spindle without chromosomes. In this case I mean, that centrosomes are necessary elements to get spindles.

Bajer: This is one of the most interesting problems in mitosis. In plant material there exist to some extent comparable structures. In the endosperm in a syncytium type phragmoplasts are formed between both sister and non-sister nuclei. Also, chromosomal fibres in plant mitosis can be formed in exceptional cases from the phragmoplast.

Taylor: Do you have evidence or do you know any evidence that would suggest what kind of changes are occuring in the shortening of chromosomes?

Bajer: No, I do not have. It is very difficult to determine the length of chromosomes before nuclear membrane disappears. They seem to decrease in length not very much. They start to decrease in length very rapidly immediately after nuclear membrane breaks or actually a short time before. I have the feeling that they are dehydrated during progress of mitosis.

Taylor: What happens, if one changes the salt concentration, for example outside the cell, to make the medium hypo- or hypertonic? Does this change the rate of condensation?

Bajer: I was performing such experiments but I was never sure to what extent the medium influences the chromosomes. In hypotonic medium, chromosomes swell, and they are not clearly visible as the refractive index changes. Swollen chromosomes do not divide, and this is followed often by death of the cell. The chromosome volume increases up to metaphase and then decreases very rapidly during anaphase and telophase. Dry mass, as seen from studies in interference microscope changes in the same way.

Taylor: Dr. STOECKENIUS, would you tell us, whether it is conceivable that enough dehydration can occur to change the form of DNA from the B-form to the A-form in the cell?

Stoeckenius: No, I don't think so.

It is conceivable, that changes in the binding of histones could effect the coiling. WILKINS has once speculated on that, but I don't think, we have any solid experimental basis, so far.

Taylor: What type of change occurs to produce the uncoiling?

Bajer: I think it is not known. I would only like to mention one point. The relational coils in prometaphase uncoil at some time before anaphase starts. However, if the start of anaphase is delayed chromatids continue to rotate in the same direction and as a result they are coiled but in opposite directions: if they were coiled counter clockwise at first, they are now coiled clockwise. The rotation continues when anaphase starts and in the case mentioned above the arms would rotate clockwise. Also there is often a reversal of coil in long arms during prometaphase: part of the arms are coiled clockwise and part counter clockwise. It is evident therefore that coiling is rather complicated. Also, we might discuss whether the term coil is a proper one at all.

Taylor: I might remark, that I agree that the coiling is very complicated. I think, that we may have been mislead by the textbooks and the early descriptions with respect to coiling cycle, particularly in mitosis. We see a helix in the early stages only. Then it disappears, and we see one after division. But in the meantime there may be no helix at all.

As to the apparent doubleness of chromosomes at anaphase, I would like to add very briefly an idea that fits the model I have proposed in my paper. If the central organizing piece of the chromosome is in the middle there would be found a series of rings at right angles to the axis stacked on top each on the other. At various degrees of twisting or swelling they might overlap somehow to form a single as-sembly, at other times two parallel assemblies.

Enzymatische Aspekte der Mitose

Von

FRANZ DUSPIVA, Heidelberg

Mit 3 Abbildungen

I. Einleitung

Obwohl der Mitose nur die relativ einfache Frage zugrunde zu liegen scheint, wie das Erbgut der Zelle in Chromosomen verpackt und auf die Tochterzellen verteilt werden kann, türmt sich ein kaum noch übersehbarer Komplex von Problemen auf, sobald man nach den zellphysiologischen Voraussetzungen fragt und den Mechanismus erfahren möchte, der den Vorgang der Mitose kontrolliert und in den Lebensablauf der Zelle einordnet. Das umfangreiche, heute über die Mitose und ihre Beziehungen zur Physiologie der Zellteilung vorliegende Wissensgut wurde bereits von D. MAZIA [47] erschöpfend behandelt. Über die Kontrolle der Zellteilung liegt eine umfangreiche und alle wesentlichen Fragen besprechende Abhandlung von M. M. SWANN [74] vor. Die Kontrolle der *DNA-Biosynthese* hat erst kürzlich K. G. LARK [40] unter Berücksichtigung aller einschlägigen Fragen vorzüglich dargestellt. Im vorliegenden Referat können daher nur noch einige erst in allerletzter Zeit erarbeitete enzymatische Aspekte des in der Vorbereitung zur Mitose ablaufenden cellulären Geschehens behandelt werden mit besonderer Berücksichtigung der Frage, wie *enzymatische Prozesse* reguliert werden können. Da aber enzymatische Umsetzungen in nahezu allen Teilstrecken des genannten Prozesses enthalten sind, muß dem Referat eine noch weitergehende Beschränkung auferlegt werden. Es sollen nur die an der DNA-Synthese direkt und indirekt beteiligten Enzyme behandelt werden; eine vollständige Erfassung der Literatur ist infolge des beschränkten Umfangs des Referats nicht möglich.

II. Vorbemerkungen zur quantitativen und zeitlichen Kontrolle der DNA-Synthese

Die Verdopplung der chromosalen Substanz ist ein normales Ereignis auf dem Wege zur Mitose und Zellteilung in dem Sinn, daß sich Zellen nicht teilen, wenn sie nicht zuvor ihren DNA-Gehalt vermehrt haben. Aber nicht alle Zellen, die den DNA-Gehalt verdoppelt haben, müssen sich notwendigerweise auch teilen. Auf dem Wege zur Teilung finden mehrere metabolische und auch strukturelle Ereignisse statt. Diese treten meist in einer genau festgelegten Reihenfolge ein. Diese ist aber nicht zwingend. In gewissen Fällen können einzelne Ereignisse ausfallen oder verspätet stattfinden. Die DNA-Synthese ist im allgemeinen eine Voraussetzung für die Teilung der Zelle, nicht für ihre Erhaltung.

Hochdifferenzierte, im Stoffwechsel sehr aktive Zellen synthetisiren häufig keine DNA. Die Biosynthese der DNA ist also mehr ein Charakteristikum der Zellvermehrung als das eines aktiven Zellstoffwechsels.

a) Der Teilungscyclus

Während eines normalen Teilungscyclus wird der DNA-Gehalt einer Zelle im chemischen und quantitativen Sinn exakt verdoppelt. Ein großes Beobachtungsgut, das an Zellen aus allen Reichen des Organischen stammt, legt dar, daß DNA nur während einer zeitlich wohlumschriebenen Periode im Vermehrungscyclus der Zelle synthetisiert werden kann. Nach autoradiographischen Untersuchungen läßt sich der Teilungscyclus in eine *postmitotische Periode* (G 1), *DNA-Syntheseperiode* (S) und ein *prämitotisches Intervall* (G 2) einteilen, dem die Mitoseperiode (M) folgt [*31, 39, 29, 77, 15, 58, 52*]. Säugerzellen wie z. B. Zellen des Mäuseohres verweilen in G 1 10—20 Std, in der S-Periode 6—8 Std, im G 2-Intervall 1—4 Std und vollziehen die Mitose in 1 Std [*18*]. Es gibt aber auch Zellen, die in G 2 mehrere Tage verweilen können. Physiologisch unterschiedliche Kerne innerhalb einer Zelle treten manchmal zu verschiedenen Terminen in die S-Phase ein. Bei *Tetrahymena* starten Groß- und Kleinkern die DNA-Synthese zu verschiedenen Terminen im Zellcyclus, brauchen aber ungefähr gleichlang, um sie durchzuführen [*48*]. Es ist bewiesen, daß der Replikation des Chromosomensatzes ein geordnetes und sequentielles Muster zugrunde liegt [*50*]. Daß dies nicht nur für Säugerzellen gilt, zeigen Untersuchungen über die Dauer der DNA-Synthese bei 3 Stämmen von *Tetrahymena*, die sich untereinander im DNA-Gehalt ihrer Macronuclei unterscheiden. Nach gegenwärtig herrschender Auffassung setzt sich der Macronucleus der Ciliaten aus einer größeren Anzahl von Genomen zusammen. Wenn jedes der Genome die gleiche Zeitspanne zur Verdopplung beansprucht, so sollten die an Genomen reicheren Macronuclei keine Verlängerung der S-Periode bedingen. Nach CAMERON u. STONE [*7*] ist dies bei *Tetrahymena* tatsächlich der Fall: Ein Kontrollmechanismus regelt die Verdopplungsdauer des Genoms. Bei *Paramecium* hingegen beginnt die DNA-Synthese in Macro- und Micronuclei gleichzeitig, wird aber in diesem früher abgeschlossen als in jenem [*81*]. Bei *Euplotes* setzt die DNA-Synthese in dem großen, hufeisenförmigen, hochpolyploiden Großkern gleichzeitig an beiden Enden ein und schreitet von hier aus in Form zweier Wellen zur Mitte fort [*17*]. Bei dem coenocytischen Myxomyceten, *Physarum polycephalum*, treten die zahlreichen Kerne innerhalb eines Plasmodiums *synchron* in die Mitose ein [*32*]. Die aufgeführten Beispiele mögen genügen um zu zeigen, daß es in der Zelle neben einer präzisen Kontrolle der DNA-*Menge* auch eine Kontrolle des Zeitpunktes des Synthese*beginns* und der Synthese*dauer* von DNA gibt. Aber hinsichtlich der Termine und Reihenfolge der Ereignisse im Zellcyclus herrscht eine große Variabilität. Der feinere

Mechanismus, der diesen Kontrolltypen zugrunde liegt, ist heute noch fast gänzlich unbekannt. Ansätze zu einer Analyse sind in den folgenden Abschnitten enthalten.

b) Das Prinzip biochemischer Kontrollsysteme

Der Übergang von einer Phase des Zellteilungscyclus zur anderen ist biochemisch durch die Synthese neuer spezifischer Makromoleküle charakterisiert. Die Biosynthese der DNA steht im Vordergrund des Interesses.

Die Kontrolle der DNA-Synthese kann auf verschiedenen Stufen einsetzen:

1. Umwandlung von Ribonucleotiden zu Desoxyribonucleotiden [60] und die Bildung von Desoxythymidylat.

2. Phosphorylierung der Desoxyribonucleotide in die entsprechenden Triphosphate [41].

3. Polymerisationsreaktion auf dem Niveau der Desoxyribonucleosidtriphosphate [37].

Die genannten Prozesse werden durch eine Serie von Enzymen katalysiert [69], deren Aktivität und Bildungsrate in der Zelle der Wirkung steuernder Mechanismen unterliegen.

Man kennt heute *Steuerungsmechanismen* enzymatischer Wirksamkeit, die auf folgenden Ebenen cellulärer Organisation wirken:

1. Makromolekül (Enzymmolekül)

Die Reaktionsgeschwindigkeit ist bei konstant gehaltener Enzymkonzentration von der Substratkonzentration, der H^+-Konzentration und der Temperatur abhängig. Sie kann von Stoffen beeinflußt werden, die nicht zur katalysierten Reaktion gehören. Diese Stoffe *(Effektoren)* können die Reaktionsgeschwindigkeit erhöhen *(Aktivatoren)* oder senken *(Hemmstoffe, Inhibitoren)*. Kompetitive Hemmstoffe konkurrieren mit dem Substrat um die gleiche Stelle des Enzyms. Der enzymatisch katalysierte Stoffdurchsatz in der Zelle kann durch Schwankungen in der Substratkonzentration und durch Effektoren gesteuert werden [43].

2. Makromolekularer Verband (Enzymkette)

Bei Enzymketten nach dem Typ: $A \xrightarrow{E_1} B \xrightarrow{E_2} \ldots \xrightarrow{E_{x-1}} X$ beobachtet man fallweise eine negative Rückkopplungsreaktion (feed back) durch das Endprodukt (X) der Enzymkette. Im allgemeinen ist nur die Aktivität des Enzyms (E_1), das den ersten Schritt der Kette ($A \rightarrow B$) katalysiert, durch das Produkt (X) hemmbar [79, 68, 76]. Der Hemmung liegt ein *allosterischer Effekt* [49] zugrunde; das regulierbare Enzymmolekül dürfte neben der Bindungsstelle für das Substrat eine zusätzliche spezifische Stelle für den Effektor besitzen. Der Effektor hat eine hochgradige Spezifität [8]. In manchen Fällen ist der allosterische Effekt durch Moleküle kompetitiv hemmbar, die mit dem Effektor um die gleiche Stelle am Enzymprotein konkurrieren und die Wirksamkeit der Enzymkette wieder herstellen [19, 8].

3. System der Proteinsynthese

Bei Mikroorganismen (Bakterien, Hefen) kann aufgrund einer heute bereits weitgehend anerkannten Hypothese von PARDEE, JACOB u. MONOD die *Neubildung von Enzymmolekülen*, damit die in einer Zelle vorhandene Enzymmenge, von *Regulatorgenen* gesteuert werden. Eine Gruppe von *Strukturgenen*, welche die Information zur Bildung der Proteine einer Serie von Enzymen besitzen, ist einem *Operatorgen* unterstellt, das die Wirkung der Strukturgene blockieren oder freigeben kann *(Operon)*. Das Operatorgen wird von *Regulatorgenen* kontrolliert, die außerhalb der Kopplungsgruppe liegen, der das *Operon* angehört [54]. Am Regulatorgen werden *Repressoren* gebildet, deren chemische Natur z. Z. noch sehr umstritten ist. Nach den an Mikroorganismen gewonnenen Erfahrungen läßt sich die *genetische Regulation der Enzymsynthese* durch Allosterie (doppelte Spezifität) des Repressors erklären. Mit einer seiner beiden Bindungsstellen kann der Repressor mit spezifischen Operatoren reagieren, mit der *anderen* einen spezifischen Effektor anlagern. Man kann sich vorstellen, daß die Zelle nach diesem einfachen Prinzip *zwei* verschiedene *Typen von Regelkreisen* aufbauen kann: 1. *Katabolischer Regelkreis:* Das Substrat (oder ein sterisch verwandtes Produkt) eines bislang repressierten Enzyms (nicht oder in Spuren wirksam) gelangt in die Zelle. In seiner Eigenschaft als Effektor lagert es sich an den spezifischen Repressor an und *inaktiviert* ihn hierdurch: Das Operon wird freigegeben und das zum Substrat passende Enzym wird synthetisiert. Der Effekt des katabolischen Regelkreises: „*Das Substrat induziert die Synthese des passenden Enzyms*". 2. *Anabolischer Regelkreis:* Das Endprodukt (X) einer anabolisch wirksamen Enzymkette wirkt als Effektor, lagert sich an den spezifischen Repressor an und *aktiviert* ihn. Als Folge wird das zugehörige Operon blockiert; die Synthese der dieser Kette angehörigen Enzyme wird eingestellt. Dieser Regeltyp ist bei Bakterien weit verbreitet und gut bekannt, z. B. bei der Synthese von Aminosäuren (z. B. Arginin hemmt die Produktion der Ornithin-Transcarbamylase, welche die Bildung von Citrullin katalysiert). Der Effekt des anabolischen Regelkreises: *Das Endprodukt einer biosynthetischen Kette repressiert die Produktion aller beteiligten Enzyme, einschließlich jenem, das den ersten biosynthetischen Schritt dieser Kette katalysiert* [5].

Es ist zu vermuten, daß ein ähnlicher Regelmechanismus — und sei es nur als elementarer Baustein in einem sehr viel komplizierteren System — auch in der Zelle höher organisierter Lebewesen enthalten ist.

III. Regelung der Aktivität einiger an der Bildung von Vorstufen der DNA beteiligten Enzyme nach dem Prinzip der Endprodukthemmung

Die DNA-Synthese erfordert die Bereitstellung der Triphosphate dATP, dGTP, dCTP und dTTP. In den Zellen ist stets ein mehr oder weniger umfangreicher Pool von Ribo- und Desoxyribonucleosiden vorhanden. Eine Gruppe von Enzymen katalysiert ein Netz von Reaktionen

(Umbau und Phosphorylierung), das zur Bereitstellung der 4 obengenannten Triphosphate dienlich ist. Davidson hat oft darauf hingewiesen, daß die DNA-Synthese ein delikates Gleichgewicht in der Konzentration dieser 4 Triphosphate erfordert und daß eine — auch nur vorübergehende — Verknappung an einem von diesen zu einem Abstoppen der DNA-Replikation führen muß; besonders gefährdet scheint die laufende Bereitstellung von dTTP zu sein. Es ist daher ein Regulationsmechanismus notwendig, der ein kooperatives Zusammenwirken der Glieder des oben genannten Multienzymsystems durch Steuerung der aktuellen Aktivität gewisser Schrittmacher-Enzyme ermöglicht.

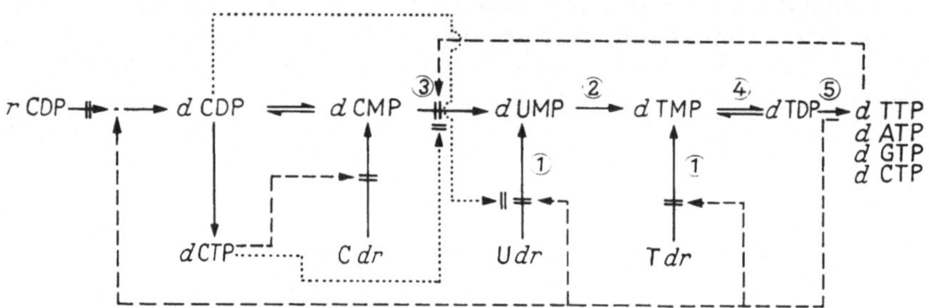

Abb. 1. Schema des Reaktionsweges und der Regelkreise bei der Bildung der Vorstufen von *DNA* (*1*) Thymidinkinase; (*2*) *dTMP*-Synthetase; (*3*) *dCMP*-Desaminase; (*4*) Thymidylatkinase; (*5*) Thymidin-diphosphokinase. — — — ▶ allosterische Hemmung; ·······▶ Ausschaltung einer Hemmung. [*33*]; etwas verändert

Reichard u. a. [*60*] fiel als erstem auf, daß der über wenig bekannte Zwischenstufen laufende Reaktionsweg von Ribocytidindiphosphat (rCDP) zu Desoxyribocytidindiphosphat (dCDP) durch dGTP, dATP und dTTP gehemmt werden kann. Es handelt sich hierbei nicht um die übliche dem Massenwirkungsgesetz gehorchende Hemmung einer enzymatischen Reaktion durch Anhäufung des Produktes derselben. Denn in diesem Fall hemmen Stoffe ein Enzym, die keine Komponenten der von diesem Enzym katalysierten Reaktion sind. Beachtlich ist aber, daß diese Inhibitoren das Endprodukt der Wirkung einer *höheren biologischen Organisationseinheit (Enzymkette)* sind, in der das durch sie hemmbare Enzym das Anfangsglied bildet. Diese Inhibitoren üben also einen negativen Feedback-Effekt aus, der dazu führt, daß der Stoffdurchsatz bereits am Anfang des Fließbandes abgedrosselt wird, wenn das Endprodukt desselben sich anzuhäufen beginnt.

Weitere Fälle von negativer Rückkopplung wurden bald darauf entdeckt. Einen Überblick vermittelt Abb. 1.

Desoxythymidin ist kein normales Zwischenglied im Verlauf der Biosynthese von DNA. Es ist daher erstaunlich, daß lebhaft proliferierende Gewebe regelmäßig ein Enzym besitzen, welches Desoxythymidin mittels ATP in Desoxythymidin-5'-phosphat verwandelt [*61,16*]. Die biologische Bedeutung der ATP: Thymidin-5'-phosphotransferase EC 2.7.1.21 (Thymidinkinase) ist unbekannt. Sie übt vermutlich nur eine Hilfsfunktion im Stoffwechsel aus, nämlich Desoxyribonucleoside, die aus dem katabolischen Stoffwechsel stammen, einer Wiederverwendung zuzuführen [3, 80]. Sie wäre dadurch imstande, eine mangelhafte Anlieferung von Desoxythymidylat aus Desoxyuridylat (bei Substratmangel oder Hemmung der Desoxythymidylat-Synthetase) zu korrigieren. Die Tatsache, daß die potentielle Aktivität der Thymidinkinase stets mit dem Mitoseindex des Gewebes in einer engen Korrelation steht [46, 80, 2, 36, 30, 11], spricht jedoch für eine *wichtige* biologische Bedeutung. Dazu kommt, daß ein besonderer Mechanismus existiert, der die *aktuelle Aktivität* der in der Zelle bereits vorhandenen Enzymmoleküle reguliert. Dieser Mechanismus beruht auf der sehr starken aktivitätshemmenden Wirkung von dTTP und in geringerem Maße von dTDP [45, 33, 4, 6].

OKAZAKI u. KORNBERG [53] gelang es kürzlich, die Thymidinkinase aus *Escherichia coli* bis zu einer 1200fachen spez. Aktivität zu reinigen und damit die Vorbedingung für ein eingehendes Studium zu schaffen.

rATP ist „in vitro" als Phosphat*donator* brauchbar, aber optimal erst bei der unphysiologisch hohen Konzentration von 10^{-2} M. Die Kinetik spricht für die Beteiligung von 2 rATP-Molekeln je Enzymkomplex. Aber in Gegenwart von dCDP, welches selbst nicht als Phosphatdonator geeignet ist, ist rATP bereits in viel geringerer Konzentration optimal verwertbar, und zwar bei gewöhnlicher Michaelis-Menten-Kinetik [53]. Als Phosphatdonatoren sind auch dGTP, dCTP und dATP wirksam; keine Wirkung haben hingegen rCTP und rUTP. Aber im Gegensatz zu rATP wirkt dATP bei normaler Michaelis-Menten-Kinetik. Als Phosphat*acceptor* dient nur Desoxyuridin und dessen am C-Atom 5 des Pyridins substituierte Derivate (Desoxythymidin, und 5-F(Cl, Br, J)-d-Uridin). Das Enzym wird ganz spezifisch durch dTTP, dem Endglied der biologischen Thymidinnucleotidsynthese, gehemmt, und zwar in Anwesenheit des Aktivators dCDP eher stärker als ohne diesen. Der molekulare Mechanismus des Hemmeffektes von dTTP ist noch nicht ganz geklärt; es ist wahrscheinlich, daß ein allosterischer Übergang des Enzymproteins unter dem Einfluß von dTTP stattfindet. Die *Thymidinkinase* ist als regelbares Enzym durch das ungewöhnliche kinetische Verhalten gegenüber den Substraten, durch den Bedarf an Aktivatoren und die Empfindlichkeit gegenüber einem sehr spezifischen Hemmstoff ausgezeichnet.

In analoger Weise wird die Desoxycytidinkinase durch dCTP gehemmt [45]. Einer komplizierten Kontrolle unterliegt auch der im Hauptweg der Biosynthese von DNA liegende Reaktionsschritt vom Desoxycytidylat zu Desoxyurydilat [45]. Die *Desoxycytidylat-Aminohydrolase* konnte von SCARANO, GERACI u. ROSSI [64] gereinigt und untersucht werden. In Gegenwart von Substrat ohne irgendwelche Zusätze resultiert eine normale Michaelis-Menten-Kinetik. Aber bei Anwesenheit von Phosphat tritt im Bereich niederer Substratkonzentration eine multimolekulare Kinetik auf. Die Gegenwart von dCTP (0,06 m M) normalisiert die Kinetik im ganzen prüfbaren Bereich der Substratkonzentration bei Anwesenheit von Phosphat und setzt die Km von 0,9 m M auf 0,4 m M herab. dTTP und dTMP hemmen stark, aber auf unterschiedliche Weise. Während die durch dTTP bewirkte Hemmung mittels dCTP aufgehoben werden kann, bleibt die durch dTMP bewirkte bestehen.

dTMP erwies sich als kompetitiver Inhibitor. dCTP kann hingegen als allosterischer Gegenspieler von dTTP aufgefaßt werden, das als allosterischer Inhibitor wirkt.

Am Beispiel der Thymidinkinase erläutert [53], funktioniert ein negativer Feedback-Mechanismus in der Zelle wie folgt:

Solange der Nachschub an dTMP im Hauptwege der DNA-Biosynthese ausreicht, bleibt die Aktivität der in der Zelle vorhandenen Thymidinkinasemoleküle unter dem Einfluß der herrschenden überschwelligen stationären Konzentration an dTTP allosterisch gehemmt. Wird die Anlieferung von dTTP aber langsamer, so wird auch weniger DNA synthetisiert und die normal angelieferten Di- und Triphosphate der 3 übrigen Desoxynucleoside häufen sich an. Manche von diesen, wie dATP und dGTP sind wirksame Phosphatdonatoren, andere, wie dCDP, Aktivatoren der Thymidinkinase. Die Aktivierung wirkt sich außerdem noch durch eine Herabsetzung der Km aus, wodurch nach Wegfall der allosterischen Hemmung sehr niedrige Konzentrationen von Thymidin in der Zelle wirksam phosphoryliert und der DNA-Synthese zugeführt werden können. Das Ungleichgewicht im gegenseitigen Verhältnis der 4 Desoxyribonucleosid-Triphosphate löst mithin einen molekularen Prozeß am Enzymprotein der Thymidinkinase aus, der die gestörte Balance der Vorstufen der DNA in „sinnvoller Weise" einzuregulieren geeignet ist.

IV. Regelung der Neusynthese und des Abbaues von Enzymprotein

Es gibt gute Gründe für die Vermutung, daß es neben dem im Abschnitt III besprochenen Regelmechanismus, der nur während der S-Phase des Zellcyclus zum Zuge kommt und am Enzymprotein selbst angreift, noch einen zweiten geben muß. Nach dem bekannten Sparsamkeitsprinzip der Zelle bezüglich Eiweiß ist ein Mechanismus zu erwarten, der die *Produktion* von anabolischen Enzymen der DNA-Synthese in der G 1, G 2 und M-Phase sperrt und nur während der S-Phase freigibt. Wenn dieser Mechanismus existiert, so müßte er sich an starken Schwankungen bezüglich der *potentiellen Aktivität* erkennen lassen, die in entsprechender zeitlicher Korrelation zum Teilungscyclus stehen.

Die atypische Kinetik der allosterisch regelbaren anabolischen Enzyme erschwert die Ermittlung ihrer Menge in den Zellen beträchtlich. Außer einer möglichst quantitativen Extraktion der Enzyme aus den Zellstrukturen, Substratüberschuß, pH-Optimum u. dgl. ist besonders die Beachtung eines optimalen Konzentrationsverhältnisses von Magnesium zu ATP sowie Ausschaltung von Inhibitoren wichtig, die heute ihrer chemischen Natur nach noch wenig bekannt sind.

a) Enzymaktivität und Teilungscyclus

Da man zur Analyse i. a. Kollektive von Zellen braucht, eignen sich zur Prüfung der vermuteten Beziehung Objekte, die aus Zellen bestehen, welche von Natur aus einen oder mehrere aufeinanderfolgende Teilungsschritte *synchron* durchlaufen. Beispiele hierfür sind der reifende Pollen in den Antheren der Blütenknospen gewisser Lilienarten, die Plasmodien des Schleimpilzes *Physarum polycephalum* sowie die ersten Furchungsteilungen der Eier von Amphibien und Seeigel. Ein vielfach untersuchtes Objekt ist Lebergewebe nach partieller Hepatektomie, dessen Zellen 24—36 Std p. o. eine synchrone Zunahme des DNA-Gehaltes aufweisen.

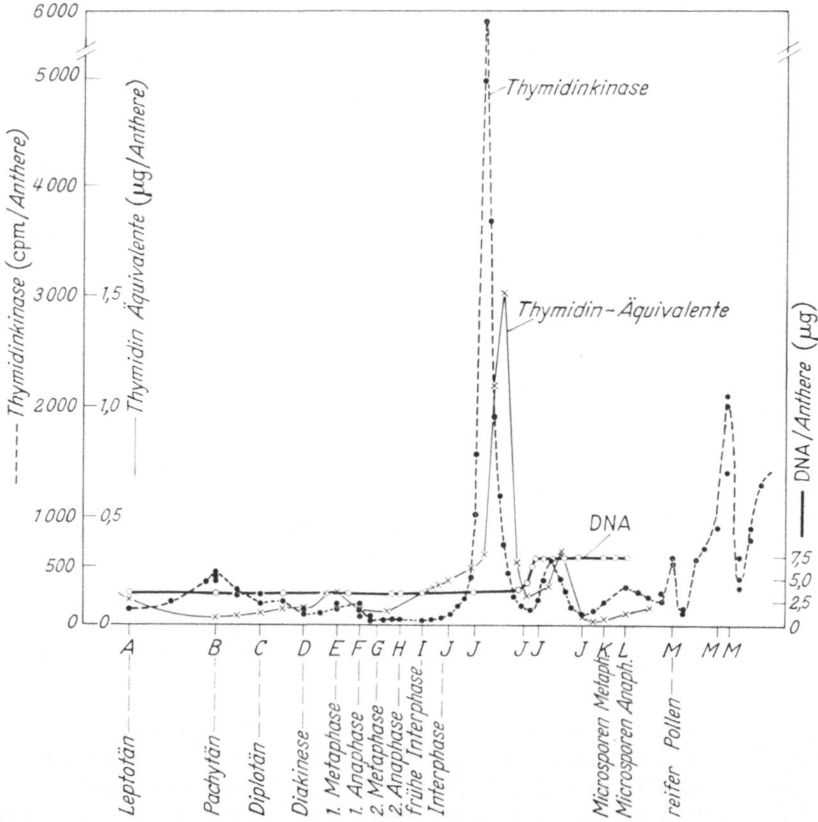

Abb. 2. Der Aktivitätsverlauf der Thymidinkinase und der *DNA*-Gehalt der Mikrosporocyten und Mikrosporen sowie die Desoxyribosidbildung in den Antheren von *Trillium erectum*. [Nach Hotta u. Stern (1961), kombiniert aus Abb. 3 u. 4]

Hotta u. Stern [*30*] untersuchten den Wandel der Thymidinkinaseaktivität in den exakt phasengleich reifenden Microsporocyten von *Lilium longifolium* und *Trillium erectum* während eines langen Zeitraumes, der die Meiose inklusive der nachfolgenden Mitose umschließt. Die Autoren beobachteten kurz von Beginn der DNA-Reduplikation während der an die Meiose anschließenden Interphase einen sprunghaften Anstieg der Thymidinkinaseaktivität auf den etwa 10 fachen Grundwert, dem unverzüglich ein ebenso steiler Abfall folgt. Kurz nach Verdoppelung der DNA und noch vor der bald danach beginnenden Mitose folgt eine zweite allerdings viel kleinere Aktivitätsausschüttung nach (Abb. 2).

Beim Myxomyceten *Physarum polycephalum* entspricht jedem synchronen Teilungsschritt der Energiden ein sprunghafter Anstieg der Thymidinkinaseaktivität, welcher mit der Kernteilung beginnt und während der an die Mitose sofort anschließenden S-Phase das Optimum erreicht. Der Abfall setzt im späteren Verlauf der S-Phase ein. Die Form und zeitliche Zuordnung des Aktivitätsverlaufes der Thymidinkinase

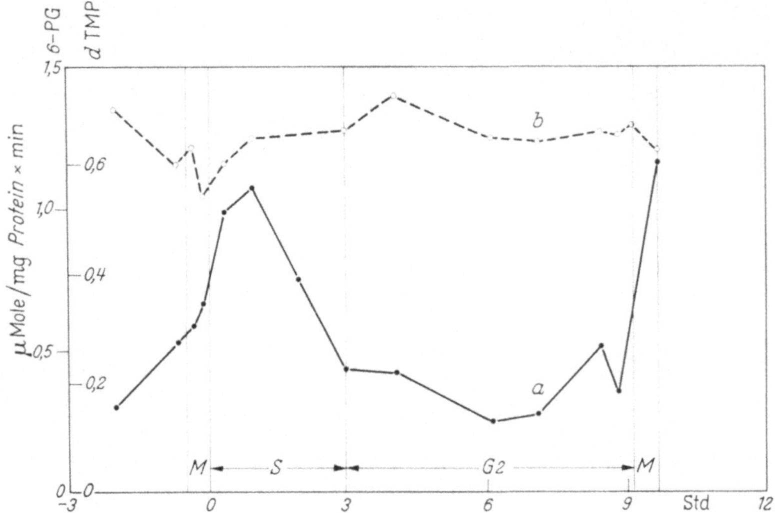

Abb. 3. Der Aktivitätsverlauf der Thymidinkinase (a) und Glucose-6-phosphatdehydrogenase (b) im Teilungscyclus synchroner Makroplasmodien von *Physarum polycephalum*. *M*: Mitoseabschnitt; *S*: *DNA*-Syntheseabschnitt; *G 2*: prämitotisches Intervall. (Sachsenmaier u. Ives; noch unveröffentlicht)

sprechen für eine engere Beziehung zum DNA-Cyclus und sind nicht zufällig oder durch einen allgemeinen dem Mitosegeschehen zugeordneten Rhythmus des Stoffwechsels bedingt. Die vergleichsweise mituntersuchte Aktivität der Glucose-6-phosphat-dehydrogenase bleibt nämlich während des gesamten Mitosecyclus konstant [*63*][1] (Abb. 3).

Amphibieneier besitzen unbefruchtet bis kurz nach der Besamung eine relativ hohe Thymidinkinaseaktivität. Aber noch vor der ersten Furchungsteilung sinkt die Aktivität stark ab [*11*]. Kleinere Schwankungen der Aktivität in der Folgezeit stehen mit den Furchungsteilungen in keinem deutlichen zeitlichen Zusammenhang. Beim Seeigelei sind Aktivitätsmaxima im Vorkernstadium und während der späten Anaphase der 1. Furchungsmitose angedeutet (Hansen-Delkeskamp u. Duspiva, noch unveröffentlicht).

---

[1] Für die freundliche Überlassung des Manuskripts zur Einsichtnahme noch vor der Drucklegung dankt der Referent Herrn Dr. W. Sachsenmaier sehr herzlich.

In der *regenerierenden Rattenleber* beginnt die DNA-Synthese 18 Std nach partieller Hepatektomie, erreicht 24—30 Std p. o. ein Maximum Maximum und fällt anschließend stark ab [*51, 25*]. Die normale Leber enthält einen Satz von Enzymen, die dAMP, dGMP und dCMP in die entsprechenden Triphosphate verwandeln. Enzyme, welche Thymidin phosphorylieren, fehlen jedoch oder sind nur in Spuren nachweisbar [*46, 23*]. Die Aktivität dieser Enzyme erscheint im Überstand zentrifugierter Zellextrakte erst im Verlauf der Regeneration [*3, 46, 26, 80*]. In der Leber Neugeborener sind die Enzyme noch vorhanden, aber verschwinden im Verlauf des Wachstums [*26*]. Nach Untersuchungen von WEISSMAN u. a. [*80*] dürfte das Auftreten der 3 Thymidin zu dTTP phosphorylierenden Kinasen succedan erfolgen. Die exakte Zuordnung der Thymidinkinaseaktivität zum Verlauf der DNA-Synthese stößt infolge heute noch unklarer Lokalisation und Bindungsverhältnissen von Enzym und Hemmfaktoren an verschiedene Zellbestandteile auf große Schwierigkeiten analytischer Art und ist z. Z. noch mit einer erheblichen Unsicherheit behaftet [*35, 13*]. In den rohen Zellextrakten sind heute ihrer chemischen Natur nach unbekannte Cofaktoren enthalten, die den Test stören. Adulte Leber enthält mehr Störfaktoren als regenerierende Leber oder Tumoren. In Zellkulturen aus Leber konnte EKER [J. biol. Chem. **240**, 2607 (1965)] zur Zeit der exponentiellen Wachstumsphase eine erhöhte Aktivität der Thymidinkinase und gleichzeitig eine verminderte Aktivität der Mono- und Diphosphonucleotid-spaltenden Phosphatasen feststellen. In der stationären Phase herrschten umgekehrte Aktivitätsverhältnisse. Daher dürften auch die genannten Phosphatasen dem Regulationssystem angehören. Nach FAUSTO u. VAN LANCKER [*12*] gewinnt die *Thymidylatkinase* bereits zwischen 12—24 Std nach partieller Hepatektomie merklich an Aktivität, um zwischen 24 und 36 Std den vierfachen Wert zu erreichen. Ähnlich verhält sich die Thymidinkinase. Der Anstieg der *DNA-Polymerase* ist innerhalb der ersten 24 Std unbedeutend, aber signifikant zwischen 24 und 36 Std p. o. Die *Thymidindiphosphat-Kinase* ändert ihre Aktivität innerhalb der ersten 24 Std nicht; bezogen auf den N-Gehalt der Extrakte ist aber ihre Aktivität von vornherein 500—1000mal höher als die der Thymidylatkinase. Diesem auf der Mittelstrecke des DNA-Syntheseweges wirkenden Enzym kommt offensichtlich im Regelkreis keine Bedeutung zu. Diese Ergebnisse sind zum Teil mit älteren Befunden im Einklang [*3, 21*]. *Desoxycytidylat-Desaminase* und Thymidylat-Synthetase, die in Normalleber nur spurenweise vorkommen, steigen in ihrer Aktivität 14 bis 18 Std nach partieller Hepatektomie scharf an [*45*]. Wir finden also, daß die neueren, mit verfeinerten Methoden an der regenerierenden Leber durchgeführten Untersuchungen bestätigen, daß mit Beginn der DNA-Verdopplung ein deutlicher Aktivitätsanstieg gerade solcher Enzyme erfolgt, die an entscheidenden Pforten der Thymidylatsynthese stehen.

b) Aktivierung oder Enzymsynthese

Der im Enzymtest erfaßte oft sprunghafte Anstieg an Aktivität kann grundsätzlich als Folge einer *Aktivierung* eines latenten Enzymdepots, einer *Ausschaltung von Hemmstoffen* oder einer *Neubildung* von Enzymmolekülen aufgefaßt werden. Die experimentelle Ausschaltung der Proteinsynthese innerhalb der Zelle bietet die Möglichkeit einer Entscheidung. Zwei Typen von Hemmung der Biosynthese von Proteinen sind anwendbar: a) Entzug von essentiellen Aminosäuren, und b) Zugabe eines Inhibitors der Proteinsynthese.

HOTTA u. STERN [*30*, 1963a] wandten an ihrem Objekt, den Mikrosporen der Lilienblüte, die unter b) genannte Methode an. Es zeigte sich, daß alle Stoffe, welche die Proteinsynthese, nach welchem Wirkungsmechanismus auch immer, hemmen, den in der Interphase auftretenden Aktivitätsanstieg der Thymidinkinase, aber zugleich auch die normalerweise nachfolgende Mitose unterbinden. Wichtig ist, daß diese Mittel *zeitgerecht*, d. h. 5 Tage vor der normalerweise auftretenden Enzymausschüttung, geboten werden. Zur Anwendung kamen: 8-Azaguanin und 5-Fluoruracil, welche nach herrschender Auffassung die RNA-Synthese hemmen; Chloramphenicol, das die Informationsübermittlung der mess-RNA verhindert, und 5-Methyltryptophan sowie Äthionin, die mit dem Aminosäuremuster der Zellproteine interferieren. Die Autoren sehen in dem Resultat eine Bestätigung ihrer Vermutung, daß ein Teil des Mechanismus, der die plötzliche Ausschüttung enzymatischer Aktivität bewirkt, den Charakter einer *induzierten Enzymsynthese* aufweist.

Der steile Aktivitäts*abfall* von Thymidinkinase kurz vor oder während der DNA-Synthese wird jedoch durch die genannten Zellgifte nicht beeinflußt [*30*, 1963b]. Ein Ausschluß von Sauerstoff oder Zugabe von Dinitrophenol vermögen jedoch den Abbau von Enzymen aufzuhalten. Offensichtlich enthält der Prozeß des Abbaues ein aerobes, energiebedürftiges Glied.

In letzter Zeit sind Erfahrungen gesammelt worden, die wahrscheinlich machen, daß auch der in der regenerierenden Leber beobachtete Aktivitätsanstieg auf eine echte Proteinsynthese zurückgehen dürfte. Werden Ratten 3 Std nach partieller Hepatektomie mit Actinomycin behandelt, so findet man in der partikelfreien Fraktion eines 24-stg. Regeneratleberhomogenats ebensowenig Thymidylatkinaseaktivität wie bei Normalleber. Die Aktivität der DNA-Polymerase läßt sich hingegen durch Actinomycin viel weniger deutlich beeinflussen [*12*]. Auf dem Wege über eine Actinomycin- und Puromycinbehandlung von Ratten zu verschiedenen Terminen nach partieller Hepatektomie ließ sich der Nachweis führen, daß die *genetische Information* zur Synthese von Desoxycytidylat-Desaminase bis zu 14 Std nach der Operation noch nicht verfügbar ist. Sie ist also als Ganzes oder in voll funktioneller Form *nicht vor Beginn der DNA-Synthese* vorhanden. Äthionin und p-Fluorphenylalanin verhindern die Enzymsynthese nur dann, wenn sie sofort nach der Operation injiziert werden. Ihr Wirkungsmechanismus ist ein anderer, offensichtlich kommt es in ihrer Gegenwart zur Bildung inaktiver Enzymmoleküle und nicht zu einer gestörten Transskription der Strukturgene [*44*].

Interessant, aber heute noch unerklärbar, ist die Feststellung von MALEY u. a. [*44*], daß der Aktivitätsanstieg der Thymidylat-Synthetase zwar durch Äthionin und p-Fluorphenylalanin, nicht aber durch Actinomycin oder Puromycin unterbunden werden kann. Solche Beobachtungen mahnen zur Vorsicht bei der Beurteilung der Versuchsergebnisse mit Antibiotica und Antimetaboliten.

Aufgrund der hier referierten Versuchsergebnisse ist unter den eingangs aufgeführten Möglichkeiten einer Erklärung für den sprunghaften Anstieg mancher Enzyme zu Beginn der S-Phase des Teilungscyclus die Annahme einer *Neusynthese* am wahrscheinlichsten.

c) Die Natur der Induktion

In Anlehnung an die Vorstellungen, die auf dem Gebiete der Mikrobiologie entwickelt wurden, kann man die minimale Aktivität vieler an der DNA-Synthese beteiligter Enzyme während der G 1- und G 2-Phase als Resultat einer *Repression* der zugehörigen Strukturgene, den sprunghaften Ausstoß an Aktivität zu Beginn der S-Phase als Ausdruck einer *Induktion* auffassen, derzufolge eine Nettosynthese der betreffenden Enzyme ausgelöst wird. Dann ist es zunächst naheliegend, Schwankungen in der Konzentration an gewissen Metaboliten, wie Ribo- und Desoxyribonucleosiden, als Induktoren von Thymidinkinase und verwandten Enzymen in Betracht zu ziehen.

Diese Vermutungen hat erstmalig STERN [zit. n. *30*, 1961] geäußert, bestärkt durch die Beobachtung, daß sich in der Flüssigkeit im Innern der Antheren von Lilien, in der die Mikrosporen heranreifen, kurz vor Eintritt der DNA-Synthese und Mitose Desoxyriboside von Guanin, Adenin, Cytosin und Thymin anhäufen; diese Stoffe verschwinden bei Beginn der DNA-Synthese (Abb. 2). Die Tatsache, daß die Mikrosporen unmittelbar nach dem Auftreten des Pools an Desoxynucleosiden die Fähigkeit erwerben, Thymidin zu phosphorylieren, drängt die Vorstellung eines *Induktionsvorganges* direkt auf [*30*, 1961]. Bei einer genaueren Untersuchung hat sich herausgestellt, daß dieser Pool im Innern der Anthere aus dem Tapetum beliefert wird, welches gerade zu dieser Zeit zu degenerieren beginnt. Eine Schwierigkeit erwächst dieser Deutung jedoch aus der Beobachtung von TAKATS [75], daß das aus dem Tapetum stammende Material von den Microsporen gar nicht zur DNA-Synthese verwendet wird. Es könnte natürlich sein, daß dieser außerhalb der Mikrosporen liegende Pool einen zweiten Pool innerhalb der Mikrosporen zu bilden anregt, der die Induktion bewirkt. Aber eine solche Hilfshypothese hat doch nur wenig Glaubwürdigkeit.

In einer neueren Arbeit untersuchten HOTTA u. STERN [*30*, 1965] die Frage, ob normalerweise nur deshalb eine so überaus kurze Zeitspanne der G 1-Phase mit Enzymbildung betraut ist (nur etwa 5% der Gesamtdauer), weil der „Induktor" nur während dieser Zeit aus dem Tapetum freigesetzt wird. Sie behandelten deshalb die Antheren zu verschiedenen Terminen innerhalb der G 1-Periode der Mikrosporen mit Tymidinlösungen (0,05—2,50 mg/ml), hatten aber nur während einer schmalen Periode des Zellcyclus Erfolg, der in der Nähe des Termins lag, zu dem normalerweise Thymidinkinase gebildet wird und etwa 10% der G 1-Phase einnahm. Die Beschränkung der stimulierenden Wirkung von

exogenem Thymidin auf diese kurze Zeitspanne zeigt, daß im Falle der
Mikrosporen sehr komplizierte Verhältnisse vorliegen, die mit dem Phänomen der Enzyminduktion bei Mikroorganismen nicht direkt vergleichbar sind. Denn im Verlauf der Entwicklung des Pollenkorns gibt es nur
eine wohlumschriebene kritische Periode, während welcher exogenes
Thymidin die Thymidinkinasesynthese effektvoll stimulieren kann.
Außerhalb dieser ist Thymidin ohne Wirkung, obwohl exogenes Thymidin
während der gesamten G 1-Phase von den Mikrosporen aufgenommen
werden kann. Hotta u. Stern sind der Meinung, daß bei der Induktion
der Thymidinkinase ein spezieller Regulationsmechanismus eine besondere
Rolle spielt, der die Zeitspanne der Induzibilität absteckt.

Es ist bekannt, daß Eier und junge Embryonen von Amphibien und
Seeigel einen umfangreichen extranucleären Pool von DNA-ähnlichen
Stoffen enthalten [28, 24, 14]. Ein erheblicher Teil dieses Pools besteht
aus Purin- und Pyrimidinnucleotiden [22, 38]. Das Material ist in einem
so großen Überschuß vorhanden, daß es fraglich erscheint, ob seine biologische Bedeutung ausschließlich darin liegt, einen Speicher von Bau-
stoffen für die DNA-Synthese zu bilden. Da jedoch genauere Informationen über den Weg dieser Stoffe sowie über ihren Konzentrations-
wandel im Verlauf der Entwicklung noch fehlen, kann man über die
Beziehung dieses Pools zur DNA-Synthese keine näheren Aussagen
machen.

Adulte Gewebe enthalten hingegen (bezogen auf Trockengewicht) nur
sehr geringe Mengen an Desoxynucleosiden und Desoxynucleotiden. In
der Leber steigt der Gehalt an diesen Stoffen zwar bei der Regeneration
an, aber die gefundenen Mengen waren stets sehr klein [66, 56, 62]. Es ist
heute noch nicht einmal sicher, daß diese Stoffe *nur* als Vorstufen der
DNA Verwendung finden [25, 55]. Eine ausführliche Diskussion der mit
dem Pool an Desoxyribonucleosiden zusammenhängenden Probleme bei
ruhenden und wachsenden Zellen findet sich bei Lark [40], auf die hier
verwiesen werden soll. Es scheint dem Referenten, daß die Ermittlung
des Pools, bezogen auf Zellen und Gewebe, als Ganzes keinen befriedigenden Einblick in eine mögliche regulative Funktion dieser Stoffe geben
kann, da die hochgradige Kompartimentation der Zelle, die der biochemische Ausdruck ihrer strukturellen Organisation ist, hierbei keine Berücksichtigung findet. Wie wichtig aber dieser Gesichtspunkt ist, zeigen
bereits die ersten tastenden Versuche in dieser Richtung. Die Untersuchung der Verteilung von leicht löslichen Stoffen auf einzelne Zellfraktionen stößt auf große technische Schwierigkeiten. Unter Anwendung
einer nicht-wäßrigen Methode zur Auftrennung von Zellbestandteilen
wurde die Verteilung der säurelöslichen Desoxyribonucleoside und Desoxyribonucleotide auf Zellkerne und Gesamtzellen unter Verwendung des
mikrobiologischen Tests bei Normalleber, regenerierender Leber und
Novikoff-Hepatom der Ratte untersucht [1]. Es stellte sich heraus, daß
die *Zellkerne* aller 3 Gewebearten eine mehrfach höhere Konzentration an
Desoxyribonucleosiden und -tiden enthalten als das Cytoplasma, und daß
es gerade die Desoxyribonucleotide sind, die in den Kernen der rasch sich
teilenden Zellen stark angereichert anzutreffen sind. Rechnet man die

gewonnenen Daten auf den DNA-Gehalt der Fraktionen als Bezugsgröße um, so ergibt sich, daß mehr als die Hälfte der Desoxyribonucleoside des Gewebes im Zellkern sitzt; von den Desoxyribonucleotiden bei der regenerierenden Leber oder dem Hepatom sogar 71 bzw. 87%. Diese Zahlen demonstrieren mit aller Deutlichkeit, daß der *Zellkern* im Umsatz der niedermolekularen Vorstufen der DNA eine führende Rolle spielen muß.

Ein faszinierendes Ergebnis wurde jüngst mit einer neuartigen autoradiographischen Methode bei *Tetrahymena* erzielt [72], die den Nachweis wasserlöslicher Stoffe ermöglicht. Die Autoren fanden, daß *Tetrahymena* ^{3}H-Thymidin dem Kulturmedium nur während der S-Periode entnehmen kann. Weder in der G 1- noch G 2-Phase findet ein Einbau von Vorstufen in den Pool löslicher Thymidinderivate statt. Während der G 1-Phase besitzt aber dieser Ciliat einen kleineren Pool löslicher Thymidinderivate; er befindet sich im Großkern. Dieser Pool wurde während der letzten vorausgehenden S-Phase aufgebaut; er hat aber in den folgenden G 2- und G 1-Phasen keinen Turnover und steht mit dem Thymidin des Kulturmediums nicht im Austausch. STONE u. PRESKOTT [73] haben gezeigt, daß eine experimentell erzwungene Verarmung der Zellen an essentiellen Aminosäuren, die kurz vor einer S-Periode manifest wird, den Ablauf der in dieser Periode stattfindenden DNA-Synthese nicht mehr verhindern kann. Aber die nächstfolgende S-Phase ist nur noch zu einer unvollständigen DNA-Synthese fähig. Spektroskopische Messungen (Feulgenfärbung) ergaben einen DNA-Zuwachs von nur 20% der Norm. ^{3}H-Thymidin wird während dieser unvollständigen DNA-Synthese nicht aus dem Medium aufgenommen und eingebaut. Die Autoren vermuten, daß der Einbau von extracellulärem Thymidin mangels der hierzu nötigen Enzyme unterbleiben muß, weil die Zellen an gewissen essentiellen Aminosäuren verarmt sind. Es konnte jedoch festgestellt werden, daß die auf 20% begrenzte DNA-Synthese allein auf Kosten des persistierenden Pools löslicher Thymidinderivate erfolgt und abstoppt, sobald dieser aufgebraucht ist.

Sehr bedeutsam ist, daß diese Untersuchungen geeignet sind, ein Licht auf die Sequenz der Ereignisse im Verlauf der DNA-Reduplikation im Makronucleus zu werfen [72].

(a) Die DNA-Synthese bei *Tetrahymena* beginnt auf Kosten eines Pools von Vorstufen, der im Makronucleus während der letzten S-Phase gebildet und seither gespeichert wurde.

(b) Der Umsatz des Pools stimuliert die Synthese von zusätzlichen Vorstufen; dadurch wird auch die Aufnahme von Thymidin aus dem Nährmedium in dem intracellulären Pool bewirkt.

Der Anschauung dieser Autoren entsprechend nimmt nicht etwa der persistente Pool von Vorstufen einen Einfluß auf den Beginn der DNA-Synthese, sondern umgekehrt: Die DNA-Synthese gibt den Anreiz für den Beginn des Poolturnovers. Es könnte nun sein, daß mit dem Einsetzen der DNA-Synthese auch Thymidinkinase in Erscheinung tritt, wobei das Auftreten dieses Enzyms die Voraussetzung für den Verbrauch des Pools sein könnte, was wahrscheinlich auch die Meinung der Autoren ist.

Der Referent möchte aber noch auf eine andere Denkmöglichkeit hinweisen, daß nämlich gerade das *Versiegen des Pools im Zellkern das auslösende Moment für die Induktion der Thymidinkinase sein könnte.*

Unter diesem Aspekt hätte der von Stone u. a. [72] aufgefundene, während der Interphase im Zellkern persistierende Pool von thymidinhaltigen Vorstufen die Funktion eines Effektors, der einen hypothetischen Repressor *aktiviert.* Der während der Interphase aktive Repressor hält eine Gruppe von Strukturgenen gesperrt mit dem Erfolg, daß die Zelle außerhalb der S-Phase nur verschwindend kleine Mengen an Thymidinkinase zu bilden vermag. Das rasche Aufzehren des intranucleären Pools bei beginnender DNA-Replikation bedeutet zunächst ein Verschwinden des Effektors, in der Folge aber eine *Inaktivierung des Repressors.* Das bisher gesperrte Operon wird frei und die plötzlich einsetzende Enzymsynthese führt zu einer sprunghaft ansteigenden Kinaseaktivität. Mit dem neuerlichen Aufbau eines intranucleären Pools von löslichen Vorstufen gegen Ende der S-Phase beginnt das Spiel von neuem, das Operon wird wieder gesperrt. Diese z. Z. natürlich noch weitgehend auf Spekulation beruhende Hypothese stimmt formal mit dem *Schema eines anabolischen Regelkreises* überein, das in der Mikrobiologie oft realisiert ist und erwartungsgemäß auch bei der DNA-Synthese der Eunucleobionten realisiert sein dürfte. Es wäre wichtig zu erfahren, ob die von Stone u. a. an *Tetrahymena* erhobenen Befunde einen Spezialfall — die Ciliatenzelle — betreffen oder von allgemeiner Bedeutung sind.

Über die primäre Ursache der DNA-Replikation ist damit nur soviel ausgesagt, daß sie offenbar nicht im Wandel der Aktivität von Thymidinkinase und ähnlichen Enzymen zu suchen ist. Offensichtlich ist die Primärursache in dem Faktorenkomplex enthalten, der mit der Primer-Funktion der DNA zusammenhängt. Da es sich aber hierbei nicht um einen enzymatischen Prozeß handelt, steht die Diskussion dieses Problems außerhalb des Rahmens dieser Abhandlung.

Schlußbemerkungen

In Anbetracht der Tatsache, daß ^{3}H-Thymidin während der S-Periode ausschließlich im *Raum des Zellkerns* eingebaut wird, wie alle autoradiographischen Befunde darlegen, ist es erstaunlich, daß ein so wesentlich an der DNA-Synthese beteiligtes Enzym wie die Desoxyribonucleotid-Polymerase im Cytoplasma lokalisiert sein soll. Jedenfalls erscheint das Enzym nach Homogenisation hauptsächlich im partikelfreien Überstand [3, 9, 46, 59, 70, 71]. Wenn man nicht an die Möglichkeit glaubt, daß die DNA-Synthese im Cytoplasma begonnen und im Kern zu Ende geführt wird, so muß man eine Verlagerung von Enzymmolekülen aus dem Grundcytoplasma in den Raum des Zellkerns während der S-Periode des Zellcyclus sowie eine Abgabe von Enzym nach Beendigung der DNA-Synthese an das Cytoplasma für möglich halten. Befunde von Littlefield, McGovern u. Margeson [42] deuten in diese Richtung. Dieser Enzymaustausch könnte durch besondere Kontrollfaktoren geregelt sein, wie z. B. solche, welche die Permeabilität der Kernmembran bzw. die

Porengröße beeinflussen. SIEBERT [67] hat besonders nachdrücklich darauf hingewiesen, wie leichtlöslich manche Enzyme in wäßrigen Medien sind; es besteht daher die Gefahr, daß die so überwiegende Anwesenheit der Polymerase im Cytoplasma ein methodisches Artefakt sein könnte. Tatsächlich enthalten Zellkerne aus regenerierender Rattenleber in einem nichtwäßrigen Medium präpariert eine mehrfach höhere DNA-Polymeraseaktivität als das Cytoplasma [34]. Während die Zellkerne aus regenerierender Leber eine mäßige Erhöhung der Polymeraseaktivität gegenüber Normalleber zeigten, waren Kerne aus Novikoff-Hepatom etwa 4mal aktiver. Auf DNA bezogen, gewinnt man aus Zellkernen der Normalleber etwa 10%, aus Hepatom-Kernen 54% der Totalaktivität des Gewebes. In nichtwäßrigen Medien präparierte Kerne erweisen sich [7] bedeutend enzymreicher als die nach üblichen Methoden gewonnenen Kerne. Wie SCHNEIDER [65] selbst betont, ist aber auch diese Methode nicht frei von Artefakten, da durch die Gefriertrocknung des Gewebes und Behandlung mit nichtwäßrigen Lösungsmitteln gewisse Änderungen in den Eigenschaften der Enzymproteine unvermeidbar sind. Das Problem der Lokalisation der Enzyme befriedigend zu lösen, bleibt der Zukunft vorbehalten.

Summary

The exact time at which DNA synthesis starts in the mitotic cycle, the duration of this process and the amount of newly synthesized DNA are determined by a special control mechanism. This control probably works by influencing three different phases of DNA synthesis:

(1) the transformation of ribonucleotides into deoxyribonucleotides (especially into deoxythymidilate);

(2) the reaction in which deoxyribonucleotides are phosphorylated into nucleotide triphosphates;

(3) the polymerizing reaction in which the nucleotide triphosphates build up the DNA molecule.

The occurrence of a system which regulates the activities of the enzymes catalysing the above mentioned reactions as well as the rate of their formation is well known. This regulating principle operates at different levels of cellular organization:

(1) *at the level of the macromolecules* (enzymes). The rate of an enzymatic reaction is regulated by the concentrations of the substrates, hydrogen ions and effectors, respectively, by direct interaction with the enzyme molecule;

(2) *at the level of macromolecular units* (enzyme chains). There are sequences of enzymatic reactions leading to the synthesis of biological compounds in which the enzyme catalysing the first step of the sequence is inhibited allosterically by the final product. In this way, the stationary concentration of the final product regulates the flow substrate through the whole reaction chain;

(3) at the *level of the protein synthesizing system*. In some instances, there is a complete repression of the formation of all enzymes taking part in the synthesis of a certain compound by the concentration of this same final product. Thus, the formation of the enzyme ATP-thymidine-5'-phosphotransferase (thymidine kinase) is regulated by d-TTP in the system of DNA synthesis. The formation of deoxycytidilate-aminohydrolase is regulated in a similar way.

It is demonstrated in some examples that thymidine kinase formation occurs only at a particular point in the mitotic cycle preceding the beginning of the S-phase. The increase in activity of this enzyme does not take place in the presence of substances which specifically inhibit protein synthesis (antimetabolites and some antibiotics). Therefore, an extensive synthesis of thymidine kinase must be induced at a certain point of the mitotic cycle. The chemical basis for this induction is still unknown.

Finally, the necessity is emphasized to attempt to localize the intermediary products and the enzymes taking part in DNA synthesis within the fine structure of the nucleus.

Literatur

[1] Behki, R. M., and W. C. Schneider: Biochim. biophys. Acta (Amst.) 61, 663 (1962); 68, 34 (1963).
[2] Bianchi, P. A., J. A. V. Butler, A. R. Crathorn, and K. V. Shooter: Biochim. biophys. Acta (Amst.) 48, 213 (1961); 53, 123 (1961).
[3] Bollum, F. J., u. V. R. Potter: J. biol. Chem. 233, 478 (1958); — Cancer Res. 19, 561 (1959).
[4] Breitman, T. R.: Biochim. biophys. Acta (Amst.) 67, 153 (1963).
[5] Bresch, C.: Klassische und molekulare Genetik. Berlin-Heidelberg-New York: Springer 1965.
[6] Bresnick, E., U. B. Thompson, H. P. Morris, and A. G. Liebelt: Biochem. biophys. Res. Comm. 16, 278 (1964).
[7] Cameron, I. L., and G. E. Stone: Exp. Cell Res. 36, 510 (1964).
[8] Changeux, J. P.: J. molec. Biol. 4, 220 (1962).
[9] Chiga, M., and G. W. E. Plaut: J. biol. Chem. 235, 3260 (1960).
[10] Crathorn, A. R., and K. V. Shooter: Nature (Lond.) 187, 614 (1960).
[11] Duspiva, F., u. E. Hansen-Delkeskamp: Z. Naturforsch. 20b, 582 (1965).
[12] Fausto, N., and J. L. van Lancker: J. biol. Chem. 240, 1247 (1965).
[13] Fiala, S., A. Fiala, G. Tobar, and H. McQuilla: J. nat. Cancer Inst. 28, 1269 (1962).
[14] Finamore, F. J., and E. Volkin: Exp. Cell Res. 15, 405 (1958).
[15] Firket, H., and W. G. Verly: Nature (Lond.) 181, 274 (1958).
[16] Friedkin, M., D. Tilson, and D. Roberts: J. biol. Chem. 220, 627 (1956).
[17] Gall, J. G.: J. biophys. biochem. Cytol. 5, 295 (1959).
[18] Gelfant, S.: Exp. Cell Res. 26, 395 (1962).
[19] Gerhart, J. C., and A. B. Pardee: J. biol. Chem. 237, 891 (1962).
[20] Giudice, G., and G. D. Novelli: Biochem. biophys. Res. Comm. 12, 383 (1963).
[21] Gottlieb, L. I., N. Fausto, and J. L. van Lancker: J. biol. Chem. 239, 555 (1964).
[22] Grant, P.: J. cell comp. Physiol. 52, 227 (1958).
[23] Gray, E. D., S. M. Weissman, J. Richards, D. Bell, H. M. Keir, R. M. S. Smellie, and J. N. Davidson: Biochim. biophys. Acta (Amst.) 45, 111 (1960).
[24] Gregg, J. R., and S. Løvtrup: Biol. Bull. 108, 29 (1955).
[25] Hecht, L. I., and V. R. Potter: Cancer Res. 16, 988 (1956); 18, 186 (1958).
[26] Hiatt, H. H., and T. B. Bojarski: Biochem. biophys. Res. Comm. 2, 35 (1960).
[27] Hinegardener, R., and D. Mazia: Science 136, 326 (1962).
[28] Hoff-Jørgensen, E., and E. Zeuthen: Nature (Lond.) 169, 245 (1952).
[29] Hornsey, S., and A. Howard: Ann. N. Y. Acad. Sci. 63, 915 (1956).
[30] Hotta, Y., and H. Stern: J. biophys. biochem. Cytol. 9, 279 (1961a); 11, 311 (1961b); — Proc. nat. Acad. Sci. (Wash.) 49, (a) 648, (b) 861 (1963); — J. Cell Biol. 25, 99 (1965).
[31] Howard, A., and S. R. Pelc: Heredity 6, (Suppl.) 261 (1953).
[32] Howard, F. L.: Ann. Bot. 46, 461 (1932).
[33] Ives, D. H., P. A. Morse jr., and V. R. Potter: J. biol. Chem. 238, 1467 (1963).
[34] Keir, H. M., R. M. S. Smellie, and G. Siebert: Nature (Lond.) 196, 752 (1962).
[35] Kielley, R.: Cancer Res. 23, 801 (1963); — Biochem. biophys. Res. Comm. 10, 249 (1963).
[36] Kit, S., L. J. Piekarski, and D. R. Dubbs: J. molec. Biol. 6, 22 (1963); 7, 497 (1963).

[37] KORNBERG, A.: Enzymatic synthesis of DNA. New York: J. Wiley a. Sons 1962.
[38] KURIKI, Y., and R. OKAZAKI: Embryologia 4, 337 (1959).
[39] LAJTHA, L. G., R. OLIVER, and F. ELLIS: Brit. J. Cancer 8, 367 (1954).
[40] LARK, K. G.: Cellular Control of DNA Biosynthesis in J. H. TAYLOR, "Molecular Genetics". New York: Academic Press 1963.
[41] LEHMAN, I. R., M. J. BESSMAN, E. S. SIMMS, and A. KORNBERG: J. biol. Chem. 233, 163 (1958).
[42] LITTLEFIELD, J. W., A. P. McGOVERN, and K. B. MARGESON: Proc. nat. Acad. Sci. (Wash.) 49, 102 (1963).
[43] LUMPER, L.: Grundlagen der Kinetik enzymatisch katalysierter Reaktionen; in HOPPE-SEYLER/THIERFELDERs Hdb. physiol. u. pathol.-chem. Analyse. 6. Bd. Teil A, p. 17. Berlin-Göttingen-Heidelberg-New York: Springer 1964.
[44] MALEY, G. F., M. G. LORENSON, and F. MALEY: Biochem. biophys. Res. Comm. 18, 364 (1965).
[45] MALEY, F., and G. F. MALEY: Biochemistry 1, 847 (1962a); — J. biol. Chem. 237, PC 3311 (1962b); — Fed. Proc. 22, 292 (1963).
[46] MANTSAVINOS, R., and E. S. CANELLAKIS: J. biol. Chem. 234, 628 (1959).
[47] MAZIA, D.: Mitosis and the Physiology of Cell Division; in J. BRACHET a. A. MIRSKY "The Cell". Vol. III. p. 77 ff. New York: Academic Press 1961.
[48] McDONALD, B. B.: J. Cell Biol. 13, 193 (1962).
[49] MONOD, J., J. P. CHANGEUX, and F. JACOB: J. molec. Biol. 6, 306 (1963).
[50] MOORHEAD, P. S., and V. DEFENDI: J. Cell Biol. 16, 202 (1963).
[51] NYGAARD, O., and H. P. RUSCH: Cancer Res. 15, 240 (1955).
[52] OEHLERT, W., u. TH. BÜCHNER: Beitr. path. Anat. u. allg. Path. 125, 374 (1961).
[53] OKAZAKI, R., and A. KORNBERG: J. biol. Chem. 239, 269, 275 (1964).
[54] PARDEE, A. B., F. JACOB, and J. MONOD: J. molec. Biol. 1, 165 (1959).
[55] POTTER, R. L., and T. BUETTNER-JANUSCH: J. biol. Chem. 233, 462 (1958).
[56] —, S. SCHLESINGER, V. BUETTNER-JANUSCH, and L. THOMPSON: J. biol. Chem. 226, 381 (1957).
[57] — Exp. Cell Res., Suppl. 9, 259 (1963).
[58] PRESCOTT, D. M.: Exp. Cell Res. 19, 228 (1960).
[59] —, F. J. BOLLUM, and B. C. KLUSS: J. Cell Biol. 13, 172 (1962).
[60] REICHARD, P., Z. N. CANELLAKIS, and E. S. CANELLAKIS: J. biol. Chem. 236, 2514 (1961).
[61] —, and B. ESTBORN: J. biol. Chem. 188, 839 (1951).
[62] ROTHERHAM, J., and W. C. SCHNEIDER: J. biol. Chem. 232, 853 (1958).
[63] SACHSENMAIER, W., u. D. H. IVES: Biochem. Z. (im Druck).
[64] SCARANO, E., G. GERACI, and M. ROSSI: Biochim. biophys. Res. Comm. 16, 239 (1964).
[65] SCHNEIDER, W. C.: Exp. Cell Res., Suppl. 9, 430 (1963).
[66] —, and L. W. BROWNELL: J. nat. Cancer Inst. 18, 579 (1957).
[67] SIEBERT, G.: Exp. Cell Res., Suppl. 9, 389 (1963).
[68] STADTMAN, E. R., G. N. COHEN, G. LE BRAS, and H. DE ROBICHON-SZULMAJSTER: Cold Spring Harbour Symp. quant. Biol. 26, 319 (1961).
[69] SMELLIE, R. M. S.: Exp. Cell Res., Suppl. 9, 245 (1963).
[70] —, and R. EASON: Biochem. J. 80, 39 P (1961).
[71] —, H. M. KEIR, and J. N. DAVIDSON: Biochim. biophys. Acta (Amst.) 35, 389 (1959).
[72] STONE, G. E., O. L. MILLER JR., and D. M. PRESCOTT: J. Cell Biol. 25, 171 (1965).
[73] —, and D. M. PRESCOTT: J. Cell Biol. 21, 275 (1964).
[74] SWANN, M. M.: The Control of Cell Division: A Review. I. General Mechanisms. Cancer Res. 17, 727—757 (1957). II. Special Mechanisms. Cancer Res. 18, 1118—1160 (1958).
[75] TAKATS, S.: Amer. J. Bot. 49, 748 (1962).
[76] TOMKINS, G. M., and K. L. YIELDING: Cold Spring Harbour Symp. quant. Biol. 26, 331 (1961).

[*77*] Taylor, J. H., P. S. Woods, and W. L. Hughes: Proc. Nat. Acad. Sci.
   (Wash.) **43**, 122 (1957).
[*78*] Tsukada, K., and I. Lieberman: J. biol. Chem. **239**, 2952 (1964).
[*79*] Umbarger, H. E.: Science **123**, 848 (1956).
[*80*] Weissman, S. M., R. M. S. Smellie, and J. Paul: Biochim. biophys. Acta
   (Amst.) **45**, 101 (1960).
[*81*] Woodard, J., B. Gelber, and J. Swift: Exp. Cell Res. **23**, 258 (1961).

Diskussion

Vorsitz: *Grundmann*

Klamerth: Die Thymidinkinase scheint tatsächlich völlig unabhängig vom Nucleotidpool zu sein. Nach Infektion eines Gewebes mit DNA-Viren (Herpes-, Pseudorabiesvirus) steigt die Thymidinkinase enorm an, auch wenn vorher die DNA-Synthese in der Zelle selbst blockiert war. Anscheinend wird vom Genom des Virus über eine vielleicht sekundäre Proteinsynthese die Thymidylatkinase erst aktiviert oder neu geschaffen.

Schaller: But there are cases known, where DNA-viruses contain their own gens for kinases (This has not been studied for herpes simplex). So we can't tell whether it is a property of the virus genome or the host genome.

Klamerth: Die Thymidylatkinase entsteht lange vor der Virussynthese. Die Mengen, die mit den Viren in die Wirtszellen gebracht werden, sind zu gering.

Klima: Another possibility to explain this behaviour of thymidine kinase would be, that we have a two-switch-system. First of all we have to switch on the first switch, and then thymidine would induce the synthesis of thymidine-kinase.

Duspiva: What do you mean by the first switch?

Klima: That is very hard to say: perhaps starting events in the replication of the DNA.

Karlson: Ich möchte Herrn Klima beipflichten. Die im Vortrag klargelegten Regulationsmechanismen dienen lediglich zur Erhaltung eines bestimmten Nucleotidpools. Es ist aber nicht anzunehmen, daß z. B. zu einer bestimmten Höhe aufgestautes Thymidintriphosphat die DNS-Synthese induziert, diese also über die Poolgröße reguliert würde (es sei denn in negativem Sinne durch einen Mangel an Bausteinen).

Schaller: There is no need to induce a synthesis of thymidine kinase by TTP or something like this, but just by time regulation of the triphosphates; you don't need to talk about compartments of the cell. It may be just a synthesis of some compound which is determined by the time of cell state. Do you know anything about the pool size of ribonucleoside triphosphates in *Tetrahymena*? In *E. coli* there are between 400 and 1000 molecules of each of the triphosphates, it is very low and therefore very difficult to detect.

Duspiva: Stone and Prescott have not communicated the pool size of triphosphates as such. They photometrically determined the relative increase of the DNA-content after blocking protein synthesis.

Klima: The grade of polyploidy of the macronucleus of *Tetrahymena* is nearly 1000. Therefore it would be necessary to have a higher pool for the high number of chromosomes.

Grundmann: Ist eigentlich über die Repressorsubstanzen etwas Näheres bekannt?

Duspiva: Es ist möglich, aber noch nicht sicher bewiesen, daß es sich um Histone handelt. Auch das System von Jacob und Monod läßt in dieser Hinsicht noch verschiedene Deutungsmöglichkeiten zu.

Parthier: I would like to mention the mechanism of the action of nucleoside antibiotics (nucleosides with either normal bases and abnormal sugars or vice versa), which are suggested to be competitive for specific enzyme sites. For instance cordycepin inhibits the formation of deoxy-ATP, psicofuranin the formation of GTP. Unfortunately so far no antibiotics are known for the specific inhibition of the thymidine kinase.

Analyse des Zellcyclus durch Eingriffe in die Makromolekül - Biosynthese[1]

Von –

Wilhelm Sachsenmaier, Heidelberg

Mit 11 Abbildungen

Einleitung

Nur ein kurzer Abschnitt des Zellcyclus entfällt auf den eigentlichen Teilungsvorgang. Dieser ist, insbesondere bei höheren Organismen, durch eindrucksvolle cytologische Veränderungen charakterisiert. Demgegenüber erscheint die Interphase, die auch bei rasch wachsenden Zellen meist mehr als 90% des Teilungscyclus beansprucht, cytologisch als Ruhestadium. Gleichwohl ist anzunehmen, daß gerade in diesem Stadium molekulare Prozesse vor sich gehen, die als Vorbereitung für die nächste Teilung von entscheidender Bedeutung sind.

Die Frage, welche Vorgänge während der Interphase wachsender Zellen spezifisch mit dem Teilungsgeschehen korreliert sind, kann auf verschiedene Weise untersucht werden. In manchen Fällen kann der zeitliche Verlauf bestimmter Stoffwechselprozesse durch direkte Messung verfolgt werden, wie z. B. die Synthese der Desoxyribonucleinsäure (DNS)[2] durch Cytophotometrie oder Autoradiographie. Versuche dieser Art haben zur Unterteilung der Interphase in die Abschnitte G_1 (postmitotische Ruhephase), S (DNS-Synthesephase) und G_2 (prämitotische Ruhephase) geführt [45]. Ein anderer Weg ist die gezielte Blockierung bestimmter Prozesse mit Hemmstoffen oder durch physikalsche Eingriffe. Sofern der Wirkungsmechanismus des Agens bekannt ist, können auf diese Weise eindeutige Informationen über die Natur, die Bedeutung und den zeitlichen Ablauf biochemischer Prozesse gewonnen werden, die die Länge des Teilungscyclus und damit den Zeitpunkt der Mitose kontrollieren. Leider ist die Spezifität chemischer oder physikalischer Eingriffe in den Zellstoffwechsel selten so streng, daß dieser Idealfall ohne weiteres realisiert werden kann. Nachfolgend sollen an Hand einiger experimenteller Beispiele die Möglichkeiten zur Analyse des Zellcyclus durch Hemmstoffe der Makromolekül — Biosynthese und

---

[1] Mit Unterstützung der Deutschen Forschungsgemeinschaft.

[2] Folgende Abkürzungen werden verwendet: BUDR = 5-Brom-2′-desoxyuridin; CR = Cytidin; DNS = Desoxyribonucleinsäure; FU = 5-Fluoruracil; FUDR = 5-Fluor-2′-desoxyuridin; FUR = 5-Fluoruridin; G_1 = postmitotische Ruhephase; G_2 = prämitotische Ruhephase; RNS = Ribonucleinsäure; S = DNS-Synthesephase; TDR = Thymidin; UR = Uridin; UV = Ultraviolett.

durch UV-Strahlung aufgezeigt werden. Hierbei sollen vorwiegend Untersuchungen an einem natürlich synchronen System, dem Schleimpilz *Physarum polycephalum*, berücksichtigt und mit einigen Befunden an anderen Systemen verglichen werden.

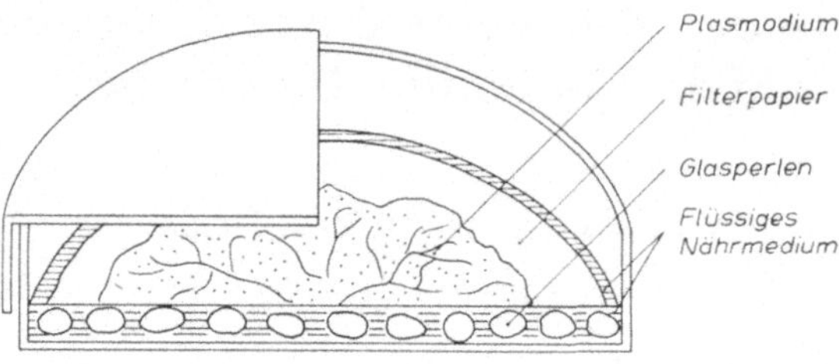

Abb. 1. Schnitt durch eine Petrischalen-Kultur von *Physarum pol.* [76]

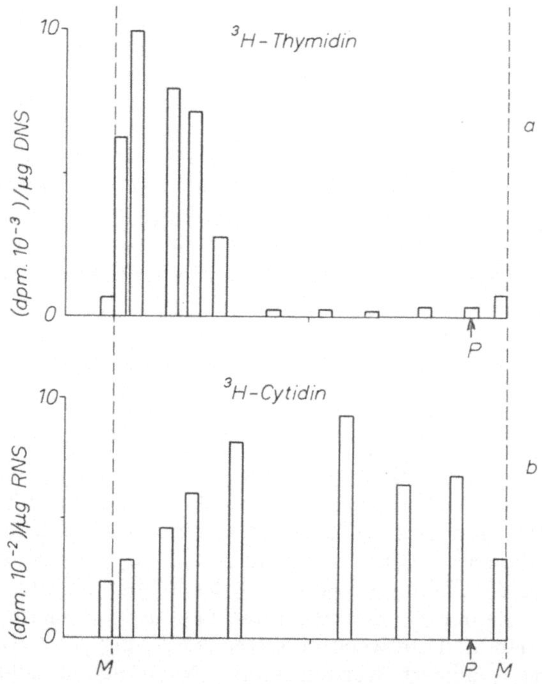

Abb. 2. DNS- und RNS-Synthese in synchronen Plasmodien von *Phys. pol.* [84]. Pulsmarkierung in verschiedenen Stadien des Teilungscyclus mit a) 5 µc/ml ³H-TDR, b) 5 µc/ml ³H-CR. Makroplasmodien wurden in etwa 2 cm² große Segmente geschnitten, in verschiedenen Stadien des Teilungscyclus 15 min auf radioaktivem Medium inkubiert und anschließend mehrmals mit kalter 0,25 M Perchlorsäure sowie mit 8%iger Trichloressigsäure in Aceton (zur Entfernung des gelben Pigments) extrahiert. Die spezifische Radioaktivität der Nucleinsäuren wurde nach Extraktion der Rückstände mit heißer 0,5 N Trichloressigsäure (*DNS*) bzw. nach *RNS*-ase Behandlung (*RNS*), durch Szintillationsmessung und colorimetrische Bestimmung der Desoxyribose und Ribose ermittelt. *M*: Synchrone Mitosen (Telophase); *P*: Prophase. Länge des Teilungscyclus: 9 Std

Teilungs-Synchronie bei Physarum polycephalum

Synchrone Zellsysteme sind aus naheliegenden Gründen besonders gut zum Studium molekularer Vorgänge des Teilungscyclus geeignet. Wir verwenden in unserem Laboratorium vielkernige Plasmodien von *Physarum polycephalum* (Klasse: Myxomycetae), die natürlich synchrone Kernmitosen durchführen [*44*]. Diese Plasmodien können frei von Fremdorganismen auf einer mit flüssigem Nährmedium [*14, 15*] getränkten Filterpapierunterlage als scheibenförmige Syncytien gezüchtet werden (Abb. 1) [*15, 76*]. Die Syncytien bestehen aus einem dichten Netzwerk kommunizierender Plasmastränge und enthalten Tausende von Zellkernen, die sich bei 26° C etwa alle 9 Std vollsynchron durch Mitose teilen. In vieler Hinsicht repräsentiert ein derartiges Makroplasmodium eine riesige Einzelzelle, die z. B. bei einem Durchmesser von 6 cm 30 mg Protein, 3 mg Ribonucleinsäure (RNS) und 0,3 mg DNS enthält. Die Verdoppelung der DNS in den Kernen (Abb. 2a) erfolgt ebenfalls synchron, und zwar während des ersten Drittels der Interphase unmittelbar im Anschluß an die Telophase [*76, 84*]. Im Teilungscyclus dieses Organismus fehlt demnach eine G_1-Periode. RNS wird während des gesamten Zellcyclus synthetisiert (Abb. 2b), jedoch mit einem ausgeprägten Minimum während der Mitose [*84, 87*]. Ein zweites Minimum der RNS-Syntheserate wird von MITTERMAYER, BRAUN und RUSCH [*69*] um die Mitte der Interphase angenommen. Die natürliche Synchronie des Systems erlaubt es, Hemmstoffe in definierten Stadien des Teilungscyclus zu applizieren und deren Effekte auf bestimmte Stoffwechselprozesse sowie auf den Mitoserhythmus zu studieren. Besonders vorteilhaft ist hierbei die Größe der Plasmodien, die es gestattet, ausreichend Material für biochemische Untersuchungen zu gewinnen, das bestimmte Stadien des Teilungscyclus repräsentiert.

Hemmung der DNS-Synthese

Die Verdoppelung des DNS-Gehaltes vor der Zellteilung scheint generell eine Voraussetzung für den normalen Ablauf einer Mitose zu sein. In Übereinstimmung mit dieser allgemeinen Regel führt eine Hemmung der DNS-Synthese mit 5-Fluor-2'-desoxyuridin (FUDR) [*18*] bei *Physarum polycephalum* zu einer Hemmung des Beginns der nächsten Mitose [*85*]. FUDR blockiert nach Phosphorylierung zum Monophosphat das Enzym Thymidylatsynthetase [*13, 37, 39*] und damit die de-novo-Synthese der Thymidylsäure. Hierdurch entsteht ein „Thymin-Mangel", der zum Erliegen der DNS-Synthese führt. Die Spezifität von FUDR wird etwas beeinträchtigt dadurch, daß der Hemmstoff durch eine Nucleosidphosphorylase teilweise zu 5-Fluoruracil (FU) bzw. 5-Fluoruridin (FUR) umgewandelt wird. Letztere führen durch Einbau in RNS zur Bildung abnormaler RNS-Moleküle und vermindern auch die RNS-Synthesegeschwindigkeit [*9*]. Der Nebeneffekt auf die RNS-Synthese kann bei *Physarum polycephalum* durch Zusatz von Uridin, das als kompetitiver Hemmstoff der Nucleosidphosphorylase wirkt [*7*], weitgehend ausgeschaltet werden. Abb. 3 zeigt, daß FUDR in Gegenwart von Uridin erwartungsgemäß nur bei Zugabe vor oder während der S-Periode die nächste Mitose hemmt. Zugabe des Hemmstoffes während der G_2-Periode beeinflußt die nächste Mitose nicht, wohl aber die zweite Mitose. Unter den gewählten Bedingungen ist die DNS-Synthese nicht völlig blockiert (Abb. 4), weshalb auch bei dauernder

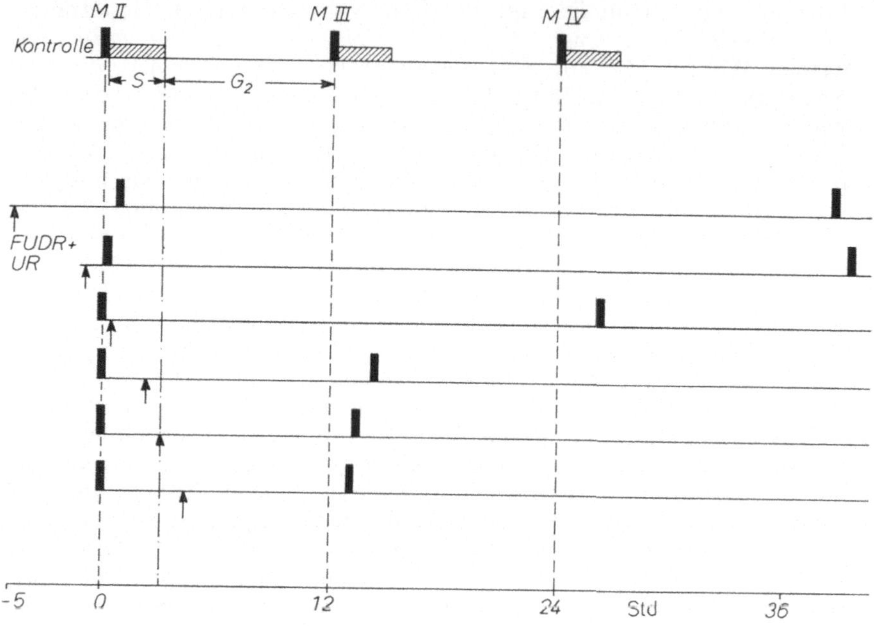

Abb. 3. Wirkung von $FUDR$ auf die synchrone Mitose von $Phys.$ $pol.$ (vgl. [85]). ↑: Zugabe von $2 \cdot 10^{-5}$ M $FUDR$ + $4 \cdot 10^{-4}$ M UR. ▮: Synchrone Mitosen

Gegenwart von FUDR schließlich der DNS-Gehalt die normale Höhe erreicht und anschließend eine Mitose erscheint.

Wie weiter aus Abb. 4 zu ersehen ist, genügt bereits eine kurzzeitige Unterbrechung der DNS-Synthese, um den Beginn der nächsten Mitose zu verzögern. In diesem Versuch wurde nach 4 Std Einwirkung von FUDR durch Zusatz von Thymidin die DNS-Synthese wieder in Gang gesetzt. Die nächste Mitose erschien unter diesen Bedingungen mit einer Verzögerung von ebenfalls etwa 4 Std. In weiteren Versuchen dieser Art [85] ergab sich eine gute Korrelation zwischen der Dauer der DNS-Synthesehemmung im Bereich von 0,5 bis 10 Std und der Verzögerung des Beginns der nächsten Mitose, was für eine weitgehende Konstanz der G_2-Periode spricht. DNS-Synthese und Mitose sind demnach zumindest bei $Physarum$ $polycephalum$ zeitlich streng koordiniert, obwohl beide Vorgänge durch eine relativ lange G_2-Periode getrennt sind. Es wäre denkbar, daß die DNS-Synthese einen geschwindigkeitsbestimmenden Schritt in einer Kette von Prozessen darstellt, die mit der Auslösung der Mitose endet. Jede Störung der DNS-Synthese unterbricht oder verändert diese Sequenz und beeinflußt dadurch auch die nächste Mitose.

Es erhebt sich die Frage, ob die straffe Koordinierung von DNS-Synthese und Mitose einem allgemeinen Prinzip entspricht oder ob die Verhältnisse bei *Physarum polycephalum* einen Spezialfall darstellen. Hierzu sollen einige Ergebnisse aus der Literatur diskutiert werden.

LARK [56] beobachtete in Hemmversuchen mit Desoxyadenosin an einem künstlich synchronisierten Bakteriensystem mit stufenweiser

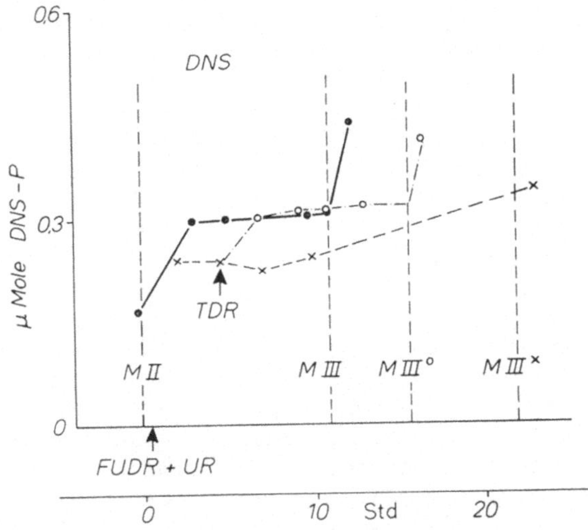

Abb. 4. Hemmung der *DNS*-Synthese in synchronen Plasmodien von *Phys. pol.* durch *FUDR + UR* und Wiederaufhebung des Syntheseblocks durch *TDR* [85]. 16 identische Makroplasmodien wurden gleichzeitig aus asynchronen Suspensionskulturen hergestellt (Methodik: [76, 84, 85]). 6 Kulturen dienten als Kontrollen, während 10 Kulturen 30 min nach Mitose (*II*) auf Medium mit $2 \cdot 10^{-5}$ M *FUDR* + $4 \cdot 10^{-4}$ M *UR* übertragen wurden. Jeder Meßpunkt entspricht dem *DNS*-Gehalt eines gesamten Plasmodiums, das zur angegebenen Zeit isoliert und analysiert wurde. (*M III*): Mitose in den Kontrollen; (*M III°*): Mitose in den mit *FUDR* behandelten Kulturen, die bei ↑ auf frisches Medium mit $2 \cdot 10^{-4}$ M *TDR* übertragen wurden; (*M III×*): Mitose in den kontinuierlich mit *FUDR* behandelten Kulturen

DNS-Synthese eine simultane Hemmung von DNS-Synthese und Zellteilung, die für eine funktionelle Koppelung beider Prozesse spricht. Diese Koppelung scheint jedoch nicht so straff zu sein wie bei *Physarum polycephalum*, da nach Aufhebung des DNS-Syntheseblocks die Teilung zeitlich weitgehend unabhängig von der DNS-Synthese erfolgt. Mitomycin C hemmt nach WILLIAMSON und SCOPES [104] die Zellteilung in synchronisierten Hefekulturen, sofern die Substanz während der DNS-Syntheseperiode zugegen ist. Da unter diesen Bedingungen die DNS-Synthese als solche nicht gehemmt ist, wird angenommen, daß in Gegenwart von Mitomycin C eine abnormale DNS entsteht, die ihren funktionellen Einfluß auf die Zellteilung nicht ausüben kann. In synchronisierten *Tetrahymena*-Kulturen kann nach CERRONI und ZEUTHEN [12] durch Hemmung der DNS-Synthese mit FUDR kein Effekt auf die ersten beiden Zellteilungen erzielt werden. Im Gegensatz hierzu hemmt FUDR die Teilung logarithmisch wachsender *Tetrahymena*-Zellen bereits vor Ablauf eines Generationscyclus. Diese Diskrepanz wird dadurch erklärt, daß während der Synchronisierung durch Temperaturschockbehandlung [*81, 90, 91*] die DNS-Synthese weiterläuft und die sich bis zum Ende der Behandlung im Überschuß ansammelnde DNS für mindestens zwei Teilungen ausreicht.

Experimente mit Zellsystemen höherer Organismen deuten ebenfalls auf eine strenge Korrelation von DNS-Synthese und Mitose hin. So schließen SISKEN und KINOSITA [*94*] aus ihren Beobachtungen an Säugetierzellen in der Gewebekultur auf eine Konstanz der S- und G_2-Phasen, bei weitgehender Variabilität der G_1-Phase. TAYLOR beobachtete in Wurzelspitzen von *Vicia faba* einen scharfen Abfall der Mitoserate etwa 2—3 Std nach Zugabe von FUDR [*100*] oder Amethopterin [*99*]. Die Ergebnisse sprechen dafür, daß nur jene Zellen in die Mitose eintreten können, die vor Zugabe der Hemmstoffe die DNS-Synthese beendet haben oder im allerletzten Abschnitt der S-Phase stehen. Die in Gegenwart der Hemmstoffe zuletzt erscheinenden Mitosen zeigen Chromosomenbrüche, die auf eine unvollständige DNS-Synthese in einzelnen, spät replizierenden Chromosomenabschnitten schließen lassen. Dies stimmt gut mit der Annahme überein, daß die Auslösung der Mitose weniger vom Beginn als vom Ende der DNS-Synthese diktiert wird.

Eine scheinbare Ausnahme der Regel, daß eine Mitose nur nach Beendigung der vorausgehenden DNS-Synthese erfolgen kann, wird von LINDNER [*62*] beschrieben. Hiernach können sich Ehrlich-Ascitestumorzellen in Gegenwart von FU trotz gehemmter DNS-Synthese und ohne Änderung des Ploidiegrades teilen. Da der DNS-Gehalt pro Zelle bei diesem Tumor 2—4mal höher liegt als in normalen diploiden Mäusezellen, hat es den Anschein, als ob in unbehandelten Tumorzellen die Zellteilung hinter der DNS-Synthese nachhinkt. Offenbar ist in diesen Zellen ein auf die DNS-Synthese folgender Schritt (eventuell die Verdoppelung der Kinetochorenregion) geschwindigkeitsbestimmend für die Auslösung der Zellteilung. Es ist bekannt, daß Tumorzellen häufig polytaene oder polyneme Chromosomen enthalten [*5, 6*], was den unveränderten Ploidiegrad nach einer Teilung in Gegenwart von FU erklären könnte.

Trotz einer gewissen Variabilität der Korrelation von DNS-Synthese und Zellteilung weisen die vorstehend diskutierten Befunde aus der Literatur, ebenso wie die Versuche mit *Physarum polycephalum*, auf eine grundlegende funktionelle Beziehung zwischen diesen beiden Vorgängen hin, derzufolge eine Zelle, die einmal mit der DNS-Synthese begonnen hat, in den meisten Fällen auch zur Teilung schreitet. Der Zeitpunkt der Teilung wird offenbar durch eine Sequenz von Stoffwechselschritten determiniert, die gegen Ende der DNS-Replikation in Gang gesetzt wird. Dieser regulative Einfluß der DNS-Synthese auf die Zellteilung könnte so verstanden werden, daß ein für die Steuerung der Zellteilung wesentlicher Abschnitt des Genoms erst nach dessen Replikation funktionell aktiv wird. Dieser Genomanteil liegt möglicherweise in Chromosomenabschnitten, die erst am Ende der S-Phase replizieren. Im Einklang hiermit stehen Beobachtungen über eine sequentielle Replikationsweise des genetischen Materials bei verschiedenen Organismen [7a, 11a, 49, 57, 67a, 73, 98] einschließlich *Physarum polycephalum* [10].

Die Tatsache, daß z. B. bei Zellen in ruhenden tierischen Geweben [25] sowie in anderen speziell gelagerten Fällen (Polytaenie) nach Abschluß der DNS-Synthese nicht notwendigerweise auch eine Zellteilung erfolgen muß, läßt darauf schließen, daß die von der DNS-Synthese kontrollierte Sequenz von Stoffwechselschritten unterbrochen bzw. modifiziert werden kann. Die Annahme einer Konstanz der G_2-Periode bezieht sich demnach streng genommen lediglich auf die Mindestlänge dieser Periode.

Die relative Länge der Interphasen-Abschnitte G_1, S und G_2 variiert erheblich unter der Vielzahl von Organismen und Zelltypen (vgl. Übersicht bei MAZIA [65]). Dies läßt vermuten, daß die zur Zellteilung führenden Stoffwechselschritte in gewissen Grenzen zeitlich verschieden korreliert sein können. *Physarum polycephalum* stellt insofern einen Spezialfall dar, als hier überhaupt keine G_1-Phase existiert. Hier scheint nicht nur die DNS-Synthese den Zeitpunkt der nächsten Mitose zu bestimmen, sondern auch umgekehrt die Mitose den Beginn der DNS-Synthese zu diktieren. Bisher ist es noch nicht gelungen, durch irgendwelche Eingriffe bei diesem Organismus den Zeitpunkt der Mitose zu verschieben ohne auch gleichzeitig den Beginn der DNS-Synthese entsprechend zu beeinflussen. Bei anderen Organismen erfolgt die Auslösung der DNS-Synthese offenbar mehr oder weniger unabhängig von der vorausgehenden Zellteilung. Hierfür sprechen u. a. die bereits erwähnten Beobachtungen über eine variable G_1-Phase bei Säugetierzellen [94] sowie die z. B. bei *Tetrahymena* durch Temperaturschockbehandlung mögliche Dissoziation des Teilungscyclus vom DNS-Synthesecyclus [12].

Hemmung der RNS- und Proteinsynthese

Die mehr oder weniger strenge Korrelation zwischen DNS-Synthese und Mitose läßt vermuten, daß beide Vorgänge nach einem gemeinsamen Prinzip reguliert werden. Es mehren sich die Anzeichen, daß die DNS-Synthese durch Genregulation gesteuert wird [23, 29, 60, 61, 92] (vgl. Übersicht von BASERGA [2]). Dies würde bedeuten, daß ein bestimmter Abschnitt des Genoms durch Bildung von messenger-RNS und Induktion bestimmter Enzymproteine die Selbstverdoppelung der DNS kontrolliert. Derselbe Mechanismus könnte auch bei der Steuerung der Zellteilung wirksam sein. Die zeitliche Koordinierung von DNS-Synthese und Zellteilung könnte dadurch geschehen, daß jene Genabschnitte, die zur Steuerung der beiden Vorgänge dienen, funktionell gekoppelt sind (siehe vorstehendes Kapitel).

Die Frage, inwieweit die Länge des Zellcyclus und damit der Zeitpunkt einer Mitose durch Genregulation bestimmt wird, kann u. a. mit Hilfe von Hemmstoffen der RNS- und Proteinsynthese untersucht werden. Zumindest ergeben sich auf diese Weise Anhaltspunkte dafür, bis zu welchem Zeitpunkt im Zellcyclus RNS und Protein zur Vorbereitung einer Teilung synthetisiert werden müssen. Zur Hemmung der RNS-Synthese bei *Physarum polycephalum* verwendeten wir das Antibioticum Actinomycin D, das bei zahlreichen Organismen und in vitro spezifisch die Synthese DNS-abhängiger RNS durch Komplexbildung mit Guanin-haltigen Abschnitten der DNS unterdrückt [*30, 31, 35, 36, 38, 47, 54, 82,*

Tabelle 1. *Hemmung der RNS-Synthese und des Mitosebeginns durch verschiedene Konzentrationen von Actinomycin D* [87]

Actinomycin D (γ/ml)	0,11		0,55	0,85	Cyclusstadium[1]
100	32		25	53	Hemmung der RNS-Syntheserate (%)[2]
200	[3]		82	66	
100	>3[5]	<15	0[4]	0[4]	Verzögerung des Mitosebeginns (Std)
200	[3]		>14	0,3	

[1] Stadium des Teilungscyclus zum Zeitpunkt der Hemmstoffzugabe. Gesamtlänge des Cyclus von Telophase zu Telophase = 1,0 (entsprechend 8,7 Std).

[2] % Hemmung des Einbaues von ^{3}H-CR in die RNS 30 min nach Hemmstoffzugabe.

[3] nicht untersucht.

[4] Dauer der Metaphase 1—2 Std (Kontrolle $\sim$ 10 min).

[5] Kerne in Metaphase arretiert.

In früher, mittlerer und später Interphase wurde je ein Makroplasmodium einer „empfindlichen" Linie von *Phys. pol.* in 3 gleich große Sektoren geteilt und letztere auf frisches Medium mit 0 (Kontrolle), 100 und 200 γ/ml Actinomycin D übertragen. Nach 30 min wurden von allen 3 Sektoren etwa 2 cm^2 große Stücke abgetrennt. 15 min auf entsprechenden radioaktiven Medien mit 2 μc/ml ^{3}H-CR inkubiert und anschließend die spezifische Radioaktivität der RNS ermittelt (vgl. Legende zu Abb. 2). Die Wirkung auf den Mitosebeginn wurde in den restlichen, nicht radioaktiven Plasmodienstücken beobachtet.

83]. Im Gegensatz zu tierischen und menschlichen Zellen in der Gewebekultur ist *Physarum polycephalum* relativ unempfindlich gegen Actinomycin D. Zudem variiert die Empfindlichkeit zwischen verschiedenen Unterlinien unseres Stammes. Mit hohen Konzentrationen (100—250 γ/ml) kann jedoch auch bei diesem Organismus eine selektive Hemmung der RNS-Synthese erzielt werden [*86, 87*]. Die Wirkung auf die Kernteilung kommt in einer Verzögerung oder in einer völligen Unterdrückung des Mitosebeginns sowie unter bestimmten Bedingungen auch in einer Arretierung der Kerne in Metaphase zum Ausdruck [*87*]. Das Ausmaß der Mitosehemmung ist sowohl von der Hemmstoffkonzentration als auch vom Zeitpunkt der Hemmstoffzugabe innerhalb des Teilungscyclus abhängig. Zur Erzielung gleicher Effekte sind in mittlerer oder später Interphase höhere Konzentrationen erforderlich als in früher Interphase (Tab. 1, Abb. 5). Abb. 5 B zeigt, daß bei Plasmodien einer „empfindlichen" Linie unseres Stammes mit der höchsten angewandten Konzentration von 250 γ/ml Actinomycin D der Mitosebeginn noch bis 2 Std vor der erwarteten Prophase durch Zugabe des Hemmstoffes verhindert werden kann. Daraus möchten wir schließen, daß zumindest während der ersten $^3/_4$ der Interphase DNS-abhängige RNS zur Vorbereitung der nächsten Mitose gebildet werden muß.

Der kritische Zeitpunkt, bis zu dem durch Zugabe von Actinomycin D die nächste Mitose gehemmt werden kann, variiert stark mit der an-

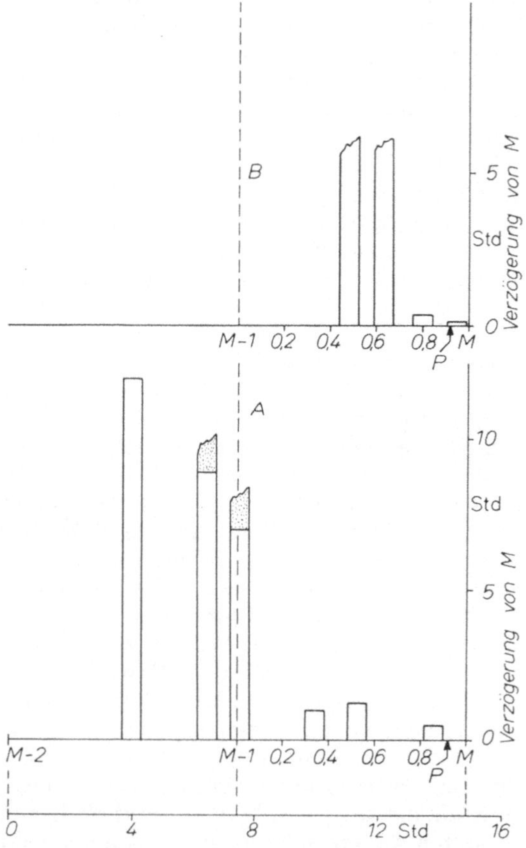

Abb. 5. Wirkung von Actinomycin *D* auf die synchrone Mitose von *Phys. pol.* [*87*]. Abszisse: Zeitpunkt der Hemmstoffzugabe; Ordinate: Verzögerung des Beginns der Mitose (*M*) gegenüber den Kontrollen. A) Einwirkung von 200 γ/ml Actinomycin *D* auf Kulturen einer „wenig empfindlichen" Linie des Organismus; punktierte Säulenabschnitte sollen andeüten, daß ein großer Teil der Kerne nach Eintritt in die Mitose in Metaphase arretiert wurde. B) Einwirkung von 250 γ/ml Actinomycin *D* auf Kulturen einer „empfindlichen" Linie des Organismus. Die Beobachtung wurde 6 Std nach Ablauf der Mitose (*M*) in den Kontrollen beendet. (*M-2*), (*M-1*), (*M*): Synchrone Mitosen in den Kontrollen (Telophase); *P*: Prophase

gewandten Hemmstoffkonzentration (Tab. 1). Dies ist wahrscheinlich darauf zurückzuführen, daß der Hemmstoff nur langsam die Plasmodienmembran durchdringt und erst nach einer gewissen Verzögerung eine wirksame intracelluläre Konzentration erreicht. Bei erhöhter Konzentration im Medium wird die Permeation des Hemmstoffes offenbar beschleunigt. In Übereinstimmung damit kann nach Beobachtungen von Mittermayer, Braun und Rusch [70] die RNS-Synthese in isolierten Zellkernen von *Physarum polycephalum* bereits durch 1 γ/ml Actinomycin D vollständig gehemmt werden.

Eine je nach Hemmstoff-Konzentration mehr oder weniger starke Verzögerung der Wirkung von Actinomycin D auf die Zellteilung wurde u. a. auch bei Gewebekulturen von Säugetier-Zellen [83, 55], an Wurzelzellen von *Allium cepa* [1] sowie an synchronisierten *Tetrahymena*-Zellen [58, 68] beobachtet. Die Autoren geben hierfür unterschiedliche Deutungen. So wird die Möglichkeit diskutiert, daß die zur Vorbereitung einer Teilung benötigte RNS schon in einem frühen Stadium des Zellcyclus, zum Teil bereits vor der vorausgehenden Teilung [71], synthetisiert wird. Die in späteren Stadien des Cyclus durch höhere Actinomycin D-Konzentrationen erzielbare Hemmung sei möglicherweise durch Abbau der in einem früheren Stadium synthetisierten RNS bedingt [58]. Für das Auftreten einer gegen Actinomycin D „empfindlichen" und einer „refraktären" Phase vor der 1. Teilung in synchronisierten *Tetrahymena*-Kulturen werden u. a. allerdings auch Permeabilitätsänderungen der Zellmembran als Ursache diskutiert [74].

Auf Grund unserer Beobachtungen an *Physarum polycephalum* möchten wir annehmen, daß die mitosehemmende Wirkung von Actinomycin D erst dann in Erscheinung tritt, wenn die RNS-Syntheserate unter eine kritische Grenze gesenkt wird. Wie aus Tab. 1 zu ersehen ist, bewirken 100 γ/ml Actinomycin D bei Zugabe zur Mitte der Interphase bereits nach 30 min eine deutliche Hemmung der RNS-Synthese, ohne daß der Beginn der nächsten Mitose beeinflußt wird. Die doppelte Hemmstoffkonzentration, die zu einer sehr viel stärkeren Hemmung der RNS-Synthese führt, unterdrückt hingegen die nächste Mitose völlig. Dies könnte bedeuten, daß die Synthese der zur Vorbereitung einer Mitose benötigten RNS gegen Actinomycin D weniger empfindlich ist als die Synthese anderer RNS-Moleküle. Mit Rücksicht auf Beobachtungen über eine unterschiedliche Synthesehemmung der drei RNS-Typen durch Actinomycin D [78] scheint es denkbar, daß in späteren Stadien des Teilungscyclus zur Vorbereitung einer Mitose zwar keine ribosomale RNS, wohl aber noch messenger-RNS neu synthetisiert werden muß. Die Ergebnisse könnten andererseits auch so gedeutet werden, daß ein Überschuß an mitosespezifischen RNS-Molekülen gebildet wird. Eine partielle Hemmung der RNS-Synthese würde dann die Vorbereitung zumindest der ersten Mitose noch nicht signifikant beeinträchtigen.

Im Einklang mit der Annahme, daß ein überwiegender Teil des Zellcyclus durch Genregulation gesteuert wird, stehen Beobachtungen über die Bedeutung der Proteinsynthese für die Vorbereitung einer Zellteilung [46, 55]. Unter anderem sprechen eingehende Untersuchungen an *Tetrahymena* [22, 34, 40, 79, 81, 102, 106, 107, 108] und *Paramaecium* [80] für die Bildung eines „Teilungsproteins" im Verlauf des Zellcyclus. Dieses Protein scheint einem ständigen „Turnover" zu unterliegen und kann durch Temperaturschock, Anaerobiose oder Hemmung der Proteinsynthese mit Antimetaboliten inaktiviert werden, sofern diese Behand-

lungen vor einem kritischen Zeitpunkt im letzten Viertel des Teilungscyclus („Transition point") erfolgen. Die Temperaturempfindlichkeit
des „Teilungsproteins" bildet die Grundlage für die künstliche Synchronisierung von *Tetrahymena*-Kulturen [*81, 90, 91*] durch eine Serie von
Temperaturschocks.

In Plasmodien von *Physarum polycephalum* kann durch Zusatz von
Puromycin [*8, 72, 75, 97, 105*], das die Proteinsynthese durch eine Blok-

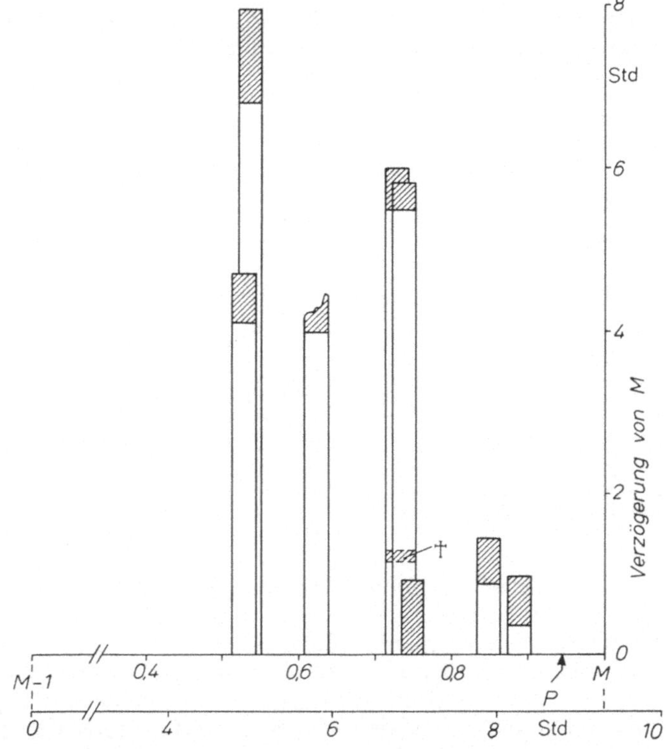

Abb. 6. Wirkung von Puromycin auf die synchrone Mitose von *Phys. pol.* (Orig.). Abszisse:
Zeitpunkt der Hemmstoffzugabe (1 mg/ml); Ordinate: Verzögerung des Beginns der
Mitose (*M*) gegenüber den Kontrollen. Die Synchronie der mit Puromycin behandelten
Kulturen war durchwegs schlechter als in den Kontrollen; die schraffierten Säulenabschnitte
geben den Zeitraum an, in welchen sich die Kerne verschiedener Plasmodienfelder teilten.
†: in einem kleinen Feld teilten sich die Kerne bereits wesentlich früher als im Hauptteil
der Kulturen. (*M-1*), (*M*): Synchrone Mitosen in den Kontrollen (Telophase);
P: Prophase

kierung des Peptisierungsschrittes an den Ribosomen hemmt [*105*],
noch bis zu einem relativ späten Stadium des Teilungscyclus die Vorbereitung einer synchronen Mitose unterbrochen werden (Abb. 6).
Die kritische Grenze für die mitosehemmende Wirkung von Puromycin
liegt bei diesem Organismus etwa 2 Std vor Prophase, also etwa am selben
Punkt der G_2-Periode, bis zu dem auch Actinomycin D den Mitosebeginn
verhindern kann (vgl. Abb. 5). Die Zeitspanne von 2 Std muß als ein

Maximalwert angesehen werden. Puromycin ist bei *Physarum polycephalum*, wahrscheinlich infolge Permeationsschwierigkeiten, nur in hohen Konzentrationen wirksam, so daß mit einem verzögerten Wirkungseintritt, ähnlich wie bei Actinomycin D, zu rechnen ist.

Die Versuche mit Hemmstoffen der RNS- und Proteinsynthese weisen übereinstimmend darauf hin, daß vor jeder Teilung bestimmte Proteine synthetisiert werden müssen. Diese Proteine können offenbar nur zur Vorbereitung *einer* Zellteilung dienen und müssen vor der nächsten Teilung erneut gebildet werden. Über die Natur und die Funktion dieser Proteine ist noch so gut wie nichts bekannt. Wahrscheinlich handelt

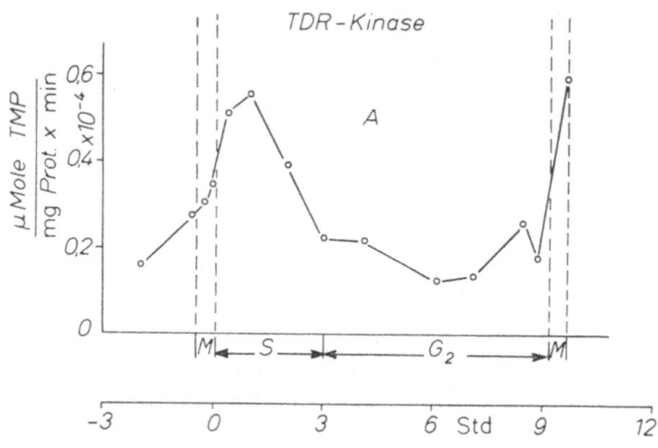

Abb. 7. Aktivität der Thymidinkinase in verschiedenen Stadien des Teilungscyclus von Phys. pol. [*88*]. Zur Enzymbestimmung [*11*] wurden die Überstände von Homogenaten ganzer Plasmodien nach 30 min Zentrifugation bei 120 000 g verwendet

es sich um eine größere Gruppe „mitosespezifischer" Proteine, zu denen u. a. auch das am Aufbau des mitotischen Apparates [*50, 63, 64, 65, 66, 67, 109*] beteiligte Spindelprotein zählen dürfte. Die Funktion dieser Proteine tritt offenbar nur periodisch in Erscheinung. Dies wäre mit der Annahme gut vereinbar, daß die Bildung dieser Proteine durch periodische Induktions- und Repressionsvorgänge auf genetischer Ebene gesteuert wird.

Rhythmische Änderungen von Enzymaktivitäten im Verlauf des Teilungscyclus, die auf einem derartigen Regelmechanismus beruhen könnten, sind in einigen Fällen beobachtet worden. So berichten HOTTA und STERN [*41, 42, 43*] über eine mit der DNS-Synthese streng korrelierte Periodizität der Thymidinkinase in Mikrosporen von Blütenpflanzen, die alle Anzeichen eines durch Genregulation [*48*] gesteuerten Vorganges trägt. Die Länge des Zellcyclus beträgt bei diesem System allerdings mehrere Wochen, so daß es sich hierbei auch um einen speziellen Fall von „Langzeitregulation" handeln könnte. Wie Abb. 7 zeigt, erfolgt jedoch auch bei *Physarum polycephalum*, das mit einer Cycluslänge von 9 Std ein relativ rasch wachsendes System darstellt, ein periodischer Anstieg und Wiederabfall der Thymidinkinase-Aktivität im Rhythmus der synchronen Kernteilungen. Bei befruchteten Kröten- und Seeigeleiern,

die einen noch weit kürzeren Teilungscyclus aufweisen, konnten Du-
SPIVA und HANSEN-DELKESKAMP [*19*] ebenfalls rhythmische Aktivi-
tätsänderungen der Thymidinkinase beobachten. Diese Beobachtungen
sprechen dafür, daß die Aktivität der Thymidinkinase nach einem für
verschiedene Organismen gemeinsamen Regulationsprinzip kontrolliert
wird. Die bereits mehrfach diskutierte Korrelation von DNS-Synthese
und Zellteilung legt die Vermutung nahe, daß die Aktivitäten „mitosespezifischer" Enzyme in ähnlicher Weise periodisch reguliert werden,
wie die Aktivität des am Nucleinsäurestoffwechsel beteiligten Enzyms
Thymidinkinase. Dies könnte vielleicht als Hinweis für die Identifizierung dieser noch unbekannten Enzyme dienen.

Mitosehemmung durch UV-Strahlung

Unter den biologischen Effekten der UV-Strahlung steht die Hemm-
wirkung auf die Zellteilung besonders im Vordergrund (vgl. Übersicht
von GIESE [*28*]). Zum Unterschied von den vorstehend besprochenen
Synthesehemmstoffen ist der Mechanismus der teilungshemmenden
Wirkung von UV-Strahlung noch nicht restlos geklärt. Viele Befunde
deuten jedoch darauf hin, daß eine Wechselwirkung mit den Nucleinsäuren, wahrscheinlich mit der DNS, als primäre Ursache für die Teilungs-
hemmung sowie auch für andere biologische Effekte der UV-Strahlung,
verantwortlich ist. So ist das Aktionsspektrum der teilungshemmenden
Wirkung identisch mit dem UV-Absorptionsspektrum der Nucleinsäuren [*26, 16*]. Die Empfindlichkeit der Biosynthese von Makromole-
külen nimmt in der Reihenfolge DNS, RNS, Protein ab [*16, 93*]. Allerdings wird eine Hemmung der Zellteilung bei Bakterien bereits durch
Strahlendosen erzielt, die noch keinen nachweisbaren Effekt auf die
DNS-Synthese zeigen [*16*]. Die durch UV-Bestrahlung verursachte Tei-
lungshemmung wird ebenso wie die mutagene und cancerogene Wirkung
[*53*] durch sichtbares Licht teilweise wieder aufgehoben. Diese Photo-
reaktivierung [*52*] wird auf die Wirkung eines Enzyms zurückgeführt,
das unter Lichteinfluß die durch UV veränderte DNS wieder restauriert.
Die primären strahlenchemischen Veränderungen an der DNS könnten
nach den Untersuchungen von BEUKERS et al. [*3, 4*] sowie von WACKER
et al. [*101*] auf einer Dimerisierung von Basenbausteinen, insbesondere
von Thymin, beruhen.

Die mitosehemmende Wirkung von UV-Strahlung bei *Physarum
polycephalum* hängt in charakteristischer Weise vom Zeitpunkt der Bestrahlung im Teilungscyclus ab (Abb. 8). Die mit einer konstanten UV-
Dosis erzielbare Mitoseverzögerung ist am stärksten bei Bestrahlung
etwa um die Mitte der Interphase, also bereits nach Abschluß der DNS-
Synthese. Dies spricht dafür, daß der Effekt auf die Mitose weniger auf
einer Hemmung der DNS-Synthese als vielmehr auf der Störung eines
bis weit in die G_2-Periode reichenden Vorganges zurückzuführen ist.
Dies schließt jedoch nicht aus, daß primär eine Wechselwirkung mit der
DNS, auch nach Abschluß der DNS-Replikation, für die Mitosehemmung
verantwortlich ist. Argumente hierfür sind die Photorevertierbarkeit des

Strahleneffektes [89] sowie Befunde, nach denen durch Einbau von
5-Brom-2′-desoxyuridin (BUDR) in die DNS synchroner Makroplasmo-

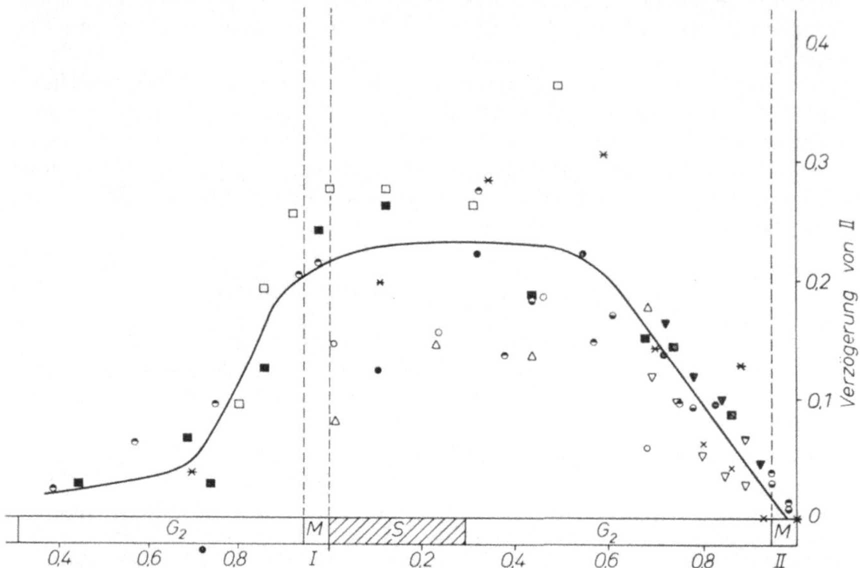

Abb. 8. Wirkung von UV-Strahlung auf die synchrone Mitose in *Phys. pol.* (Orig.). Makroplasmodien wurden in 4 bis 8 gleich große Sektoren (2—4 cm²) getrennt und die Teilstücke
in verschiedenen Stadien des Cyclus mit einer konstanten Dosis UV-Licht (2600 erg/mm²;
λ max: 258 mμ) bestrahlt. Ein Sektor diente jeweils als Kontrolle. Zusammenstellung der
Ergebnisse von 4 getrennt durchgeführten Versuchen. Gleiche Symbole kennzeichnen Teilstücke einer Ausgangskultur. Abszisse: Zeitpunkt der Bestrahlung; Ordinate: Verzögerung
des Beginns der Mitose (*M II*) gegenüber den Kontrollen. (*M I*), (*M II*): Synchrone
Mitosen in den Kontrollen (Prophase bis Telophase)

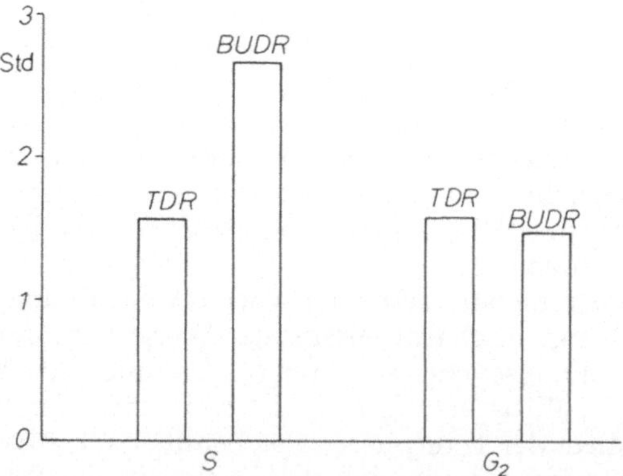

Abb. 9. Strahlensensibilisierende Wirkung von *BUDR*. (Orig.) Je ein Plasmodium wurde
in der *S*-Phase bzw. während 3 Std der G_2-Phase in Gegenwart von $2 \cdot 10^{-4}$ M *BUDR* +
+ $2 \cdot 10^{-5}$ M *FUDR* + $4 \cdot 10^{-4}$ M *UR* inkubiert und anschließend auf normalem Medium
weitergezüchtet. Zu Beginn der nächsten G_2-Phase wurden beide Kulturen sowie 2 Kontroll-
plasmodien, die anstelle von *BUDR* unter sonst gleichen Bedingungen mit $2 \cdot 10^{-4}$ M
TDR inkubiert worden waren, mit UV-Licht (1750 erg/mm²; λ max 258 mμ) bestrahlt.
Die Höhe der Säulen gibt die Verzögerung der nächsten Mitose gegenüber unbestrahlten
Kulturen an

dien während der S-Periode der mitoseverzögernde Effekt einer in der G_2-Periode applizierten UV-Dosis verstärkt wird (Abb. 9 [86]). Die radiosensibilisierende Wirkung von BUDR [17, 20, 21, 32, 33, 51, 77, 95, 96] ist nach SZYBALSKI [96] ein starkes Argument dafür, daß UV-Strahlung primär mit der DNS in Wechselwirkung tritt. Inwieweit daneben auch Reaktionen mit anderen Zellkomponenten (RNS, Protein) an

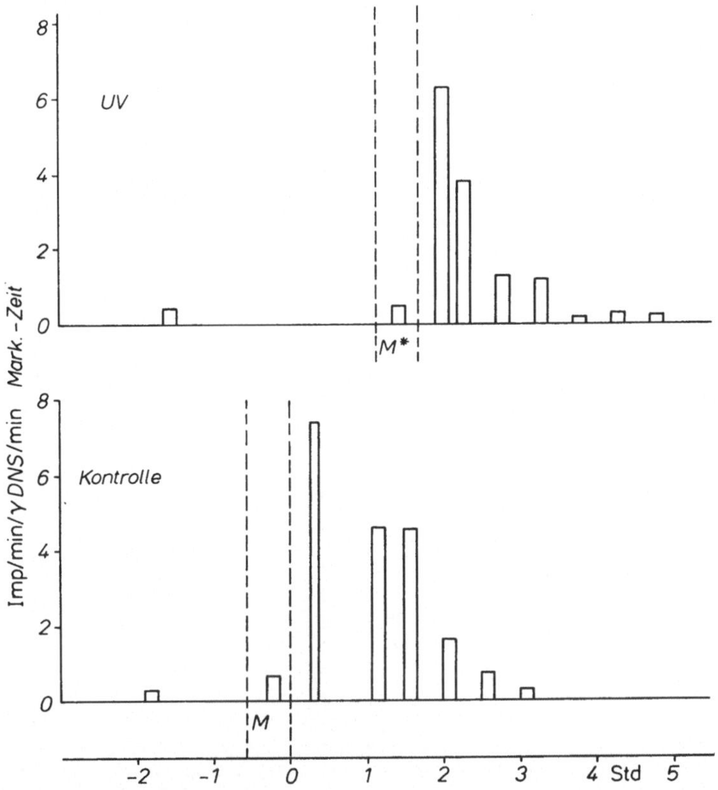

Abb. 10. Wirkung von UV-Strahlung auf die *DNS*-Synthese in *Phys. pol.* (Orig.) Pulsmarkierung (10 min) mit 5 μC/ml ^{3}H-TDR (vgl. Abb. 2). A) Kontrollen; B) 5,4 Std vor der erwarteten Mitose mit UV-Licht (2600 erg/mm^2; λ max 258 mμ) bestrahlt. M: Synchrone Mitose in den Kontrollen (Prophase bis Telophase); M^*: verzögerte Mitose in den mit UV bestrahlten Kulturen

der teilungshemmenden Wirkung von UV-Strahlung beteiligt sind, muß dahingestellt bleiben. In diesem Zusammenhang sei auf die Untersuchungen von GAULDEN und PERRY [24] hingewiesen, in denen durch UV-Bestrahlung des Nucleolus von Heuschrecken-Neuroblasten mittels eines UV-Mikrostrahlenbündels eine bevorzugte Hemmung der Zellteilung erzielt wurde. Da Nucleolen neben RNS und Protein auch DNS (vgl. R. LETTRÉ [59]) enthalten, kann hieraus jedoch nicht auf die chemische Natur des empfindlichen Trefferbereiches geschlossen werden.

Gegen Ende der Interphase von *Physarum polycephalum* ist ein starker Abfall der UV-Empfindlichkeit hinsichtlich der nächsten Mitose zu verzeichnen, der zeitlich mit einem Abfall der RNS-Syntheserate (Abb. 2) parallel geht. Andererseits nimmt der Effekt auf die zweite folgende Mitose durch Bestrahlung gerade in dieser Phase stark zu. Dies könnte so gedeutet werden, daß eine Schädigung der DNS am Ende der G_2-Periode die nächste Mitose nur mehr wenig beeinflußt, da die zur Vorbereitung

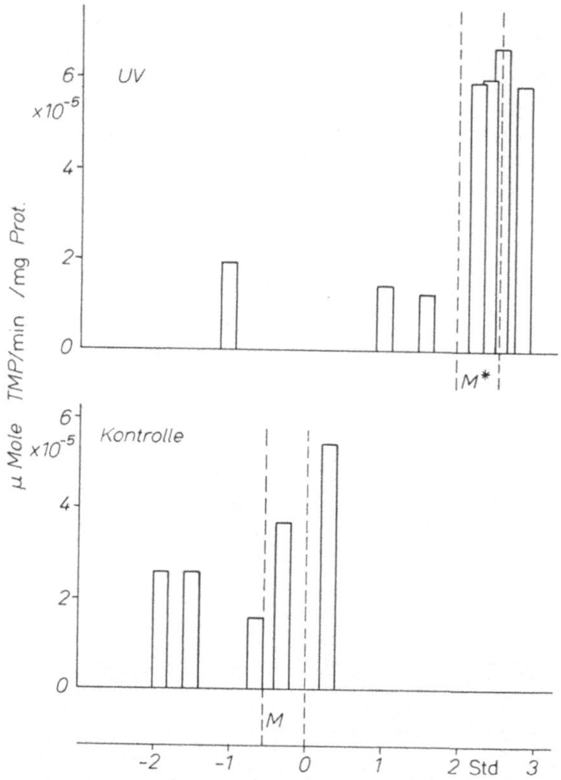

Abb. 11. Wirkung von UV-Strahlung auf die Thymidinkinase-Aktivität von *Phys. pol.* (Orig.) A) Kontrollen (vgl. Abb. 7). B) 5,5 Std vor der erwarteten Mitose mit UV-Licht (2600 erg/mm²; λ max 258 mµ) bestrahlt. *M*: Synchrone Mitose in den Kontrollen (Prophase bis Telophase); *M**: verzögerte Mitose in den mit UV bestrahlten Kulturen

dieser Mitose benötigte RNS bereits größtenteils fertig gebildet ist. Die geschädigte DNS-Matrize wird jedoch in den nächsten Teilungscyclus übernommen und stört dann die zur Vorbereitung der zweiten Mitose notwendige RNS-Synthese. UV-Bestrahlung von *Physarum polycephalum* führt nicht nur zu einer Verzögerung des Mitosebeginns, sondern auch zu einer entsprechenden Verschiebung der DNS-Syntheseperiode (Abb. 10) und des Aktivitätsanstieges der Thymidinkinase (Abb. 11). Die gleichsinnige Beeinflussung dieser drei periodischen Vorgänge durch

UV-Strahlung spricht erneut für eine gekoppelte Regulation von DNS-Synthese und Mitose durch periodische Genaktivierung.

Zusammenfassung

Zur Vorbereitung einer Mitose sind bei *Physarum polycephalum* und auch bei einer Reihe anderer Organismen in verschiedenen Stadien des Zellcyclus die Synthese von DNS, RNS und Protein erforderlich. Versuche mit Hemmstoffen und mit UV-Strahlen deuten darauf hin, daß eine strenge zeitliche Korrelation zwischen DNS-Synthese und Mitose besteht, die allerdings zwischen verschiedenen Zelltypen variiert und unter dem Einfluß von Milieufaktoren gelegentlich modifiziert werden kann. Wenn es auch noch kaum direkte Hinweise auf die Existenz und die Natur von spezifisch mit der Regulation der Zellteilung verknüpften Prozessen gibt, so sprechen doch einige Argumente für die Annahme, daß der Rhythmus des Zellcyclus durch Genregulation gesteuert wird. Unter Berücksichtigung dieser vorstehend besprochenen Argumente sei daher folgende Arbeitshypothese zur Diskussion gestellt.

Die Auslösung einer Zellteilung ist das Endergebnis mehrerer Prozesse, die teils in Sequenz, teils parallel zueinander im Zellcyclus ablaufen (vgl. MAZIA [65]). Die Länge des Teilungscyclus und damit der Zeitpunkt der Teilung wird durch eine Folge von Stoffwechselschritten bestimmt, in der die Replikation der DNS geschwindigkeitsbestimmend wirkt. Durch die Auslösung der DNS-Synthese wird demnach die nächste Teilung determiniert. Die zeitliche Koordinierung der zur Teilung führenden Prozesse, einschließlich der DNS-Replikation, geschieht durch periodische Induktion und Repression bestimmter Enzyme. Jene Enzyme, die zur Vorbereitung der Zellteilung während der G_2-Periode benötigt werden, können offenbar erst dann induziert werden, wenn die Synthese der DNS bzw. eines am Ende der S-Phase replizierenden DNS-Abschnittes, beendet ist. Der Ablauf des Zellcyclus würde nach diesem Modell durch eine stufenweise Aktivierung und Inaktivierung bestimmter Gengruppen gesteuert werden, in der Weise, daß das Signal zum nächsten Schritt erst nach Beendigung des vorausgehenden Schrittes gegeben werden kann.

Summary

The role of DNA-, RNA- and protein synthesis on the timing of synchronous mitosis in *Physarum polycephalum* was studied by means of metabolic inhibitors and UV-light. The myxomycete *Physarum polycephalum* forms multinuclear plasmodia exhibiting naturally synchronous mitoses every 8 to 10 hours at 26° C. S-period lasts for about 3 hours immediately following telophase. There is no G_1-period.

Treatment with FUDR + UR during S-period inhibits or delays the onset of the next mitosis. The length of the delay corresponds to the length of inhibition of DNA-synthesis, suggesting that a constant minimum G_2-period has to elaps prior to mitosis. It appears that DNA synthesis functions as a rate limiting step in the preparation for mitosis.

Inhibition of RNA synthesis with 250 γ/ml actinomycin D prevents the next mitosis if added to the medium at least 2 hrs prior to prophase corresponding to about 70% of the total length of interphase. The critical point within the cell cycle, however, up to which addition of actinomycin D interfers with preparation of the next mitosis depends largely on the concentration of the inhibitor. At intermediate levels (100 γ/ml) RNA synthesis may be partially inhibited without affecting the onset of the following mitosis. This could mean that formation of particular RNA molecules pertinent to the control of mitosis is somewhat less sensitive to treatment with actinomycin D than the formation of other RNA molecules. Inhibition of protein synthesis with puromycin (1 mg/ml) also delays the onset of the next mitosis if the substance is added during a period of the division cycle which roughly coincides with the actinomycin D-sensitive phase.

Irradiation with UV-light causes maximum delay of mitosis when applied during a portion of the cell cycle ($\sim 60\%$) which includes S-period and the first half of G_2-period. During the second half of G_2-period UV sensitivity gradually drops to zero. Mitotic delay after UV irradiation is accompanied by a corresponding delay of DNA synthesis and of the increase of thymidinekinase activity. The latter normally rises sharply just prior to and during mitosis.

These results as well as related data from the literature seem to be compatible with the idea that control of periodic events of the cell cycle, like DNA synthesis and mitosis, basically occurs by gene regulation.

Literatur

[1] Bal, A. K., and P. R. Gross: Science 139, 584 (1963).
[2] Baserga, R.: Cancer Res. 25, 581 (1965).
[3] Beukers, R., J. Ijlstra, and W. Berends: Rec. Trav. Chim. 77, 729 (1958).
[4] — — — Rec. Trav. Chim. 79, 101 (1960).
[5] Biesele, J. J.: Cancer Res. 4, 540 (1944).
[6] — Cancer Res. 5, 179 (1945).
[7] Birnie, G. D., H. Kröger, and C. Heidelberger: Biochemistry 2, 566 (1963).
[7a] Bonhoeffer, F., and A. Gierer: J. molec. Biol. 7, 534 (1963).
[8] Bosch, L., and H. Bloenendahl: Biochim. biophys. Acta (Amst.) 5, 613 (1961).
[9] —, E. Harbers, and C. Heidelberger: Cancer Res. 18, 335 (1958).
[10] Braun, R., C. Mittermayer, and H. P. Rusch: Proc. nat. Acad. Sci. (Wash.) 53, 924 (1965).
[11] Bresnick, E., and R. J. Karjala: Cancer Res. 24, 841 (1964).
[11a] Cairns, J.: J. molec. Biol. 6, 208 (1963).
[12] Cerroni, R. E., and E. Zeuthen: C. R. Carlsberg 32, 499 (1962).
[13] Cohen, S. S., J. G. Flaks, H. D. Barner, M. R. Loeb, and J. Lichtenstein: Proc. nat. Acad. Sci. (Wash.) 44, 1004 (1958).
[14] Daniel, J. W., and H. P. Rusch: J. gen. Microbiol. 25, 47 (1961).
[15] —, and H. H. Baldwin: In: Methods in Cell Physiology I, p. 9. New York: Academic Press 1964.
[16] Deering, R. A.: J. Bact. 76, 123 (1958).
[17] Djordjevic, B., and W. Szybalski: J. exp. Med. 112, 509 (1960).
[18] Duschinsky, R., E. Pleven, J. Malbica, and C. Heidelberger: 132nd Meeting Amer. Chem. Soc., Wash., Abstr. p. 19 C (1957).

[19] DUSPIVA, F., u. E. HANSEN-DELKESKAMP: Persönliche Mitteilung.
[20] ERIKSON, R. L., and W. SZYBALSKI: Biochem. biophys. Res. Commun. 4, 258 (1961).
[21] — — Radiation Res. 20, 252 (1963).
[22] FRANKEL, J.: C. R. Carlsberg 33, 1 (1962).
[23] FUJIOKA, M., M. KOGA, and I. LIEBERMAN: J. Biol. Chem. 238, 3401 (1963).
[24] GAULDEN, M. E., and R. P. PERRY: Proc. nat. Acad. Sci. (Wash.) 44, 553 (1958).
[25] GELFANT, S.: Exp. Cell. Res. 15, 423 (1958).
[26] GIESE, A. C.: Physiol. Zool. 26, 12 (1953).
[27] — Physiol. Zool. 26, 14 (1953).
[28] — In: Cell Physiology, p. 177. Philadelphia, London: W. B. Saunders Comp. 1962.
[29] GIUDICE, G., F. T. KENNEY, and G. D. NOVELLI: Biochim. biophys. Acta (Amst.) 87, 171 (1964).
[30] GOLDBERG, I. H., and M. RABINOWITZ: Science 136, 315 (1962).
[31] — —, and E. REICH: Proc. nat. Acad. Sci. (Wash.) 48, 2094 (1962).
[32] GREER, S.: J. Gen. Microbiol. 22, 618 (1960).
[33] —, and S. ZAMENHOF: Abstr. Amer. Chem. Soc., 131st Meeting, p. 3 C (1957).
[34] HAMBURGER, K.: C. R. Carlsberg 32, 359 (1962).
[35] HARBERS, E., u. W. MÜLLER: Biochem. biophys. Res. Commun. 7, 107 (1962).
[36] — —, u. R. BACKMANN: Biochem. Z. 337, 224 (1963).
[37] —, N. K. CHAUDURI, and C. HEIDELBERGER: J. biol. Chem. 234, 1255 (1959).
[38] HARTMANN, G., u. U. COY: Angew. Chemie 74, 501 (1962).
[39] HARTMANN, K.-U., and C. HEIDELBERGER: J. biol. Chem. 236, 3006 (1961).
[40] HOLZ JR., G. G., L. RASMUSSEN, and E. ZEUTHEN: C. R. Carlsberg 33, 289 (1963).
[41] HOTTA, Y., and H. STERN: Proc. nat. Acad. Sci. (Wash.) 49, 648 (1963).
[42] — — J. Cell Biol. 16, 259 (1963).
[43] — — Proc. nat. Acad. Sci. (Wash.) 49, 861 (1963).
[44] HOWARD, F. L.: Ann. Bot. 46, 461 (1932).
[45] HOWARD, A., and S. R. PELC: Heredity 6, 261 (1953).
[46] HULTIN, T.: Experientia (Basel) 17, 410 (1948).
[47] HURWITZ, J. J., J. FURTH, M. MALAMY, and M. ALEXANDER: Proc. nat. Acad. Sci. (Wash.) 48, 1222 (1962).
[48] JACOB, F., and J. MONOD: J. molec. Biol. 3, 318 (1961).
[49] KAJIWARA, K., and G. C. MUELLER: Proc. Amer. Ass. Cancer Res. 5, 33 (1964).
[50] KANE, R. E.: J. Cell Biol. 12, 47 (1962).
[51] KAPLAN, H. S., K. C. SMITH, and P. A. TOMLIN: Nature (Lond.) 190, 794 (1961).
[52] KELNER, A.: Proc. nat. Acad. Sci. (Wash.) 35, 73 (1949).
[53] —, and E. B. TAFT: Cancer Res. 16, 860 (1956).
[54] KERSTEN, W.: Biochim. biophys. Acta (Amst.) 47, 610 (1961).
[55] KISHIMOTO, S., and I. LIEBERMAN: Exp. Cell Res. 36, 92 (1964).
[56] LARK, K. G.: Biochim. biophys. Acta (Amst.) 45, 121 (1960).
[57] —, T. REPKO, and E. J. HOFFMAN: Biochim. biophys. Acta (Amst.) 76, 9 (1963).
[58] LAZARUS, L. H., M. R. LEVY, and O. H. SCHERBAUM: Exp. Cell Res. 36, 672 (1964).
[59] LETTRÉ, R., u. W. SIEBS: Z. Krebsforsch. 60, 564 (1955).
[60] LIEBERMAN, I., R. ABRAMS, and P. OVE: J. biol. Chem. 238, 2141 (1963).
[61] — —, N. HUNT, and P. OVE: J. biol. Chem. 238, 3955 (1963).
[62] LINDNER, A.: Cancer Res. 19, 189 (1959).
[63] MAZIA, D.: Harvey Lectures, 1957—1958, 53, 130 (1959).
[64] — In: Sulfur in Proteins (Edit.: R. BENESCH et al.), p. 367. New York: Academic Press 1959.
[65] — In: The Cell III, J. BRACHET, and A. E. MIRSKY (Edit.), p. 77. New York: Academic Press 1961.

[66] — J. cell. comp. Physiol. 62, Suppl. 3, 123 (1963).
[67] —, J. M. MITCHISON, H. MEDINA, and P. HARRIS: J. biophys. biochem. Cytol. 10, 467 (1961).
[67a] MESELSON, M., and F. W. STAHL: Proc. nat. Acad. Sci. (Wash.) 44, 671 (1958).
[68] MITA, T.: Biochim. biophys. Acta (Amst.) 103, 182 (1965).
[69] MITTERMAYER, C., R. BRAUN, and H. P. RUSCH: Biochim. biophys. Acta (Amst.) 91, 399 (1964).
[70] — — — J. Cell Biol. 23, 61 A (1964).
[71] — — — Exp. Cell Res. 38, 33 (1965).
[72] MUELLER, G. C., J. GORSKI, and Y. AIZAWA: Proc. nat. Acad. Sci. (Wash.) 47, 164 (1961).
[73] —, K. KAJIWARA, E. STUBBLEFIELD, and R. R. RUECKERT: Cancer Res. 22, 1084 (1962).
[74] NACHTWEY, D. S.: Zit. bei [58].
[75] NATHANS, D., and F. LIPMANN: Proc. nat. Acad. Sci. (Wash.) 47, 497 (1961).
[76] NYGAARD, O. F., S. GÜTTES, and H. P. RUSCH: Biochim. biophys. Acta (Amst.) 38, 298 (1960).
[77] OPERA-KUBINSKA, Z., Z. LORKIEWICZ, and W. SZYBALSKI: Biochem. biophys. Res. Commun. 4, 288 (1961).
[78] PERRY, R. P.: Nat. Cancer Inst. Monogr. 14, 73 (1964).
[79] RASMUSSEN, L.: C. R. Carlsberg 33, 53 (1963).
[80] — Biol. Bull. 127, 386 (1964).
[81] —, and E. ZEUTHEN: C. R. Carlsberg 32, 333 (1962).
[82] RAUEN, H., H. KERSTEN, u. W. KERSTEN: Z. Physiol. Chem. 321, 139 (1960).
[83] REICH, E., R. M. FRANKLIN, A. J. SHATKIN, and E. L. TATUM: Proc. nat. Acad. Sci. (Wash.) 48, 1238 (1962).
[84] SACHSENMAIER, W.: Biochem. Z. 340, 541 (1964).
[85] —, and H. P. RUSCH: Exp. Cell. Res. 36, 124 (1964).
[86] —, and J. E. BECKER: VI. Intern. Congr. Biochem., New York, 1964, Abstr. p. 238.
[87] — — Mh. Chem. 96, 754 (1965).
[88] —, u. D. H. IVES: Biochem. Z. 343, 399 (1965).
[89] —, u. H. RUPFF: Unveröffentlichte Versuche.
[90] SCHERBAUM, O. H., and E. ZEUTHEN: Exp. Cell Res. 6, 221 (1954).
[91] — — Exp. Cell Res. 3 (Suppl.), 312 (1955).
[92] SCHWARTZ, H. S., S. S. STERNBERG, and F. S. PHILIPS: Cancer Res. 23, 1125 (1963).
[93] SIBATANI, A., and N. MIZUNO: Biochim. biophys. Acta (Amst.) 76, 188 (1963).
[94] SISKEN, J. E., and B. KINOSITA: J. biophys. biochem. Cytol. 9, 509 (1961).
[95] STAHL, F. W., J. M. CRASEMANN, L. OKUN, E. FOX, and C. LAIRD: Virology 13, 98 (1961).
[96] SZYBALSKI, W., and Z. LORKIEWICZ: In: Strahleninduzierte Mutagenese — Erwin Bauer Gedächtnisvorlesungen II, 1961 (Edit.: H. STUBBE); Abhandl. Dtsch. Akad. Wiss. Berlin, Kl. Med. No. 1, p. 63 (1962).
[97] TAKEDA, Y., S. HAYASHI, H. NAKAGAWA, and F. SUZUKI: J. Biochem. 48, 169 (1960).
[98] TAYLOR, J. H.: J. biophys. biochem. Cytol. 7, 455 (1960).
[99] — Proc. Am. Soc. Cell Biol. 2, 186 (1962).
[100] —, W. F. HAUT, and J. TUNG: Proc. nat. Acad. Sci. (Wash.) 48, 190 (1962).
[101] WACKER, A., H. DELLWEG, u. D. WEINBLUM: Naturwissenschaften 47, 477 (1960).
[102] WATANABE, Y., and M. IKEDA: Exp. Cell Res. 38, 432 (1965).
[103] WHITSON, G., and G. PADILLA: Exp. Cell Res. 36, 667 (1964).
[104] WILLIAMSON, D. H., and A. W. SCOPES: Proc. Intern. Union of Physiol. Sci. 1, 758 (1962).
[105] YARMOLINSKY, M. B., and G. L. DE LA HABA: Proc. nat. Acad. Sci. (Wash.) 45 , 1721 (1959).

[*106*] ZEUTHEN, E.: In: Biological Structure and Function II (Edit.: T. W. GOODWIN, and O. LINDBERG), p. 537. London, New York: Academic Press 1961.

[*107*] — In: Growth in Living Systems (Edit.: M. X. ZARROW), Proc. Intern. Symp. on Growth, Purdue Univ., June 1960, p. 135. New York: Basic Books 1961.

[*108*] — In: Synchrony in Cell Division and Growth (E. ZEUTHEN Edit.). New York: Interscience Publishers 1964.

[*109*] ZIMMERMAN, A. M.: In: The Cell in Mitosis (L. LEVINE Edit.), p. 159. New York, London: Academic Press 1963.

Diskussion

Vorsitz: *Grundmann*

Grundmann: The myxomycete you have investigated is of course a very interesting object for several reasons. Perhaps I should repeat one special point: In the most somatic cells after mitosis the DNA content is at a half, and then we have a G_1-phase, then a S-phase and then the G_2-phase, and then the next mitosis. In this myxomycete after mitosis the DNA synthesis begins, then we have the G_2-phase and then the next mitosis; but there is no G_1-phase.

Taylor: In your experiments the actinomycin D block is given two hours before mitosis. I recall that work by GUTTES had indicated that there was a delay of one division cycle before the actinomycin had an effect. Is that correct? And if so, why the difference here? One of the surprising things to me is, that the delay is as much as two hours, because there is certainly RNA synthesis in the last part of the cycle.

Sachsenmaier: In our earlier experiments we have also found a delay of about one complete cycle before mitosis was inhibited. However, the length of the delay depends on the concentration of the inhibitor and also on the sensitivity of the organism. There are some sublines which show different sensitivities to actinomycin. We were lucky to find a subline in which 250 µg/ml (the highest concentration we have used) are effective after a delay of only 2 hours. If the concentration is reduced in the same line down to 100 µg/ml, then we observed much longer delays similar to the findings of GUTTES (Exper. [Basel] *20*, 269 (1964)) and others [*71*]. With the lower concentration there is much less effect on RNA synthesis too.

Concerning the question whether this 2 hour period may be shortened further: this seems to be possible because actually there is still RNA synthesis going on. My feeling is that actinomycin does not penetrate fast enough to reach an effective concentration within a few minutes. So this two-hour period is certainly a maximum period. It might be even shorter.

Grundmann: Do you know anything about the time or the intensity of the histone reduplication in your system? Does it run parallel to the reduplication of DNA?

Sachsenmaier: Some work on this is carried out in the laboratory of Dr. RUSCH in Madison where I have also started some of my own experiments with *Physarum*. I just heard from Dr. RUSCH a few weeks ago that, according to recent results of his group, histone synthesis occurs simultaneously with DNA synthesis. It continues if DNA synthesis is blocked with 5-fluorodeoxyuridine.

Grundmann: And if you block the histone synthesis?

Sachsenmaier: This has not been studied yet.

Grundmann: Is it possible at all to block specifically the histone synthesis?

Sachsenmaier: You can block protein synthesis.

Grundmann: Yes, but I don't know any specific inhibitor for histone synthesis.

Klamerth: You mentioned a certain rhythm in enzymatic activity before mitosis. Did you check some protein building enzymes like transaminase?

Sachsenmaier: We have not studied yet too many enzymes. I have shown the data of a dehydrogenase („Zwischenferment") which did not exhibit a significant periodicity. In the meantime we have also studied (together with Dr. PETTE, München) several other enzymes including different dehydrogenases and condensing enzyme. None of them showed periodicity like thymidinekinase. I would expect that enzymes related to DNA- and perhaps also to RNA-metabolism may show periodicity. Protein synthesis in this organism occurs quite continuously throughout the cycle.

Schaller: Have you tested for thymidine-kinase just in crude extracts? In this case the periodicity in activity may be due to a feedback inhibition. When you have no DNA synthesis, you have — at least in bacteria — a constant pool size of DNA precursors which inhibit synthesis of precursors by feedback inhibition.

Sachsenmaier: We have studied this in particular, and we have dialysed the enzyme preparation prior to analysis. So any pool present in the extract would be removed. The dialysed preparations show exactly the same periodicity like undialysed extracts.

Schaller: Have you extracted RNA- or protein fractions from that late state before mitosis, and put them back onto normal cultures to find out whether they had any effect on mitosis?

Sachsenmaier: Yes, we have tried this, and we did not find any such effects. However, I should mention: If two plasmodia, which are in a different state of the division cycle (so that mitosis in one plasmodium occurs at 9 o'clock and in the second one at 12) are fused, the next mitosis occurs synchronously at an intermediate time. So there is apparently some factor, which is exchanged between both plasmodia, and which is pertinent to the timing of mitosis.

Grundmann: Is it possible to elongate the G_2-phase by low temperature and to shorten it by temperatures higher than 26° C? And have you any possibility, to get G_1-phase by any method, for in fairly all the differentiated tissues, the G_1-phase is a very important one within the generation cycle, and is easily lengthened or shortened by several methods.

Sachsenmaier: It is possible, to change the length of G_2-period by changing the temperature. If we cultivate the organism at 22° C then we observe a total length of the cycle of about 13 hours instead of 9 hours at 26° C. This is mainly due to an elongation of G_2-period. On the other hand, raising the temperature up to 28° C still shortens the G_2-period. As to the other question about introducing a G_1-period: the only mean until now is to block DNA synthesis by antimetabolites.

Beobachtung von Vorgängen im Submikroskopischen — Möglichkeiten und Grenzen

Von

Hellmuth Sitte, Homburg/Saar

Mit 5 Abbildungen

Die Vorgänge bei der Reduplikation und Weitergabe der genetischen Information können derzeit nur bei relativ begrenzter Auflösung beobachtet werden. Einer Rekonstruktion zeitlicher Abläufe nach elektronenmikroskopischen Einzelaufnahmen fixierter Zellen bei höherer Auflösung sind ebenfalls Grenzen gesetzt. Viele Probleme können infolge dieser Beschränkungen im Augenblick nicht oder nur teilweise geklärt werden. Es scheint daher gerechtfertigt, die vorhandenen Möglichkeiten und die derzeit erkennbaren prinzipiellen Grenzen zu diskutieren.

A. Vorbemerkungen und Definitionen

Die klassische submikroskopische Morphologie klammert den Zeitbegriff aus [46]: Das Studium zeitlicher Abläufe fällt in den Bereich von Physiologie und physiologischer Chemie, welche auf den morphologischen Fakten aufbauen und die Variation der Strukturen als Funktion der Zeit registrieren. Die reine statische Morphologie kann demnach beim Beobachten wie Rekonstruieren von Vorgängen nur Mittel zum Zweck sein, da für eine allgemeine Cytologie oder Biologie letztlich nur Kenntnisse von Interesse sind, die das gesamte raum-zeitliche System erfassen. Das Thema des Referates schließt daher Morphologie und Physiologie ein. Eine Kombination dieser Disziplinen wird oftmals als „funktionelle" oder „dynamische Morphologie" bezeichnet. Diese Begriffe sind im Hinblick auf die klare Abgrenzung der Fachgebiete nicht glücklich gewählt [46], kennzeichnen aber die Willkür der Trennung ebenso wie die augenblickliche Situation der Cytomorphologie. Bei der praktischen Arbeit müssen die künstlich errichteten Grenzen überschritten werden, da erst das Zusammenspiel von Morphologie und Physiologie — also das Studium der Lebensvorgänge — unsere Kenntnisse erweitern kann.

Beim Einführen des Zeitfaktors in den Rahmen morphologischer Studien muß Einiges berücksichtigt werden. Dies gilt vor allem für relativ rasch ablaufende Vorgänge, welche im folgenden Abschnitt B vorzugsweise betrachtet werden (langsame Vorgänge: Vgl. Abschnitt C 1). Zunächst kann man ein und denselben Vorgang auf sehr verschiedene

Weise beobachten und analysieren. Als Modell kann die gleichmäßig beschleunigte Bewegung dienen. Ein Körper weist zum Zeitpunkt t_0 (Start) eine Geschwindigkeit $v_0 = 0$ auf und steigert diese bis zum Zeitpunkt t nach Durchmessen der Strecke s auf den Wert v. Der Vorgang läßt sich entweder fortlaufend beobachten oder aus Einzelbildern rekonstruieren. *Beim fortlaufenden Beobachten* wird der Vorgang (= die Bewegung) nicht unterbrochen — der Körper bewegt sich, bis er aus dem gegebenen Beobachtungsfeld verschwindet. Mit einer einzigen derartigen Beobachtung kann man den Einzelvorgang bereits klar erfassen. Dabei erscheint zunächst eine Trennung in ein *kontinuierliches visuelles* und ein *diskontinuierliches kinematographisches Beobachten* sinnvoll. Im erstgenannten Fall erhielte man einen direkten Eindruck von der Kontinuität des Ablaufes — im zweiten lediglich Einzelbilder, welche eine kontinuierliche Bewegung in kleine Einzelabschnitte zerhacken. Überprüft man die beiden Begriffe „kontinuierlich" und „diskontinuierlich" in diesem Zusammenhang, so kommt man zwangsläufig zum Schluß, daß ein kontinuierliches Beobachten nur in sehr bescheidenem Umfange möglich ist. Abgesehen vom begrenzten Auffassungs- und Reproduktionsvermögen gestattet unser Gesichtssinn nur eine Trennung von etwa 15 Bildern pro Sekunde. Diese „Verschmelzungsfrequenz" bildet die Grundlage jeder Kinematographie: Ruckartige Verschiebungen „verschmelzen" zu einem scheinbar kontinuierlichen Ablauf, können also zeitlich nicht mehr getrennt wahrgenommen werden. Demgegenüber gestattet eine normale kinematographische Kamera noch eine Trennung von 65 Bildern, die Hochfrequenzkinematographie im heute erreichten Grenzfall noch eine Separation von etwa 10^7 Bildern pro Sekunde. Das diskontinuierliche kinematographische ist damit dem kontinuierlichen visuellen Beobachten wesentlich überlegen, da wir rasch ablaufende Vorgänge (Flügelschlag der Insekten, Bewegungen eines Geschosses) mit unserem Auge nicht mehr erfassen können.

Neben dem fortlaufenden Beobachten kann eine *Rekonstruktion eines Vorganges aus Einzelbildern*, welche das Objekt jeweils statisch in einem fixierten Zustand zeigen, auch zum Ziel führen. Wichtig ist dabei dreierlei:

a) Der Vorgang muß wiederholbar sein und muß im Rahmen dieser Repetitionen statistisch reproduzierbar ablaufen.

b) Der zeitliche Abstand t_{ae} des Einzelbildes vom Startzeitpunkt muß hinreichend genau festliegen und hinreichend klein gehalten werden können.

c) Die Einzelbilder zu verschiedenen Zeitpunkten nach dem Start müssen sich voneinander in einer Weise unterscheiden, welche die Richtung des Ablaufes klar erkennen läßt (vgl. Abschnitt C. 3).

Im Rahmen des gegebenen Beispieles könnte man den Körper in einem ersten Experiment zum Zeitpunkt t_1 nach einer Zeit $t_{ae} = (t_1 - t_0)$ abstoppen (= „fixieren"), in einem zweiten Experiment zum Zeitpunkt t_2 usw.. Der definitionsgemäß beliebig oft wiederholbare Vorgang wäre auf diese Weise aus Einzelbildern zu rekonstruieren.

Man kann einen Vorgang umso genauer erfassen, je kürzer der zeitliche Abstand $t_b = 1/\nu_b$ der einzelnen Aufnahmen bei der fortlaufenden kinematographischen Beobachtung ist, bzw. je kleiner die Zeitspanne t_{ae} zwischen Start (Auslösung) und Ende (Fixation) eines Ablaufes bei der Rekonstruktion nach Einzelbildern gehalten werden kann. Neben der Längenauflösung $_L\delta_{min}$ des optischen Systems spielt daher für das Beobachten von Vorgängen die *Zeitauflösung* $_T\delta_{min}$ (Trennschärfe) eine entscheidende begrenzende Rolle. Die Zeitauflösung soll wie folgt definiert werden:

$$_T\delta_{min} = \frac{1}{\nu_b} = t_b = t_{ab} \tag{1}$$

Hierbei entspricht ν_b der kinematographischen Bildfrequenz, t_b der Zeitspanne zwischen zwei aufeinanderfolgenden kinematographischen Einzelaufnahmen beim fortlaufenden Beobachten. t_{ab} gibt den Minimalabstand zwischen Auslösen und Ende (Fixieren) eines Vorganges beim Rekonstruieren nach Einzelaufnahmen am fixierten Material an; je kürzer diese Zeitspanne gehalten werden kann, desto günstiger (= kleiner) wird die Zeitauflösung $_T\delta_{min}$. Die Zeitauflösung besitzt die Dimension einer Zeiteinheit – im cgs-System beispielsweise [sec].

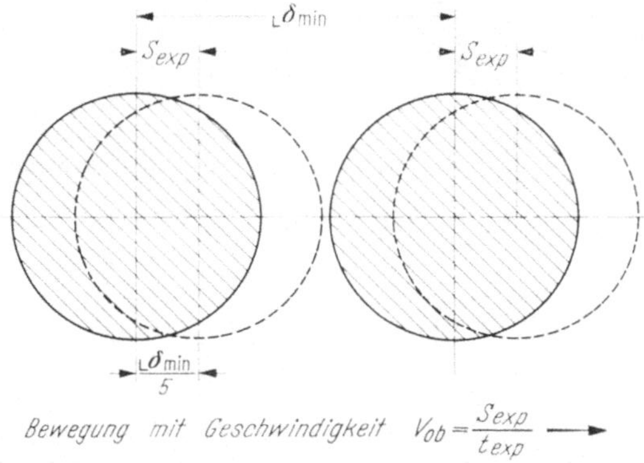

Abb. 1. Zur Abhängigkeit der Punktauflösung $_L\delta_{min}$ bewegter Objekte von der Belichtungszeit t_{exp} und der Objektgeschwindigkeit $v_{ob} = s_{exp}/t_{exp}$. Es wird angenommen, daß sich beide Objektpunkte mit identischer Geschwindigkeit in die gleiche Richtung bewegen

Beim statischen Betrachten, bzw. bei der Momentaufnahme eines fixierten Objektes ist die Längenauflösung $_L\delta_{min}$ bei hinreichender Auflösung der Netzhaut bzw. der Photoemulsion ausschließlich durch die Qualität des optischen Systemes gegeben. Beim Aufnehmen von Vorgängen treten zusätzlich *Belichtungszeit* t_{exp} und *Objektgeschwindigkeit* v_{ob} als Faktoren auf, welche die Längenauflösung $_L\delta_{min}$ limitieren. Unabhängig vom verwendeten optischen System, sowie unabhängig vom Abbildungsmaßstab (Vergrößerung des Endbildes) ist eine Auflösung zweier Scheibchen im Abstand $_L\delta_{min}$ nur dann denkbar, wenn die *Verschiebung* s_{exp} *des Objektes während der Belichtungszeitspanne* t_{exp} wesentlich unter der gewünschten und optisch erreichbaren Grenzauflösung liegt ($s_{exp} \ll {}_L\delta_{min}$). Für die Praxis darf man nach Abb. 1 eine Auflösung

$_L\delta_{min}$ noch erwarten, wenn die Verschiebung s_{exp} während des Belichtens der Photoemulsion unter einem Fünftel von $_L\delta_{min}$ liegt. Man kann daher folgende Richtlinien aufstellen:

$$v_{ob} = \frac{s_{exp}}{t_{exp}} \text{ bzw. } s_{exp} = v_{ob} \cdot t_{exp}$$

$$_L\delta_{min} > 5 \cdot v_{ob} \cdot t_{exp} \tag{2}$$

$$\frac{_L\delta_{min}}{5 \cdot v_{ob}} > t_{exp} \tag{3}$$

Die oben formulierten Relationen zwischen der Längenauflösung $_L\delta_{min}$ (Trennen zweier Punkte), der Belichtungszeit t_{exp} und der Objektgeschwindigkeit v_{ob} geben naturgemäß nur eine Richtlinie. Sie gelten für

Tabelle 1 *Geschwindigkeit von Vorgängen im Submikroskopischen. Abhängigkeit der Belichtungszeit bei vorgegebener Auflösung*

Vorgang	Geschwindig-keit[1] [μ/sec]	Maximalbelich-tung für 100 Å Auflösung nach Formel (3)	Basis der Angaben
Gasmolekel $\{$ H$_2$ . . .	$1{,}7 \cdot 10^9$	—	
(0° C) $\{$ H$_2$O . . .	$5{,}7 \cdot 10^8$	—	[93]
$\{$ CO$_2$ . . .	$3{,}6 \cdot 10^8$	—	
Flugmuskel von Insekten	$2{,}5 \cdot 10^3$	—	Frequenz rd. 500 Hz Kontraktion im Bildfeld von 5 μm rd. 2,5 μm
Plasmaströmung (*Physarum* Plasmodienstrang) . . .	bis $1{,}3 \cdot 10^3$	—	[69]
Blutströmung (Arteriolen)	280	7 μ sec	[101]
Plasmaeinschlüsse in Pflanzenhaaren (*Viola*)	bis 60	35 μ sec	[54]
Blutströmung (Capillaren)	50	40 μ sec	[101]
Cilienbewegung	40	50 μ sec	Frequenz 15 Hz. Länge rd. 8 μm. Halbkreisbahn mit 25 μm pro Schlag [17]
Plasmaströmung in Pflanzen (Normwerte)	2 bis 10	1,0 bis 0,2 msec	[69]
Herzmuskelkontraktion	2,5	0,8 msec	Frequenz rd. 1 Hz. Bildfeld rd. 5 μm. Kontraktion im Bildfeld rd. 2,5 μm
Kontraktion glatter Muskelzellen	1	2 msec	Verkürzung um 2 μm im Blickfeld binnen 2 sec
Mitochondrienbewegung	1	2 msec	Bewegung um 5fache Breite in $^1/_2$ sec

[1] Soweit nicht nach Literatur, rohe Schätzwerte, welche auf eine Bildfeldgröße von rd. 5 μm $\varnothing$ bezogen werden.

die gleichsinnige Bewegung der aufzulösenden Punkte mit gleichen Geschwindigkeiten v_{ob} nach Abb. 1. Jedes Abgehen von diesem Schema muß man entsprechend berücksichtigen. Dies gilt beispielsweise für ungleiche Geschwindigkeiten, Bewegung in verschiedenen Richtungen usw. Immerhin ermöglicht die Formel (3) ein Abschätzen der maximalen Belichtungszeiten t_{exp}, welche bei der gegebenen Größenordnung der Objektgeschwindigkeiten v_{ob} noch eine vorgegebene Auflösung $_L\delta_{min}$ zulassen; vgl. Tab. 1. Man erkennt sofort, daß beim Beobachten von Vorgängen im Submikroskopischen extrem niedrige Belichtungszeiten im Größenordnungsbereich unter 1 msec notwendig sind, wenn man scharfe Abbildungen erhalten will.

Im Weiteren muß man berücksichtigen, daß die Geschwindigkeit v_{ob} der Objektbewegung im Endbild um den Betrag der *Vergrößerung V* vergrößert abgebildet wird. Eine Objektbewegung von der Größe 1 mm/sec wird dadurch bei $V = 30000:1$ auf rund 100 km/h „vergrößert" und ist nur dann entsprechend zu analysieren, wenn folgende Faustregel eingehalten wird:

$$\frac{V}{v_b} = V \cdot t_b \sim \text{const} \tag{4}$$

Der Bruch der Vergrößerung V durch die Bildfrequenz v_b, bzw. das Produkt Vergrößerung mal Bildfolgezeit t_b soll beim kinematographischen Aufnehmen eines bewegten Objektes etwa konstant gehalten werden. Genügt beispielsweise bei der Vergrößerung $V = 10:1$ zum Analysieren einer Bewegung mit $v_{ob} = 1$ mm/sec eine Bildfrequenz $v_b = 1$ Bild/sec (bzw. $t_b = 1$ sec), so müßte man bei der 100mal gesteigerten Vergrößerung $V = 1000:1$ die Bildfrequenz v_b ebenfalls mit dem Faktor 100 multiplizieren ($v_b = 100$/sec). Darüber hinaus ist zu bedenken, daß die *Endbildgröße bei gegebener Vergrößerung V* durch das Bildfeld des Okulars, bzw. die Größe von Mattscheibe, Leuchtschirm oder Photomaterial (Platte, Film) limitiert ist. Bei Vorgängen, welche mit hoher Geschwindigkeit v_{ob} ablaufen, wird daher die *Zeitspanne der Beobachtung mit steigender Vergrößerung V immer kürzer* und ist im allgemeinen eine Funktion des Betrages $1/V$. Anders ausgedrückt: Bewegt sich das Objekt mit einer Geschwindigkeit $v_{ob} = 1$ mm/sec von einem zum anderen Rand eines Bildfeldes bestimmter Größe (z. B. $\varnothing = 100$ mm), so ist seine Bewegung bei $V = 10:1$ über eine Zeitspanne von 10 sec, bei $V = 10000:1$ lediglich über eine Zeit von 10 msec zu beobachten.

Schließlich ist man aus der reinen, statischen Morphologie gewöhnt, normalerweise *zweidimensionale, ebene Bilder* zu beobachten und auszuwerten. Vorstellungen über den räumlichen Aufbau der Objekte leitet man entweder aus Einzelbildern auf der Basis stereologischer Überlegungen [61] oder aus Serienschnitten ab [107, 128]. Demgegenüber muß man bei der fortlaufenden Beobachtung *Vorgänge im räumlichen Objekt dreidimensional* erfassen, da sich diese Abläufe normalerweise nicht auf eine Ebene beschränken. Es muß daher ein genügend großer Tiefenbereich des Objektes scharf abgebildet werden.

B. Vitalbeobachtung im Submikroskopischen
1. Allgemeines und Auswahl des Instrumentes

Ein fortlaufendes Beobachten von Lebensvorgängen mit vergrößernden optischen Systemen wird üblicherweise als *Vitalmikroskopie* bezeichnet. Jeder Vitalmikroskopie sind bei der Auswahl geeigneter Objekte relativ enge Grenzen gesetzt. Dies gilt nicht nur für eine Arbeit bei höherem und höchstem Auflösungsvermögen im Submikroskopischen, sondern bereits für die Lichtmikroskopie.

Die *Durchlicht-Vitalmikroskopie* muß sich auf gut durchstrahlbare Objekte beschränken. Diese weisen unter normalen Lebensbedingungen im Hellfeld sehr geringe Kontraste auf. Da Vitalfarbstoffe in vielen Fällen Artefakte erzeugen, brachte erst die Kombination der *Gewebekulturmethode* mit der *Phasenkontrastoptik* Fortschritte auf einer breiteren Basis. Die Auflichtvitalmikroskopie an Oberflächen größerer undurchstrahlbarer Organe und Organismen (u. A.: Capillarmikroskopie, Kolpomikroskopie) bleibt auf schwächere Vergrößerungen beschränkt und ist in vielen Fällen wiederum auf Vitalfarbstoffe angewiesen[1].

Es ist bekannt, daß man mit *indirekten Methoden* (Polarisations- und Interferenzmikroskopie, Röntgenbeugung usw.) Aufschlüsse über Strukturelemente erhalten kann, deren Größe im submikroskopischen Bereich liegt; derartige Verfahren scheinen daher für eine Vitalbeobachtung im Submikroskopischen geeignet. Bei genauerem Betrachten zeigt sich jedoch, daß ihre Einsatzbreite äußerst gering ist. Insbesondere können sie kein geschlossenes submikroskopisch-morphologisches Bild einer (lebenden) Zelle liefern[2]. Alle genannten Verfahren ermöglichen lediglich sehr genaue Längenmessungen in einer vorgegebenen Richtung (eindimensional). Die Punktauflösung im mikroskopisch-optischen Sinn (zweidimensional in der Fläche) bleibt nach wie vor durch die Lichtwellenlänge beschränkt. Man befindet sich dadurch in der Lage eines Geographen, der zwar sehr exakte Höhenmessungen ausführen kann, die Höhen in einer zweidimensionalen Karte aber nur mit sehr viel geringerer Genauigkeit eintragen kann. Die Genauigkeit in der einen Achsenlage (Höhe) wird durch Ungenauigkeit in den beiden restlichen Achsen (Fläche) entwertet. Ein geschlossenes räumliches Bild kann daher nur durch ein direktes Beobachten mit submikroskopischer Auflösung erreicht werden. Die häufig in diesem Zusammenhang diskutierte „*Ultramikroskopie*" nach Siedentopf und Zsigmondy kann zwar die Sichtbarkeitsgrenze weiter hinausschieben, nicht aber die Auflösung des lichtoptischen Systemes verbessern. Dunkelfeldsysteme sind daher nur dort nützlich, wo mangelnde Kontraste eine Abbildung vereiteln. Damit verbleiben für eine Vitalmikroskopie im submikroskopischen Bereich nach

---

[1] Die Auflicht-Phasen- und Interferenzkontrastsysteme der Metallmikroskopie wurden nach Wissen des Ref. in der Biologie bislang nicht eingesetzt. Über ihre Brauchbarkeit kann wahrscheinlich nur der Versuch entscheiden.

[2] Dies soll nicht bedeuten, daß sie in allen Fällen unbrauchbar sind. Man kann indessen mit ihrer Hilfe stets nur günstig gelagerte Detailfragen klären, niemals aber ein geschlossenes morphologisches Bild erhalten.

dem heutigen Stand der Technik nur zwei Geräte, welche bei der geforderten direkten Abbildung eine bessere Auflösung als das Lichtmikroskop erreichen können: das Röntgenstrahlen-Schattenmikroskop (hinfort RSM) und das Elektronenmikroskop (hinfort EM).

Das *Röntgenstrahlen-Schattenmikroskop* [*26, 27, 37, 38, 86*] besteht im wesentlichen aus einem elektronenoptischen Feinfokussystem, das auf einer dünnen Folie einen sehr feinen Brennfleck erzeugt. Die von diesem Brennfleck ausgehenden Röntgenstrahlen können für eine vergrößerte Schattenabbildung eines Objektes verwendet werden. Das Objekt wird hierzu in unmittelbarer Nähe des Brennfleckes angebracht. Die Vergrößerung im Schattenbild ergibt sich aus den Abständen des Objektes und des Leuchtschirmes (bzw. der Photoplatte) vom Brennfleck. Das Auflösungsvermögen hängt im wesentlichen vom Minimaldurchmesser des Brennfleckes und vom Minimalabstand des Objektes vom Brennfleck auf der Folie ab, welche die Röntgenstrahlen emittiert. Das RSM hat heute noch nicht den Stand technischer Vollkommenheit erreicht, welchen die Durchstrahlungs-Elektronenmikroskope aufweisen. Für das RSM scheint jedoch der Umstand zu sprechen, daß die Objekte außerhalb des evakuierten Elektronenstrahlrohres unter normalen atmosphärischen Bedingungen liegen. Auf diese Weise könnte man offenbar feuchte Präparate ohne besondere Vorkehrungen untersuchen. Darüber hinaus spräche das beliebig große und innerhalb weiter Grenzen variable Durchdringungsvermögen der Röntgenstrahlen für das RSM. Auch dickere Objektschichten könnten mühelos durchstrahlt werden. Die angeführten Eigenschaften wären für das Beobachten lebender Zellen unter normalen Bedingungen äußerst vorteilhaft. Leider trügen aber diese scheinbaren Vorteile des RSM. Das elegante Gerät scheint derzeit gerade für eine Vitalmikroskopie aus zwei Gründen kaum geeignet: Einerseits kann man hinreichende Bildkontraste nur unter bestimmten Bedingungen am trockenen Objekt erzielen. Andererseits darf man beim derzeit gegebenen Stand der Gerätetechnik hinreichend kurze Belichtungszeiten nach Tab. 1 nicht erwarten. Die Frage der Strahlenbelastung ist daher zunächst nur von sekundärem Interesse, da der Einsatz des RSM für eine Vitalbeobachtung im Submikroskopischen vorerst am mangelnden Bildkontrast und an den zu hohen Belichtungszeiten scheitert. Das RSM wird daher in diesem Referat nicht weiter diskutiert. Trotzdem muß darauf verwiesen werden, daß die Entwicklung dieses Instrumentes noch nicht abgeschlossen ist. Neue technische Lösungen können unter Umständen in der Zukunft eine Röntgenstrahlen-Vitalmikroskopie in den Bereich des Möglichen rücken.

Von allen verfügbaren Geräten kommt daher derzeit per exclusionem nur das *Elektronenmikroskop* (EM) für eine Vitalmikroskopie im Submikroskopischen in Frage, obwohl man gerade diesem Gerät auf Grund seiner normalen Betriebsbedingungen (Vakuum, Erwärmung und Belastung des Objektes mit ionisierenden Strahlen) die Eignung für diesen Zweck immer wieder abgesprochen hat. Ohne Rücksicht auf diese Vorurteile haben bereits in frühesten Entwicklungsstadien der Elektronenmikroskopie mehrere Autoren mit verschiedenen Mitteln versucht, die

Hindernisse zu überwinden. Es soll hier besonders auf die Arbeiten und Überlegungen von v. Ardenne [6—8] sowie v. Ardenne und Friedrich-Freksa [9] hingewiesen werden. Die *Frage des Überlebens nach dem Elektronenbeschuß* stand bei diesen Untersuchungen und Berechnungen im Vordergrund des Interesses. So wurden beispielsweise Bakteriensporen im EM bei minimal gehaltener Belichtungszeit photographiert und danach auf ihre Keimfähigkeit untersucht, welche zum Teil erhalten blieb. Die Arbeiten im EM wurde auf relativ trockene Objekte beschränkt, welche Ruheformen ohne nennenswerten Stoffwechsel darstellen. Damit entfällt ein Spezifikum des Lebens und die Vitalmikroskopie verliert ihren Sinn, da jede Dynamik fehlt. Im Folgenden soll daher als *Vitalmikroskopie lediglich das Beobachten von Vorgängen (Bewegungen) bezeichnet werden, die sich in stark hydratisierten Zellen oder Lebewesen abspielen*. Das Interesse an einer derartigen reellen Vitalmikroskopie hat in dem Augenblick nachgelassen, in dem man durch das Einführen geeigneter Fixations- und Einbettungsverfahren mit speziellen Schneidegeräten und Messern [89] im Stand war, Ultradünnschnitte gut erhaltener Objekte im Elektronenstrahl zu studieren. Die angeführten Verfahren haben binnen eines Jahrzehntes ein fast geschlossenes Bild der submikroskopischen Struktur der Zellen und Gewebe ergeben [130]. Der Vorwurf einer groben Artefizierung des Untersuchungsgutes konnte durch die Identität der Resultate nach verschiedenartiger Fixation [89, 97] und Einbettung [89], insbesondere aber durch die elegante Gefrierätztechnik nach Moor und Mühlethaler [82—84] weitgehend entkräftet werden. Nachdem man zudem in vielen Fällen Vorgänge im Submikroskopischen nach Einzelbildern am fixierten Objekt verbindlich rekonstruieren konnte (vgl. Abschnitt C), schien der große Aufwand einer Vitalbeobachtung im EM kaum mehr gerechtfertigt. Inzwischen haben verschiedene grundlegende Versuche und Entwicklungen eine neue Ausgangsposition geschaffen, die im Folgenden diskutiert werden soll. Diese Diskussion beschränkt sich auf das Beantworten der grundlegenden Frage, ob ein Überwinden der Schwierigkeiten nach dem heutigen Stand der Technik und der Kenntnisse physikalisch denkbar erscheint oder nicht.

2. Gasdruck, relative Feuchtigkeit und Temperatur in der Objektkammer

Bei den heute gebräuchlichen EM-Modellen liegt das Objekt im Hochvakuum des Strahlrohres (10^{-4} bis 10^{-5} Torr). Die unelastische Streuung von Elektronen am Objekt führt zu einer teilweise beträchtlichen Erwärmung. Vakuum und Wärme stören bei den normalen Untersuchungen von anorganischem Material, Abdrucken, trockenen organischen Objekten und Ultradünnschnitten in einem viel geringeren Umfang, als man zunächst prophezeite und befürchtete. Es ist aber unmöglich, normal hydratisierte lebende Zellen in derartigen Geräten ohne Zusatzeinrichtung zu untersuchen. Bereits 1942 führte E. Ruska [95] Versuche mit einer *Objektkammer für höhere Drucke* durch; vgl. Abb. 2a. Feine

Öffnungen erlauben einen Durchtritt der Elektronen durch eine 0,4 bis
0,9 mm hohe Objektkammer, welche mit Gas gefüllt werden kann. Die
kleinen Gasmengen, welche durch die Ein- und Austrittsöffnungen für
den Elektronenstrahl in den Tubus eindringen, werden durch das

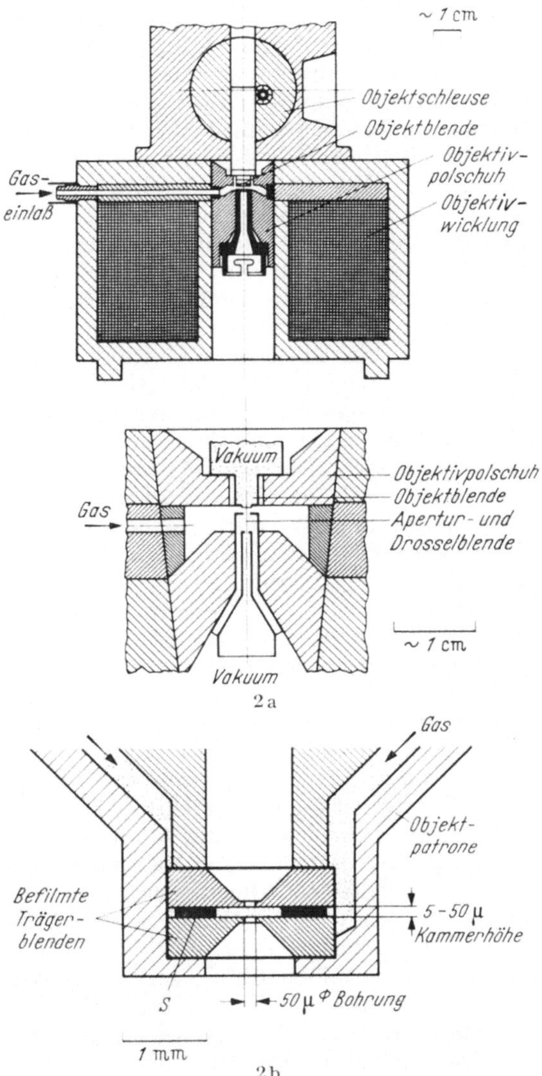

Abb. 2. (a) Magnetisches Polschuhobjektiv mit Druckkammer nach E. Ruska [95]. Die
Druckkammer ist auf der einen Seite durch die Objektblende, auf der anderen Seite durch
die Aperturblende des Objektives bzw. eine zusätzliche Drosselblende abgeschlossen.
(b) Objektpatrone mit Gaseinlaß nach Heide [62, 63]. Die Druckkammer ist beidseitig
durch befilmte Objektträgerblenden abgeschlossen und wird über ein Ventilsystem kontrolliert mit Gas beliefert

Pumpenaggregat ohne Schwierigkeit beseitigt und stören nicht. Mit dieser Kammer konnte nachgewiesen werden, daß Gasdrucke bis maximal 5 Torr (Luft), bzw. 20 Torr (H_2) die Bildqualität nicht beeinflussen. Es war nach diesen Resultaten zu erwarten, daß bei höheren Gasdrucken zwar die Wechselwirkung zwischen Elektronen und Gasmolekeln ein größeres Ausmaß annimmt, daß aber trotz reduzierter Bildqualität noch eine submikroskopische Auflösung erhalten werden kann.

Eine im Prinzip ähnliche Anordnung entwickelte etwa ein Jahrzehnt später HEIDE [62] zum Studium der Objektverschmutzung. In weiterer Folge wurde diese Druckkammer zum Beobachten von Objekten unter kontrollierten Gasdrucken verfeinert [63]. Der Objektraum wird nach der schematischen Abb. 2b durch zwei Objektträgerblenden begrenzt, deren Öffnungen durch Trägerfilme verschlossen sind. Der Elektronenstrahl tritt durch einen derartigen Film in die Kammer ein und verläßt sie wiederum durch einen derartigen Film. Im Vergleich zu einer normalen elektronenoptischen Untersuchung muß ein zusätzlicher Trägerfilm durchstrahlt werden. Die Verhältnisse entsprechen damit der „Sandwichpräparation", welche für Ultradünnschnitte in Plexiglaseinbettung entwickelt wurde [123]. Der Nachteil der erhöhten Trägerfilmstärke wird durch das Fehlen der Objektverschmutzung weitgehend kompensiert und fällt bei den relativ dicken Objektschichten im Rahmen einer Vitalbeobachtung kaum ins Gewicht. Der Gasdruck muß der Höhe der Kammer angepaßt werden. Bei einer Höhe vom maximal 50 μm kann man mit normaler Strahlspannung (100 kV) ohne Nachteil unter Atmosphärendruck und bei normaler Feuchtigkeit arbeiten. Die Chancen für eine Vitalbeobachtung dürften sich beim Arbeiten mit höheren Strahlspannungen [25, 28, 72, 92, 115, 120] wesentlich verbessern, da hierbei auch größere Objektkammern ohne Einbuße an Bildqualität unter Atmosphärendruck gehalten werden können.

Beim Kultivieren und Untersuchen lebender Zellen muß bekanntlich neben der Feuchtigkeit auch die *Temperatur innerhalb enger Grenzen konstant* gehalten werden. Im EM wird man daher zunächst versuchen, die Strahlstromstärke und damit die Wärmeentwicklung am „Objektwiderstand" [14, 122] möglichst gering zu halten. Dies erreicht man bekanntlich dadurch, daß man jeweils nur dasjenige Objektfeld ausleuchtet, das man im Endbild gerade beobachtet; man arbeitet im „Feinstrahl" [73, 96]. Die unter diesen Bedingungen freigesetzte Wärme muß abgeführt werden. Man erreicht mit Kühlvorrichtungen unterschiedlicher Bauart und Funktionsweise [74, 102, 105, 121, 133] Objekttemperaturen, die weit unterhalb −100° C liegen und damit den für eine Vitalbeobachtung sinnvollen Bereich weit unterschreiten. Für diesen speziellen Zweck würde wahrscheinlich ein kleines Peltier-Element vollständig genügen. Das technische Problem wäre weniger das Abführen großer Wärmemengen, sondern eher das Erreichen einer guten Temperaturkonstanz. Es müßte jedoch beim Verwenden massiver Objektträgerblenden mit gutem Wärmeleitvermögen (z. B. aus Cu, Ag oder Au) bei der vorgegebenen Dimension der darauf liegenden Objekte ohne weiteres technisch lösbar sein.

3. Durchstrahlen dickerer Objektschichten, Schärfentiefe und Überlagerung von Strukturelementen

Die Schichtdicke, welche bei einem Beobachten von Lebensvorgängen im Submikroskopischen durchstrahlt werden muß, ist durch die Größenordnung der Zellen[1] gegeben. Sie beträgt etwa 1–5 μm und liegt damit wesentlich über der Dicke der üblichen EM-Präparate. In einem kommerziellen EM mit 100 kV Strahlspannung kann man etwa 0,2 μm trockenes organisches Material durchstrahlen[2]. Zum Durchstrahlen dickerer Schichten sind schnellere Elektronen notwendig, welche man mit höheren Beschleunigungsspannungen erhält. Die Spitze beim *Bau von Höchstspannungs-Elektronenmikroskopen* [*24, 25, 28, 72, 92, 115, 120*] halten derzeit DUPOUY und PERRIER vom CNRS in Toulouse, die ein Gerät für 1500 kV = 1,5 Million Volt = 1,5 MV Beschleunigungsspannung entwickelt und 1960 in Betrieb genommen haben [*31–36*]. Ein derartiges EM erfordert nicht nur einen besonderen Elektronenbeschleuniger mit entsprechenden Stabilisierungselementen. Die Amperewindungszahl der magnetischen Elektronenlinsen muß zum Ablenken der härteren Elektronen wesentlich gesteigert werden; nur dadurch erhält man genügend kleine Brennweiten und in weiterer Folge die gewünschten starken Vergrößerungen. Darüber hinaus muß man die starken Röntgenstrahlen, welche 1,5 MV-Elektronen beim Auftreffen auf Blenden und Leuchtschirme erzeugen, durch entsprechende Schutzelemente abfangen. Die einzelnen Bauteile erhalten dadurch gewaltige Dimensionen. So weist das Objektiv beispielsweise einen Durchmesser von einem halben Meter und ein Gewicht von rd. 700 kg auf. Der genannten Arbeitsgruppe ist es gelungen, die großen apparativen Schwierigkeiten zu überwinden und mit diesem Großgerät bereits eine Vielzahl interessanter Studien durchzuführen. So war es unter anderem möglich, 8 μm dicke Metallfolien zu durchstrahlen. Da die Dichte der hydratisierten Biokolloide wesentlich geringer ist, kann man mit derartigen Beschleunigungsspannungen auch lebende Zellen durchstrahlen. In Toulouse gelang es bereits, bei 650–750 kV Elektronenbilder von Bakterien in feuchtem Milieu bei 760 Torr herzustellen [*31, 33*]. Dies erreichte man durch einen Einschluß zwischen vakuumdichten Plastikfolien. Die Bakterien waren nach dem Beobachten im EM noch lebensfähig und konnten weiter gezüchtet werden. Die elektronenoptischen Aufnahmen selbst sind im Vergleich zum heute üblichen Schnittpräparat, wie zu Gefriertrocknungs- und Kritischen-Punkt-Präparationen [*3, 127*] nach Beschatten nicht besonders ansprechend, zeigen aber zumindest trotz der hohen Strahlspannung hinreichende Kontraste. Es soll in diesem Zusammenhang an die ersten Elektronenbilder von biologischen Objekten erinnert werden, welche MARTON 1934/35 [*76*] publizierte. Auch diese Bilder

---

[1] Beim Untersuchen kleinerer Mikroorganismen könnte man Schichtdicken unter 1 μm erreichen. Diese Werte sollte man jedoch nicht als Norm betrachten, sondern als günstig gelagerte Ausnahmefälle. Eine Vitalmikroskopie auf breiterer Basis müßte Schichtdicken $\geqq$ 1 μm einschließen.

[2] Die Dichte dieses trockenen Materials ist sicherlich höher, als die Dichte im hydratisierten Cytoplasma.

mußten zunächst im Vergleich zu den Resultaten der perfektionierten lichtoptischen Mikromethodik enttäuschen, standen aber trotzdem am Anfang einer äußerst fruchtbaren Entwicklung. So sollte man auch diese ersten Versuche am Toulouser 1,5 MV-EM nicht zu kritisch beurteilen, da noch viele Möglichkeiten derzeit ungenützt sind, welche die Bildqualität wesentlich verbessern können.

Schichtdicken von 1—5 μm sind nach dem eben Dargelegten mit 1,5 MV-Elektronen sicher durchstrahlbar. Es bleibt trotzdem die Frage, ob man hierbei brauchbare Bilder erwarten darf. Die Anwort hierauf steht mit der *Schärfentiefe des EM* in engem Zusammenhang. Vergleicht man Lichtmikroskop und EM, so fällt die äußerst geringe Apertur der Elektronenlinsen auf. Sie führt dazu, daß jedes elektronenoptische System eine sehr viel größere Schärfentiefe als das Lichtmikroskop aufweist. Man kann im EM also sehr viel größere Tiefenbereiche des Objektes im Bild scharf erfassen. Der „optische Schnitt" des Lichtmikroskopes, der.beispielsweise eine Zellschichte eines dickeren Objektes scharf erfaßt und die ebenfalls in ihm enthaltenen darüber und darunter liegenden Ebenen verschwimmen läßt, ist beim EM normalerweise nicht zu beobachten. Diese Sachlage läßt sich infolge des großen Öffnungfehlers jeder Elektronenlinse derzeit kaum ändern und bringt für die praktische Arbeit am EM Vor- und Nachteile. Es wurde bereits einleitend in Abschnitt A darauf hingewiesen, daß man Vorgänge dreidimensional beobachten muß, da die dritte Dimension im Gegensatz zur statischen Morphologie am fixierten Objekt nicht rekonstruiert werden kann. Da ein räumliches Beobachten eine beträchtliche Schärfentiefe voraussetzt, mag sie unter diesem Gesichtspunkt als Vorteil gewertet werden. Auf der anderen Seite darf man nicht übersehen, daß eine große Schärfentiefe zwangsläufig dazu führt, daß sich viele Objektdetails überlagern. Damit ändert sich der Bildhabitus — jede Interpretation wird schwieriger. Der Einfluß der Schnittdicke auf das Bild, seine Deutung und morphometrische Analyse wurde unter den verschiedensten Gesichtspunkten diskutiert [*47, 60, 107, 128, 132*]. Abb. 3a zeigt als Modell einen Querschnitt durch ein Objekt, das Membranen (M) und kugelförmige Partikel (K) enthält. Wird der Gesamtbereich der Dicke (D) scharf abgebildet, so überlagern sich die Kugeln und geben auch dann einen homogenen Hintergrund ohne Details, wenn der Kugelabstand über der Auflösungsgrenze des EM liegt. Die Membranen werden nur dann scharf abgebildet, wenn sie über größere Strecken parallel zum Elektronenstrahl verlaufen [*80, 107, 128*] und einen hinreichenden Dichteunterschied zur umgebenden Matrix aufweisen. Dieser Umstand ist ohne weiteres verständlich, wenn man sich vor Augen hält, daß die Helligkeit einzelner Bildstellen im wesentlichen von der Massendicke (= Dicke $\times$ Dichte) der entsprechenden Objektpartien abhängt. Abb. 3b zeigt das Massendickediagramm zweier Membranen in einem Ultradünnschnitt durch ein OsO_4-fixiertes Gewebe. Da bei der Osmiumfixation im Membranbereich Osmiumverbindungen niedergeschlagen werden, ist die Dichte der Membranen in diesem Fall wesentlich höher, als die Dichte der umgebenden Plasmapartien. Man erhält daher in diesem Fall trotz der starken Krümmung und der teilweisen Überlagerung eine scharfe Abbildung der beiden Membranen an derjenigen Stelle, an der sie annähernd parallel zum Elektronenstrahl verlaufen. In analoger Weise wäre eine Abbildung entsprechend orientierter Bereiche in dickeren Schnitten denkbar. Vergleicht man diese Sachlage bei einem dicken Schnitt, bzw.

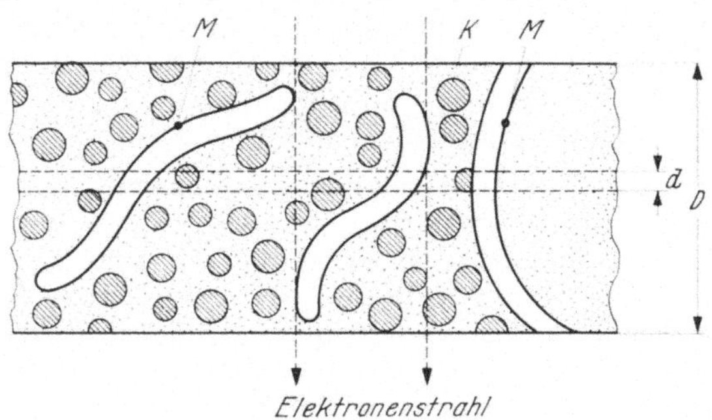

Abb. 3. (a) Schnittpräparat in Seitenansicht. Im dicken Schnitt (D) überlagern sich bei entsprechend großer Schärfentiefe die Kugeln (K) und können nicht mehr einzeln wahrgenommen werden. Demgegenüber können sie bei geringerer Schnittdicke (d) oder geringerer Schärfentiefe (= „optischer Schnitt") gut aufgelöst werden. Die Membranen (M) sind bei größerer Schichtdicke nur an denjenigen Stellen wahrzunehmen, an denen sie vom Elektronenstrahl tangential getroffen werden. (b) Abbildung der Membranen im dicken Schnitt. MD = Massendickeverlauf

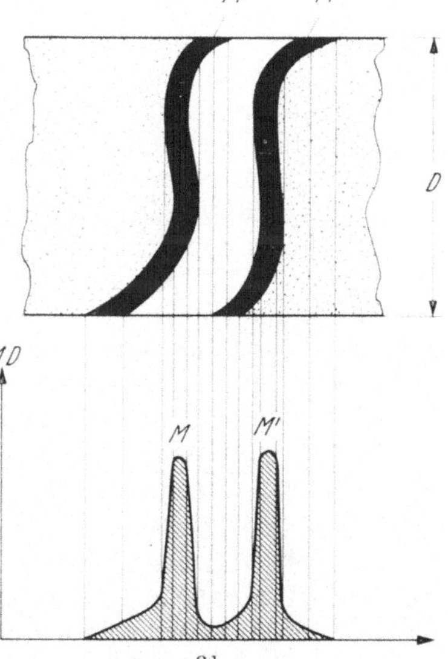

einem dicken Objekt mit den Gegebenheiten bei einem Ultradünnschnitt der Dicke (d) oder bei einem „optischen Schnitt" mit einem Abbildungssystem geringer Schärfentiefe, so ergeben sich wesentliche Unterschiede. Membranen können bis zu einer gewissen Winkellage auch im Schrägschnitt abgebildet und getrennt werden [80, 128]. Ihr Verlauf ist daher im Gegensatz zum dicken Schnitt durch Serienschnitte, bzw. durch kontinuierliches Verstellen des Fokus zu rekonstruieren. Die Kugeln sind einzeln zu erkennen. Man muß daher beim Beobachten von Vorgängen an dickeren Objektschichten im EM Bilder erwarten, die in jedem Fall wesentlich von den gewohnten Ultradünnschnittbildern abweichen. Es mag in dieser Hinsicht von Vorteil sein, daß die zunächst vorzugsweise in Frage kommenden Zellen tierischer oder menschlicher Gewebekulturen weitgehend entdifferenziert und ausgesprochen strukturarm sind, sodaß die Überlagerung von Strukturen bei ihnen unter Umständen nicht besonders stört.

4. Bildkontrast

Der Kontrast im elektronenoptischen Bild hängt im Gegensatz zur Lichtoptik vorwiegend von den Massendickedifferenzen im Objekt und von der Beschleunigungsspannung im EM ab. Der Bildkontrast nimmt mit steigenden Massendickeunterschieden zu, mit steigender Beschleunigungsspannung demgegenüber ab. Aus diesem Sachverhalt ergeben sich schwerwiegende Probleme für eine Vitalbeobachtung im EM, da man ausschließlich unfixierte, praktisch schwermetallfreie Objekte mit hohen Strahlspannungen untersuchen kann.

Vergleicht man Ultradünnschnitte von formolfixierten Objekten ohne Schwermetallimprägnation (Kontrastierung) mit Ultradünnschnitten nach Osmiumfixation oder Kontrastierung, so konstatiert man bei identischer Schnittdicke und Optik sehr große Kontrastunterschiede. Das schwermetallfreie Präparat zeigt bedeutend geringere Kontraste. In ähnlicher Weise unterscheiden sich Präparate, die man durch Auftrocknen von Viren- oder Bakteriensuspensionen auf Trägerfolien herstellt. Derartige Suspensionspräparate weisen nur dann gute Bildkontraste auf, wenn man durch Beschatten, Negativkontrastieren oder Imprägnieren Schwermetalle an- oder einlagert. Dieser Umstand beruht auf der hohen Dichte der Schwermetalle, die bereits in relativ dünnen Schichten oder geringen Konzentrationen eine hohe Massendicke erzeugen. Im lebenden Gewebe weisen nur sehr wenige Strukturen eine große Dichte, bzw. Schwermetalle auf (Pigmente, Blutfarbstoffe, mineralische Einschlüsse). Im allgemeinen sind die Dichtegradienten im wäßrigen Cytoplasma sehr gering. Die größere Dicke der einzelnen Strukturelemente im Totalpräparat und die daraus resultierenden Dickendifferenzen können die Dichtedifferenzen voraussichtlich nur zum Teil kompensieren, sie fallen insbesondere nur bei größeren Strukturen ins Gewicht. Man muß daher erwarten, daß die *fehlenden Dichteunterschiede* in hydratisierten Zellen zu ausgesprochen kontrastarmen Elektronenbildern führen. Eine Verbesserung dieser Situation darf man sich sowohl auf optischem, wie auf photographischem Weg erwarten.

Jedem Elektronenmikroskopiker sind Phasenphänomene bekannt, die bei leichter Defokussierung auftreten und den Bildkontrast teilweise beträchtlich verstärken [*2, 15, 39, 49, 111, 117*]; sie treten vor allem im Bereich der Auflösungsgrenze der heute gebräuchlichen Hochleistungsgeräte, also in der Größenordnung ≤ 20 ÅE auf und dürften demnach bei einer Vitalbeobachtung im EM weder besonders nützen noch die Arbeit stärker behindern. Will man in der hier interessierenden Größenordnung über 100 ÅE die Bildkontraste wesentlich erhöhen, so muß man andere Wege beschreiten und ein *Phasenkontrast-Elektronenmikroskop* verwenden. Nach diversen Vorarbeiten [*15, 70, 71, 75*] ist es dem Arbeitskreis um CH. FERT vor einigen Jahren gelungen, das Phasenkontrastprinzip nach dem Konzept von ZERNIKE [*134*] auf die Elektronenmikroskopie zu übertragen [*39, 40*]. Dabei wird eine gestufte Kohlefolie als Phasenplatte verwendet. Ein Verschmutzen dieser Folie während des Betriebes wird durch ein Erwärmen auf 250° C vermieden. Auf diese

Weise bleibt die Elektronentransparenz der Phasenplatte über längere Zeit konstant. Die derzeit erreichte Auflösung im Phasenkontrast-EM liegt bei etwa 20 ÅE — also jenseits der Grenze von ≥ 100 ÅE, die zunächst für eine Vitalelektronenmikroskopie in Frage käme.

Auf gänzlich andersartige Weise scheint zusätzlich ein *Steigern des Kontrastes auf photographischem Weg* durch das Ausarbeiten des Negatives mit einem elektronisch gesteuerten Vergrößerungsapparat möglich. Derartige Geräte werden beispielsweise von der Firma Logetronics hergestellt und haben sich beim Ausarbeiten von Elektronenbildern bereits vorzüglich bewährt [*119*].

5. Vergrößerung und Durchmesser des Endbildes

Für eine kinematographische Aufnahme eines Vorganges spielen Endbildvergrößerung und -durchmesser eine wichtige Rolle, da von ihnen bei vorgegebener Strahlstromstärke die Bildhelligkeit und in weiterer Folge die Zeitauflösung $_T\delta_{min}$ wie die Belichtungszeit t_{exp} abhängen. Es ist in dieser Hinsicht sehr günstig, daß in der Elektronenmikroskopie im Gegensatz zur Lichtmikroskopie stets nur ein einziges Objektiv annähernd konstanter Apertur verwendet wird, welches bei einer Eigenvergrößerung im Bereich von $100-200:1$ bereits die optimale Auflösung gewährleistet. Mit Hilfe weiterer Elektronenlinsen (Zwischenlinse und Projektiv) erzeugt man ein Endbild mit einer Vergrößerung im Bereich zwischen rd. 1000 und $200000:1$. Die hohen Endbildmaßstäbe sind derzeit zum Erreichen der Auflösungsgrenze unentbehrlich, da man über genügend feinkörnige Leuchtschirme noch nicht verfügt und damit ein hinreichend exaktes Fokussieren unmöglich ist.

Eine Reduktion des Endbildmaßstabes bei gleichbleibender Auflösung $_L\delta_{min}$ würde bei identischer Objektbelastung wesentlich *kürzere Belichtungszeiten* t_{exp} ermöglichen. Für die in diesem Referat behandelten Fragen scheint die damit erzielbare *bessere Zeitauflösung* $_T\delta_{min}$ ebenso wichtig wie *das scharfe Abbilden rasch bewegter Details*. Bei linearer Reduktion der Vergrößerung nimmt die Bildhelligkeit quadratisch zu, bzw. die Belichtungszeit mit dem Quadrat ab. Verringert man also beispielsweise den Endbildmaßstab auf 1/100 von $10000:1$ auf $100:1$, so nimmt die Bildhelligkeit am Endbildschirm bei identischer Bestrahlung des Objektes (Gleicher Strahlstrom, gleiche Kondensoreinstellung) auf das 10000fache zu. Umgekehrt kann die Belichtungszeit in diesem Fall auf 1/10000 des ursprünglichen Wertes reduziert werden. Benötigt man beispielsweise für eine Aufnahme bei 10000facher Endbildvergrößerung eine Belichtungszeit von 1 sec, so genügt bei $100:1$ bereits 0,1 msec. Damit gelangt man in den Belichtungszeitbereich, der nach Tab. 1 zum scharfen Abbilden rasch ablaufender Vorgänge im Submikroskopischen gefordert werden muß. Zudem erzielt man eine Zeitauflösung $_T\delta_{min}$ im Millisekundenbereich.

Normalerweise ist man — wie bereits oben erwähnt — am *Erfassen eines möglichst großen Objektbereiches* interessiert. Da man ohne weiteres Ultradünnschnitte in Flächen von 1 mm² herstellen und beobachten kann, ist dies sinnvoll. Man könnte auf einem Endbild der Größe 6×6 cm² bei einem Maßstab 100:1 ein Objektfeld von $0,6 \times 0,6 = 0,36$ mm² aufnehmen. Beim Beobachten relativ rasch ablaufender Vorgänge wird man jedoch unter Umständen die gegebene Situation in anderer Weise nützen. Man kann dabei vom Gedanken ausgehen, daß der interessierende Bereich in der cellulären Größenordnung liegt. Eine Objektfläche von 10×10 μm² würde also in vielen Fällen ausreichen. Diese nimmt bei 100facher Vergrößerung 1 mm² ein. Diese *geringe Endbildgröße* vereinfacht sicherlich die kinematographische Aufnahmetechnik. Die Fläche des Negativmateriales bleibt in diesem Fall sehr klein — Geschwindigkeiten und Wege beim Verschieben des Filmes von Aufnahme zu Aufnahme erreichen keine allzu großen Werte. Man wird in jedem einzelnen Fall beide Möglichkeiten — d. h. großes Objektfeld bei normaler Endbildgröße, bzw. kleines Objektfeld bei minimaler Endbildgröße — im Auge behalten und jeweils entscheiden, was besser entspricht.

Das Rechnen mit kleinen Endbildmaßstäben bleibt natürlich so lange utopisch, als man noch nicht über ein genügend feinkörniges Leuchtschirm- und Negativmaterial verfügt. Das *Leuchtschirmmaterial* [10] spielt unter den heute gegebenen technischen Voraussetzungen die entscheidende Rolle, da man ohne visuelle Kontrolle nicht hinreichend genau fokussieren kann und ohne exaktes Fokussieren keine brauchbare Auflösung erreicht. Im Moment ist die Auflösung der Leuchtschirme in den kommerziellen Elektronenmikroskopen schlechter, als die Auflösung der normalen, handelsüblichen Negativmaterialien. Da man im allgemeinen nicht auf die feinsten Strukturdetails scharf einstellt, gelingt es dem Geübten trotzdem, die Auflösungsgrenze des photographischen Negativmateriales annähernd zu nützen. Es besteht aber kein Zweifel darüber, daß feinkörnigere Leuchtschirme auch im Rahmen der normalen Routinearbeit am EM vorteilhaft wären. Man befaßt sich daher immer wieder mit diesem Konstruktionselement. Erste konkrete Resultate konnten BROSER-WARMINSKY und E. RUSKA vorlegen [16]. Einkristalle von ZnS und CdS liefern im Gegensatz zu den üblichen polykristallinen Leuchtschirmen ein kornloses Bild. Man erhält aber leider nicht die gleiche Bildhelligkeit, da das Luminescenzlicht zum großen Teil durch Totalreflexion nach den Seitenkanten abgeleitet wird. Bei CdS ist die Ausbeute an sichtbarem Luminescenzlicht zudem temperaturabhängig und erreicht erst bei Temperaturen unter $-100°$ C brauchbare Werte. Man verwendet daher ein System, das mit flüssigen Gasen auf sehr tiefe Temperaturen gekühlt werden kann. Die Leuchtschirme liegen im Hinblick auf die richtungsabhängige Lumineszenzlicht-Emission exakt senkrecht zur optischen Achse des Mikroskopes. Die ersten Resultate mit diesen Einkristall-Leuchtschirmen berechtigen durchaus zu gewissen Hoffnungen. Der Aufwand zum Kühlen der Leuchtschirme scheint zwar für die normale Routinearbeit an kommerziellen Geräten derzeit zu groß, ist für spezielle Zwecke aber sicherlich nicht übertrieben und technisch zu beherrschen. Weitere Möglichkeiten bieten *Bildwandler*, welche das Elektronenbild direkt oder über einen feinkörnigen Leuchtschirm aufnehmen [55, 85]. Nach den ersten Resultaten besteht auf diesem Wege innerhalb der prinzipiellen physikalischen Grenzen die Möglichkeit zu einem Regeln der Bildhelligkeit und des Bildkontrastes.

Die *Auflösungsgrenze des Photomaterials* scheint heute noch nicht im vollen Umfang genützt. Es existieren zahlreiche Untersuchungen und Vergleiche über das handelsübliche Negativmaterial. Bereits auf dieser Basis lassen sich Fortschritte erzielen [*1, 4, 29, 104, 113*]. Es liegen demgegenüber nach Wissen des Referenten noch keine Untersuchungen mit extremen Feinkorn-Emulsionen vor. Goldberg-Emulsionen auf LiCl-$AgNO_3$-Citronensäure-Basis liefern nach Entwickeln in Metol-Hg und fixieren in KCN eine Strichauflösung von 1 μm [*4, 50*]. Im Verein mit einer geeigneten Fokussierungshilfe wäre damit eine Reduktion des Endbildmaßstabes im Verhältnis 100:1 unter Umständen möglich. Damit könnte den Anforderungen der Vitalmikroskopie entsprochen werden. Es sind jedoch sicherlich noch umfangreiche Experimente und Entwicklungsarbeiten notwendig, um diese Grenzwerte zu erreichen.

6. Strahlenbelastung, Auflösungsgrenze und Beobachtungsdauer

Es ist einleuchtend, daß man *zum Erzeugen eines Bildes von einer bestimmten Objektfläche eine Mindestanzahl von Elektronen* benötigt. Befinden sich beispielsweise in der Mitte dieser Objektfläche zwei Punkte größerer Dichte oder Dicke, so werden drei Elektronen sicherlich noch nicht genügen, um diese Punkte vom Hintergrund abzuheben. Umgekehrt wird man sie um so klarer darstellen können, je mehr Elektronen man zum Abbilden verwendet. Dabei ist es nicht gleichgültig, in welchem Zeitraum diese Elektronen die interessierende Objektstelle penetrieren. Betrachtet man das Bild visuell, so ist der Maximalzeitraum einerseits durch die Nachleuchtzeit der Leuchtmasse, andererseits durch die Zeitauflösung, bzw. durch die bereits in Abschnitt A erwähnte Verschmelzungsfrequenz von rd. 15 Hz gegeben. Wir bezeichnen diesen Maximalzeitraum unverbindlich als *Beobachtungszeit*. Stellt man eine photographische Aufnahme her, so entspricht dieser Zeitraum der *Belichtungszeit*. In jedem Fall muß der Strahlstrom bzw. die Beleuchtungsintensität so gewählt werden, daß die Objektfläche in der Beobachtungs- bzw. Belichtungszeit von der zum Auflösen der interessierenden Strukturdetails notwendigen Elektronenzahl getroffen wird. Je höher die geforderte Auflösung $_L\delta_{min}$ ist, um so höher muß auch die Zahl dieser Elektronen liegen. Der Informationsgehalt eines Elektronenbildes hängt in jedem Fall von der Zahl der Elektronen ab, die zum Entstehen dieses Bildes beitragen [*7, 8*]. Eine Parallele hierzu ist das Abbilden eines Gegenstandes mit einem Rasterdruck. Ein objektähnliches Bild erfordert eine Minimalanzahl einzelner Rasterpunkte, die in keinem Fall unterschritten werden kann. Diese Minimalzahl ist unabhängig von der Bildgröße. So kann man mit der gleichen Punktzahl den gleichen Gegenstand auf Flächen von 1 m², 1 cm² oder 1 mm² gleich deutlich abbilden. Überträgt man diesen Sachverhalt auf die Elektronenmikroskopie, so erkennt man, daß einer Reduktion der Belichtungszeit bei gleichbleibendem Strahlstrom Grenzen gesetzt sind. Unterschreitet man diese, so muß man ein schlechteres Auflösungsvermögen in Kauf nehmen. Nach dem derzeitigen Stand unserer Kenntnisse wird diese Grenze für die Punktauflösung $_L\delta_{min}$ nur in Bereichen $\leqq 10$ ÅE bei der normalen Arbeit spürbar. Verwendet man mittlere elektronenoptische Vergrößerungen zum Erreichen eines Auflösungsvermögens im Bereich von $30-40$ ÅE, so penetrieren unter den üblichen Aufnahmebedingungen wesentlich mehr Elektronen durch die Objektfläche, als zum Erzielen dieser Auflösung notwendig wären. Es kann aber sehr wohl möglich sein, daß man durch eine *extreme Reduktion des Endbildmaßstabes* und der damit verbundenen verkürzten Belichtungszeit bei gleichbleibendem Strahlstrom (vgl. Abschnitt B. 5) ebenfalls in den kritischen Bereich vorstößt und als Folge ein reduziertes Auflösungsvermögen $_L\delta_{min}$ hinnehmen muß. Die Belichtungszeit t_{exp} bestimmt bekanntlich die Zeit- und Längenauflösung ($_T\delta_{min}$ und $_L\delta_{min}$) und kann aus diesem Grund nicht willkürlich verändert werden; vgl. Abschnitt A. Bei einem vorgegebenen Negativmaterial muß daher die Bildhelligkeit so festgelegt werden, daß mit der Belichtungszeit t_{exp} optimal belichtete Aufnahmen entstehen. Dadurch ist die Bildhelligkeit ebenfalls festgelegt. Die Bildhelligkeit selbst ist durch den Strahlstrom pro Flächeneinheit des Objektes („Strahlstromdichte") und den Endbildmaßstab gegeben. Da die Strahlstromdichte im Verein mit der Belichtungszeit t_{exp} nach dem oben Dargelegten die Auflösung $_L\delta_{min}$ bestimmt, kann sie nicht unter ein bestimmtes Maß gesenkt werden. Soweit man sich nicht mit der zufällig erreichbaren Auflösung $_L\delta_{min}$ zufrieden gibt, bleibt daher im vorliegenden Fall der Maßstab des Endbildes die einzige Variable. Er allein kann so verändert werden, daß man mit einem Photomaterial vorgegebener Empfindlichkeit die gewünschten Auflösungsgrenzen $_L\delta_{min}$ und $_T\delta_{min}$ erreicht.

Man kann demnach ein bestimmtes Auflösungsvermögen nur mit einer bestimmten Strahlbelastung des Objektes erzielen. V. Ardenne hat bereits frühzeitig erkannt, daß der *ionisierende Elektronenstrahl die entscheidende Grenze* bei jeder Vitalbeobachtung im EM setzt und daß demgegenüber andere begrenzende Faktoren von untergeordneter Bedeutung sind. Es ist relativ einfach, die Minimalbelastung für eine Einzelaufnahme bestimmter Auflösung $_L\delta_{min}$ (z. B. 100 ÅE) zu berechnen. Einzelaufnahmen „lebender Zellen" sind jedoch für den Biologen vollkommen uninteressant, da sie wiederum nur eine statische Morphologie ermöglichen, über Lebensvorgänge im Zeitkontinuum aber nicht das Geringste aussagen. Zudem ermöglichen Ultradünnschnitte durch fixiertes und eingebettetes Material ebenso lebenstreue Bilder bei wesentlich besserer Bildqualität. Im Rahmen einer Vitalmikroskopie sind daher nur Aufnahmeserien interessant. Bei derartigen Reihenaufnahmen muß die Strahlenbelastung natürlich auf das unumgänglich notwendige Mindestmaß beschränkt werden. Dies geschieht unter anderem dadurch, daß das Objekt zwischen den einzelnen Aufnahmen durch eine Sektorenblende abgeschattet wird, welche gleichzeitig die Belichtungszeit festlegt. Die Fokussierung wird nicht im Beobachtungsbereich, sondern in der unmittelbaren Nachbarschaft durchgeführt [9]. Die Minimalbelastung pro Einzelaufnahme muß bei diesem Vorgehen mit der Aufnahmezahl in der Serie multipliziert werden, welche durch die Dauer des Vorganges und die Geschwindigkeit der dabei ablaufenden Bewegungen festgelegt wird. Die danach kalkulierte auflösungsabhängige Maximalbelastung beim Aufnehmen des interessierenden Vorganges kann man mit der maximalen Belastbarkeit des Objektes in bezug setzen.

V. ARDENNE berechnet die Relation „tragbare Elektronenbelastung-Auflösung" für Einzelbilder auf der Basis der radiologischen Fachliteratur [6–8]. Gemeinsam mit FRIEDRICH-FREKSA führte er die erste experimentelle Arbeit zur Empfindlichkeit von Bakteriensporen im EM durch [9]. DUPOUY, PERRIER und DURRIEU stellten fest, daß Bakterien eine Elektronenbestrahlung überleben, welche für EM-Aufnahmen ausreicht [31, 33]. Leider erlauben aber diese interessanten Einzelresultate keine verbindlichen Schlüsse oder Voraussagen. Es ist im Moment noch nicht klargestellt, welche Schäden eine Vitalmikroskopie im Elektronenstrahl begrenzen. Üblicherweise wird zum *Beurteilen eines Strahlenschadens* der genetische Test herangezogen (Veränderte Teilungsrate; Stillstand der Vermehrung). Da ein Beobachten von Lebensvorgängen in jedem Fall auf Zeitspannen beschränkt werden muß, die weit unterhalb der Interphasezeit zwischen zwei Zellteilungen liegen, erscheint der genetische Test nicht adäquat. Strahlungsbedingte Letalfaktoren am genetischen Material der Zelle schließen eine erfolgreiche Vitalbeobachtung im EM nicht aus. Andere Strahlenschäden treten ebenfalls erst nach einer gewissen Latenzzeit auf. Die Strahlen setzen die physikochemischen Veränderungen an verschiedensten Molekeln. Ein Großteil dieser Prozesse dürfte harmlos verlaufen. Werden jedoch Schlüsselstellen getroffen, so führen diese Treffer mit der Zeit zu einem qualitativ und quantitativ verändertem Stoffwechsel. Es ist von vorneherein anzunehmen und durch die radiologische Praxis belegt, daß diese Veränderungen nicht schlagartig, sondern erst nach einer gewissen *Latenzperiode* morphologisch wahrnehmbar sind. Die Länge dieser Latenzperiode hängt von der Intensität des Stoffwechsels ab, ist also von Zelltyp zu Zelltyp verschieden. Diese Unterschiede zeigen sich naturgemäß auch in einer unterschiedlichen Strahlenempfindlichkeit verschiedener Zellen, bzw. Gewebe. Faßt man diese Fakten zusammen, so kommt man zu folgenden Schlüssen:

a) Verbindliche Unterlagen zum Berechnen der tragbaren Strahlenbelastung im Rahmen einer Vitalbeobachtung im EM fehlen derzeit.

b) Die Latenzperiode zwischen dem Elektronenbeschuß und dem Auftreten morphologisch faßbarer Strahlenschäden kann im Kurzzeitversuch eine Strahlendosis erlauben, welche die normale Letaldosis unter Umständen wesentlich übersteigt.

c) Die Dauer der Latenzperiode (b) ist wahrscheinlich ebenso wie die Strahlenempfindlichkeit vom Zellstoffwechsel abhängig und von Zelltyp zu Zelltyp verschieden. Die Auswahl geeigneter, möglichst strahlenresistenter Objekte ist daher äußerst wichtig.

Im einzelnen ist das Problem wohl so komplex, daß man die Verhältnisse wie die gegebenen Möglichkeiten nur empirisch abklären kann. Erst danach wird man wahrscheinlich eine bessere Definition für die im

EM kurzfristig über $^1/_2$ bis 2 min zulässige Strahlendosis für verschiedene Objekte bzw. Objektpartien geben können[1]. In jedem Fall wäre es aber falsch, wenn man Versuche zu einer Vitalmikroskopie im EM im Hinblick auf die Schäden unterlassen würde, welche der ionisierende Elektronenstrahl in lebenden Zellen hervorruft.

---

[1] Bahr et al. [Lab. Invest. **14**, 1115 (1965)] studierten in einer Spezialapparatur Strahlenschäden an Filmen aus organischem Material. Nach ihren Befunden treten bereits nach $2 \cdot 10^{-2}$ Coul $\cdot$ cm^{-2} deutliche Defekte auf. Zum Herstellen einer EM-Aufnahme (Einstellen des Focus und Exponieren des Photomaterials) wird das Objekt normalerweise mit rd. $1 \cdot 10^{-1}$ Coul $\cdot$ cm^{-2} entsprechend rd. $6 \cdot 10^{17}$ Elektronen $\cdot$ cm^{-2} belastet. Demgegenüber liegt nach Heide [Lab. Invest. **14**, 1135 (1965)] die Minimaldosis für ein Elektronenbild mit $_L\delta_{min} \sim 150$ ÅE nur bei rd. $1 \cdot 10^{13}$ Elektronen $\cdot$ cm^{-2}; dies entspricht wiederum nach Ennos [Brit. J. appl. Physics **4**, 101 (1953); zit. nach Heide] bei biologischem Material mit 100 kV Strahlspannung etwa $1 \cdot 10^6$ r. Es ist im Moment sehr schwierig, Modellversuche wie Gegebenheiten bei der normalen EM-Arbeit an trockenen Objekten [vgl. Bahr et al., l. c. sowie Reimer, Lab. Invest. **14**, 1083 (1965); dortselbst weitere Literaturhinweise] mit Objektschäden in bezug zu setzen, die beim kurzfristigen Bestrahlen von lebenden Zellen mit höchsten Elektronendosen bei hohen Beschleunigungsspannungen $\geqq 500$ kV auftreten. Dis bislang als Testobjekte verwendeten Bakterien [*31*, *33*] oder Bakteriensporen [*9*] lassen ebenfalls keine bindenden Rückschlüsse auf normale Zellen zu, da ihre Strahlenresistenz kaum gleiche Werte aufweisen dürfte. Zudem würde man beim Beobachten von Bakterien aller Voraussicht nach Auflösungen $_L\delta_{min} \ll 150$ ÅE benötigen. Einen Anhaltspunkt geben jedoch u. U. die Befunde von Schneider [Protoplasma **53**, 530 (1961)], welcher Paramecien mit 10^6 r $\cdot$ min^{-1} ein bis vier Minuten lang bestrahlte. Nach $4 \cdot 10^6$ r zeigen 20% der Tiere eine erhöhte Aktivität, der Rest eine Strahlenlähmung. Im Plasma und in den Zellorganellen sind zwar deutliche Veränderungen zu erkennen — von einer totalen Destruktion kann aber keine Rede sein. So sind beispielsweise die Mitochondrien noch deutlich zu erkennen. Dabei bleibt zu berücksichtigen, daß geschädigte Objekte meist fixierungslabil sind. Darüber hinaus setzt die Fixation nicht schlagartig ein; weitere strahlenpathologische Veränderungen liegen demnach ebenfalls im Bereich der Möglichkeit. Das Objekt dürfte daher beim Abschluß der Bestrahlung besser erhalten sein, als nach dem Ultradünnschnittbild zu erwarten wäre. Schließlich besteht die Möglichkeit, nicht das gesamte Areal, sondern lediglich strahlenresistente Teilgebiete (z. B. Ausläufer) einer Zelle im Feinstrahl zu untersuchen; dadurch könnte die Dosis unter Umständen beträchtlich erhöht werden. Dies wird beispielsweise durch Resultate gezielter Röntgenbestrahlungen der kernhaltigen bzw. kernlosen Hälfte von *Drosophila*-Eiern nahegelegt [Ulrich, H., zit. nach Scherer, E.: In: E. Scherer u. H.-S. Stender: Strahlenpathologie der Zelle. Beiträge zur Cyto- und Histopathologie der Strahlenwirkung. Stuttgart: G.

C. Rekonstruktion von Vorgängen nach Einzelaufnahmen
1. Zeitauflösung beim Rekonstruieren

Im einleitenden Abschnitt A wurde die Bedeutung der Zeitauflösung $_T\delta_{min}$ für die Rekonstruktion von relativ rasch ablaufenden Vorgängen im Submikroskopischen eingehend diskutiert. Besonders wichtig scheint im vorliegenden Problemkreis daher ein sinnvolles Abstimmen der Zeitauflösung auf die Geschwindigkeit des interessierenden Prozesses. In lebenden Zellen gibt es Vorgänge, die in wenigen Millisekunden ablaufen. Auf der anderen Seite kennt der Pathologe natürliche und experimentell induzierte Krankheitsabläufe, die von der Ursache bis zum vollen Entfalten aller Symptome Monate bis Jahre benötigen. Zwischen beiden Extremen gibt es fließende Übergänge. Alle genannten Vorgänge verschiedener Geschwindigkeit und Dauer fallen in den Rahmen dieses Referates. Es ist daher nicht zu umgehen, die Eignung der verschiedenen Verfahren für bestimmte Vorgänge kurz zu diskutieren.

Eine *Vitalbeobachtung im Submikroskopischen* muß sich — soweit sie in Zukunft überhaupt realisierbar ist — aller Voraussicht nach auf *kurze Perioden rasch ablaufender Prozesse* beschränken; vgl. vorangehenden Abschnitt B. Dafür scheinen sehr kurze Belichtungszeiten und damit eine Zeitauflösung $_T\delta_{min}$ im Bereich von Sekundenbruchteilen prinzipiell möglich. Demgegenüber ist eine Vitalbeobachtung langsamer Vorgänge im EM kaum sinnvoll. Dies gilt bereits für *Vorgänge, welche eine Zeitspanne von mehreren Minuten überschreiten*. In diesen Fällen führt eine Rekonstruktion nach Einzelaufnahmen mit der heute gängigen Methodik der Ultramikrotomie auf einfachere Weise und schneller zu wesentlich besseren Resultaten, als man sie zunächst bei einer Vitalbeobachtung im EM erwarten darf.

Die Zeitauflösung einer Rekonstruktion hängt in den meisten Fällen davon ab, wie präzise man den interessierenden Vorgang einleiten und abstoppen kann, bzw. wie genau man den Zeitpunkt der Fixation auf bestimmte natürliche Vorgänge abstimmen kann. Eine Ausnahme bilden lediglich die Rekonstruktionen an Hand einer Leitstruktur; vgl. den folgenden Abschnitt C. 2. Soweit man einen Vorgang experimentell einleitet, ist die *zeitliche Präzision dieses Startes* ein limitierender Faktor. Nur beim Anwenden von mechanischen oder elektrischen Reizen können größere Objektbereiche annähernd gleichzeitig erfaßt werden. Bei den meisten experimentell-pathologischen oder pharmakologischen Studien müssen demgegenüber zumindest Diffusionsgeschwindigkeit und Permeabilität von Zellgrenzen in den Kreis der Betrachtung einbezogen

---

Thieme-Verlag 1963, p. 47]: Das Abtöten der kernlosen Hälfte erfordert eine über 180fach höhere Dosis, als das Abtöten des kernhaltigen Teiles. Naturgemäß können diese Befunde lediglich Hinweise geben. Diese zeigen aber immerhin, daß Chancen bestehen, mit „letalen Elektronenstrahlendosen" unter den angeführten Bedingungen eine „Vitalelektronenmikroskopie" zu realisieren. (Fußnote anläßlich der Korrektur angefügt)

werden. Soweit man die Pharmaka nicht gezielt applizieren kann, sondern auf Resorption aus dem Darm, Peritoneum oder Blutkreislaufsystem angewiesen ist, muß man die dadurch eingeführte Verzögerung und zeitliche Unsicherheit ebenfalls einkalkulieren. Wird beim medikamentösen Induzieren eines Vorganges ein hohes Zeitauflösungsvermögen verlangt, so kommen nur bestimmte Objekte in Frage. Hierzu zählen vor allem Mikroorganismen, isolierte Zellen von Gewebekulturen und Organoberflächen [*77, 88*]. Darüber hinaus kann man durch Perfusion [*87, 108*] und Mikroperfusion [*116*] in Einzelfällen exakte Startbedingungen erzielen. Es ist jedoch auch in diesen günstig gelagerten Fällen kaum möglich, die Zeitauflösung $_T\delta_{min}$ unter 10 sec zu senken.

Auf analoge Probleme trifft man beim *Fixieren eines bestimmten Zustandes*. Auch dieser Endpunkt des Versuches kann nicht beliebig genau festgesetzt werden. Die übliche *Fixation mit Lösungen* (OsO_4, $KMnO_4$, Aldehyde) beruht auf einem Diffusionsvorgang mit einer Geschwindigkeit in der Größenordnung von 10 μm/sec [*94, 108, 110, 131*]. Auch sehr kleine Gebilde werden von der Front des eindiffundierenden Fixans nicht schlagartig, sondern erst allmählich erfaßt. Für das Durchdringen einer freien Einzelzelle mit einem Durchmesser von 10 μm muß man bei allseitigem Angriff des Fixans etwa 30 sec rechnen. Darüber hinaus muß man berücksichtigen, daß beim Eintreffen des Fixans die Lebensvorgänge nicht augenblicklich zum Stillstand kommen. Man wird also nicht weit fehlen, wenn man bei der Fixation mit verschiedenen Lösungen eine Zeitauflösung $_T\delta_{min} \geqq 30$ sec schätzt. Dieser Wert gilt für freie Einzelzellen aus Kulturen, Protozoen, Organoberflächen bis zur Tiefe einer Zellage, Perfusion und Mikroperfusion. Bei der üblichen Blöckchenfixation wird $_T\delta_{min}$ in tieferen Zonen bedeutend größer (z. T. > 30 min), bei kleinen Mikroorganismen (Bakterien) kann man wahrscheinlich Werte in der Größenordnung $_T\delta_{min} \geqq 1$ sec erzielen. Neben dem Fixieren in Lösungen besteht noch eine weitere Möglichkeit: Man kann die *Objekte rasch einfrieren* [*43, 56, 57, 82—84, 94, 103, 112*]. Hier bestimmt der Temperaturgradient die Geschwindigkeit dieser „physikalischen Fixation". Man wählt daher möglichst tiefe Umgebungstemperaturen. Da die meisten verflüssigten Gase um wärmere Körper augenblicklich einen wärmeisolierenden Gasmantel bilden (Leidenfrostsches Phänomen), kommt ein direktes Einbringen der Objekte in diese Medien nicht in Frage. Man muß entweder eine Flüssigkeit mit entsprechend niedrigem Gefrierpunkt (z. B. Isopentan) zwischenschalten oder ein Metall mit gutem Wärmeleitvermögen einführen, das mit dem verflüssigten Gas über eine größere Wärmeaustauschfläche in Kontakt steht [*57*]. Schließlich kann man flüssiges Helium-II verwenden, das infolge seines besonderen physikochemischen Zustandes eine minimale Viskosität und ein optimales Wärmeleitvermögen aufweist und daher das Leidenfrostsche Phänomen nicht zeigt [*18, 44*]. Es ist jedoch auch mit diesem Verfahren nicht möglich, größere Objektbereiche — beispielsweise die üblichen Blöckchen in Größen von 1—10 mm³ — schlagartig zu fixieren, wie häufig angenommen wird. Der Grund ist die rasche Abnahme des Temperaturgradienten. Dies zeigt ein Gedankenexperiment: Unmittelbar nach dem

Einbringen in das Kühlmedium möge der Temperaturgradient in der Randzone beispielsweise 200° C/μm betragen. Der von ihm abhängige Wärmetransfer ist in diesem Fall sehr rapid, die äußerste Randzone friert daher tatsächlich sehr rasch ein. Nach kurzer Zeit ist die „Kältefront" etwa 10 μm tief in das Objekt eingedrungen. Der Gradient beträgt nur mehr 200° C/10 μm = 20° C/μm. Dies ist nur mehr ein Zehntel des Startwertes im vorliegenden Fall und entspräche damit dem Anfangswert bei einem Kühlmedium mit einer Temperatur von lediglich −20° C. Der Wärmetransfer ist entsprechend geringer, die Kältefront dringt wesentlich langsamer vor. Man darf demnach nicht vergessen, daß die Geschwindigkeit der Kältefront in einem größeren Objekt nicht allein von der Temperatur, Viscosität und dem Wärmeleitvermögen des Kühlmediums abhängt. Der Effekt relativ einfacher konventioneller Methoden und des aufwendigen He-II-Verfahrens ist in derart gelagerten Fällen praktisch der gleiche [57].

2. Rekonstruktion nach Leitstrukturen

Häufig sind Vorgänge im Submikroskopischen mit deutlich wahrnehmbaren Veränderungen im mikro- oder makroskopischen Bereich gekoppelt. In diesen Fällen besteht die Möglichkeit einer zeitlichen Orientierung auch dann, wenn eine Fixation zu einem bestimmten Zeitpunkt nicht sicher durchführbar ist. Die Zeitauflösung $_T\delta_{\min}$ im Submikroskopischen ist hier im wesentlichen durch die Zeitauflösung im mikro- oder makroskopischen Bild gegeben und kann damit sehr gute Werte im Sekundenbereich oder sogar darunter erreichen.

Ein typisches Beispiel liefern elektronenoptische Untersuchungen am Nephridialapparat von Paramecium. Dieser zeigt bereits im lichtoptischen Bild einen zeitlich genau faßbaren rhythmischen Formwandel; SCHNEIDER [100] war an Hand dieser lichtoptisch klar darstellbaren Leitstrukturen in der Lage, die Vorgänge im submikroskopischen Bereich zeitlich genau einzuordnen. Während der Systole der contractilen Vacuole ist das endoplasmatische Reticulum an die erweiterten Radialkanäle angeschlossen — während der Diastole sind keine derartigen Verbindungen zu beobachten. Innerhalb weniger Sekunden findet also jeweils ein kompletter Umbau dieses Systems statt. Wäre die Leitstruktur nicht vorhanden, so wäre es kaum möglich gewesen, beide Zustände so klar voneinander zu trennen und zeitlich zu erfassen. In prinzipiell ähnlicher Weise konnten FLOREY und GRANT [45] die submikroskopischen Vorgänge bei der Leukodiapedese im Entzündungsfeld an Hand des lichtoptisch faßbaren und gut bekannten Ablaufes darstellen. Dieses Verfahren ist so einfach und elegant, daß man es jeder anderen Methode zum Beobachten von Vorgängen im Submikroskopischen vorziehen wird, soweit das Objekt Leitstrukturen aufweist.

3. Vorgänge im Fließgleichgewicht — Rekonstruktion mit Markierungsmethoden

Vorgänge im Rahmen eines Fließgleichgewichtes („Steady state") kommen nicht nur häufig vor — sie sind geradezu typisch für lebende

Organismen und wurden an diesen erstmals erkannt und analysiert [*12, 30, 66*]. Es ist daher von Interesse, daß man derartige Vorgänge *ohne zusätzliche Hilfsmittel nach Einzelbildern nicht rekonstruieren* kann. Das Modell in Abb. 4 gibt hierfür ein Beispiel. In einem Rohr befindet sich eine Flüssigkeit, in der Kugeln (K) enthalten sind. Fixiert man den augenblicklichen Zustand im Objektfeld (a), so kann man aus diesem Einzelbild keine Aussage darüber ableiten, ob die Flüssigkeit und die Kugeln im Rohr still stehen, sich von links nach rechts oder umgekehrt von rechts nach links bewegen. Wiederholt man die Fixation an einem anderen gleichartigen Rohr, so erhält man ein analoges Bild. Eine Vielzahl derartiger Bilder erlauben zwar eine Aussage über die durchschnittliche Anzahl der Kugeln pro Volumeinheit der Flüssigkeit (= Konzentration), sagen aber nichts über Richtung und Geschwindigkeit der Flüssigkeitsbewegung aus. Der am meisten interessierende Vorgang läßt sich also nicht fassen. Dies beruht darauf, daß die im Abschnitt A angeführte Voraussetzung (a) nicht erfüllt ist. Der Vorgang ist nicht wiederholbar, sondern verläuft im Rahmen des Gleichgewichtes kontinuierlich. Ein Beispiel für ein derartiges System gibt die Druckfiltration im Nierenkörperchen. Das Lumen der glomerulären Blutcapillare ist gegen das Lumen der Bowmanschen Kapsel durch eine dreischichtige Capillarwand abgegrenzt. Diese besteht aus einer Basalmembran, welche von einem dünnen Endothelhäutchen auf der einen, einer speziell ausgebildeten Epithellage auf der anderen Seite bedeckt wird. Endothel und Epithel weisen charakteristische Poren auf. Der Primärharn wird aus der Blutcapillare durch die Capillarwand in den Kapselraum abgepreßt. Größere Molekel (z. B. Proteine des Blutplasmas) bleiben dabei im Blutraum zurück. Morphologisch ist lediglich der Aufbau der Capillarwand faßbar. Cytochemisch kann am Schnittpräparat der Eiweißgehalt des Blutplasmas, bzw. die Eiweißfreiheit des Primärharnes nachgewiesen werden. — Eine Aussage über den Vorgang der Harnbildung ist nach Einzelaufnahmen vom fixierten Präparat nicht möglich. Hierzu wäre — wie auch im dargestellten Modellfall der Abb. 4 — ein fortlaufendes Beobachten im Sinne der gegebenen Definitionen nötig; vgl. Abschnitte A und B. Demgegenüber ist eine *Veränderung der Fließgleichgewichtslage faßbar*. Sie ist ein Vorgang, welcher der Bedingung (a) im Abschnitt A entspricht. So ist es beispielsweise auf verschiedene Art möglich, die Capillarwand zu schädigen. Dabei kann sich die Basalmembran verändern, es kann das Epithel quellen usw. Zugleich mit diesem morphologisch faßbaren Vorgang verschiebt sich auch die Funktion, oder anders ausgedrückt, die Lage des Fließgleichgewichtes [*13, 106*]. Aus der Blutcapillare tritt jetzt ein Primärharn aus, der sich nach Qualität und Quantität vom normalen Primärharn unterscheidet. Findet im Laufe der Zeit eine restitutio ad integrum statt, so kann auch diese Rückbildung der Strukturen an Hand von Einzelaufnahmen von fixierten Objekten erfaßt und rekonstruiert werden. Über die Vorgänge im Rahmen der jeweils eingestellten Fließgleichgewichte läßt sich aber zu keinem Zeitpunkt aus Einzelaufnahmen nach einer Fixation irgendeine verbindliche Aussage ableiten.

Es existiert jedoch ein Kunstgriff: Eine *Markierung* ermöglicht Einblicke in den Vorgang, der im Rahmen eines Fließgleichgewichtes abläuft. Abb. 4 gibt dafür ein Beispiel: Man spritzt zum Zeitpunkt t_0 an der Stelle (A) eine Suspension in das Rohr, die als Markierung (M) kleine Kügelchen enthält. Bewegt sich die Flüssigkeit im Rohr in Pfeilrichtung von rechts nach links, so werden diese Teilchen mitgeführt und mögen

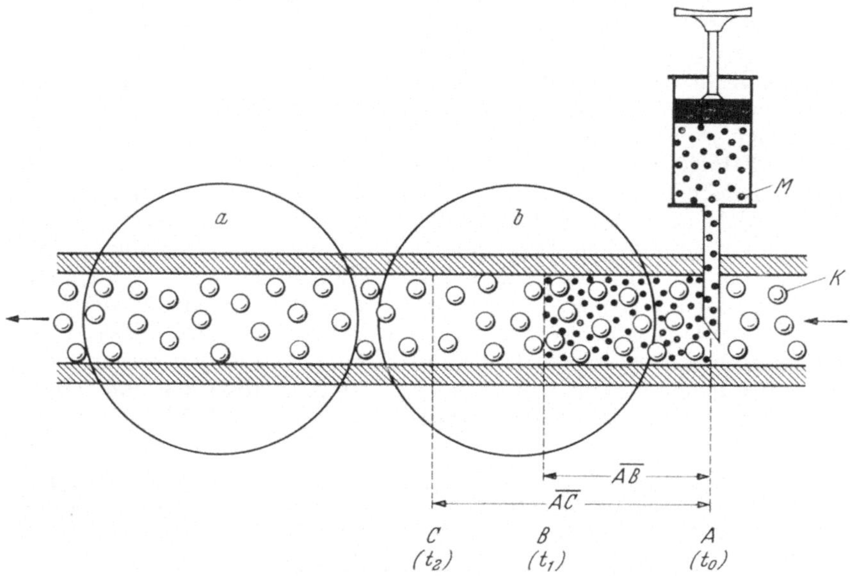

Abb. 4. Studium von Vorgängen im Fließgleichgewicht (steady state) durch Überlagern eines zweiten Prozesses (Einführen einer Markierung = *M*). Vgl. Text

zum Zeitpunkt t_1 die Stelle (B), zum Zeitpunkt t_2 die Stelle (C) im Gesichtsfeld (b) erreichen. Fixiert man zum Zeitpunkt t_1 oder t_2, so kann man die Längen $\overline{AB}$ und $\overline{AC}$ bestimmen. Kennt man Entfernungen und Zeitspannen ($\overline{AB}/t_1$ bzw. $\overline{AC}/t_2$), so kann man daraus Richtung und Geschwindigkeit der Flüssigkeitsbewegung und damit die entscheidenden Daten des Fließgleichgewichtes ermitteln. Dies ist jetzt möglich, weil das Markieren einen zusätzlichen Vorgang einleitet, der zwar mit dem Vorgang im Fließgleichgewicht gekoppelt ist, trotzdem aber beliebig oft wiederholt werden kann. Der Bedingung (a) im Abschnitt A ist damit entsprochen. Die zeitliche Auflösung $_T\delta_{\min}$ ist auch in diesem Fall durch die zeitliche Präzision der Markierung (Einleitung des Vorganges) und der Fixation gegeben. Sie liegt daher in den Grenzen, die in Abschnitt C. 1 besprochen wurden.

Als Indikatoren (Markierungssubstanzen) kamen zunächst vorzugsweise Partikel in Frage, welche Schwermetalle enthalten und dadurch im EM gute Kontraste liefern. Hierbei wählte man einerseits *schwermetallhaltige organische Makromolekel*, beispielsweise Hämoglobin [*81*], Ferritin [*41*], Myofer [*99*], sowie Porferrin [*23*], andererseits *anorganische*

Kolloide bzw. Kristalloide wie Quecksilbersulfid [78], Thorotrast [23, 91] oder Gold [51]. Man konnte auf diese Weise über viele Prozesse wertvolle Aufschlüsse erhalten, welche mit anderen Methoden nicht analysiert werden konnten. Als Beispiele seien hier die Untersuchungen über den Durchtritt von Kolloiden durch die glomeruläre Capillarwand [41], die Resorption von Eiweiß aus den Tubuli der Säugerniere [81], der Austritt von Plasmaeiweiß aus den Blutcapillaren im Entzündungsfeld [78, 91], die Resorption aus dem Darmlumen [99, 126] und die Verdauungsprozesse in Gewebekulturzellen [51] genannt. Die Zahl derartiger Beispiele ließe sich vermehren. Sie demonstrieren die umfangreichen Möglichkeiten, welche diese Markierungssubstanzen bei der Analyse von Fließgleichgewichten offerieren.

Die Erfolge mit kolloidalen oder makromolekularen Indikatoren können jedoch nicht darüber hinwegtäuschen, daß eine derartige Markierung nicht allen Ansprüchen gerecht wird. Die Indikatorpartikel sind in gewissem Umfang unphysiologisch — verhalten sich also anders, als die Biokolloide und die anderen, am Fließgleichgewicht beteiligten kleineren Molekel. Weitere Fortschritte waren nur von der in der Biochemie bereits vielfach erprobten und bewährten *radioaktiven Markierung* zu erwarten. Es ist daher sehr wichtig, daß in den vergangenen Jahren Methoden für eine *Autoradiographie im EM* entwickelt und bereits mit Erfolg angewendet werden konnten [11, 20—22, 52, 53, 79, 90, 98].

Tabelle 2. *Verteilung der Silberkörner (%) über verschiedenen Zellstrukturen der exokrinen Pankreas zu unterschiedlichen Zeiten nach der Injektion einer markierten Aminosäure* (DL-Leucin-4,5-H³). *Ergebnisse einer autoradiographischen Studie im EM.* Nach CARO u. PALADE [21]

Zeit [min]	Endoplasmatisches Reticulum + RNS + Umgebung	Golgifeld	Zymogengrana
4	67	27	1
6	53	39	2
20	11	73	10
240	11	10	73

Anstelle der Kolloide appliziert man nunmehr radioaktiv markierte Molekel und behandelt die Objekte daraufhin nach den gängigen Methoden der Ultramikrotomie (Fixation — Einbetten — Schneiden [89])[1]. Die Ultradünnschnitte werden mit einer feinkörnigen Emulsion überzogen, die nach dem Exponieren und Entwickeln im Bereich der radioaktiven Einlagerungen Silberkörner zeigt. Die Trennschärfe dieser Methode liegt noch nicht im Bereich der Auflösungsgrenze des EM, ist aber für sehr viele Zwecke bereits vollkommen ausreichend. Ein gutes Beispiel geben Studien über den Sekretionsprozeß in der Pankreas, die CARO und

---

[1] Die wesentlichste Grenze der EM-Autoradiographie liegt in einem ortsgetreuen Fixieren der radioaktiven Indikatoren; dies gilt insbesondere für niedermolekulare Markierungssubstanzen.

Palade ausgeführt haben [20, 21]. Die Autoren konnten zeigen, daß die markierten Aminosäuren zunächst im Bereich des Ergastoplasmas (RNS-haltiges = rauhes endoplasmatisches Reticulum), darauf im Golgiapparat und schließlich in verschiedenen Reifungsstadien der Sekretgrana zu finden sind; vgl. Tab. 2. Damit konnte der bislang nicht einwandfrei belegte Gang der Sekretbildung verbindlich rekonstruiert werden.

4. Morphometrische Verfahren zum Rekonstruieren von Vorgängen

Beim Rekonstruieren ist man stets auf Einzelaufnahmen angewiesen, welche verschiedene zeitlich aufeinanderfolgende Zustände an verschiedenen Objekten wiedergeben. Da die Vorgänge im cellulären Bereich meist aus einem Verschieben von Größenverhältnissen bestehen, ist ein quantitatives Auswerten der Schnittbilder oft vorteilhaft — bei geringen Veränderungen teilweise unerläßlich. Bei Vorgängen im Submikroskopischen trifft man neben einer Bewegung der verschiedenen Strukturen häufig eine Veränderung der Volumanteile (Quellung — Entquellung) oder Flächenanteile (Membrantransformation). Es ist daher von großem Interesse, Volum- und Flächenanteile im räumlichen Objekt nach Schnittbildern zu ermitteln.

Man verfügt heute über „*stereologische Verfahren*" [61, 64, 65, 109, 124, 125], die jeden Praktiker in die Lage versetzen, ohne größeres Instrumentarium das notwendige morphometrische Auswerten selbst vorzunehmen. Ein kompliziertes Planimetrieren ist hierbei ebenso wenig notwendig, wie zeitraubende Längenmessungen mit dem Kurvimeter. Man verwendet im allgemeinen die *Punktzählverfahren*, die zunächst von Geologen und Mineralogen entwickelt und in den letzten Jahren für biologische Zwecke adaptiert wurden [59, 65, 109, 118, 125].

Der große Nutzen dieser quantitativen Verfahren liegt darin, daß man auch Vorgänge rekonstruieren kann, die mit relativ geringen morphologischen Veränderungen verbunden sind. Führt beispielsweise ein experimentell-pathologisches Experiment zu einem Anschwellen der Mitochondrien um 5%, so ist diese Differenz zur Norm nur morphometrisch zu erfassen. Beim normalen Betrachten der EM-Aufnahmen kann man derartige Veränderungen unter Umständen übersehen. Trotzdem können sie für den Zellstoffwechsel — soweit es sich um Schlüsselpunkte handelt — sehr wichtig sein. Das gleiche gilt für die vieldiskutierte Membrantransformation [130], die mit der Nadelmethode klar registriert werden kann. Man wird daher in derart gelagerten Fällen am besten routinemäßig eine quantitativ-morphometrische Analyse des Vorganges durchführen. Die Resultate erlauben an Hand der Verlaufskurven meist klare, statistisch gesicherte Rückschlüsse.

D. Diskussion

Für ein Beobachten von Vorgängen im Submikroskopischen wurden zwei grundsätzlich verschiedene Möglichkeiten diskutiert: Das fortlaufende Beobachten des Ablaufes (Vitalmikroskopie) auf der einen —

die Rekonstruktion des Vorganges nach Einzelaufnahmen von fixierten Objekten auf der anderen Seite. Relativ langsam ablaufende Vorgänge, die Minuten oder noch längere Zeiten in Anspruch nehmen, kann man nach Einzelaufnahmen rekonstruieren, soweit sie gewissen Bedingungen entsprechen (Abschnitt A). Damit sind sehr viele Prozesse den heute gebräuchlichen Methoden der Elektronenmikroskopie ohne weiteres zugänglich. Soweit der interessierende Prozeß mit einem charakteristischen Strukturwandel im lichtmikroskopischen oder makroskopischen Bereich verknüpft ist, kann man sich zeitlich nach der Veränderung dieser „Leitstrukturen" orientieren. Andernfalls muß man den Ablauf durch genaues Festlegen des Anfangs- und Endzeitpunktes (Fixation) rekonstruieren. Die hierbei erreichbare Genauigkeit bestimmt das „Zeitauflösungsvermögen" $_T\delta_{min}$ der Analyse (Abschnitt A). Dieses kann beim Rekonstruieren von Vorgängen im cellulären Bereich nach Einzelbildern normalerweise kaum wesentlich unter 30 sec gesenkt werden (Abschnitt C. 1). Diese Grenze scheint prinzipieller Natur. Sowohl das Eindringen von fixierenden Molekeln, wie der Wärmetransfer beim Einfrieren folgt den Gesetzen der Diffusion. Den Anfangsgradienten der Konzentration (Chemische Fixation) und Temperatur (Physikalische Fixation) sind durch Sättigung (Löslichkeit) und absoluten Nullpunkt Grenzen gesetzt. Die Gradienten nehmen zudem im Laufe des Fixationsvorganges schnell ab. Man kann daher weder durch die maximale Konzentration der Fixierungslösung, noch durch Verwenden eines Kühlmittels minimaler Temperatur und besonderer Beschaffenheit räumlich ausgedehnte Objekte schlagartig fixieren (Abschnitt C. 1). Ebenso scheint zunächst eine Rekonstruktion von Vorgängen im Rahmen eines Fließgleichgewichtes nicht möglich. Dieses Problem kann jedoch durch einen Kunstgriff gelöst werden: Man führt markierte Partikel in das System ein (Abschnitt C. 3). Alle genannten Vorgänge können im Rahmen der Rekonstruktion mit den stereologischen Punktzählverfahren auf einfache Weise quantitativ-morphometrisch erfaßt werden. Dies ist besonders bei geringgradigen Strukturänderungen nützlich, da man diese beim qualitativen Betrachten und Auswerten der Elektronenbilder häufig übersieht (Abschnitt C. 4).

Im Gesamten sind die damit bereits heute gegebenen Möglichkeiten beim Beobachten von Vorgängen im Submikroskopischen so weitreichend und faszinierend, daß man sehr leicht den großen Bereich des derzeit Unerreichbaren übersieht, der ausschließlich durch ein fortlaufendes Beobachten erschlossen werden kann.

Alle Verfahren der Fixation unterbrechen einen Vorgang irreversibel und schließen damit ein Fortsetzen der Untersuchung am gleichen Objekt aus. Dies gilt für die chemische Fixation ebenso wie für die physikalische: Gefrorene Zellen, von denen man bereits Anschnitte und Abdrucke hergestellt hat, sind weder lebend, noch sind sie weiter lebensfähig. Man kann die gleichen Zellen nicht weiter beobachten − Fixation und fortlaufendes Beobachten lassen sich scheinbar grundsätzlich nicht vereinen[1]. Nachdem der Fixationsvorgang die Zeitauflösung $_T\delta_{min}$ limitiert, können schneller ablaufende Prozesse (vgl. Tab. 1) in jedem Fall nur mit grundlegend anderen Methoden erfaßt werden. Neben diesen rasch ablaufenden Prozessen gibt es noch weitere Vorgänge, die man am fixierten Material kaum erfolgreich studieren wird, weil der hierfür notwendige Aufwand das erträgliche Maß übersteigt. Es handelt sich um kurzdauernde Prozesse, welche auf längere Ruheperioden folgen. Ist (T) die Zeitdauer der Ruhe, (t) diejenige des Prozesses und $t \ll T$, so wird es mit steigendem (T) und sinkenden (t) immer unwahrscheinlicher, daß man den Vorgang im fixierten Objekt erfaßt. Dabei braucht der Prozeß nicht besonders selten aufzutreten, wie folgendes Beispiel zeigt: Beobachtet man eine instabile Emulsion, so sieht man im mikroskopischen Bild (Abb. 5a) immer wieder die Konfluenz einzelner Tröpfchen zu größeren Aggregaten. Diese Konfluenz verläuft schlagartig — wir nehmen nach Abb. 5b hierfür eine Zeit von 0,1 sec an. Sie ist aber einwandfrei zu beobachten und im mikroskopischen Bild beim fortlaufenden Beobachten kaum zu übersehen. Man kann im Rahmen einer derartigen Beobachtung die durchschnittliche Lebensdauer eines Tröpfchens von einer Konfluenz zur nächsten bestimmen — wir nehmen sie mit 30 min an. (T) ist in diesem Fall die Lebensdauer mit $T = 30$ min, (t) wäre die Konfluenzzeit mit $t = 0,1$ sec. Die Wahrscheinlichkeit, in einer Momentaufnahme mit einer Belichtungszeit $t_{exp} \ll t$ ein Tröpfchen im Konfluenzprozeß zu erfassen, wäre 1:18000. Trotzdem sähe man im mikroskopischen Bildfeld (Abb. 5a) mit 30 Tropfen durchschnittlich jede zweite Minute eine Konfluenz. Die Wahrscheinlichkeit nimmt noch weiter ab, wenn man zum Rekonstruieren des Vorganges Schnitte verwendet. Bezieht man den Einfluß der Schnittlage ein, so lassen maximal 25% der Schnitte eine tatsächlich vorhandene Konfluenz erkennen (Abb. 5c). Die Wahrscheinlichkeit einer Darstellung verringert sich damit auf $^1/_4$ oder 1:70000. Nimmt man schließlich an, daß man auf einer Mikroaufnahme im Schnittpräparat 30 Tropfen sieht und daß an einer Konfluenz zwei Tropfen teilnehmen, so ergibt sich für das Darstellen des Konfluenzvorganges eine Wahrscheinlichkeit 1:4700. Anders ausgedrückt: Auf 4700 Aufnahmen dürfte man im Durchschnitt eine erkennbare Konfluenz erwarten. Schließt man noch die notwendige statistische Sicherung ein, so erkennt man die hoffnungslose Situation, in der man sich bei derartigen Prozessen mit einem Studium am fixierten Objekt befindet. Auch in diesem Fall kann nach dem Dargelegten nur ein fortlaufendes Beobachten helfen. Eine Vitalmikroskopie höherer Auflösung wäre daher sicher nicht sinn- und zwecklos.

---

[1] Eine Ausnahme von dieser Regel könnte höchstens ein temporäres Einfrieren oder eine vorübergehende Hypothermie einer kompletten Zelle im Rahmen einer Vitalbeobachtung am EM bilden. Durch ein derartiges Vorgehen könnte man unter Umständen sehr rasch ablaufende Vorgänge etwas verlangsamen. Dadurch könnte man in Grenzfällen mit einer etwas schlechteren Zeitauflösung auskommen. Derartige Arbeiten erfordern jedoch den vollen Aufwand einer Vitalelektronenmikroskopie und ermöglichen damit wiederum nicht den angestrebten Kompromiß.

Sucht man nach einem Gerät für Vitalbeobachtungen im Submikroskopischen, so scheiden sowohl die indirekten Verfahren wie auch das Röntgenschattenmikroskop aus (Abschnitt B. 1): Die indirekten Verfahren offerieren nur in einer einzigen räumlichen Achse eine submikroskopische Meßgenauigkeit (nicht Auflösung!), in den beiden anderen Achsen übersteigt ihre Auflösung nicht den lichtmikroskopischen Bereich. Sie liefern aus diesem Grund nie ein vollständiges Bild, sondern stets nur Detailaussagen. Das Röntgenschattenmikroskop hat bei seinem heute gegebenen Entwicklungsstand eine schlechte Zeitauflösung (große

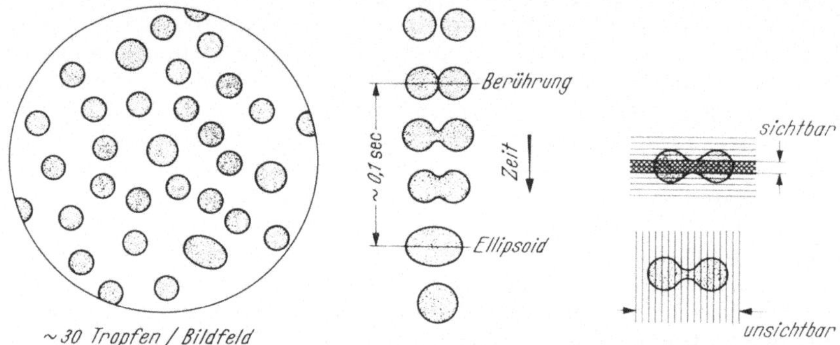

Abb. 5. Erfassen eines kurzdauernden Konfluenzvorganges. Vgl. Text

Belichtungszeit) und unzureichende Bildkontraste bei feuchten Objekten. Es bleibt daher nur das Durchstrahlungs-Elektronenmikroskop — also ein Gerät, dessen Hauptmangel nach allgemeiner Ansicht eben darauf beruht, daß es zum Studium lebender Objekte vollkommen ungeeignet ist.

Der Verfasser fühlt sich nicht berufen, kompetent über alle apparativen oder radiologischen Probleme zu referieren und zu urteilen. Trotzdem erscheint es notwendig, neuerlich eine Diskussion über die prinzipiellen Möglichkeiten und Grenzen einer Vitalelektronenmikroskopie in Gang zu bringen. Nur auf diesem Weg dürfte es möglich sein, Physiker, Techniker und Biologen zur notwendigen Zusammenarbeit und zum Ausräumen immer wieder auftauchender Mißverständnisse zu bewegen. Will man Vorarbeiten für eine Vitalelektronenmikroskopie in Angriff nehmen, so befindet man sich in einer ähnlichen Situation, wie sie zu Beginn der Elektronenmikroskopie bestand: Es ist vollkommen unmöglich, sichere Prognosen zu stellen. Hierfür gibt es nur zwei Ausnahmen: Sicher ist einerseits, daß der notwendige apparative Aufwand sehr groß sein wird; sicher ist andererseits, daß die Ergebnisse zunächst im Vergleich mit den Resultaten der perfektionierten Ultramikrotomie ebenso unansehnlich anmuten werden, wie die ersten elektronenoptischen Aufnahmen biologischer Objekte von Marton aus den Jahren 1934/35 gegenüber der damals hochentwickelten Mikrotomhistologie. Die Skepsis an einer Vitalelektronenmikroskopie wird daher zunächst ebenso berechtigt erscheinen, wie seinerzeit die Skepsis gegenüber der Elektronenmikroskopie. Inzwischen sind seit Martons ersten Aufnahmen zwanzig

Jahre vergangen. Man hat das Beschatten der Objekte mit Schwermetall, die Abdruckmethode, die Ultramikrotomie und das Negativkontrast-Verfahren entwickelt und damit ungeahnte Fortschritte erzielt. Es wird gut sein, wenn man diese Parallele im Auge behält. GABOR leitet seine ausgezeichnete retrospektive Betrachtung über die Entwicklungsgeschichte des Elektronenmikroskopes mit folgenden Sätzen ein:

„ . . . . Niemand konnte an Hand der damals (1932; d. Ref.) verfügbaren Unterlagen voraussehen, daß das Elektronenmikroskop so erfolgreich sein würde, wie es sich später erwies. Dies wird ein Teil der nachstehend zu erzählenden Geschichte sein. Nur wollen wir im Voraus die Moral aus der Geschichte ziehen: Selbst in der Wissenschaft ist es oft besser, Mut zu besitzen, als gescheit zu sein! . . .‟

Naturgemäß bedarf der „Mut‟ irgendeines Rückhaltes. Fehlt dieser, so tritt an die Stelle eines mutigen Vorgehens ein Hasardspiel, ein sinn- und planloses Umherirren. Der Mut zu weiteren Bemühungen um eine Vitalelektronenmikroskopie scheint teils durch apparative Fortschritte der Elektronenmikroskopie in den letzten Jahren, teils durch prinzipielle Überlegungen und teils durch Fakten der Radiologie gerechtfertigt. So ist es beim Arbeiten mit dem Feinstrahl möglich, die Erwärmung des Objektes auf das notwendige Minimum zu beschränken (Abschnitt B. 2); diese Mindestenergiemenge kann durch ein Kühlsystem abgeleitet werden (Abschnitt B. 2). Objektkammern können eine Arbeit bei 760 Torr und normaler relativer Feuchtigkeit ermöglichen. Die Höhe dieser Objektkammern kann mit steigender Strahlspannung vergrößert werden (Abschnitt B. 2). Dickere Objektschichten sind ebenfalls mit höheren Strahlspannungen durchstrahlbar. Man darf nach den Ergebnissen von DUPOUY u. Mitarb. am 1,5 MV-EM vom CNRS Toulouse erwarten, daß feuchte Objekte in einer Dicke bis etwa 10 μm bei höheren Strahlspannungen im Elektronenstrahl hinreichend transparent sind. Die große Schärfentiefe des EM führt bei dickeren Objekten zwar zu einer Überlagerung übereinanderliegender Strukturdetails, ermöglicht aber auf der anderen Seite das notwendige Erfassen der dritten Dimension (Abschnitt B. 3). Man muß die Vitalelektronenmikroskopie aus diesem Grund zunächst wahrscheinlich auf einfache, relativ strukturarme Objekte beschränken, wie sie die weitgehend entdifferenzierten Zellen tierischer und menschlicher Gewebekulturen darstellen (Abschnitt B. 3). Darüber hinaus muß man relativ geringe Bildkontraste erwarten, da größere Dichteunterschiede im Detail fehlen. Es besteht jedoch berechtigte Hoffnung, daß die Sachlage mit Phasenkontrastsystemen sowie auf phototechnischer Basis verbessert werden kann (Abschnitt B. 4). Ein wesentliches Reduzieren des Leuchtschirmkornes würde es ermöglichen, den Abbildungsmaßstab im Endbild wesentlich zu verringern. Ein um 1:10 verringerter Abbildungsmaßstab gestattet ein Senken der Belichtungszeit auf $^1/_{100}$. Man könnte damit die Zeitauflösung $_T\delta_{min}$ wesentlich verbessern und unter Umständen bis in den Millisekundenbereich vorstoßen. In diesem Fall wäre der Einsatz besonders feinkörniger Photoemulsionen gerechtfertigt. Auch in diesem Sektor „Abbildungsmaßstab – Bildhelligkeit – Belichtungszeit‟ liegen bereits erste positive Resultate von Versuchen vor (Abschnitt B. 5). *Die bislang aufgeführten*

Probleme scheinen durchwegs bei entsprechendem Einsatz technisch lösbar, wenn man sich für den Anfang mit einer mittleren elektronenoptischen Auflösung im Bereich von ewa 100—200 ÅE zufrieden gibt und die Arbeit auf Einzelzellen oder Mikroorganismen mit Durchmessern unter 10 μm beschränkt. Damit befindet man sich bei der Vitalelektronenmikroskopie in der gleichen Lage wie beim Rekonstruieren am fixierten Objekt (Abschnitt C. 1) und in der Vital-Lichtmikroskopie (Abschnitt B. 1).

Entscheidend für eine Vitalbeobachtung im EM ist die Frage der *Strahlenschäden.* Zum Erreichen eines bestimmten Auflösungsvermögens $_L\delta_{min}$ in einer Mikrophotographie ist eine bestimmte Elektronenzahl pro Objektflächeneinheit notwendig (Abschnitt B. 6) — je höher die Zahl der Elektronen, um so größer ist der Informationsgehalt des Bildes, bzw. um so besser ist die Auflösung $_L\delta_{min}$. Hier trifft man auf eine prinzipielle Limitierung der Möglichkeiten, die durch keinen Kunstgriff zu umgehen ist. Dies hat bereits in der Anfangszeit der Elektronenmikroskopie v. Ardenne klar herausgestellt. Trotzdem werden nach Ansicht des Referenten zwei grundlegend wichtige Fakten scheinbar generell übersehen. Diese sollen im Folgenden ausführlicher diskutiert werden.

Untersucht man ein bestimmtes Objekt bei stetig zunehmenden Vergrößerungen, so zwingt das Steigern des Abbildungsmaßstabes zu einer Reduktion des Beobachtungsfeldes. So kann man beispielsweise ein Feld von $10 \times 10 = 100$ cm² mit einer 10fach vergrößernden Lupe ohne weiteres nach Strukturdetails in der Größe von 0,1 mm durchmustern. Es ist aber gänzlich unmöglich, die gesamte Objektfläche bei 100 000 facher Vergrößerung nach Details von 10 ÅE durchzusehen, welche bei diesem Abbildungsmaßstab 0,1 mm groß wären. Wollte man es tun, so entspräche das einem Aufsuchen von einzelnen verstreuten Härchen auf einem Areal von $100 \times 100 = 10 000$ km². In vollkommen analoger Weise schränkt eine stets weiter gesteigerte Zeitraffung die Beobachtungsdauer ein. Dies ist jedem Hochfrequenz-Kinematographen wohl bekannt. Es ist unmöglich und sinnlos, einen zweistündigen Vorgang mit einer Bildfrequenz von 1000 Hz (1000 Einzelbilder pro Sekunde) aufzunehmen. Man erhielte dabei 7,2 Millionen Einzelbilder, deren exaktes Auswerten — Bild für Bild — kaum möglich wäre. Man müßte sich dabei vor Augen halten, daß nur ein genaues Auswerten den großen Aufwand bei der kinematographischen Aufnahme rechtfertigen könnte. Die angeführten Beispiele zeigen, daß ein Beobachten längerer Abläufe mit bester Zeitauflösung $_T\delta_{min}$ ebenso ad absurdum führt, wie ein Durchmustern größerer Objektflächen bei optimaler Zweipunktauflösung $_L\delta_{min}$. In beiden Fällen ist man gezwungen, das Ausmaß der Beobachtung auf das unbedingt notwendige Mindestmaß einzuschränken. Dieses wird beim fortlaufenden Beobachten von Vorgängen im wesentlichen durch die Zeitauflösung $_T\delta_{min}$ gesetzt, die man beim Rekonstruieren nach Einzelaufnahmen erreicht. *Es genügt daher zunächst, wenn man in der Vitalelektronenmikroskopie Vorgänge im Minutenbereich beobachten kann* — z. B. in einer Zeitspanne zwischen 30 sec und 2 min. Größere Beobachtungszeiten sind im Augenblick weder notwendig noch sinnvoll, da man größere Zeitspannen ohne weiteres durch ein Rekonstruieren erfassen kann.

In engem Zusammenhang mit der Beobachtungsdauer steht eine weitere Tatsache, die meist übersehen wird. Die *morphologische Manifestation von Strahlenschäden erfolgt nicht sofort, sondern erst nach einer gewissen Latenzzeit.* Diese Latenzzeit ist aller Voraussicht nach größer als die maximale Dauer einer einzelnen Vitalbeobachtung. Zusätzlich könnte man für die Vitaluntersuchung am EM Objekte oder Objektteile auswählen, deren Latenzzeit besonders groß und deren Strahlenempfindlichkeit relativ gering ist (Abschnitt B. 6); unter Umständen ermöglicht eine Teilbestrahlung höhere Strahlendosen als eine Ganzzellbestrahlung.

Faßt man alle diese Fakten und Überlegungen zusammen, so kann man den Bemühungen, die auf das Entwickeln einer Vitalelektronenmikroskopie ausgerichtet sind, weder einen Sinn und Zweck, noch eine reelle Chance absprechen. Eine bessere Zeitauflösung $_T\delta_{min}$ kann man nach dem heutigen Stand unserer Kenntnisse nur auf diesem Weg erreichen. Prinzipielle Hindernisse, welche in physikalischen Gesetzmäßigkeiten begründet wären, sind nicht zu erwarten. Es ist danach gerechtfertigt, es einmal zu versuchen und den notwendigen Einsatz zu riskieren, ohne im Detail die daraus erwachsenden Entdeckungen im Voraus zu kennen oder auch nur zu ahnen. Einer der Altmeister der submikroskopischen Morphologie, A. FREY-WYSSLING, bemerkt zu den Möglichkeiten einer direkten Beobachtung submikroskopischer Strukturen im Elektronenmikroskop

".... The discovery of the electron microscope ... suddenly brought the submicroscopic regions within reach. ... the precipices and gullies which had hitherto been hidden have become accessible to the investigator, who is now equipped with the means whereby he can move about in this difficult province. Submicroscopic morphology has accordingly lost some thing of its mysterious charme. The unravelling of its secrets no longer wholly depends upon an ingenious combination of partial evidence obtained indirectly, as it still does in the study of the constitution of organic molecules in structural chemistry. There is now a direct means of checking the conceptions developed so far ..."

Dem Referenten scheint, daß der reizvolle Zustand beim *direkten Beobachten von Vorgängen* im Submikroskopischen noch einige Zeit erhalten bleiben wird. Auch einer relativ gut entwickelten, apparativ und methodisch perfekten Vitalelektronenmikroskopie der Zukunft werden enge Grenzen gesetzt sein. Der Biologe und Mediziner wird auch mit diesem Hilfsmittel nur einzelne, hierfür besonders geeignete Vorgänge beobachten können und darüber hinaus weiterhin auf der Basis von Indizien Hypothesen formulieren müssen.

Summary

Two different methods permit the study of a biological process: A continuous visual or cinematographic observation of the living organisms (vitalmicroscopy) or a reconstruction based on material fixed in different periods of the process. Every study is limited not only by the well known point resolution $_L\delta_{\min}$ of the optical system used. It is limited also by the time resolving power. This time resolution $_T\delta_{\min}$ mainly depends on the frequency ν_b of the cinematographic pictures, respectively the minimal time interval t_b between two pictures following one another or the minimal interval t_{ae} between the starting point of the process and the stop by the fixation; see equation (1), Chapter A. The faster a process the better $_T\delta_{\min}$ is needed to resolve all details.

In the moment only a reconstruction is practicable. The process must be stopped at a certain time t_e by chemical or physical fixation. The further treatment of the fixed object is a conventional one [89]: Embedding in synthetic resin, ultrathin sectioning and observation in the EM (= Electron microscope). It must, of course, be presupposed that the process can be repeated. Different objects are then fixed at different time intervals t_{ae}. The minimal amount of t_{ae} (= $_T\delta_{\min}$) in all cases is given by the diffusion of heat [57] or chemical fixing agent [110, 131]. The velocity of this diffusion process is strictly limited, since the maximal temperature gradient is given by normal temperature and $0°$ K and the maximal concentration gradient of any solution is limited by saturation. Both gradients are rapidly dropping down during the fixation of any threedimensional object [57]. Therefore and for other reasons any fixation generally is a time consuming process, even on the smallest object obtainable. Objects in the size of single cells (10 μm in diameter) scarcely can be fixed within time intervalls below 30 sec. In the case of reconstruction based on fixed material time resolution therefore normally remains in the range $_T\delta_{\min} \geqq 30$ sec.

The reconstruction of a process at the submicroscopic level will succeed very easy, if any structural changes at the microscopic or macroscopic level are connected with it. Such structures („Leitstrukturen") allow allways a correct timing and an optimal $_T\delta_{\min}$ [45, 100]. If such structures are not visible, exact starting and fixing points are needed. The study of processes involed in a "steady-state-system" [11, 30, 66] — e. g. diffusion processes between blood capillaries and the interstice — need additional methods, since they are running continuously without any limit or variation. A reconstruction of such processes using only the normal methods is naturally impossible. But it is possible to superimpose a second (normal) process by introducing of a marker — e. g. radioactive tracer molecules or heavy metal containing particles, like ferritin [41], Thorotrast [23, 91], HgS [78] or gold colloid [51]. Especially radioactive tracers in connexion with EM-autoradiography offer lots of possibilities [11, 20—22, 52, 53, 79, 80, 88]. Finally pointcounting of areas [64, 65] and volums [59,64,65] are very useful. These simple stereological methods [61, 109, 124, 125] allow binding conclusions from the plane EM-picture, respectively the flat ultrathin section to the object in space. This is especially valuable in cases concerning little structural variations during the event observed. Correct and statistical significant values hereby only are obtainable by means of such morphometric methods.

Fast running processes frequently need a better time resolution $_T\delta_{\min}$ as given by reconstruction on fixed material. In the moment the only possible solution of this difficult problem seems to be a continuous observation of living organisms. Every observation of this kind depends mainly on the optical system chosen. Indirect methods — e. g. polarizing microscopy, interference microscopy, X-ray diffraction — seem not to be very profitable, since they offer a high precision measurement only in one direction whereas the localization of any point in the object plane is limited by the light microscopic resolution. It is therefore impossible to obtain a complete submicroscopic picture with such a method. The efficiency of the projection X-ray microscope (Shadow X-ray m.) [5, 26, 27, 37, 38, 86] is limited by the intensity of the X-ray point source. Therefore the exposure time t_{exp} is very high and the time resolution $_T\delta_{min} \geqq t_{exp}$ very bad. Since the point resolution $_L\delta_{min}$ in the study of movements is also depending on the exposure time t_{exp} and the object velocity v_{ob} according to the approximative equation $_L\delta_{min} \geqq 5. \ v_{ob} \cdot t_{exp}$ (comp. fig. 1), a very short exposure time is urgently needed. Only the EM seems to be promising with respect to a vitalmicroscopy at submicroscopic level. Therefore the basic, the technical and the radiological problems of an electron microscopy of living cells must be discussed.

The typical processes of life only occur in highly hydrated systems. Therefore any observation of living cells must be carried out at normal pressure (760 Torr) and normal humidity. This seems to be possible, since different object chambers are known (comp. e. g. fig. 2), which allow higher pressures [58, 62, 63, 67, 68, 85, 114, 133]. Another requirement is to keep the temperature constant at the normal level. In this respect one has to reduce the energy loss of the electron beam into the object [13, 122] to the amount given by the resolving condition for the object field under observation. It is possible to fulfil this condition by means of a double condensor, which provides a small area illumination [96]. The heat produced in this area [73] must be lead away by a cooling device. Such systems also are known [74, 102, 105, 121] and surely can be adapted for the special purpose. Another serious problem is the penetration of living objects by electrons. The normal thickness of cells ranges between 2 and 10 micron. Such objects are only transparent, if higher accelerating voltages > 400 kV are used [25, 92, 115]. First results with the 1,5 MV-EM from the CNRS Toulouse [44—49] show clearly that also heavy objects — e. g. metal foils [34] — in relatively great thicknesses are transparent, further that living bacteria can be investigated [31, 33] and last not least that there is a considerable image contrast. As a result of the minimal numerical aperture of electron lenses it is possible to get a sharp picture including the whole depth of the object. This circumstance has the advantage that every observation is including the third dimension of space. But it has also the disadvantage that different structural details may be invisible on account of overlaping (Fig. 3). This fact restricts the range of observation to objects with a minimal structural index like human or animal tissue culture cells. It is of course not possible to enhance the contrast of details by incorporation of heavy metal atoms in this living systems. With respect to the mainly uniform specific weight of different cell structures, the contrast in the EM-picture may contrarily to ultrathin sections be a cause of different thicknesses. Additional contrast could be obtained by phase contrast systems [14, 39, 40, 70, 71, 75] based on the fundamental idea of ZERNIKE [134] or by photographic methods [119]. Since the amelioration of time resolution is only possible by restriction of the exposure time t_{exp}, it must be worked with the smallest magnification obtainable. A 100-fold restriction of the magnification on the screen causes a reduction 1:10000 of t_{exp}, which lies then in the acceptable range of milliseconds. First experiments with grainless single crystal screens for such purposes [10, 16] seem to be successful. Every further approach in this direction is very important. It may be combined with image intensifying systems [55, 85] and high resolution photographic recording [1, 29, 50, 104, 113].

The more electrons contribute to an electron micrograph the better is the resolution $_L\delta_{min}$. With other words: A minimal amount of electrons must penetrate the observed object area to from a picture of a given resolution. This minimal amount per picture can not be restricted in any way — it is a fundamental limit. Therefore, the very important question arises, if a series of pictures for a cinematographic study of a process can be gained without serious artifacts by the electron irradiation needed for submicroscopic resulution [6—9]. In this respect two facts seem to be of the greatest importance: (1) Any amelioration of the resolving power $_L\delta_{min}$ or $_T\delta_{min}$ causes a restriction of the observation. It is impossible to observe a field of 10×10 cm² with a resolution of $_L\delta_{min} = 10$ ÅE — as it is impossible to observe a time interval of 2 hours with a resolution $_T\delta_{min} = 0,1$ msec. In every case one is obliged to restrict the observation area or the observation time considerably. Slow processes can be reconstructed with conventional methods based on fixed material. On the other hand studies on living organisms must not exceed time intervals of 30 sec to 2 min. — (2) The morphological manifestation of irradiation damage occurs normally not immediately but after a characteristic time („Latenzperiode").

It seems to be very probably, that even letal doses of electrons do not cause a serious morphological alteration within a sufficient short observation time. Every irradiation damage test based on genetics (cultivation) in this respect seems to be unprofitable and without any obligation.

Resuming all facts and considerations discussed above, electron microscopy of living cells and microorganisms at moderate resolutions of about 100 ÅE $< _L\delta_{min} < 200$ ÅE and 0,01 sec $< _T\delta_{min} < 1$ sec seems to be possible. The technical and methodical efforts may be enormous, the results obtainable may be obscure — it remains at the present state of our knowledge the only practicable way to observe with a better time resolution. One should therefore not hesitate to take the unavoidable risk and to go consequently in this direction.

Literatur

[1] AKASHI, K., T. MASUDA, H. TOCHIGI, K. YAMANOUCHI u. E. IGUCHI: In:
    G. MÖLLENSTEDT et al. (Edit.): 4. Int. Kongr. Elektronen Mikr., Berlin 1958
    Bd. 1, p. 131. Berlin-Göttingen-Heidelberg: Springer-Verlag 1960.

[2] ALBERT, L, R. SCHNEIDER u. H. FISCHER: Z. Naturforsch. 19a, 1120 (1964).
[3] ANDERSON, T. F.: Trans. N. Y. Acad. Sci. 16, 242 (1959).
[4] ANGERER, E. v.: Wissenschaftliche Photographie. Eine Einführung in Theorie und Praxis. 5te. erw. Aufl., herausg. von G. Joos. Leipzig: Akademische Verlagsges. Geest und Portig KG, 1952.
[5] ARDENNE, M. v.: Naturwissenschaften 27, 485 (1939).
[6] — Elektronenübermikroskopie. Physik — Technik — Ergebnisse. Berlin: Julius Springer, 1940. Vgl. insbesondere Abschnitt III: Die Empfindlichkeit lebender Substanzen gegen Elektronenstrahlung, pp. 105 ff.
[7] — Z. techn. Physik 20, 239 (1940); Z. Physik 117, 657 (1941; sowie in: G. MÖLLENSTEDT et al. (Edit.): Verh. 4. Int. Kongr. Elektronenmikroskopie, Bd. 1, p. 112, Berlin 1958. Berlin-Göttingen-Heidelberg: Springer-Verlag 1960.
[8] — Tabellen zur angewandten Phsyik. Bd. 1. Elektronenphysik, Übermikroskopie, Ionenphysik. 2. Aufl.. Berlin: VEB Deutscher Verlag der Wissenschaften, 1962. Vgl. insbesondere Abschnitt 2. 3. 10: Untersuchung lebender Substanz, pp. 460—471. Daselbst weitere Hinweise.
[9] — u. H. FRIEDRICH-FREKSA: Naturwissenschaften 29, 523 (1941).
[10] AREND, H., R. BROSER-WARMINSKY u. E. RUSKA: Z. wiss. Mikr. 62, 46 (1954).
[11] BACHMANN, L., and M. M. SALPETER: In: TITLBACH, M. (Edit.): Proc. 3d. Europ. Conf. Electron Micr., Prag 1964. Czechosl. Acad. Sc. 1964, Vol. B, p. 15.
[12] BERTALANFFY, L. v.: Naturwissenschaften 28, 521 (1940); Theoretische Biologie. Bd. II. Stoffwechsel und Wachstum. 2. Aufl., Bern 1951; Biophysik des Fließgleichgewichtes, Braunschweig: Vieweg.
[13] BOHLE, A., u. H. SITTE: In: E. WOLLHEIM (Edit.): Glomeruläre und tubuläre Nierenerkrankungen, p. 205. Int. Nierensymposion Würzburg 1960. Stuttgart: G. Thieme Verlag, 1962.
[14] BORRIES, B. v., u. W. GLASER: Kolloid Z. 106, 123 (1944).
[15] — u. F. LENZ: In: F. S. SJÖSTRAND and J. RHODIN (Edit.): Proceedings Stockholm Conf. Electron Micr. 1956, p. 60. Uppsala: Almqvist & Wiksell 1957.
[16] BROSER-WARMINSKY, R., u. E. RUSKA: In: G. MÖLLENSTEDT et al. (Edit.): Verh. 4. Int. Kongr. Elektronen Mikr. p. 104, Berlin 1958, Bd. I. Berlin-Göttingen-Heidelberg: Springer-Verlag 1960.
[17] BUCHER, O.: Histologie und mikroskopische Anatomie des Menschen, pp. 76-78. 3. Aufl. Bern u. Stuttgart: Medizinischer Verlag Hans Huber 1962.
[18] BULLIVANT, S.: J. biophysic. biochem. Cytol. 8, 639 (1960).
[19] CARLSTRÖM, D.: Micro X-ray diffraction on biological materials. In: Advances in Biol. and Med. Phys. Vol. VII. New York: Academic Press, 1960.
[20] CARO, L. G.: J. biophysic. biochem. Cytol. 10, 37 (1961).
[21] — et G. E. PALADE: C. R. Soc. Biol. (Paris) 155, 1750 (1961); J. Cell Biol. 20, 473 (1964).
[22] — and R. P. VAN TUBERGEN: J. Cell Biol. 15, 173 (1962).
[23] CHOI, J. K.: J. Cell Biol. 25, 175 (1965).
[24] COSSLETT, V. E.: In: C. R. Congr. Microscopie Electronique, Paris 1950. Rev. d'Optique (Paris) 1953, 556.
[25] — J. roy. micr. Soc. 81, 1 (1962).
[26] —, A. ENGSTRÖM, and H. H. PATTEE: X-ray microscopy and microradiography. New York: Academic Press 1957.
[27] — and W. C. NIXON: Nature (Lond.) 168, 24 (1951); 170, 436 (1952); Proc. roy. Soc. London B, 140, 422 (1952/53); J. appl. Physics 24, 16 (1953).
[28] COUPLAND, J. H.: In: R. Ross (Edit.): Proc. 3rd. Int. Conf. Electron Microscopy, London 1954. Roy. micr. Soc. London 1956, 159.
[29] D'ANS, A. M., E.-G. BERGANSKY u. G. TOCHTERMANN. In: G. MÖLLENSTEDT et al. (Edit.): 4. Int. Kongr. Elektronen Mikr. Berlin 1958. Bd. I, p. 127. Berlin-Göttingen-Heidelberg: Springer-Verlag 1960.
[30] DOST, F. H.: Mitt. Dtsch. Akad. Naturforsch. (Leopoldina), 4/5, 143 (1958/59), Halle a. d. Saale 1961; Dtsch. med. Wschr. 87, 1833 (1962), dortselbst weitere Literaturhinweise.

[31] Dupouy, G., et F. Perrier: In: S. S. Breese (Edit.): Proc. 5th. Int. Conf. Electron Micr. Philadelphia 1962. Vol. I. New York-London: Academic Press 1962, Ref. A. 2/3.

[32] — — Ann. Phys. Fr. 8, 251 (1963).

[33] — — et L. Durrieu: C. R. Acad. Sci. (Paris) 251, 2836 (1960).

[34] — — — In: M. Titlbach (Edit.): Proc. 3d Europ. Conf. Electron Micr. Prag 1964. Publ. House Czechoslovak. Acad. Sc. 1964, Vol. A., p. 103.

[35] — — et F. Fabre: C. R. Acad. Sci. (Paris) 252, 627 (1961).

[36] — — R. Uyeda, R. Ayroles et A. Mazel: In: M. Titlbach (Edit.): Proc. 3d Europ. Conf. Electron Micr. Prag 1964. Publ. House Czechoslovak. Acad. Sc. 1964, Vol. A., p. 105.

[37] Engström, A.: X-ray microanalysis in biology and medicine. Elsevier Monograph. Amsterdam-London-New York-Princetown: Elsevier Pulb. Co. 1962.

[38] — V. E. Cosslett, and H. H. Pattee: X-ray microscopy and X-ray microanalysis. Amsterdam: Elsevier Publ. Co. 1960

[39] Faget, J., M. Fagot et Ch. Fert: In: Houwink A. L. and B. J. Spit (Edit.): Proc. Europ. Conf. Electron Micr. Delft 1960. Vol. I. Nederl. Veren. Electr. Micr. Delft 1961, p. 18. Daselbst zahlreiche Literaturhinweise.

[40] — — J. Ferré et Ch. Fert: In: S. S. Breese (Edit.): Proc. 5th Int. Conf. Electron Micr. Philadelphia 1962. Vol. I. New York-London: Academic Press 1962. Ref. A. 7.

[41] Farquhar, M. G., and G. E. Palade: J. biophysic. biochem. Cytol. 7, 297 (1960).

[42] — S. L. Wissig, and G. E. Palade: J. exp. Med. 113, 47 (1961).

[43] Feder, N., and R. L. Sidman: J. biophysic. biochem. Cytol. 4, 593 (1958).

[44] Fernández-Morán, H.: Science 129, 1284 (1959); Ann. N. Y. Acad. Sci. 85, 689 (1960); Circulation 26, 1039 (1962).

[45] Florey, H. W., and Grant: J. Path. Bact. 82, 13 (1961).

[46] Frey-Wyssling, A.: Submicroscopic morphology of protoplasm. 2nd English Edit., Amsterdam: Elsevier Publ. Co. 1953.

[47] Fujiwara, T.: J. Electron Micr. (Tokyo) 6, 65 (1958).

[48] Gabor, D.: Elektrotechn. Z. Ausg. A, 78, 522 (1957).

[49] Gansler, H., u. Th. Nemetschek: Z. Naturforsch. 13b, 190 (1958).

[50] Goldberg, E.: Z. techn. Physik 7, 500 (1926); zit. nach Angerer (1952).

[51] Gordon, G. B., L. R. Miller, and K. G. Bensch: J. Cell Biol. 25, 41 (1965).

[52] Granboulan, Ph.: J. roy. micr. Soc. 81, 165 (1963).

[53] Granboulan, N., et Ph. Granboulan: In: M. Titlbach (Edit.): Proc. 3d Europ. Conf. Electron Micr. Prag 1964. Czechoslovak. Acad. Sc. Publ. House 1964, p. 31, Vol. B.

[54] Greb, W.: Die Haare der Viola-Blüten, ein neues Objekt für Plasma-Untersuchungen. Diss. Gießen.; Z. Wiss. Mikr. 53, 1 (1936); zit. nach Küster, E.: Handb. d. Protoplasmaforschung, Protoplasmalogia, Bd. II. A. 1b. Wien: Springer-Verlag 1957.

[55] Haine, M. E., and P. A. Einstein: In: Houwink, A. L. and B. J. Spit (Edit.): Proc. Europ. Conf. Electron Micr. Delft 1960. Vol. I. Nederl. Veren. Electr. Micr. Delft 1961, p. 97.

[56] Hanssen, O. E.: Acta path. microbiol. scand. 49, 280, 297 (1960).

[57] Harreveld, A. van, and J. Crowell: Anat. Rec. 149, 381 (1964).

[58] Hashimoto, H., K. Tanaka, and E. Yoda: J. Electrónmicr. (Tokyo) 6, 8 (1958); zit. nach Hiziya et al. (1960).

[59] Haug, H.: Z. Anat. Entwickl.-Gesch. 118, 302 (1955); Med. Grundlagenforsch. 4, 299 (1962).

[60] — In: Weibel, E. R. u. H. Elias (Edit.): Symposion über quantitative Methoden in der Morphologie, 8. Anat. Kongr. Wiesbaden 1965. Berlin-Heidelberg-New York: Springer-Verlag (Im Druck).

[61] — u. H. Elias: (Edit.): Verh. 1. Int. Kongr. Stereologie, Wien 1963. Wien VI., Kaunit, zg. 33, Congressprint 1963

[62] Heide, H. G.: In: G. Möllenstedt et al. (Edit.): Verh. 4. Int. Kongr. Elektronenmikr. Berlin 1958. Bd. I, p. 87. Berlin-Göttingen-Heidelberg: Springer-Verlag 1960.

[63] HEIDE, H. G.: J. Cell Biol. 13, 147 (1962); daselbst weitere Literaturhinweise.
[64] HENNIG, A.: Verh. dtsch. Anat. Ges. 54, 254 (1957).
[65] — In: HAUG, H. u. H. ELIAS (Edit.): Verh. 1. Int. Kongr. Stereologie Wien
1963. Wien VI: Congressprint 1963; Zeiss-Werkz. 6/H. 30, 78 (1958); In:
WEIBEL, E. R., u. H. ELIAS (Edit.): Symposion über quantitative Methoden
in der Morphologie, 8. Anat. Kongr. Wiesbaden 1965. Berlin-Heidelberg-
New York: Springer Verlag (Im Druck).
[66] HILL, A. V.: Adventures in Biophysics. Philadelphia 1931; zit. nach DOST
(1962).
[67] HIZIYA, K., H. HASHIMOTO, M. WATANABE, and K. MIHAMA: In: G. MÖLLEN-
STEDT et al. (Edit.): Verh. 4. Int. Kongr. Elektronenmikr. Berlin 1958. Bd. I,
p. 80, Berlin-Göttingen-Heidelberg: Springer-Verlag 1960.
[68] ITO, T., and K. HIZIYA: J. Electronmicr. (Tokyo) 6, 4 (1958); zit. nach HIZIYA
et al. (1960).
[69] KAMIYA, N.: Cytologia 15, 183, 194 (1950); sowie: Protoplasmic streaming.
In: HEILBRUNN, L. V. u. F. WEBER, Protoplasmatologia. Handbuch der
Protoplasmaforschung. Bd. VIII. 3a. Wien: Springer Verlag 1959. Vgl.
insbes. Tabelle p. 32 mit zahlreichen Literaturhinweisen.
[70] KAMIYA, Y., M. NONYAMA, and R. UYEDA: J. Phys. Soc. Japan 14, 1334 (1959).
[71] KANAYA. K., and H. KAWAKATSU: In: G. MÖLLENSTEDT et al. (Edit.): Verh.
4. Int. Kongr. Elektronen Mikr. Berlin 1958. Bd. 1, p. 308. Berlin-Göttingen-
Heidelberg: Springer-Verlag 1960.
[72] KOBAYASHI, K., E. SUITO, and S. SHIMADZU: In: E. MÖLLENSTEDT et al.
(Edit.): Verh. 4. Int. Kongr. Elektronen Mikr. Berlin 1958, Bd. I, p. 165.
Berlin-Göttingen-Heidelberg: Springer-Verlag 1960.
[73] LEISEGANG, S.: In: R. Ross (Edit.): Proc. 3d Int. Conf. Electron Micr. London
1954. Roy. micr. Soc. London 1956, 176
[74] — ebenda, p. 184
[75] LOQUIN, M.: ebenda, p. 285.
[76] MARTON, L.: Bull. Cl. Sci. Acad. Roy. Belgique 20, 439 (1934); 21, 553, 606
(1935).
[77] MAUNSBACH, A. B., S. C. MADDEN, and H. LATTA: J. Ultrastruct. Res. 6, 511
(1962).
[78] MAJNO, G., and G. E. PALADE: J. biophysic. biochem. Cytol. 11, 571 (1961).
[79] MEEK, G. A., and M. J. MOSES: J. roy. micr. Soc. 81, 187 (1963).
[80] MILLER, F.: Verh. dtsch. Ges. Path. 42, 261 (1959).
[81] — J. biophysic. biochem. Cytol. 8, 689 (1960).
[82] MOOR, H.: Z. Zellforsch. 62, 546 (1964).
[83] —, and K. MÜHLETHALER: J. Cell Biol. 17, 609 (1963).
[84] — C. RUSKA u. H. RUSKA: Z. Zellforsch. 62, 581 (1964).
[85] NIRIKOFF, V. G., J. M. KUSCHNIER, M. M. BUTSLOFF, and G. A. BORDOWSKY:
In: G. MÖLLENSTEDT et al. (Edit.): Verh. 4. Int. Kongr. Elektronen Mikr.
Berlin 1958. Bd. I, p. 103. Berlin-Göttingen-Heidelberg: Springer-Verlag
1960.
[86] ONG SING POEN: Microprojection with X-rays. Den Haag, Niederlande:
Martinus Nijhoff 1959.
[87] PALAY, S. L., S. M. McGEE-RUSSELL, S. GORDON, and M. A. GRILLO: J. Cell
Biol. 12, 385 (1962).
[88] PEASE, D. C.: Anat .Rec. 121, 723 (1955); J. Histochem. Cytochem. 3, 295
(1955).
[89] — Histological techniques for electron microscopy. 2nd Ed., New York-
London: Academic Press 1964.
[90] PELC, S. R.: J. roy. micr. Soc. 81, 131 (1963).
[91] PETERSON, D. A., and R. A. GOOD: Lab. Invest. 11, 507 (1962).
[92] POPOV, N. M.: Bull. Acad. Sci. USSR, 23, 436 (1959); zit. nach COSSLETT (1962).
[93] RÖMPP, H.: Chemie-Lexikon, p. 1818. 5. Aufl. Stuttgart: Franckh'sche Verlags-
handlung 1962.
[94] ROMEIS, B.: Mikroskopische Technik. 15. Aufl. München: Leibniz Verlag 1948.
Vgl. insbesondere §§ 146—186 (Das lebende Präparat) sowie §§ 193—197
(Eindringen von Fixantien).

[95] RUSKA, E.: Kolloid Z. **100**, 212 (1942).
[96] — In: R. Ross (Edit.): Proc. 3d Int. Conf. Electron Micr. London 1954. Roy. Micr. Soc. London W. C. 1, **1956**, 673.
[97] SABATINI, D. D., K. G. BENSCH, and R. J. BARRNETT: Anat. Rec. **142**, 274 (1962); J. Cell Biol. **17**, 19 (1963).
[98] SALPETER, M. M., and L. BACHMANN: J. Cell Biol. **19**, 63 A (1963).
[99] SCHMIDT, W.: Z. Zellforsch. **54**, 803 (1961).
[100] SCHNEIDER, L.: Verh. dtsch. Zool. Ges. Münster, Zool. Anz. Suppl. **23**, 457 (1959); Z. Protozool. **7**, 75 (1960).
[101] SCHNEIDER, M.: Einführung in die Physiologie des Menschen. 15. Aufl. der von H. REIN begründeten Einführung in die Physiologie des Menschen, p. 123. Berlin-Göttingen-Heidelberg: Springer-Verlag 1964
[102] SCHOTT, O., u. S. LEISEGANG: In: F. S. SJÖSTRAND, and J. RHODIN (Edit.): Proc. Stockholm Conf. Electron Micr. 1956, p. 27. Uppsala: Almqvist & Wiksell 1957.
[103] SENO, S., and K. YOSHIZAWA: J. biophysic. biochem. Cytol. **8**, 617 (1960).
[104] SHIMIZU, Y.: J. Electronmicr. (Tokyo) **11**, 157 (1962).
[105] *Siemens und Halske AG*, Wernerwerk für Meßtechnik: Einrichtung für Objektraumkühlung am Elmiskop-I. Techn. Beschreibung. Vorl. Information. Dr. schr. Eg1/1010/464 (1964); sowie: Objektkühleinrichtung zum Elmiskop IA und I. Dr. schr. Eg 1/6 (1964).
[106] SITTE, H.: Verh. dtsch. Ges. Path. **43**, 225 (1959).
[107] — In: BARGMANN et al. (Edit.): Verh. 4. Int. Kongr. Elektronen Mikr. Berlin 1958. Bd. II, p. 63. Berlin-Göttingen-Heidelberg: Springer-Verlag 1960.
[108] — In: WOHLFARTH-BOTTERMANN, K. E. (Edit.): 2. wissensch. Konf. Ges. Dtsch. Naturf. u. Ärzte, Schloß Reinhardsbrunn 1964, p. 343. Berlin-Heidelberg-New York: Springer-Verlag 1965.
[109] — In: WEIBEL, E. R., u. H. ELIAS (Edit.): Symposion über quantitative Methoden in der Morphologie, 8. Anat. Kongr. Wiesbaden 1965. Berlin-Heidelberg-New York: Springer-Verlag (Im Druck).
[110] SITTE, P.: Histochemie **2**, 76 (1960); daselbst weitere Literaturhinweise.
[111] SJÖSTRAND, F. S.: Science Tools, LKB-J. **2**, 25 (1955).
[112] —, and R. F. BAKER: J. Ultrastruct. Res. **1**, 239 (1958).
[113] SPEIDEL, R.: In: G. MÖLLENSTEDT et al. (Edit.): Verh. 4. Int. Kongr. Elektronenmikrosk. Berlin 1958. Bd. I, p. 110. Berlin-Göttingen-Heidelberg: Springer-Verlag 1960.
[114] STOJANOWA, I. G.: In: G. MÖLLENSTEDT et al. (Edit.): Verh. 4. Int. Kongr. Elektronen Mikr. Berlin 1958. Bd. I, p. 82. Berlin-Göttingen-Heidelberg: Springer-Verlag 1960.
[115] TADANO, B., S. KATAGIRI, K. ICHIGE, Y. SAKAKI, u. S. MARUSE: In: G. MÖLLENSTEDT et al. (Edit.): Verh. 4. Int. Kongr. Elektronen Mikr. Berlin 1958. Bd. I, p. 166. Berlin-Göttingen-Heidelberg: Springer-Verlag 1960.
[116] THOENES, W., K. HIERHOLZER, u. M. WIEDERHOLT: Klin. Wschr. **43**, 794 (1965).
[117] THON, F.: Z. Naturforsch. **20** a, 154 (1965).
[118] TOMKEIEFF, S. I.: Nature (Lond.) **155**, 24, 107 (1945).
[119] TRILLAT, J. J., L. TERTIAN et C. SELLA: In: HOUWINK, A. L., and B. J. SPIT (Edit.): Proc. Europ. Conf. Electron micr. Delft 1960. Vol. I. Nederl. Veren. Electr. Micr. Delft, 1961, p. 596.
[120] WATANABE, H., S. NAGAKURA, and N. KATO: In: Proc. 1st Conf. Electronmicr. in Asia and Oceania, Tokyo 1956. Electrotechn. Laboratory Tokyo, 1957, p. 125.
[121] WATANABE, M., I. OKAZAKI, G. HONJO and K. MIHAMA: In: G. MÖLLENSTEDT et al. (Edit.): Verh. 4. Int. Kongr. Elektronen Mikr. Berlin 1958. Bd. I, p. 90. Berlin-Göttingen-Heidelberg: Springer-Verlag 1960.
[122] — T. SOMEYA, and Y. NAGAHAMA: In: S. S. BREESE (Edit.): Proc. 5th Int. Congr. Electron microsc. Philadelphia 1962. Vol. I. Ref. A. 8. New York-London: Academic Press 1962.
[123] WATSON, M. L.: J. biophysic. biochem. Cytol. **3**, 1021 (1957).

[*124*] WEIBEL, E. R.: Morphometry of the human lung. Berlin-Göttingen-Heidelberg: Springer-Verlag 1963; sowie: Lab. Invest. **12**, 131 (1963).

[*125*] WEIBEL, E. R., u. H. ELIAS (Edit.): Symposion über quantitative Methoden in der Morphologie. 8. Anat. Kongr. Wiesbaden 1965. Berlin-Heidelberg-New York: Springer-Verlag, (Im Druck).

[*126*] WERNER, G., u. W. FORTH: Unveröffentl. Befunde am Dünndarm der Ratte (1965).

[*127*] WILLIAMS, R. C.: Exp. Cell Res. **4**, 188 (1953).

[*128*] —, and F. KALLMAN: J. biophysic. biochem. Cytol. **1**, 301 (1955).

[*129*] WOHLFARTH-BOTTERMANN, K. E.: Protoplasma **54**, 1 (1961); daselbst zahlreiche Literaturhinweise zur Protoplasmaströmung.

[*130*] — Verh. dtsch. Ges. Naturforsch. u. Ärzte **102**, 77 (1963); Naturwissenschaften **50**, 237 (1963).

[*131*] WÜSTENFELD, E.: Z. wiss. Mikrosk. **62**, 241 (1955); **63**, 7, 86 (1956).

[*132*] YAMAMOTO, T.: Tohoku J. exp. Med. **82**, 201 (1964).

[*133*] YOSHIDA, S.: J. Electronmicr. (Tokyo) **12**, 2 (1963); daselbst weitere Literaturhinweise.

[*134*] ZERNIKE, F.: Z. techn. Physik **16**, 454 (1935).

Diskussion

Vorsitz: *P. Sitte*

Sachsenmaier: Welche Vorgänge sind nach Ihrer Meinung für eine Vital-Elektronenmikroskopie besonders geeignet?

H. Sitte: Darüber läßt sich heute ebensowenig eine Angabe machen, wie zu Beginn der Elektronenmikroskopie eine Prophezeiung dessen möglich gewesen wäre, was man heute damit untersuchen kann.

Meyer: Wie ist bei 1,5 Mill. V Beschleunigungsspannung die Objektbelastung? Sind die Objekte noch lebensfähig?

H. Sitte: Die von DUPOUY untersuchten Staphylokokken waren noch lebensfähig. Diese Beobachtung steht mit Berechnungen von COSLETT (1953) in Einklang, wonach die Überlebenschance mit steigender Strahlspannung wächst, da ceteris paribus die Objektbelastung sinkt.

Karlson: Eine Überlebensrate von 10% würde m. E. wenig helfen. Hinzu kommen subletale Veränderungen der Zellen, die die Beurteilung des normalen Erscheinungsbildes sehr erschweren würden.

H. Sitte: Ganz gewiß. Bei der Realisierung der Vitalelektronenmikroskopie würde bei sehr geringen Vergrößerungen, d. h. zugleich recht großem Bildbereich gearbeitet. Dabei wächst die Chance, überlebende Organismen zu beobachten. Nach einer Überlegung von ARDENNE reicht eine Überlebensrate von 1—5% zunächst dabei aus.

P. Sitte: Darf ich ergänzend bemerken, daß das Arbeiten bei schwacher Vergrößerung beim Elektronenmikroskop im Gegensatz zur Lichtmikroskopie wegen der dabei unveränderten Apertur nicht mit einer Auflösungsminderung verbunden ist.

Stoeckenius: Man muß auch bedenken, daß im Elektronenstrahl eine statistische Verteilung von Elektronen besteht, die einen bestimmten Störuntergrund bildet. Für eine Abbildung muß man über diesem Störpegel liegen. Für eine Auflösung von 150 Å — also kaum besser als im Lichtmikroskop — ergibt dies nach Berechnungen von H. G. HEIDE eine Objektbelastung von mehr als 1 Mill. Rö.-Äquivalente. Das stimmt mich etwas pessimistisch.

H. Sitte: Das ist in der Tat der einzige Punkt, wo wir eine gewisse Grenze sicher nicht überspringen können. Andererseits ist hier auf Berechnungen von v. ARDENNE hinzuweisen, der die zur Erzielung eines gewissen Kontrastes unvermeidbare Belastung und die Überlebensrate bei einer solchen Bestrahlung in Beziehung gesetzt hat und dabei zu etwas günstigeren Ergebnissen gekommen ist. Hinzu kommt, daß die in der Literatur gemachten Angaben über Letaldosen stark untereinander differieren, so daß heute noch kein bestimmter Wert als Limit betrachtet werden kann. Die Letaldosis hängt z. B. stark von der Strahlspannung ab. M. W. hat HEIDE seine Berechnungen nicht für verschiedene Strahlspannungen angestellt.

Stoeckenius: Ein weiterer Punkt ist, daß Sie zur Erfassung eines Vorganges mindestens 2 Aufnahmen — besser aber 100 — benötigen.

H. Sitte: Bei einer Belichtungszeit von beispielsweise 0,1 msec besteht durchaus die Möglichkeit, im kleinen Bereich rasch ablaufende Vorgänge mit sehr vielen Aufnahmen zu erfassen.

Stoeckenius: Bei 0,1 msec muß aber dann der Strahlstrom und damit die Objektbelastung entsprechend erhöht werden!

H. Sitte: Die von Dupouy untersuchten Staphylokokken haben die Belastung, die wohl länger war als 2 sec, überlebt. In dieser Zeit wären sehr viele Aufnahmen bei einer Belichtungszeit von 0,1 msec möglich gewesen.

Stoeckenius: Da aber ein bestimmter Strahlstrom für eine bestimmte Auflösung notwendig ist, läßt sich die Belastungszeit bei gleichzeitig minimal gehaltenem Strahlstrom nicht verkürzen.

H. Sitte: Das stimmt; da liegt die Grenze.

Karlson: Ich glaube, mit der Belastung an ionisierender Strahlung kann man hier ruhig etwas höher gehen, da (wie von Röntgenologen gezeigt wurde) viele Schäden, die an sich letal sind, erst sehr viel später auftreten.

Bei z. B. 2 sec dauernder Beobachtung würde immer noch der normale Ablauf und nicht die späteren, letalen Folgen erfaßt.

Stoeckenius: Nicht unbedingt, da Sie nicht wissen, was sich in der molekularen Dimension abspielt, die Sie im Elektronenmikroskop beobachten.

H. Sitte: Man wird da sehr vorsichtig sein müssen. Ich möchte auch keine Prognosen stellen. Ich halte es aber dennoch für richtig, einen Weg in dieser Richtung nicht von vornherein auszuschließen.

Grundmann: 5% Überlebensrate ist für die Morphologie ungenügend, es müssen 95% gefordert werden, wobei die 5% nicht überlebenden als solche erkennbar sein müssen. Es müßte sichergestellt werden, daß die gerade beobachtete Zelle eine der überlebenden ist. Statistische Betrachtungen sind hierbei problematisch.

H. Sitte: Mit einer anschließenden lichtoptischen Untersuchung sollte es möglich sein, Überlebende und Nichtüberlebende zu erkennen.

Klima: Man kann nachträglich mit dem Präparat in Kultur gehen und feststellen, welche der beobachteten Zellen überlebt haben. Nur die an diesen gemachten Beobachtungen wertet man aus.

P. Sitte: Wenn man das Überleben sicher erkennt (vielleicht schon bei der Beobachtung im Elektronenmikroskop), dann genügen übrigens schon 5% Überlebensrate durchaus.

Beermann: Ich glaube Sie bezweifeln nicht, daß auf eine lebende Zelle im Elektronenmikroskop ein wie immer gearteter Einfluß mit bestimmten Wirkungen ausgeübt wird. Man sollte aber andererseits die Betonung nicht so sehr darauf legen, ob es sich um die Beobachtung lebender oder toter Zellen dreht, sondern darauf, ob man Abläufe studieren kann oder statische Bilder sieht.

H. Sitte: Ich möchte einen Vergleich zur Phasenkontrastmikroskopie ziehen: Ihr sind nur wenige Objekte zugänglich (z. B. Gewebekulturen, Organoberflächen usw.); dennoch ist sie nicht mehr wegzudenken. Eine Vital-Elektronenmikroskopie könnte gleichfalls nur einige wenige Fakten überprüfen, sicherlich mit sehr hohem Aufwand, dies erscheint aber zuletzt vielleicht doch gerechtfertigt.

Lettré: Auf welcher Temperatur befindet sich das Objekt im Mikroskop, wenn außen mit —160°C gekühlt wird? Ist da mit vitalen Abläufen zu rechnen?

H. Sitte: Eine Objekttemperatur von —160°C ist heute ohne weiteres zu verwirklichen. So sollte es auch möglich sein, +20°C zu realisieren; m. a. W. die Objekterwärmung läßt sich durch eine Kühlvorrichtung kompensieren. Dies ergäbe übrigens die Möglichkeit, zu rasch ablaufende Vorgänge durch Hypothermie zu verlangsamen.

Morphologie und Morphogenese der Plastiden (Chloroplasten)

Von

W. Wehrmeyer, Hannover

Mit 8 Abbildungen

Nach elektronenmikroskopischen Untersuchungen bestehen die Chloroplasten im wesentlichen aus Chloroplastenhülle, Stroma (Matrix, Grundsubstanz) und dem sehr unterschiedlich ausgebildeten Thylakoidsystem [*119*] (Abb. 1).

Der Begriff Morphologie läßt eine umfassende Beschreibung aller Plastiden und Plastidenbestandteile zu; hier mußte eine Auswahl getroffen werden. Der Begriff Morphogenese ist im strengen Sinn nur auf das Thylakoidsystem anwendbar, denn nur dieses zeigt im Zuge der Entwicklung und unter verschiedenen physiologischen Bedingungen einen im Elektronenmikroskop faßbaren Formwechsel. Angesichts der in letzter Zeit immer deutlicher werdenden Beziehungen zwischen Stroma und Thylakoidsystem erscheint jedoch eine Behandlung des Stroma im Hinblick auf seine Syntheseleistungen als Voraussetzung zum Verständnis der morphogenetischen Prozesse im Thylakoidsystem unumgänglich. Die Darstellung folgt der morphologisch vorgegebenen Gliederung des Chloroplasten in Stroma und Thylakoidsystem, mit der eine funktionelle Gliederung in Dunkel- und Lichtreaktionen im Photosynthesegeschehen vermutlich korreliert ist [*3, 191*].

A. Morphologie des Stroma und seiner Einschlüsse

Das Stroma ist bislang kaum Gegenstand eigener elektronenmikroskopischer Untersuchungen gewesen. Einmal stellt das Stroma eine polydisperse Phase im wäßrigen Dispersionsmittel dar und bietet infolge seines Flüssigkeitscharakters wenig strukturelle Ansatzpunkte; hinzu kommt, daß die Dimension der suspendierten Teilchen an der Grenze des Auflösungsvermögens einer routinemäßig betriebenen elektronenmikroskopischen Praxis liegt. Schließlich bestand bis vor kurzem auch methodisch gesehen eine Schranke in der Unzulänglichkeit der Fixationsmethoden. $KMnO_4$ fixiert das Stroma nur unzureichend, und OsO_4 liefert erst nach zusätzlicher Kontrastierung brauchbare Bilder. Die neuerdings eingeführte Fixierung mit Glutaraldehyd und Nachfixierung mit OsO_4 bzw. $KMnO_4$ [*160*] scheint dagegen gute Ergebnisse zu liefern [*63*].

Die Chloroplasten stellen nicht nur einen hohen Eiweißanteil im Blatt dar [226], sie sind auch an der Bildung dieser Proteine in hohem Maße beteiligt [u. a. 57, 153, 176, 181]. Nach den derzeitigen Vorstellungen

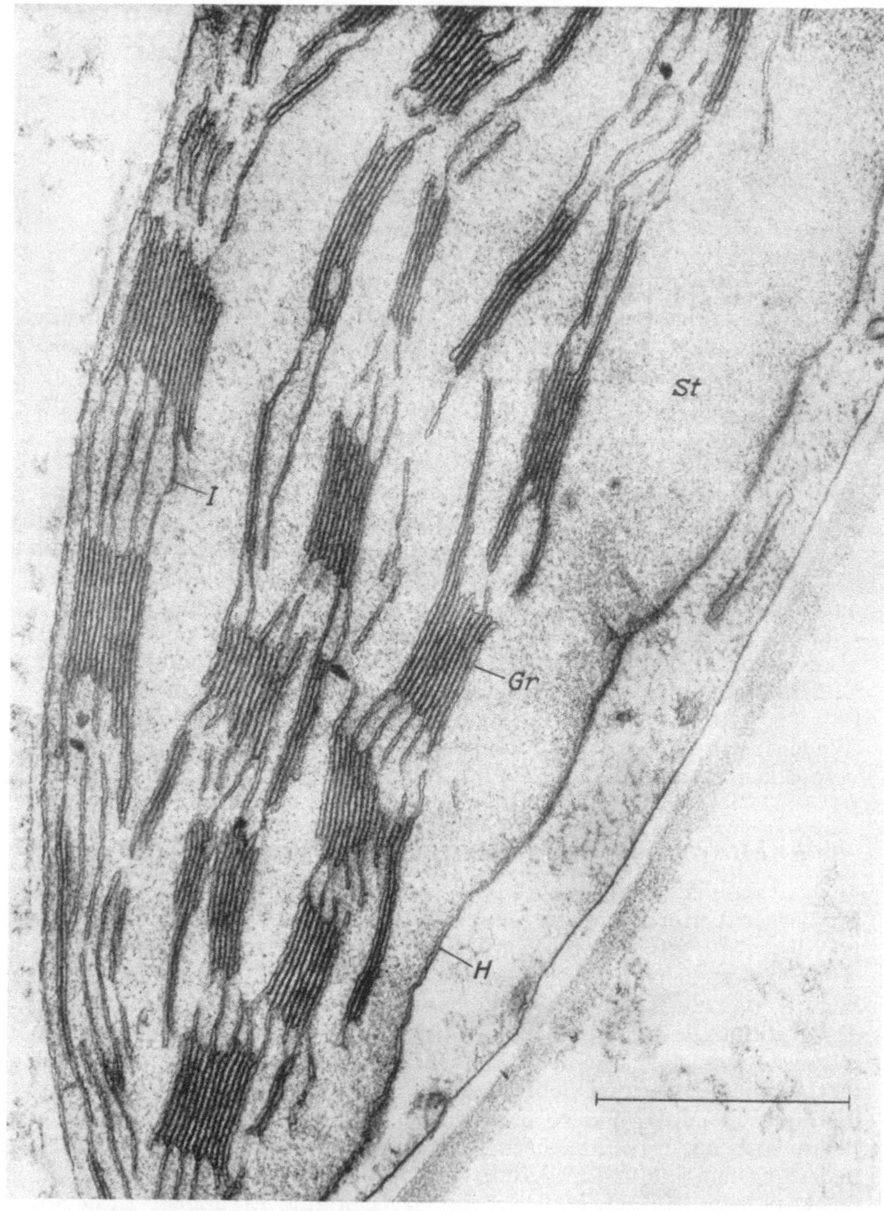

Abb. 1. Ausdifferenzierter Chloroplast mit Chloroplastenhülle (H), Stroma (St) und dem in Grana- (Gr) und Intergrana- (I) bereich gegliederten Thylakoidsystem. Original. 33 000 : 1

über den Ablauf von Proteinsynthesen [Lit. *65, 172, 219*] sind eine Reihe von Strukturen im Chloroplasten bzw. seiner Umgebung zu erwarten, unter denen die Ribosomen und DNS-Fäden eine der Elektronenmikroskopie zugängliche Dimension erreichen. Das im Elektronenmikroskop granuliert erscheinende Stroma hat als möglicher Ort der Proteinsynthese in diesem Zusammenhang neuerdings die Aufmerksamkeit auf sich gezogen.

1. Ribosomen

Von biochemischer Seite war auf Grund des RNS-Gehalts von Chloroplasten [*23*] und nachgewiesener Inkorporation von Aminosäuren [*184*] sehr frühzeitig die Existenz von Ribosomen in Chloroplasten postuliert worden [*107, 125*]. Nachhaltige Bestätigung der Befunde über Chloroplasten-RNS bei Algen [*12, 148*] und höheren Pflanzen [*68, 80, 83, 85, 158*] und über den Einbau markierter Aminosäuren in Chloroplasten [*1, 2, 40, 57, 67, 176*] ließen die Existenz von Ribosomen wahrscheinlich werden. Der Nachweis eines festen RNS-Proteinverhältnisses in bestimmten Fraktionen des Chloroplastenextraktes, die reproduzierbaren Sedimentationskonstanten und die Tatsache einer reversiblen Dissoziation der Partikel bei Verschiebung der Mg^{2+}-Konzentration sprachen für die Existenz von Ribosomen [*13, 14*]. Es gelang auch ihre Unterscheidung von den Ribosomen des Cytoplasma [*16, 41, 139, 176*]. Ihre Größe wird mit 170 Å für Mais [*80*] bzw. 180—200 Å für *Chenopodium album* und *Clivia miniata* [*139*] angegeben (vgl. jedoch [*125*] und die Angabe für die Chrysophycee *Ochromonas* mit 90—120 Å [*53*]). Nach GUNNING [*63*] erreichen die Ribosomen im Stroma von *Avena*proplastiden nur zwei Drittel der Größe von Ribosomen des Cytoplasma. Es ist bekannt, daß die Ribosomen des Cytoplasma zur Synthese spezifischer Eiweiße über messenger-RNS zu Polyribosomen (= Polysomen, Ergosomen) verknüpft werden [*56, 60, 93, 109, 147, 167, 164, 195, 225*]. Solche Polyribosomen konnten vor kurzem von CLARK u. Mitarb. [*25, 26*] auch aus Chloroplasten von *Brassica pekinensis* isoliert und von GUNNING [*63*] in *Avena*proplastiden in situ dargestellt werden. Wenn auch bislang die Elektronenmikroskopie nur sehr zögernd den von der Biochemie erschlossenen Bereich betreten hat [*53, 63, 80, 132*], so dürfte sich hier in nächster Zukunft ein Wandel vollziehen.

2. DNS

Das Vorkommen von DNS in Verbindung mit Chloroplasten ist einmal auf Grund genetischer Untersuchungen über den selbständigen Erbgang der Plastiden [Lit. *27, 55*], zum anderen auf Grund bestimmter Vorstellungen über den Mechanismus DNS-abhängiger RNS-Synthese [*86*, Lit. *65*] und schließlich im Parallelschluß über positive DNS-Befunde in Mitochondrien postuliert worden [*81, 106, 136, 165*].

Da die genetische Seite dieses Problems von W. STUBBE, die Biochemie der Nucleinsäuren von B. PARTHIER und R. WOLLGIEHN und die Frage der Nucleinsäuren in Mitochondrien von H. TUPPY und E. WINTERSBERGER im Rahmen dieses Symposiums ausführlich behandelt werden sollen, kann sich die vorliegende Darstellung auf wenige Angaben beschränken.

Positive DNS-Nachweise in Verbindung mit Chloroplasten liegen für eine Reihe von Algen vor [15, 24, 97, 149, 162], wobei die Befunde an *Acetabularia* besonders aufschlußreich erscheinen, in denen der Kern experimentell beseitigt wurde [5, 54]. Nachweise für DNS in Chloroplasten höherer Pflanzen sind dagegen spärlich [84, 159]. Die elektronenmikroskopische Analyse hat bislang nur für *Chlamydomonas* einen Direktnachweis der DNS-Fibrillen von 25 Å Dicke erbracht [154], jedoch sind durch Verbesserung der Autoradiographie von Dünnschnitten [163] indirekt weitere Hinweise gewonnen worden. Die bisherigen Befunde über den Einbau von Tritium-Thymidin in Chloroplasten [161, 185, 221, 222] konnten danach auf elektronenmikroskopischer Ebene an Maischloroplasten [105] und an *Allium*proplastiden [18] bestätigt werden. Doch geschieht die Interpretation solcher Bilder meist mit einer gewissen Vorsicht [105].

3. Stromazentren

Im Zuge intensiver Syntheseprozesse können in Proplastiden und Chloroplasten Fibrillen von 80 bis 85 Å Dicke von unterschiedlicher Länge (bis zu 1200 Å) gebildet werden, die als konzentrische Bündel Aggregate bis zu 2 μ im Durchmesser aufbauen. Diese frei im Stroma liegenden Stromazentren konnten bislang nur für *Avena sativa* nachgewiesen werden [63]. Auf Grund mangelnder Anfärbung nach OsO_4-, Uranyl- und Perjodsäure/Schiff-(PAS)-Behandlung wird die Proteinnatur der Stromazentren angenommen. Es ist jedoch auf die Ähnlichkeit in Dimension und Anordnung mit den Lipoproteingranulastrukturen hinzuweisen, wie sie in corticalen Nierentubuli der Maus vorkommen [126]; diese sind jedoch PAS-positiv.

4. Phytoferritin

Verschiedene Autoren beobachteten kontrastreiche, kristalline Aggregate globulärer Partikel im Stroma von Proplastiden [156], Leukoplasten [177] und Chloroplasten [169, 144]. Diese werden neuerdings mit den von Hyde et al. [79] erstmals für Pflanzen nachgewiesenen Phytoferritin identifiziert [6]. Bei dem Phytoferritin handelt es sich um eine an hochmolekulares Protein gebundene Eisenreserve. Das genauer untersuchte Ferritin A aus der Pferdemilz enthält 20% Eisen und erreicht bei einem Mol.-Gew. von 747000 in Übereinstimmung mit elektronenmikroskopischen Befunden [46] einen Durchmesser von 111 bis 112 Å [66]. Der eisenführende und daher allein elektronenstreuende zentrale Teil des Ferritin macht dabei etwa 55 Å aus [46, 94]. Die Größenangaben für Partikel aus den Plastiden liegen bei 40 Å [144], knapp 60 Å [177] und 80—100 Å [156]; der Abstand der Netzebenen beträgt etwa 100 Å [144, 177]. Die Eisenreserven in Plastiden sind im Zusammenhang mit dem Vorkommen von Cytochrom b u. f und der zentralen Bedeutung von Ferredoxin für die NADP-Reduktion bemerkenswert [173, Lit. 192].

5. Osmiophile Globuli

Unter den im Elektronenmikroskop sichtbaren Einschlüssen im Stroma der Plastiden sind die osmiophilen Globuli [101] besonders charakteristisch. Ihre Größe schwankt von 100 bis 5000 Å [62]. Während sie nach $KMnO_4$-Fixierung nur schlecht erkennbar sind, bleiben sie nach OsO_4-Behandlung deutlich kugelförmig erhalten und sind so in fast allen Entwicklungsstadien der Chloroplasten nachgewiesen. Gehäuft treten sie in den Chromoplasten [20, 49, 96, 138, 183], in Mutanten [211] oder nach Mangelkultur [95] unter teilweisem Abbau des Thylakoidsystems auf. Vermutlich ist ein Teil der von den älteren Autoren [28, 166] lichtmikroskopisch als Grana angesprochenen Strukturen in Chromoplasten daher mit diesen elektronenmikroskopisch nachgewiesenen Globuli zu identifizieren. Dies gilt nach Döbel

[*32*] auch für die sogenannten „Macrograna" in einigen Mutanten [*108, 217, 223*].
In Sonderfällen wurden auch in völlig normal grünen Chloroplasten in situ starke
Ansammlungen osmiophiler Globuli beobachtet [*44, 179*]. Durch die neuerdings
durchgeführte Isolierung der Globuli gelang ihre chemische Charakterisierung [*4,
104, 62*]. Während die Globuli von *Vicia faba* [*62*] und *Beta vulgaris* [*4*] Plastochinon und 2 verschiedene Galaktolipide und β-Sitosterol enthielten, bestanden die
aus Spinat isolierten in der Hauptsache aus β-Carotin (Lutein, Violaxanthin),
daneben Plastochinon A, α-Tocopherol und Chlorophyll a und b [*104*]. Es handelt
sich also stets um Bestandteile des Thylakoidsystems. Wenn auch kein Beweis
dafür vorliegt, daß die osmiophilen Globuli mit der Thylakoidvermehrung in Zusammenhang stehen [*213*], so ist doch möglich, daß sie überschüssige Lipid-
anhäufungen im Bereich aktiver Lipidsynthese darstellen [*62*]. Die Lokalisation
der Globuli im Stroma hat vielfach zu der Ansicht geführt, daß auch die Biosynthese
hier stattfinde. Die auffällige Verteilung sehr kleiner osmiophiler Tröpfchen in der
Randzone der Grana in Profil und Aufsicht in situ [*19, 180*] und das starke Vorkommen an den Oberflächen stromafreier Thylakoidpräparationen in Aufsicht nach
mechanischer Beanspruchung [*197*] läßt jedoch auch den umgekehrten Schluß
eines Austritts dieser Substanzen aus dem Thylakoidbereich in das Stroma zu.
Hier ist die Entscheidung über primären Bildungsort und sekundäre Verbreitung
zumal bei der nachgewiesenen heterogenen Zusammensetzung der Tröpfchen
schwer zu treffen.

6. Stärkekörner

Die Stärke gehört mit zu den regulären Bestandteilen des Chloroplasten und ist
nach elektronenmikroskopischen Befunden stets im Stroma anzutreffen. Die für
viele Pflanzen spezifische Form der Stärkekörner wird meist nur in den Leukoplasten beim Aufbau von Reservestärke erreicht. Die transitorische „Assimilations-
stärke" in den Chloroplasten hingegen zeigt in der Regel keinen besonders hohen
Ordnungsgrad. Im Elektronenmikroskop erweist sich die Stärke elektronen-
transparent und sehr oft ist ihre Lage nur indirekt an auseinanderweichenden
Thylakoiden zu erschließen. Bemerkenswert ist dagegen die klare, kontrastreiche
Begrenzung der Stärkekörner in nicht-wäßrig isolierten Chloroplasten oder nach
Glutaraldehyd-KMnO$_4$-Fixierung in situ [*205*]. Diese Befunde weisen auf Stoffe hin,
die vermutlich an der Biosynthese der Stärke beteiligt sind. Der in Chloroplasten
sich vollziehende rasche Auf- und Abbau der Stärke setzt eine komplette Enzym-
garnitur voraus und läßt Zwischenprodukte der Stärkesynthese in unmittelbarer
Nähe der Stärkekörner vermuten. Der Aufbau der Stärke aus D-Glucosylresten über
Uridindiphosphoglucose (UDPG) [*30*] und verstärkt über Adenosindiphospho-
glucose als Donatoren (ADPG) [*31, 50, 134, 135, 150*] hat durch den Nachweis dieser
Überträgersysteme (UDPG: [*99*], ADPG: [*82, 133, 151*]) in Verbindung mit den
Stärkekörnern große Wahrscheinlichkeit gewonnen und wird durch den Nachweis
entsprechender UDPG- und ADPG-synthetisierender Enzyme gestützt [*9, 43*].
Darüber hinaus liegen Angaben über Glykolipide in Verbindung mit Stärke vor,
die noch der Bestätigung bedürfen [*38*].

B. Morphologie und Morphogenese
des Thylakoidsystems

1. Beziehungen zwischen Stroma und Thylakoidsystem

Das auffälligste Strukturelement des Chloroplasten sind die Thylakoide. Elektronenmikroskopisch stellen sie membranumgebene Räume
dar, für die auch die Bezeichnungen discs [*51*], abgeplattete Vesikel [*129*]
und compartments [*206*] gebräuchlich sind. Auf die mit anderen Methoden gewonnenen Vorstellungen vom molekularen Bau der Thylakoide
soll angesichts der neueren zusammenfassenden Darstellungen [*8, 123,
124*] nicht näher eingegangen werden.

Die außerordentliche Vielfalt im Erscheinungsbild der Thylakoide im Pflanzenreich in Phylogenese und Ontogenese geht zum Teil auf eine noch nicht im einzelnen bekannte Eigengesetzlichkeit morphogenetischer Prozesse zurück. Zum Teil jedoch wird die Morphogenese von basalen Syntheseprozessen her gesteuert. Angesichts der komplexen chemischen Zusammensetzung der Thylakoide, die nicht nur aus analytischen Arbeiten hervorgeht [102, 103, 141, 142, Lit. 8, 121, 124], sondern auch indirekt aus den Schemata über den Ablauf primärer Lichtreaktionen im Thylakoidsystem zu folgern ist [192], überrascht die Anfälligkeit gegenüber Störungen im Stoffwechsel nicht. Wenn im folgenden einige Beziehungen zur Proteinsynthese und in der weiteren Darstellung zum Chlorophyllgehalt besonders berücksichtigt werden, so stellt dies eine nur bedingt zulässige Vereinfachung der wirklichen Verhältnisse dar. Die Abhängigkeit normaler Membranbildung von sehr verschiedenen Faktoren muß im Hinblick auf die Morphogenese des Thylakoidsystems immer wieder betont werden.

Die Notwendigkeit ungestörter Proteinsynthese zum Aufbau der Thylakoide geht aus Versuchen unter Stickstoffmangel [111] und mit Inhibitoren der Protein- bzw. RNS-Synthese mit z. T. spezifischer Wirkung hervor. Als solche haben in den letzten Jahren u. a. Streptomycin, Chloramphenicol und Actinomycin Anwendung auch im Rahmen entwicklungsgeschichtlicher und funktioneller Untersuchungen an Chloroplasten gefunden. Durch Streptomycinbehandlung ließen sich bei Cyanophyceen [157], *Euglena* [115] und höheren Pflanzen [34] Reduktionen im Thylakoidsystem bei gleichzeitiger Abnahme von Chlorophyll induzieren (vgl. aber 35]. Ein Zusammenhang mit der Proteinsynthese in Chloroplasten ist auf Grund von Befunden über die Streptomycinanlagerung an Ribosomen und Blockierung der messenger-RNS bei Mikroorganismen naheliegend [29, 37, 48]. Ähnliche Reduktionen traten nach Einwirkung von Chloramphenicol ein [34, 77], dessen Wirkung vermutlich über eine Blockierung spezifischer Orte der Ribosomenoberfläche verläuft [210]. Die Wirkung von Actinomycin besteht bei Mikroorganismen in der Hemmung DNS-abhängiger RNS-Synthese. Ob es dabei zu einer Verbindung des Actinomycin mit der DNS und Blockierung der DNS-abhängigen RNS-Polymerase kommt [58, 59, 78, 152] oder zu einer Degradation vorhandener messenger-RNS und Hemmung weiterer m-RNS-Synthese [22, 92, 182] ist noch offen. Versuche über die Wirksamkeit von Actinomycin auf die Plastiden liegen unseres Wissens noch nicht vor, wohl aber konnte ein Abfall des Chlorophyllgehalts dunkel gezogener Pflanzen nach Belichtung im Vergleich zu den Kontrollen festgestellt werden, wenn vorher Actinomycin appliziert worden war [11, 112].

Da die Thylakoide zu einem hohen Anteil von etwa 30% aus Lipiden — bezogen auf das Trockengewicht — bestehen [7, 8], darf neben dem Proteinstoffwechsel der Lipidstoffwechsel nicht völlig übersehen werden. Bereits nach kurzer Photosynthese in $^{14}CO_2$-Atmosphäre erfolgt ein rascher Einbau von ^{14}C in verschiedene Lipide bei *Chlorella* [47]. Enzyme für die Synthese von gesättigten [17, 127, 189] und ungesättigten Fettsäuren [*113*] und von Galaktolipiden [*137*] sind in isolierten Chloroplasten nachgewiesen, jedoch ist der genaue Bildungsort der Lipide nicht angegeben. Nicht einmal die endgültige Verteilung dieser oberflächenaktiven Stoffe im Thylakoidsystem ist bekannt. Nach der ursprünglich von BENSON [*7*] vertretenen Auffassung besetzen die Lipide die inneren und äußeren Oberflächen der Thylakoide [s. *129*]. Diese Konzeption wurde neuerdings zugunsten eines corpuscularen Membranaufbaues aufgegeben [*8*]. Auf Grund von röntgenstrukturanalytischen Daten über Elektronendichteverteilung begrenzen Proteinschichten das Thylakoid nach außen, während die Lipidschichten im Innern , und zwar nur dort, liegen [Lit. *123, 124*]. Elektronenmikroskopisch läßt sich ein Wechsel in der Osmiophilie der Schichten in Abhängigkeit von der Hydratur nachweisen [*145*].

2. Vesiculäre und tubuläre Phase der Morphogenese

a) Frühstadien der Plastidenentwicklung

Nach den derzeitigen Vorstellungen geht die Entwicklung des Thylakoidsystems höherer Pflanzen von Bläschen oder Tubuli aus. Diese entstehen durch Invagination der inneren Membran der Plastidenhülle und werden durch Abschnürung anschließend frei [Lit. *121*]. Die Vesikelbildung verläuft unabhängig vom Licht und von der Gegenwart von Chlorophyll [*212*]. Sie ist aber nicht auf die frühe Phase der Plastidenentwicklung beschränkt, sondern hält aus nicht geklärten Gründen vielfach noch an, wenn bereits ein ausdifferenziertes Thylakoidsystem in den Plastiden vorhanden ist [*155, 190*]. Verschiedentlich bleiben Tubuli einzeln [*168*] oder in dichtem Verband mit der Plastidenhülle in Kontakt [*169, 178, 205*]. Solche Tubulibündel zeigen im Querschnitt eine hexagonale Ordnung. Über ihre Bedeutung ist nichts bekannt. MENKE [*117*] wies solche Tubuliverbände auch frei vorkommend in Plastiden von *Elodea* nach; sie sollen als Ausgangspunkt für die Prolamellarkörperentwicklung dienen. Unterbleibt die Abschnürung der durch Invagination entstandenen Vesikel, so können zusammenhängende Bläschenreihen gebildet werden. Entsprechende Stadien der Thylakoid̓differenzierung — allerdings auf cellulärer Ebene — liegen bei einigen Purpurbakterien vor [*76*, vgl. *36*]. Im allgemeinen führt die weitere Differenzierung der Tubuli und Vesikel zu einer flächenhaften Ordnung von Thylakoiden. Selbst diese können noch mit der Plastidenhülle in Verbindung stehen [*124, 178, 155*]. Ob diese Verbindung jedoch ursprünglich oder durch sekundäre Fusion bedingt ist, muß offen bleiben.

b) Struktur und Entwicklung der Prolamellarkörper

Unter natürlich oder experimentell bedingter Dunkelheit kann es durch ein vermutlich hochgeordnetes Wachstum von Tubuli im Raum zur Ausbildung kristallgitterartiger Anordnungen kommen. Diese sind von den einzelnen Autoren unterschiedlich bezeichnet worden (Prolamellarkörper [*75*], Primärgrana [*186*], Plastidenzentren [*98, 211*], Heitz-Leyon-Kristalle [*120, 122*]). Prolamellarkörper aus tubulären

Elementen hoher Ordnung sind für *Selaginella* [70], *Picea* [212], *Pinus* [21], verschiedene dikotyle *(Eranthemum* [70], *Phaseolus* [88, 89, 91, 114, 140, 193, 216]* und vor allem monokotyle Pflanzen nachgewiesen *(Chlorophytum* [69, 120, 122, 128, 143], *Aspidistra* [100], *Hordeum* [131, 212], *Elodea* [98, 128], *Liriope* [130], *Zea* [75, 80, 87, 175], *Avena* [63, 170]*)).

Voraussetzung für einen hohen Ordnungsgrad der Prolamellarkörper scheint absolute Dunkelheit während der Entwicklung zu sein. Auch späterhin führt Lichteinwirkung in der Regel zur Desorganisation. Daß aber auch unter normalen Lichtverhältnissen Prolamellarkörper existieren können, ist für Plastiden der Drüsenzellen von *Drosophyllum* [168] und der Nektarien von *Passiflora* [169] angegeben bzw. gezeigt worden. Jedoch handelt es sich hierbei um persistierende Strukturen, deren Bildung wahrscheinlich ebenfalls im Dunkeln erfolgte [vgl. 155]. Über die Struktur der Prolamellarkörper ist vor kurzem zusammenfassend berichtet worden [204].

Tabelle

Typ	Morphologische Grundform	Nachweis
1. Prolamellarkörper geringer Ordnung	überwiegend 5 gliederige Ringe	*Phaseolus vulgaris* [204]
2. Kristalline Prolamellarkörper		
a) Zinkblendegitter	ausschließlich 6 gliederige Ringe (Sesselform)	*Phaseolus vulgaris* [201] *Chlorophytum spec.* [120, 122, 128, 143] *Avena sativa* [63]
b) Wurtzitgitter	6 gliederige Ringe (Sessel- und Wannenform)	*Phaseolus vulgaris* [140, 202]
c) Kombination von a) und b) (Polymorphie)		*Phaseolus vulgaris* [205]
3. Konzentrische Prolamellarkörper	5- und 6 gliedrige Ringe (Sessel- und Wannenform)	
a) mit 1 Zentrum		*Phaseolus vulgaris* [91, 200, 204] *Chlorophytum* spec. [143]
b) mit > 1 Zentrum		*Phaseolus vulgaris* [39, 204]

Die Vielfalt der Aspekte tubulärer Anordnungen in den Ultradünnschnitten von Prolamellarkörpern einerseits und das Fehlen eines Bezugssystems zur Orientierung der Schnitte andererseits haben zu unterschiedlichen Auffassungen von ihrem submikroskopischen Bau geführt [61, 64, 120, 122]. Vermutlich liegt jedoch den tubulären Anordnungen ein einheitliches Bauprinzip zugrunde; als eigentliches Bauelement kann rein formal das Tetraeder angesehen werden. Ausgehend von dieser morphologischen Grundstruktur des Tetraeder lassen sich ebene, 5 gliedrige Ringe mit einem Kantenwinkel von 108° und gewinkelte,

6gliedrige Ringe mit einem Kantenwinkel von etwa 109° in zwei stereoisomeren Formen als Sessel und Wanne aufbauen (Abb. 2a, b, c). Durch die Kombination dieser Grundform im Raum haben sich die bisher bekannt gewordenen Schnittbildaspekte von Prolamellarkörpern weitgehend interpretieren lassen. Wie aus der Tabelle hervorgeht, sind sehr verschiedene Kombinationen der Grundformen sterisch möglich und in der Natur wahrscheinlich verwirklicht. Während die Prolamellarkörper geringerer Ordnung (Tabelle Typ 1) mit überwiegend oder ausschließlich pentagonaler Verknüpfung ohne größere Schwierigkeiten anhand der

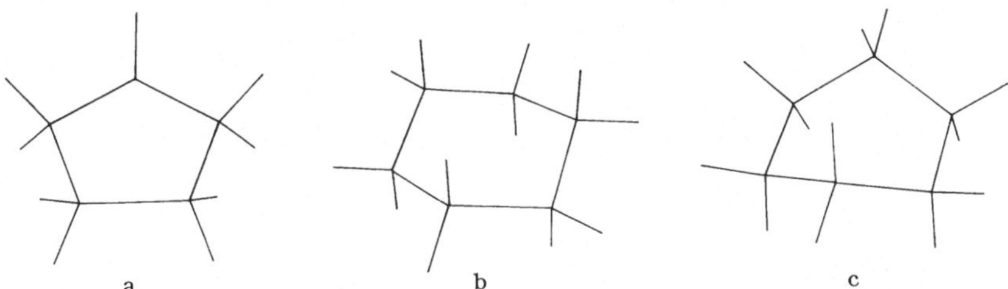

a   b   c

Abb. 2a—c. Morphologische Grundstrukturen der Prolamellarkörper: ebene, 5gliedrige Ringe (a), gewinkelte, 6gliedrige Ringe nach der Sesselform (b) und der Wannenform (c)

elektronenmikroskopischen Bilder noch unmittelbar erkannt werden können, ist der Beweis für die Existenz kristallähnlicher Prolamellarkörper nach dem Zinkblende- (Tabelle Typ 2a) bzw. Wurtzitgitter (Tabelle Typ 2b) mit einem Aufbau aus 6gliedrigen Ringen nur durch den Vergleich verschiedener Schnittebenen durch diese Kristalle mit den vorliegenden Ultradünnschnitten durch die Prolamellarkörper möglich [201, 202]. Die angenommenen Modelle können jedoch erst dann als genügend gesichert betrachtet werden, wenn sie nach mindestens 3 Richtungen des Raumes vollständig beschrieben und untersucht worden sind. Es zeigt sich bei solcher Prüfung, daß bestimmte Schnittlagen zu völlig gleichartigen Schnittbildern führen können. Dadurch wird die Anzahl der zur Kontrolle notwendigen Ebenen reduziert. Im Zinkblendegitter wurden die [111], [110], [112] und [001] Ebene senkrecht zur Tafelebene [1$\bar{1}$0] und die [1$\bar{1}$0], [110], [100], [010] Ebene senkrecht zur Tafelebene [001] untersucht und diskutiert [201], im Wurtzitgitter waren es Schnitte nahe der [0001], [1$\bar{1}$0$\bar{2}$], [1$\bar{1}$00] und [1$\bar{1}$01] Ebene senkrecht zur Tafelebene [11$\bar{2}$0] [202]. Zinkblende und Wurtzit zeigen im Aufbau identische Ebenen [111] und [0001], jedoch wird in der Zinkblende die Identität in der Schichtung erst in jeder 4. Lage in der Sequenz ABCABC erreicht, während im Wurtzit jede 3. Lage entsprechend der Sequenz ABABAB miteinander korrespondiert. Darüber hinaus ergeben sich viele Möglichkeiten wechselseitiger Kombination (Polymorphie), die die verschiedentlich sichtbaren spontanen Strukturänderungen in den Schnittbildern erklären. Unsere Vorstellungen vom Bau der Prolamellarkörper werden durch Befunde über die konzentrischen Typen (Tabelle Typ 3a, b) gestützt (Abb. 3a, b). Diese besitzen ein oder mehrere zentralständige Pentagondodekaeder bei einer im übrigen dominierend „hexagonalen" Umgebung. Da das zentralständige Pentagondodekaeder 6 fünfzählige, 10 dreizählige und 15 zweizählige Achsen besitzt, die gegenüberliegende Flächenmitten, Ecken bzw. Kantenmitten durchlaufen, ist von Ultradünnschichten durch konzentrische

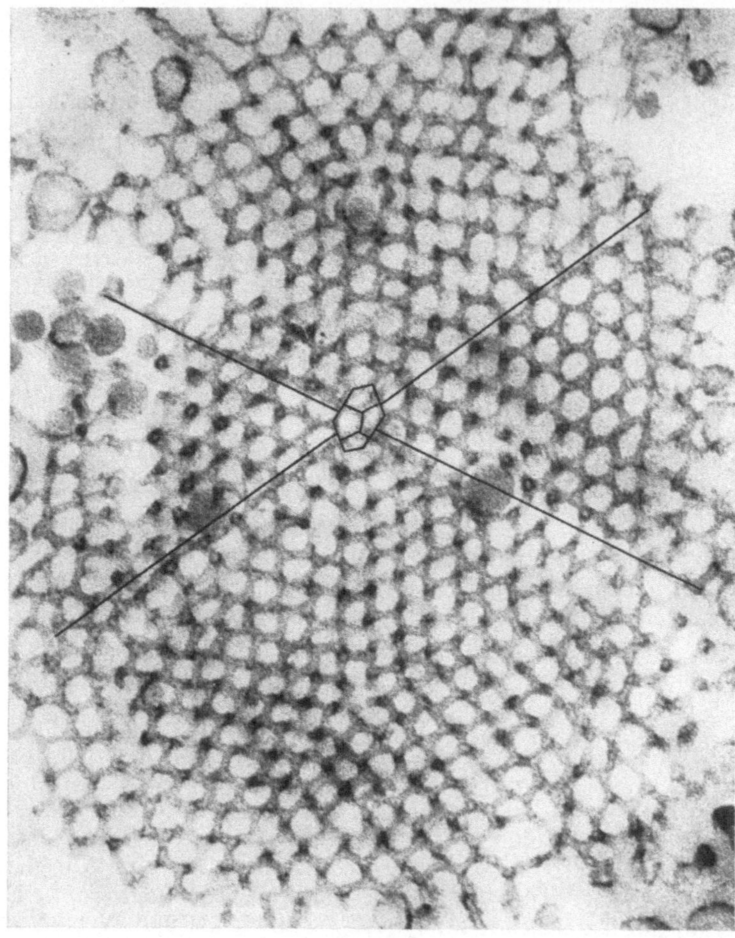

Abb. 3a

Abb. 3a u. b. Konzentrischer Prolamellarkörper mit zentralständigem Pentagondodekaeder (P. D., markiert) im Schnitt nach Wehrmeyer [200]. 99000:1. Zwei 5zählige Achsen (markiert) und zwei 2zählige Achsen liegen ungefähr in der Tafel, eine weitere 2zählige Achse steht zur Tafelebene senkrecht. Das Modell (Abb. 3b) ist nur aus Tetraedern (Fa. W. Büchi, Schweiz) aufgebaut

Prolamellarkörper eine 532-Symmetrie zu erwarten, sofern die Schnittebenen diese Achsen genau treffen. Dies konnte vor kurzem gezeigt werden [200]. Aus der weitgehenden Identität von Modellen und Schnittbildern ist nun zu folgern, daß Tubuli sich verzweigen können, daß die

Winkel zwischen 2 Tubuli in etwa konstant sind und bei 109° liegen und
daß schließlich die Verzweigungen mit einer gewissen Regelmäßigkeit
aufeinander folgen. Damit wird die Genese der Prolamellarkörper verständlich, denn unter diesen Voraussetzungen gelangen die Tubuli nach
der 4. oder 5. Verzweigung im Raum wieder zu ihrem Ausgangspunkt
zurück. Sie können dort durch Fusion oder Kombination enden. Auf
diese Weise sind tetraedrische Tubulianordnungen erklärlich [204].

Wie für die Vesikelbildung ist auch die Bildung der Prolamellarkörper unabhängig von der Gegenwart von Chlorophyll. Die fluorescenz-
mikroskopische Kontrolle [10, 188] und Pigmentanalysen [39, 89] von

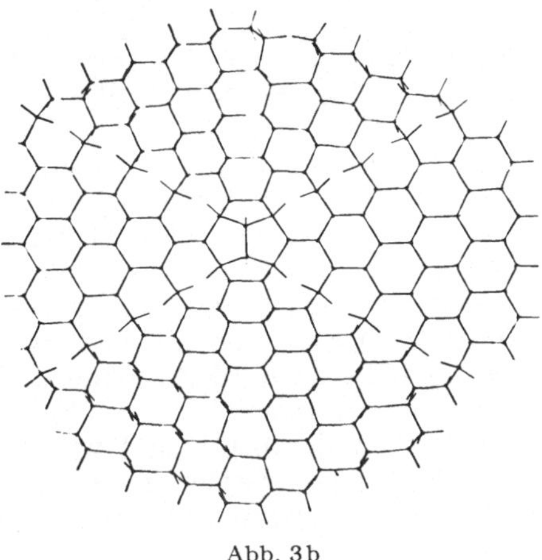

Abb. 3b

Prolamellarkörpern zeigen jedoch, daß bereits nach kurzer Belichtung
Chlorophyll in ihnen nachweisbar ist. Daraus kann auf die Existenz von
Protochlorophyll in den Prolamellarkörpern geschlossen werden [vgl.
90].

3. Übergang zur Fläche

Der Übergang von Vesikeln und Tubuli zu Thylakoidflächen ist
wohl wegen des raschen Ablaufs dieser Prozesse erst wenig untersucht.
Die Vorstellung von der Fusion kleiner Bläschen zu Thylakoiden [75]
bedarf des Beweises. Hinsichtlich der Tubuli-Thylakoid-Transformation
liegen mehrere Untersuchungen über die strukturellen Veränderungen
der Prolamellarkörper im Anschluß an Belichtung vor. Dabei standen
lichtphysiologische Gesichtspunkte [39, 90, 114, 193, 216] oder Fragen
der Biochemie [114] und insbesondere der Pigmentsynthese und -kon-
version im Vordergrund [89, 91, 193].

Die Ausbildung flächenhafter Ordnung wird fast generell im Zusammenhang
mit dem Faktor Licht betrachtet, da unter natürlicher oder experimentell bedingter
Dunkelheit die Spreitung vesiculärer oder tubulärer Elemente in der Regel unterbleibt. Das Beispiel ausgebildeter Thylakoide [212] und sogar von Grana [21] in
dunkel gehaltenen Coniferen zeigt jedoch, daß das Licht nur Auslöser ist und die
eigentliche Induktion vermutlich über Stoffe erfolgt, die in Sonderfällen auch im

Dunkeln angeliefert werden können. Da die Protochlorophyll-Chlorophyll-Konversion in Coniferencotyledonen in Dunkelheit abläuft, erhebt sich hier die Frage nach dem Zusammenhang von Chlorophyllgehalt und der Ausbildung eines typischen Thylakoidsystems. Die Tubuli-Thylakoid-Transformation läuft bei *Phaseolus* mit der Protochlorophyll-Chlorophyll-Konversion parallel und wird übereinstimmend mit einem Energiebedarf von 10^4 erg/cm^2 sec bei einer Wellenlänge von 660 mμ oder 445 mμ angegeben [*42, 193, 90*]. Demgegenüber stehen Befunde über eine vom Chlorophyllspiegel unabhängige Thylakoidbildung. So gelang es, allein durch Fütterung etiolierter Bohnen mit Rohrzucker trotz dauernder Dunkelheit die Thylakoidbildung aus Prolamellarkörpern zu induzieren. Die Protochlorophyll-Chlorophyll-Konversion unterblieb dabei [*140*, vgl. aber *220*]. Konzentrische Thylakoide sollen nach 20 tägiger Dunkelheit und in der Gerstenmutante xantha 10 auch bei absoluter Hemmung der Chlorophyllsynthese auftreten [*212*]. Durch Behandlung etiolierter Bohnenblätter mit δ-Aminolävulinsäure kann eine Störung im Chlorophyllstoffwechsel erzeugt werden, so daß nach der Belichtung ein rasches Ausbleichen der zunächst ergrünten Blätter erfolgt. Trotz der Photodestruktion der Pigmente konnte die Bildung von Thylakoidstapeln elektronenmikroskopisch nachgewiesen werden [*91*].

Bei den bisherigen Untersuchungen über Differenzierungsprozesse im Anschluß an Belichtung von Prolamellarkörpern ist die Struktur der Prolamellarkörper nicht berücksichtigt worden. Dabei ergeben sich gerade von der strukturellen Seite her ganz neue Aspekte. Für die Entwicklung von Thylakoiden in Plastiden mit Prolamellarkörpern sind offensichtlich verschiedene Wege möglich. Der erste besteht in der völligen Neubildung von Thylakoiden von der Oberfläche der Prolamellarkörper her unter weitgehender Erhaltung der übrigen Teile. Während der Thylakoidbildung persistierende Prolamellarkörper sind weit verbreitet [*70, 71, 80, 120, 122, 128, 140, 100*]. Diese werden in der Regel erst allmählich durch Schwund undeutlich. Dauernde Persistenz dagegen ist selten [*168, 169*]. Die zweite Möglichkeit liegt in dem sofortigen Zerfall des Prolamellarkörpers in Vesikeln nach kurzfristiger Belichtung und ist für *Phaseolus* angegeben [*91, 193, 216*]. Nach diesen Autoren erfolgt in einer zweiten lichtabhängigen Phase die Verteilung der Bläschen und Anordnung in primären Schichten, sogenannten „Bläschenreihen". In einem dritten Schritt soll die Granabildung durch Verschmelzen von Bläschen in Scheiben und deren Aggregation zu Grana einsetzen [*193*]. In eigenen Untersuchungen am gleichen Objekt konnten die lichtphysiologischen Angaben bezüglich des Ablaufs der Entwicklung bestätigt werden. In der Beurteilung der in den einzelnen Phasen vorliegenden Strukturen jedoch ergeben sich Differenzen. Bei genauer räumlicher Analyse erweisen sich die „Bläschenreihen" als kontinuierlich zusammenhängende Flächen netzartig verknüpfter mehr oder weniger abgeplatteter Tubuli, die unmittelbar von Prolamellarkörperschichten abzuleiten sind. Die Schichtenbildung erfolgt durch Verschmelzen der Tubuli einer Ebene. Die in den Thylakoiden stellenweise verbleibenden Löcher [vgl. *170*] sind Reste des Raumes, der ursprünglich von den Tubuli ausgespart wurde. Unter dieser Vorstellung ist der Terminus „Aussparung" geprägt worden und dem der „Perforation" vorzuziehen [*198, 199*]. Den Charakter von Aussparungen haben auch die Löcher des Thylakoidsystems, die nicht auf Prolamellarkörper zurückgeführt werden können. Wenn daher Löcher in den Thylakoiden durchaus ver-

ständlich erscheinen und als reell anzusehen sind, ist auf fixations-
bedingte nachträgliche Veränderungen insbesondere nach KMnO$_4$-
Einwirkung zu achten [45, 178].

In der Anordnung der sich entwickelnden, meist einzeln liegenden
Thylakoidschichten ergeben sich trotz aller Vielfalt bestimmte, regelmäßig wiederkehrende Typen. Dazu gehören die Anordnung in linearen

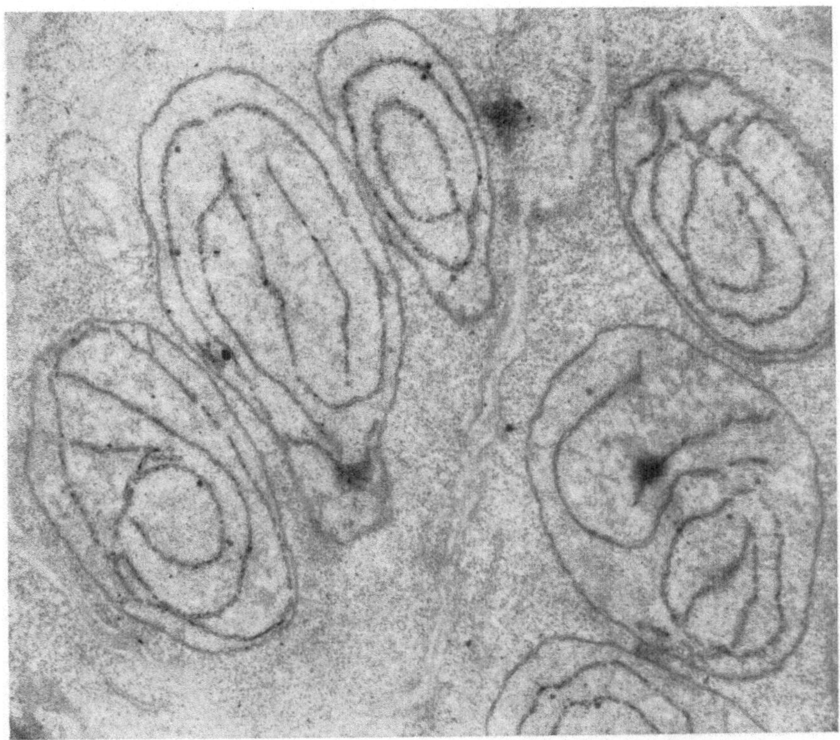

Abb. 4. Nach kurzfristiger (1 h) Belichtung etiolierter Primärblätter der Bohne aus den
Tubuli der Prolamellarkörper hervorgehende Thylakoide in einzeln liegenden Schichten.
Original. 24 500 : 1

Reihen, Schleifen oder in konzentrischen Ringen (Abb. 4). Diese Stadien
der Thylakoidentwicklung, insbesondere die konzentrischen Ringe, die
bislang als gegeben hingenommen oder auf Temperatur [87] bzw.
Schwachlichtbedingungen [39] zurückgeführt wurden, scheinen mit der
Struktur der Prolamellarkörper in Zusammenhang zu stehen.

Angesichts der sterischen Verhältnisse in der Anordnung der Tubuli sind nämlich bestimmte „Gitterebenen" in den Prolamellarkörpern für eine Fusion als
bevorzugt, andere als behindert zu betrachten. In jedem Falle muß beim Übergang der ursprünglich dreidimensionalen Ordnung in die zweidimensionale Fläche
eine Dimension abgebaut werden. Dies läßt sich nachweisen.

Es ist naheliegend, die konzentrischen „Bläschenreihen" auf konzentrische Prolamellarkörper (Tabelle Typ 3), parallele Thylakoidzüge

entsprechend auf Prolamellarkörper mit Schichtenbau zurückzuführen (Tabelle Typ 2), wobei allerdings hochgeordnete Prolamellarkörper zur Persistenz neigen. Bei der Vielfalt der darüber hinaus realisierten Prolamellarkörper ist die Uneinheitlichkeit der Aspekte in den Plastiden

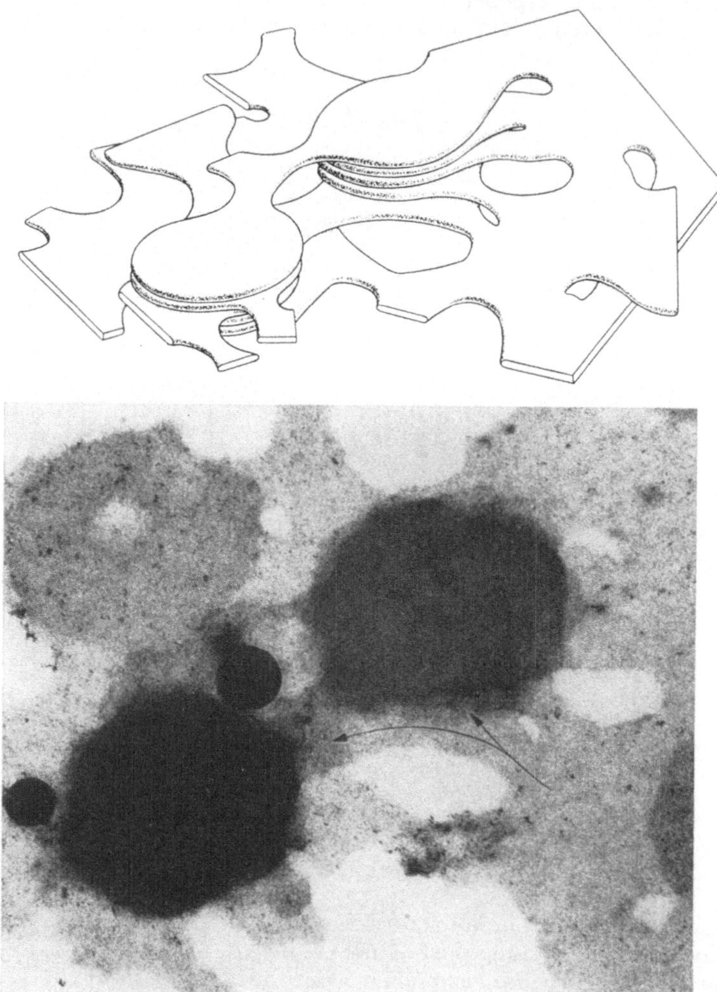

Abb. 5. Stapelbildung durch Überschiebung lateral miteinander verbundener Thylakoide in Spinatchloroplasten in Aufsicht nach Wehrmeyer [*199*]. 80 000:1

selbst in ein und derselben Zelle nicht überraschend. Sie wird durch die unmittelbar nach der Thylakoidbildung einsetzende Neubildung von Flächen noch verstärkt.

4. Stapelbildung

Das Vorkommen der Thylakoide in Stapeln ist — von wenigen Ausnahmen bei Rhodophyceen und Cyanophyceen abgesehen — fast generell nachgewiesen [Lit. *52, 124, 203*]. Während das morphologische Bauelement der Stapel in den Thylakoiden [*119*], discs [*51*] bzw. Vesikeln [*129*] erkannt werden konnte, blieben die Fragen nach der Entwicklung und Differenzierung der Schichten bislang weitgehend offen. Die Klärung der morphologischen Seite der Entwicklung der Stapel setzt eine genaue Kenntnis ihrer räumlichen Ordnung voraus. Diese ist erst wenig untersucht und kann nur anhand kompletter Schnittserien [*155, 203, 224*] oder durch kombinierte Untersuchung von Profil- und Aufsichtsbildern beurteilt werden [*73, 197, 198, 199, 206, 207, 208*]. Aus bestimmten räumlichen Gegebenheiten stehen nach unserer Vorstellung prinzipiell nur beschränkte Möglichkeiten für eine Stapelbildung offen:

1. Stapelbildung aus voneinander unabhängigen Einzelschichten.

2. Stapelbildung aus *lateral* miteinander verknüpften Schichten (Abb. 5).

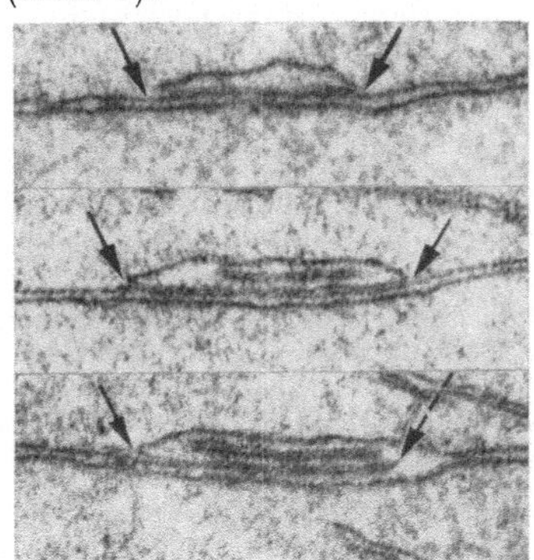

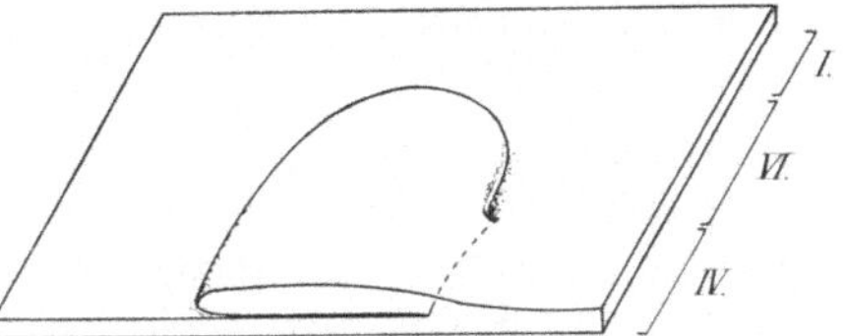

Abb. 6. Stapelbildung durch Überschiebung serial miteinander verknüpfter Thylakoide in einer chlorina-Mutante von *Arabidopsis* in Profilansicht nach Wehrmeyer u. Röbbelen [*203*]. 60 000 : 1

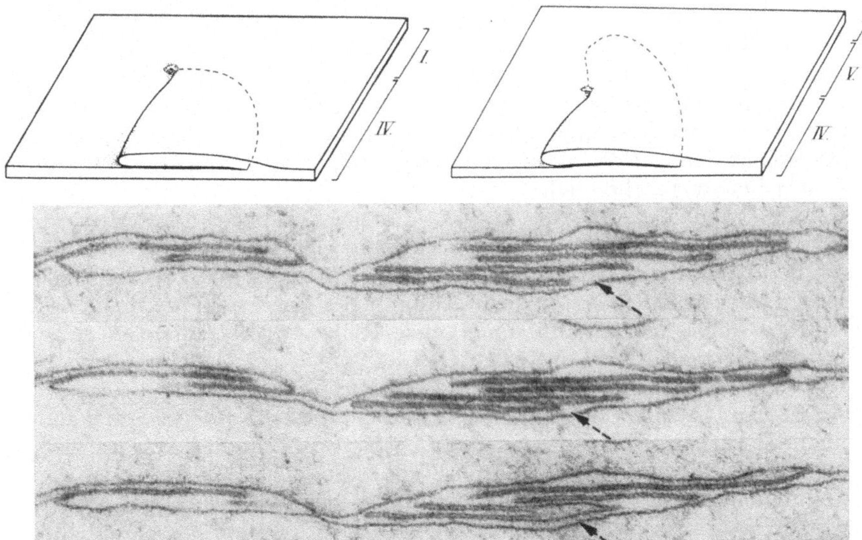

Abb. 7. Innere Aufgliederung der Thylakoide durch Invagination in einer chlorina-Mutante von *Arabidopsis* nach [*203*]. 54 000 : 1

3. Stapelbildung aus *serial* miteinander verknüpften Schichten (Abb. 6).

4. Stapelbildung durch nach *innen* fortschreitende Aufgliederung der Thylakoide, *Invagination* [*116, 121*] (Abb. 7).

Zur Beurteilung der Stapel sind hauptsächlich Profilansichten zugänglich. Thylakoidschichtungen nach Typ 1 u. 2 erscheinen dabei in Schnittserien stets oder überwiegend getrennt (disjunktiv), Stapel nach Typ 3 lassen zwischen benachbart übereinander liegenden Schichten in günstigen Fällen Verbindungen erkennen (konjunktiv). Treten konjunktive Thylakoidschichten dominierend auf, wie dies bei *Anthoceros* [*110, 119, 218*] und einigen Mutanten der Fall ist [*155, 171, 203*], so ist mit (zusätzlicher) Invagination zu rechnen.

Die Stapelbildung über isolierte Einzelstrukturen ist ursprünglich allgemein und insbesondere für die Granaordnung in Chloroplasten höherer Pflanzen angenommen worden. Die Genese solcher Stapel sollte durch identische Reduplikation der Granascheibchen erfolgen [*187*]. Neuere Befunde sprechen für die Ausbildung der Stapel über spezifische Wachstumsprozesse im Membranbereich der Thylakoide, jedoch sind viele Fragen über den Modus und Ablauf und die Richtung dieses Wachstums noch offen.

Kombinierte Untersuchung an Profil- und Aufsichtsbildern der Chloroplasten erbrachten den Nachweis eines Zusammenhangs zwischen Grana- und Intergranabereich [*196, 206, 207, 197*]. In Weiterführung dieser Befunde an Spinatchloroplasten ergaben sich kontinuierliche Verknüpfungen zwischen Granamembranen benachbarter Grana und solchen ein und desselben Granums und zwischen Granamembranen und einem durch Aussparungen aufgegliederten Thylakoidsystem [*198, 199*]. Die Interpretation dieser Befunde führt zur Ableitung der Granamembranen von ihresgleichen oder von einem vorgegebenen Thylakoidsystem. Die daraus resultierende Stapelbildung durch Überschiebung wird durch den Nachweis kontrastreicher lokaler Doppellagen von Thylakoidbereichen in der Nähe der Grana in Aufsichtsbildern, durch zum Raum hin geöffnete Überschiebungslücken und schließlich durch das Vorkommen von 2 Stegen an ein und derselben Granamembran gestützt. Diese Form der Stapelbildung ist — wenn auch nicht ausschließlich — für Spinat [*189, 199*], *Nicotiana* [*206, 207, 208*], *Cannabis* [*73*] und *Elodea* [*45*] wahrscheinlich.

Weit weniger sind die Stapelbildungen serialer Verknüpfung in ihrem räumlichen Bau und erst recht von der Entwicklung her verständlich. Konjunktive Thylakoidstapel konnten in Chloroplasten niederer Pflanzen bei 14 Arten aus 5 Abteilungen festgestellt werden [*203*]. Weitere Nachweise sind für eine ganze Reihe höherer Pflanzen zu erbringen *(Oenothera* [*116, 124, 171*], *Lycopersicum* [*33*], *Spinacia, Phaseolus* [*205*], *Arabidopsis* [*155, 203*], *Nicotiana* [*206, 207*], *Cannabis* [*73*], *Chlorophytum* [*124*], *Fritillaria* [*129*], *Hordeum* [*209*]). Serial miteinander verknüpfte Thylakoide bei *Oenothera* veranlaßten Menke [*116*] zur Aufstellung seiner Invaginationshypothese und damit zu ersten Vorstellungen über die Entwicklung der Granastapel. Es konnte vor kurzem anhand von Serienschnitten und deren Kombination zu räumlichen Modellen gezeigt werden, in welchem Umfang und Modus Invaginationen im Binnenraum der Thylakoide ablaufen können [155, 203]. Bestimmte Profilansichten lassen z. Z. eine Interpretation im Sinne einer Invagination als auch von Überschiebungsprozessen zu (Abb. 8). Hier ist von Schnittserien her in Zukunft eine Klärung zu erwarten. Wichtig erscheint in diesem Zusammenhang, daß Invaginationen mit Überschiebungen kombiniert auftreten können und sich nicht gegenseitig ausschließen [155, 203]. Dies wurde bereits an einem weiteren Objekt bestätigt [171].

Die physiologischen Bedingungen für die Stapelbildung sind nur unzulänglich bekannt. Auf der einen Seite wird die Granabildung mit einer Phase starker Chlorophyllzunahme in den Chloroplasten korreliert [193, 212, 216]; auf der anderen Seite stehen Befunde über abnehmenden oder sehr geringen Chlorophyllgehalt bei eindeutig ausgebildeter Schichtstruktur [91, 140, 53]. Versuche mit Inhibitoren des Proteinstoffwechsels

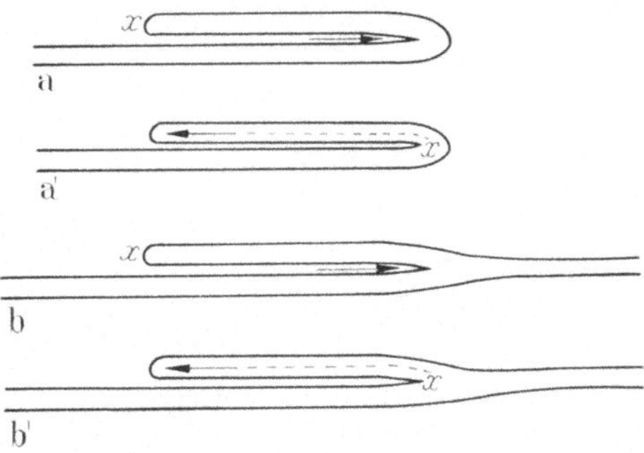

Abb. 8. Schematische Darstellung konjunktiver Thylakoidschichtung durch Invagination (a, b) bzw. Überschiebung (a', b') in Pfeilrichtung nach [203]

geben weitere Hinweise. Die nach Besprühen mit Chloramphenicol und Streptomycin neu zuwachsenden, völlig weißen Sproßabschnitte der Tomate enthalten zwar in der Entwicklung gehemmte, anomale Plastiden; es sind aber stets Stadien mit Thylakoidstapeln in Form von „Makrograna" dabei zu beobachten [34]. Ähnliches gilt für die Thiouracilbehandlung, die ebenfalls zu starker Störung in der Ausfärbung der Blätter führt, jedoch die Stapelbildung nicht völlig unterbindet [34, 72, 74]. Darüber hinaus konnten in pigmentdefekten, letalen Gerstenmutanten [194, 215] und beim Hanf [74] Makrograna nachgewiesen werden. Wenn also auch im einzelnen Abweichungen in der Zahl der Grana, in ihrem Durchmesser und der Feinstruktur ihrer Schichtenordnung vorkommen, erscheinen ihre Bildung und Aggregation trotz der Pigmentdefekte grundsätzlich möglich. Ob Pigmente an der Ausbildung der Kontaktzonen in dem vermuteten Maße oder überhaupt beteiligt sind [214, 215], ist fraglich [53]. Die Notwendigkeit höherer Lichtenergie für die Granabildung in Chloroplasten ist von verschiedenen Autoren bestätigt worden, jedoch besteht hinsichtlich des genauen Energiebedarfs noch keine einheitliche Meinung [39, 42, 90, 114, 140, 193, 216]. Dies gilt auch bezüglich der morphogenetischen Wirkung des Hellrot-Dunkelrot-Systems [90, 114].

Summary

1. The morphology of the stroma is discussed particularly with regard to the protein synthesizing system of ribosomes in the presence of DNA. Moreover, the various inclusions of the stroma, e.g. stromacentres, phytoferritin, osmiophilic globules, and starch grains are biiefly dealt with.

2. The morphology and morphogenesis of the thylakoid system is represented in three steps of development: the vesicular, the tubular, and the planar phase including the multiplication of thylakoid layers. Concerning the tubular phase recent studies on the structure and development of prolamellar bodies of the author are summarized. From the different types of prolamellar bodies being indicated in dark grown etiolated leaves, a new interpretation of the tubules — thylakoid — transformation in proplastids after illumination is possible. The so-called primary layers are not rows of vesicles, but consist of an interconnected net-like system of more or less spread tubules as seen in face view. The transition of this single-layered system to the multi-layered system of the grana-containing chloroplast is performed by stacking processes within 18 hours. Different types of stacks can be distinguished, probably due to differently directed growth in space of thylakoid membranes. Membrane growth in relation to the stroma and to the pigment content of thylakoids is discussed.

Literatur

[1] App, A., and A. T. Jagendorf: Biochim. biophys. Acta (Amst.) **76**, 286 (1963).

[2] — — Plant Physiol. **39**, 772 (1964).

[3] Arnon, D. I.: Nature **184**, 10 (1959).

[4] Bailey, J. L., and A. G. Whyborn: Biochim. biophys. Acta (Amst.) **78**, 163 (1963).

[5] Baltus, E., and J. Brachet: Biochim. biophys. Acta (Amst.) **76**, 490 (1963).

[6] Baumann, E.: Z. Pflanzenphysiol. **53**, 90 (1965).

[7] Benson, A. A.: In: Light and Life (W. D. McElroy, and B. Glass eds.), p. 392. Baltimore: Johns Hopkins Press 1961.

[8] — Ann. Rev. Plant Physiol. **15**, 1 (1964).

[9] Bird, I. F., H. K. Porter, and C. R. Stocking: Biochim. biophys. Acta (Amst.) **100**, 366 (1965).

[10] Boardman, N. K., and J. M. Anderson: Aust. J. biol. Sci. **17**, 86 (1964).

[11] Bogorad, L., and A. B. Jacobson: Biochem. Biophys. Res. Commun. **14**, 113 (1964).

[12] Brawerman, G., A. O. Pogo, and E. Chargaff: Biochim. biophys. Acta (Amst.) **48**, 418 (1961).

[13] — Biochim. biophys. Acta (Amst.) **61**, 313 (1962).

[14] — Biochim. biophys. Acta (Amst.) **72**, 317 (1963).

[15] —, and J. M. Eisenstadt: Biochim. biophys. Acta (Amst.) **91**, 477 (1964).

[16] — — J. Molec. Biol. **10**, 403 (1964).

[17] Brooks, J. L., and P. K. Stumpf: Biochim. biophys. Acta (Amst.) **98**, 213 (1965).

[18] Budd, G. C., and G. M. Mills: Nature (Lond.) **205**, 524 (1965).

[19] Buvat, R.: Ann. Sci. Nat., Bot., Sér. XI, **19**, 121 (1958).

[20] Camefort, H.: C. R. Acad. Sci. (Paris) **258**, 1017 (1964).

[21] — C. R. Acad. Sci. (Paris) **258**, 5705 (1964).

[22] Chantrenne, H.: Biochim. biophys. Acta (Amst.) **95**, 351 (1965).

[23] Chiba, Y., and K. Sugahara: Arch. Biochem. **71**, 367 (1957).

[24] Chun, E. H. L., M. H. Vaughan jr., and A. Rich: J. molec. Biol. **7**, 130 (1963).

[25] Clark, M. F., R. E. F. Matthews, and R. K. Ralph: Biochim. biophys. Acta (Amst.) **91**, 289 (1964).

[26] — Biochim. biophys. Acta (Amst.) **91**, 671 (1964).

[27] Correns, C.: Nichtmendelnde Vererbung. Hdbuch der Vererbungswiss. Bd. II, Lfg. 22. Berlin: Bornträger 1937.

[28] COURCHET, M.: Ann. Sci. Nat., Bot., Sér. VII, **7**, 263 (1888).
[29] DAVIES, J., W. GILBERT, and J. GORINI: Proc. nat. Acad. Sci. (Wash.) **51**, 883 (1964).
[30] DE FEKETE, M. A. R., L. F. LELOIR, and C. E. CARDINI: Nature (Lond.) **187**, 918 (1960).
[31] —, and C. E. CARDINI: Arch. Biochem. **104**, 173 (1964).
[32] DÖBEL, P.: Kulturpflanze **7**, 168 (1959).
[33] — Biol. Zbl. **81**, 721 (1962).
[34] — Biol. Zbl. **82**, 275 (1963).
[35] DRAWERT, H., u. M. MIX: Planta **57**, 51 (1961).
[36] DREWS, G., u. P. GIESBRECHT: Zbl. Bakt., I. Abt. Orig. **190**, 508 (1963).
[37] DUBIN, D. T., and B. D. DAVIS: Biochim. biophys. Acta (Amst.) **55**, 793 (1962).
[38] DUNCAN, H. J., and R. W. REES: Biochem. J. **94**, 18 p (1965).
[39] EILAM, Y., and S. KLEIN: J. Cell Biol. **14**, 169 (1962).
[40] EISENSTADT, J., and G. BRAWERMAN: Biochim. biophys. Acta (Amst.) **76**, 319 (1963).
[41] — — J. molec. Biol. **10**, 392 (1964).
[42] ERIKSSON, G., A. KAHN, B. WALLES u. D. v. WETTSTEIN: Ber. dtsch. bot. Ges. **74**, 221 (1961).
[43] ESPADA, J.: J. biol. Chem. **237**, 3577 (1962).
[44] FALK, H.: Planta **55**, 525 (1960).
[45] —, u. P. SITTE: Protoplasma (Wien) **57**, 290 (1963).
[46] FARRANT, J. L.: Biochim. biophys. Acta (Amst.) **13**, 569 (1954).
[47] FERRARI, R. A., and A. A. BENSON: Arch. Biochem. **93**, 185 (1961).
[48] FLAKS, J. G., E. C. COX, and J. R. WHITE: Biochem. Biophys. Res. Commun. **7**, 385 (1962).
[49] FREY-WYSSLING, A., u. E. KREUTZER: Planta **51**, 104 (1958).
[50] FRYDMAN, R. B.: Arch. Biochem. **102**, 242 (1963).
[51] GIBBS, S. P.: J. Ultrastruct. Res. **4**, 127 (1960).
[52] — J. Ultrastruct. Res. **7**, 418 (1962).
[53] — J. Cell Biol. **15**, 343 (1962).
[54] GIBOR, A., and M. IZAWA: Proc. natl. Acad. Sci. (Wash.) **50**, 1164 (1963).
[55] —, and S. GRANICK: Science **145**, 890 (1964).
[56] GIERER, A.: J. molec. Biol. **6**, 148 (1963).
[57] GOFFEAU, A., and J. BRACHET: Biochim. biophys. Acta (Amst.) **95**, 302 (1965).
[58] GOLDBERG, I. H., and M. RABINOWITZ: Science **136**, 315 (1962).
[59] — —, and E. REICH: Proc. nat. Acad. Sci. (Wash.) **48**, 2094 (1962).
[60] GOODMAN, H. M., and A. RICH: Nature (Lond.) **199**, 318 (1963).
[61] GRANICK, S.: In: The Cell (J. BRACHET et A. E. MIRSKY eds.) **2**, 489. New York, London: Academic Press 1961.
[62] GREENWOOD, A. D., R. M. LEECH, and J. P. WILLIAMS: Biochim. biophys. Acta (Amst.) **78**, 148 (1963).
[63] GUNNING, B. E. S.: J. Cell Biol. **24**, 79 (1965).
[64] — Protoplasma (Wien) **60**, 111 (1965).
[65] HARBERS, E.: Die Nucleinsäuren. Stuttgart: G. Thieme Verlag 1964.
[66] HARRISON, P. M.: J. molec. Biol. **1**, 69 (1959).
[67] HEBER, U.: Nature (Lond.) **195**, 91 (1962).
[68] — Planta **59**, 600 (1963).
[69] HEITZ, E.: Exp. Cell Res. **7**, 606 (1954).
[70] — Experientia (Basel) **12**, 476 (1956).
[71] — Experientia (Basel) **16**, 265 (1960).
[72] HESLOP-HARRISON, J.: Planta **58**, 237 (1962).
[73] — Planta **60**, 243 (1963).
[74] — Port. Acta biol. A, **8**, 13 (1963/64).
[75] HODGE, A. J., J. D. McLEAN, and F. V. MERCER: J. biophys. biochem. Cytol. **2**, 597 (1956).
[76] HOLT, S. C., and R. G. MARR: J. Bact. **89**, 1402 (1965).

[77] Hudock, G. A., G. C. McLeod, J. Moravkova-Kiely, and R. P. Levine: Plant Physiol. **39**, 898 (1964).
[78] Hurwitz, J., J. J. Furth, M. Malamy, and M. Alexander: Proc. nat. Acad. Sci. (Wash.) **48**, 1222 (1962).
[79] Hyde, B. B., A. Hodge, A. Kahn, and M. L. Birnstiel: J. Ultrastruct. Res. **9**, 248 (1963).
[80] Jacobson, A. B., H. Swift, and L. Bogorad: J. Cell Biol. **17**, 557 (1963).
[81] Kalf, G. F.: Biochemistry **3**, 1702 (1964).
[82] Kauss, M., u. O. Kandler: Z. Naturforsch. **17b**, 858 (1962).
[83] Kern, H.: Protoplasma (Wien) **50**, 505 (1959).
[84] Kirk, J. T. O.: Biochim. biophys. Acta (Amst.) **76**, 417 (1963).
[85] — Biochem. biophys. Res. Commun. **16**, 233 (1964).
[86] — Biochem. biophys. Res. Commun. **14**, 393 (1964).
[87] Klein, S.: J. biophys. biochem. Cytol. **8**, 529 (1960).
[88] —, and A. Poljakoff-Mayber: Exp. Cell Res. **24**, 143 (1961).
[89] — — J. biophys. biochem. Cytol. **11**, 433 (1961).
[90] —, G. Bryan, and L. Bogorad: J. Cell Biol. **22**, 433 (1964).
[91] —, and L. Bogorad: J. Cell Biol. **22**, 443 (1964).
[92] Korner, A., and A. J. Munro: Biochem. biophys. Res. Commun. **11**, 235 (1963).
[93] Kretsinger, R. H., G. Manner, B. S. Gould, and A. Rich: Nature (Lond.) **202**, 438 (1964).
[94] Kuff, E. J., and A. L. Dalton: J. Ultrastruct. Res. **1**, 62 (1957).
[95] Lamprecht, I.: Protoplasma (Wien) **53**, 162 (1961).
[96] Lance-Nougarède, A.: C. R. Acad. Sci. (Paris) **250**, 173 (1960).
[97] Leff, J., M. Mandel, H. T. Epstein, and J. A. Schiff: Biochem. biophys. Res. Commun. **13**, 126 (1963).
[98] Lefort, M.: Rev. gén. Bot. **66**, 461 (1959).
[99] Leloir, L. F., M. A. R. de Fekete, and C. E. Cardini: J. biol. Chem. **236**, 636 (1961).
[100] Leyon, H.: Exp. Cell Res. **7**, 609 (1954).
[101] — Exp. Cell Res. **7**, 265 (1954).
[102] Lichtenthaler, H. K., and R. B. Park: Nature (Lond.) **198**, 1070 (1963).
[103] —, and M. Calvin: Biochim. biophys. Acta (Amst.) **79**, 30 (1964).
[104] — Ber. dtsch. bot. Ges. **77**, 398 (1964).
[105] Lima-de-Faria, A., and M. J. Moses: Hereditas (Lund) **52**, 367 (1965).
[106] Luck, D. J. L., and E. Reich: Proc. nat. Acad. Sci. (Wash.) **52**, 931 (1964).
[107] Lyttleton, J. W.: Exp. Cell Res. **26**, 312 (1962).
[108] Maly, R., u. A. Wild: Z. Vererbungslehre **87**, 493 (1956).
[109] Manner, G., and B. S. Gould: Nature (Lond.) **205**, 670 (1965).
[110] Manton, I.: J. exp. Bot. **13**, 325 (1962).
[111] Marinos, N. G.: Amer. J. Bot. **50**, 998 (1963).
[112] McCalla, D. R., and R. K. Allan: Nature (Lond.) **201**, 504 (1964).
[113] McMahon, V., and P. K. Stumpf: Biochim. biophys. Acta (Amst.) **84**, 359 (1964).
[114] Mego, J. L., and A. T. Jagendorf: Biochim. biophys. Acta (Amst.) **53**, 237 (1961).
[115] — Biochim. biophys. Acta (Amst.) **88**, 663 (1964).
[116] Menke, W.: Z. Naturforsch. **15b**, 479 (1960).
[117] — Z. Naturforsch. **15b**, 800 (1960).
[118] — Experientia (Basel) **16**, 537 (1960).
[119] — Z. Naturforsch. **16b**, 334 (1961).
[120] — Z. Naturforsch. **17b**, 188 (1962).
[121] — Ann. Rev. Plant Physiol. **13**, 27 (1962).
[122] — Z. Naturforsch. **18b**, 821 (1963).
[123] — Nat. Acad. Sci.-Nat. Res. Council, Publication 1145, 1963.
[124] — Ber. dtsch. bot. Ges. **77**, 340 (1964).
[125] Mikulska, E., M. S. Odintsova, and N. M. Sissakian: Naturwissenschaften **49**, 549 (1962).
[126] Miller, F.: J. Cell Biol. **9**, 157 (1961).

[*127*] Mudd, J. B., and T. T. McManus: Plant Physiol. **39**, 115 (1964).
[*128*] Mühlethaler, K., u. A. Frey-Wyssling: J. biophys. biochem. Cytol. **6**, 507 (1959).
[*129*] — Z. wiss. Mikroskopie **64**, 445 (1960).
[*130*] Murakami, S.: J. Electronmicroscopy **9**, 91 (1960).
[*131*] — Experientia (Basel) **18**, 168 (1962).
[*132*] — Exp. Cell Res. **32**, 398 (1963).
[*133*] Murata, T., T. Minamikawa, and T. Akazawa: Biochem. biophys. Res. Commun. **13**, 439 (1963).
[*134*] —, and T. Akazawa: Biochem. biophys. Res. Commun. **16**, 6 (1964).
[*135*] —, T. Sugiyama, and T. Akazawa: Arch. Biochem. **107**, 92 (1964).
[*136*] Nass, M. M. K., S. Nass, and B. A. Afzelius: Exp. Cell Res. **37**, 516 (1965).
[*137*] Neufeld, E. F., and C. W. Hall: Biochem. biophys. Res. Commun. **14**, 503 (1964).
[*138*] Nougarède, A.: C. R. Acad. Sci. (Paris) **258**, 683 (1964).
[*139*] Odintsova, M. S., E. V. Golubeva, and N. M. Sissakian: Nature (Lond.) **204**, 1090 (1964).
[*140*] Orsenigo, M., G. Marziani e M. R. Macchi: Giornale Botan. Ital. **71**, 152 (1964).
[*141*] Park, R. B., and N. G. Pon: J. molec. Biol. **6**, 105 (1963).
[*142*] —, and J. Biggins: Science **144**, 1009 (1964).
[*143*] Perner, E. S.: Z. Naturforsch. **11b**, 560 (1956); **11b**, 567 (1956).
[*144*] — Portugal. Acta Biol., Sér. A, **6**, 359 (1962).
[*145*] — Planta **65**, 334 (1965).
[*146*] — Unveröffentlicht.
[*147*] Philipps, G. R.: Nature (Lond.) **205**, 567 (1965).
[*148*] Pogo, A. O., G. Brawerman, and E. Chargaff: Biochemistry **1**, 128 (1962).
[*149*] Ray, D. S., and P. C. Hanawalt: J. molec. Biol. **9**, 812 (1964).
[*150*] Recondo, E., and L. F. Leloir: Biochem. biophys. Res. Commun. **6**, 85 (1961).
[*151*] —, M. Dankert, and L. F. Leloir: Biochem. biophys. Res. Commun. **12**, 204 (1963).
[*152*] Reich, E.: Science **143**, 684 (1964).
[*153*] Rhodes, M. J. C., and E. W. Yemm: Nature (Lond.) **200**, 1077 (1963).
[*154*] Ris, H., and W. Plaut: J. Cell Biol. **13**, 383 (1962).
[*155*] Röbbelen, G., u. W. Wehrmeyer: Planta **65**, 105 (1965).
[*156*] Rossner, W.: Protoplasma (Wien) **52**, 580 (1960).
[*157*] — Planta **60**, 166 (1963).
[*158*] Ruppel, H. G.: Biochim. biophys. Acta (Amst.) **80**, 63 (1964).
[*159*] —, u. D. van Wyk: Z. Pflanzenphysiol. **53**, 32 (1965).
[*160*] Sabatini, D. D., K. Bensch, and R. J. Barrnett: J. Cell Biol. **17**, 19 (1963).
[*161*] Sagan, L., V. Ben-Shaul, J. A. Schiff, and H. T. Epstein: J. Cell Biol. **23**, 81 A (1964).
[*162*] Sager, R., and M. R. Ishida: Proc. nat. Acad. Sci. (Wash.) **50**, 725 (1963).
[*163*] Salpeter, M. M., and L. Bachmann: J. Cell Biol. **22**, 469 (1964).
[*164*] Schaechter, M.: J. molec. Biol. **7**, 561 (1963).
[*166*] Schimper, A.: Jb. wiss. Bot. **16**, 1 (1885).
[*167*] Schlessinger, D.: J. molec. Biol. **7**, 569 (1963).
[*168*] Schnepf, E.: Planta **54**, 641 (1960).
[*169*] — Protoplasma (Wien) **54**, 310 (1961).
[*170*] — Planta **61**, 371 (1964).
[*171*] Schötz, F.: Planta **64**, 376 (1965).
[*172*] Schweiger, H. G.: Naturwissenschaften **51**, 521 (1964).
[*173*] Shin, M., and D. I. Arnon: J. biol. Chem. **240**, 1405 (1965).
[*174*] Siegesmund, K. A., W. G. Rosen, and S. R. Gawlik: Amer. J. Bot. **49**, 137 (1962).
[*175*] Signol, M.: C. R. Acad. Sci. (Paris) **252**, 4177 (1961).
[*176*] Sissakian, N. M., I. I. Filippovich, E. N. Svetailo, and K. A. Aliyev: Biochim. biophys. Acta (Amst.) **95**, 474 (1965).

[177] SITTE, P.: Protoplasma (Wien) 53, 438 (1961).
[178] — Port. Acta. biol. A, 6, 269 (1962).
[179] — Protoplasma (Wien) 56, 197 (1963).
[180] — Protoplasma (Wien) 57, 304 (1963).
[181] SMILLIE, R. M.: Canad. J. Bot. 41, 123 (1963).
[182] STAEHELIN, T., F. O. WETTSTEIN, and H. NOLL: Science 140, 180 (1963).
[183] STEFFEN, K., u. F. WALTER: Planta 50, 640 (1958).
[184] STEPHENSON, M. L., K. V. THIMANN, and P. C. ZAMECNIK: Arch. Biochem. 65, 194 (1956).
[185] STOCKING, C. R., and E. M. GIFFORD: Biochem. biophys. Res. Commun. 1, 159 (1959).
[186] STRUGGER, S.: Naturwissenschaften 37, 166 (1950).
[187] — Ber. dtsch. bot. Ges. 64, 69 (1951).
[188] — Naturwissenschaften 41, 286 (1954).
[189] STUMPF, P. K., and A. T. JAMES: Biochim. biophys. Acta (Amst.) 70, 20 (1963).
[190] SUN, C. N.: J. Electronmicroscopy 12, 254 (1963).
[191] TREBST, A. V., H. Y. TSUJIMOTO, and D. I. ARNON: Nature (Lond.) 182, 351 (1958).
[192] — Ber. dtsch. bot. Ges. 77, 123 (1964).
[193] VIRGIN, H. I., A. KAHN, and D. VON WETTSTEIN: Photochem. Photobiol. 2, 83 (1963).
[194] WALLES, B.: Hereditas (Lund) 50, 317 (1963).
[195] WARNER, J. R., A. RICH, and C. E. HALL: Science 138, 1339 (1962).
[196] WEHRMEYER, W.: Vortr. Tagung. d. dtsch. bot. Ges. (Karlsruhe 1962).
[197] — Planta 59, 280 (1963).
[198] — Planta 62, 272 (1964).
[199] — Planta 63, 13 (1964).
[200] — Z. Naturforsch. 20b, 1270 (1965).
[201] — Z. Naturforsch. 20b, 1278 (1965).
[202] — Z. Naturforsch. 20b, 1288 (1965).
[203] WEHRMEYER, W., u. G. RÖBBELEN: Planta 64, 312 (1965).
[204] — Colloque "Croissance et Vieillissement des Chloroplastes" Gorsem-St. Trond, 13.—18. 9. 1965, Belgie.
[205] — Unveröffentlicht.
[206] WEIER, T. E.: Amer. J. Bot. 48, 615 (1961).
[207] —, and W. W. THOMSON: J. Cell Biol. 13, 89 (1962).
[208] —, C. R. STOCKING, W. W. THOMSON, and H. DREVER: J. Ultrastruct. Res. 8, 122 (1963).
[209] —, C. R. STOCKING, C. E. BRACKER, and E. B. RISLEY: Amer. J. Bot. 52, 339 (1965).
[210] WEISBERGER, A. S., S. ARMENTROUT, and S. WOLFE: Proc. nat. Acad. Sci. (Wash.) 50, 86 (1963).
[211] V. WETTSTEIN, D.: Exp. Cell Res. 12, 427 (1957).
[212] — Brookhaven Symposia Biology 11, 138 (1959).
[213] — In: Developmental Cytology (D. RUDNICK, ed.). New York: The Ronald Press Company 1959.
[214] — J. Ultrastruct. Res. 3, 235 (1959).
[215] — Hereditas (Lund) 46, 700 (1960).
[216] —, and A. KAHN: Proc. European Reg. Conf. Electron Microscopy, Delft, 1960, 2, 1051 (1961).
[217] WILD, A.: Planta 50, 379 (1958).
[218] WILSENACH, R.: J. Cell Biol. 18, 419 (1963).
[219] WITTMANN, H. G.: Naturwissenschaften 50, 76 (1963).
[220] WOLFF, J. B., and L. PRICE: J. biol. Chem. 235, 1603 (1960).
[221] WOLLGIEHN, R., u. K. MOTHES: Naturwissenschaften 50, 95 (1963).
[222] — — Exp. Cell Res. 35, 52 (1964).
[223] WOODS, M. W., and H. G. DU BUY: Amer. J. Bot. 38, 419 (1951).
[224] WYGASCH, J.: Z. Naturforsch. 18b, 827 (1963).
[225] ZIPSER, D. J.: J. molec. Biol. 7, 739 (1963).
[226] ZUCKER, M., and H. T. STINSON JR.: Arch. Biochem. 96, 637 (1962).

Diskussion

Vorsitz: P. Sitte

Karlson: Hat man Vorstellungen darüber, warum sich in den Prolamellar-körpern diese Strukturen zu solchen fünf- und sechseckigen Gebilden ordnen ? Sind das inhärente Kräfte, die in der Art des Proteins oder des Lipids oder in Wechsel-wirkung von beiden liegen, oder ist dazu irgend eine zusätzliche Information nötig ?

Wehrmeyer: Gunning [Protoplasma (Wien) **60**, 111, 1965] konnte nachweisen, daß in den Winkeln der Tubuli Ribosomen liegen, die nach seiner Vorstellung das Muster prägen. Damit wird allerdings das Problem nur in einen anderen Bereich transformiert. Gunning diskutiert in seiner Arbeit als ordnendes Agens Fermi-Kräfte. Diese kann man nach meinem Dafürhalten wohl im Bereich von wenigen Ångström geltend machen, der Dimensionsbereich der Prolamellarkörper liegt um etwa zwei Zehnerpotenzen höher.

Meine Vorstellung ist: Wenn sich die Tubuli in regelmäßigen Abständen nach konstanten Winkeln (etwa 108°) verzweigen, werden sie nach der dritten oder vierten Verzweigung wieder zum Ausgangspunkt zurückkommen; dort können sie sich mit dem Ausgangssystem entweder durch Fusion verbinden oder auch nur anlagern, wie das ja auch beim Membranwachstum der Grana erfolgt.

Bis vor kurzem waren keine ähnlichen Strukturen bekannt. Aber inzwischen sind von Eymé und Le Blanc (C. R. Acad. Sci. **256**, 4958, 1963) in Nektarien von Ranunculaceen und von Behnke (Planta **66**, 106, 1965) in Siebröhren ähnlich hoch geordnete Strukturen nachgewiesen worden. Man findet solche Proteinstrukturen hoher Ordnung also nicht nur in Proplastiden.

P. Sitte: Wir haben bei Organismen noch wesentlich größere (manchmal sogar makroskopische) Strukturen von beinahe kristallgitterartiger Regelmäßigkeit. Das Problem ist doch im Grunde immer das gleiche: Elemente bestimmter Größe — das können auch physiologische Sperrfelder sein — werden so nahe zusammengeschoben wie möglich. Stimmen nun die Größen überein, dann muß es zu einer regelmäßigen Anordnung kommen.

Stoeckenius: Lipoide und Lipoproteine bilden flüssige Kristalle, die ähnliche Dimensionen haben wie jene in den Proplastiden. Die in Plastiden gefundene An-ordnung haben wir allerdings nicht beobachtet. Aber die Kristallform hängt von den speziellen Bedingungen (Salzkonzentration, pH usw.) ab.

Ich halte die Prolamellarkörper für eine Art Vorratsform von Lipoiden und Pro-teinen für die nachfolgende Membranbildung. Sie gehen in den halbkristallinen Zustand von selbst über, wenn sie in genügender Konzentration vorhanden sind. Es gibt ähnliche Erscheinungen bei Abbauvorgängen in Zellen, wenn größere Men-gen von Lipoiden frei werden.

P. Sitte: Gunning hat ausgerechnet, daß tatsächlich die gesamte Oberfläche dieser Tubuli der Fläche der unmittelbar aus ihnen entstehenden Chloroplasten-membranen entspricht.

Parthier: Der Biochemiker fragt natürlich nach der chemischen Zusammen-setzung der Prolamellarkörper.

Bei Belichtung etiolierter Gewebe kommt es zu einer starken Protein- (und auch Ribonucleinsäure-)Synthese, die so schnell abläuft, daß man sie nach dem bekannten Schema (DNS→messenger-RNS→Ribosomen→Proteine) nicht recht erklären kann. Vielleicht kann man die Prolamellarkörper als „unterkühltes" System auffassen (DNS plus langlebige messenger-RNS enthaltend), das bei „Startschußbelichtung" sofort mit hoher Intensität Proteine bilden kann. Die messenger-RNS für die Photo-synthese-Enzymproteine würde erst nach der Belichtung gebildet.

Wehrmeyer: Die Transformation der Tubuli in primäre Thylakoidflächen geht tatsächlich rasch (innerhalb von etwa 3 min) vor sich. Deshalb verdient der Gedanke von Gunning Beachtung, daß die Tubuli das erforderliche Material prä-formiert enthalten. Viele Systeme liegen ja in photolabilen Vorstufen vor, die zur Umwandlung in die entsprechenden aktiven Formen nur einer Lichtgabe bedürfen (z. B. Chlorophyll).

3. Symposion Naturforscher   15

Dennoch würde ich die Annahme von Gunning nicht überbewerten, daß die Oberflächen im Prolamellarkörper den Flächen der entstehenden Thylakoide entsprechen. Nach unserer Vorstellung wird ein Teil der Tubuli zugunsten der flächigen Strukturen abgebaut.

Quantitative, biochemische Untersuchungen sind schwierig, weil einerseits Proplastidenfraktionen noch nicht genügend rein gewonnen werden können, zum andern das Problem der Bezugsgröße besteht: Hill-Reaktion und Photophosphorylierung werden gewöhnlich auf den Chlorophyllgehalt bezogen; das geht hier natürlich nicht.

Karlson: Welchen Einfluß hat das Chlorophyll selbst auf die Ausbildung der Granastruktur ? Wie sind die Verhältnisse bei chlorophylldefekten Mutanten ?

Wehrmeyer: Viele Autoren halten einen hohen Chlorophyllspiegel für die normale Morphogenese der Chloroplasten für notwendig. Andere fanden, daß in pigmentfreien oder pigmentarmen „Chloroplasten" durchaus übereinandergestapelte Thylakoide vorliegen können, wenn auch in reduziertem Ausmaß (z. B. Döbel bei mit Streptomycin behandelten, farblosen Tomatenblättern. Ähnliche Befunde wurden neuerdings aus der Brachetschen Schule für actinomycinbehandelte *Acetabularia* mitgeteilt).

Bei chlorophylldefekten Mutanten ist insgesamt die Ausbildung des Thylakoidsystems reduziert; eine restlose Hemmung aller Membranbildungsprozesse besteht jedoch nicht.

Parthier: Man hat guten Grund anzunehmen, daß bei Antibiotica-Versuchen mit chloroplastenführenden Geweben primär die Proteinsynthese — im Falle des Actinomycins die RNS-Synthese — gehemmt wird. Unter diesen Umständen ist das normal gebildete Chlorophyll instabil, weil es nicht strukturgebunden vorliegt; aber das ist eine sekundäre Erscheinung.

Hagemann: Wenn die Genetiker von Chlorophyll-Mutanten sprechen, sind meist solche gemeint, die das Chlorophyll nicht stabilisieren können.

Beermann: Welche Strukturen charakterisieren die Plastiden eigentlich in morphologischer Hinsicht ? Was ist sozusagen die „Minimumdefinition" für diese semi-autonomen Systeme ?

Wehrmeyer: Doppelte Hülle und etwas Grundsubstanz.

P. Sitte: Und darin wohl spezifische DNA.

Wehrmeyer: Ja, die biochemische Analyse hat bislang erheblich mehr Informationen über Plastiden erbracht als die Elektronenmikroskopie.

Über die Vermehrung von Plastiden und Mitochondrien während der Oogenese von Sphaerocarpus

Von

Lothar Diers

Mit 9 Abbildungen

Die Frage nach der Genese der Plastiden ist bis heute nicht endgültig geklärt. Zahlreiche Untersuchungen an niederen und höheren Pflanzen sprechen für eine morphologische Kontinuität, die umfassend zum ersten Mal von Schmitz [16], Schimper [15] und Meyer [10] als Hypothese aufgestellt wurde. Nach ihr gehen Plastiden nur aus ihresgleichen durch Teilung hervor. Obwohl Lebendbeobachtungen diese Vermehrungsart bei einigen Arten unmittelbar beweisen, wird sie als die ausschließlich vorkommende Vermehrungsform der Plastiden nicht allgemein anerkannt. So entwickeln sich die Plastiden in meristematischen Zellen einiger Angiospermen nach der Deutung elektronenmikroskopischer Befunde aus 200–500 Å großen Partikeln, die entsprechend der Ansicht von Mühlethaler und Frey-Wyssling [11] in der Zelle *de novo* entstehen können. Hier wird also der Gedanke einer morphologischen Kontinuität aufgegeben. Auch die Interpretationen, die Mühlethaler und Bell [12] ihren in den letzten Jahren veröffentlichten Untersuchungen geben, weisen auf eine Neubildung der Plastiden hin. Nach der Deutung dieser Befunde degenerieren alle Plastiden und Mitochondrien während der Oogenese bei dem Farn *Pteridium aquilinum* und entstehen neu aus Abschnürungen des Eikerns. Die zuletzt erwähnte Arbeit ist besonders anregend, weil in ihr die Vermutung geäußert wird, „daß diese Rekonstitution des Plasmas nicht nur für die Farne, sondern auch für die niederen und höheren Pflanzen sowie Tiere gilt." Die Folgerungen, die sich aus dieser Aussage für die Plastidengenetik ergeben, seien hier nicht ausführlich erwähnt (vgl. [18], [17], [8]).

Trifft nun die von Mühlethaler und Bell geäußerte Ansicht zu, dann sollte man annehmen, daß sie sich auch bei den Moosen, die wie die Farne zu den Archegoniaten gehören, bestätigen läßt. Daher wurde bei der elektronenmikroskopischen Untersuchung des Lebermooses *Sphaerocarpus donnellii*, über die hier im folgenden näher zu berichten ist, u. a. auch geprüft, ob die vermutete vollständige Degeneration von Plastiden und Mitochondrien sowie die Neubildung dieser Organellen vom Eikern aus ebenfalls für die vorliegende Art zutrifft. Die Archegonentwicklung von *Sphaerocarpus donnellii* wurde bereits ausführlich dargestellt [4, 5].

15*

Daher genügt hier die Erwähnung einiger wesentlicher elektronenmikroskopischer Beobachtungen über die Plastiden und Mitochondrien. In den
jungen Archegonien enthalten alle Zellen Mitochondrien und Plastiden,
die den gleichen Feinbau zeigen (Abb. 1). Erst nach der Ausbildung der

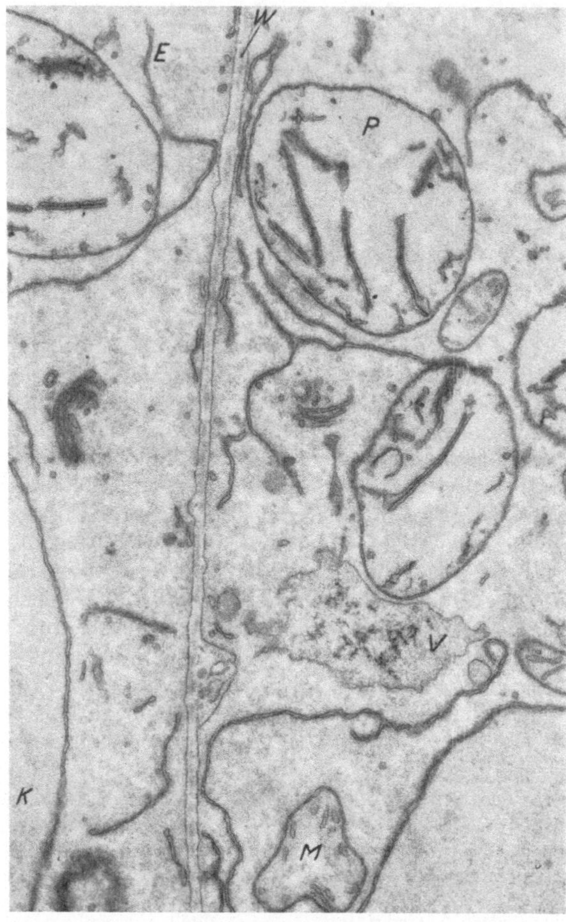

Abb. 1. Ausschnitt aus Archegonanlage. Links Primärzelle der axialen Zellreihe, rechts an-
grenzende Wandzelle. Aus der Primärzelle wird über die primäre und sekundäre Zentralzelle
die Eizelle hervorgehen. E = endoplasmatisches Reticulum; K = Kern; M = Mitochondrium; P = Plastide; V = Vacuole; W = Zellwand. Das Thylakoidsystem in den Plastiden
beider Zellen ist gleichartig ausgebildet. Vergr. ~20 000 mal

sekundären Zentralzelle treten geringfügige Änderungen auf. Das Thylakoidsystem in den Plastiden dieser Zelle und der beiden Halskanalzellen erscheint schwächer entwickelt als in den sich immer mehr zu
Chloroplasten differenzierenden Plastiden der benachbarten Archegonwandzellen. Während des weiteren Wachstums wird dieser Unterschied

immer deutlicher. In der kurz vor der Mitose stehenden sekundären
Zentralzelle und in der jungen Eizelle besitzen die Plastiden schließlich

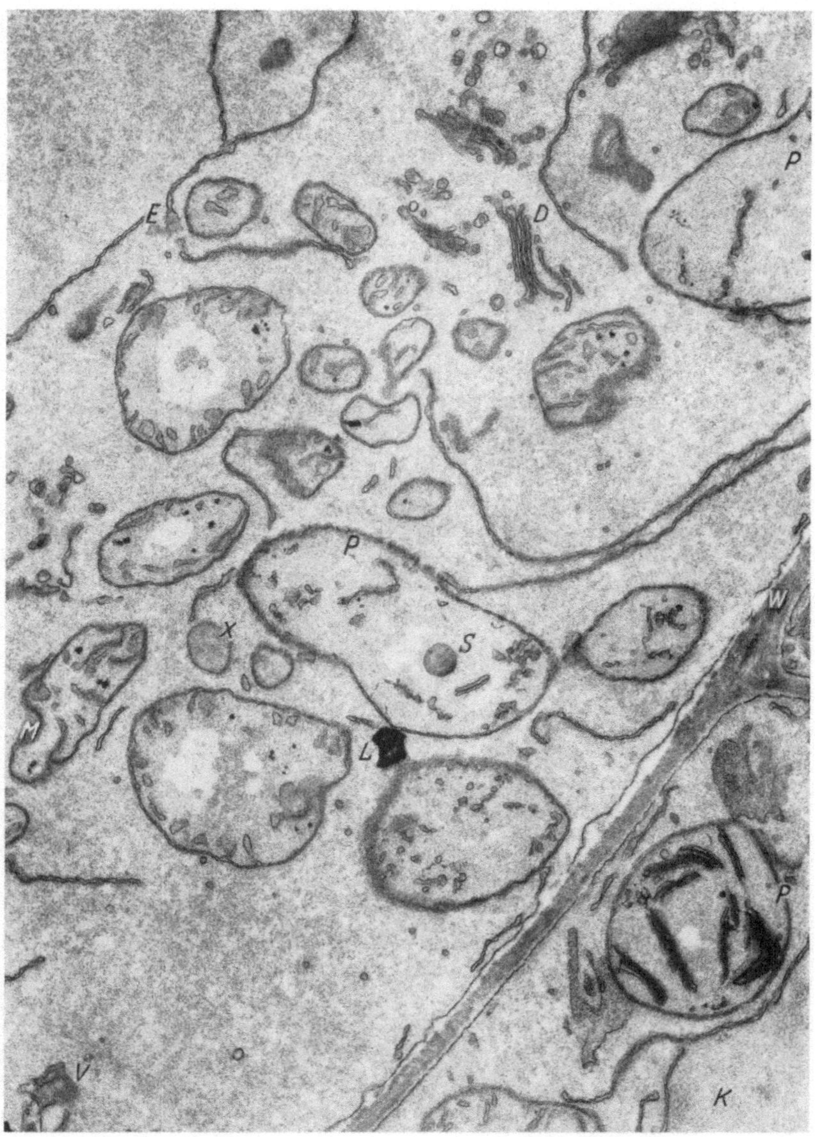

Abb. 2. Ausschnitt aus sich teilender sekundärer Zentralzelle und benachbarter Archegon-
wandzelle (rechts unten). D = Dictyosom; E = endoplasmatisches Reticulum; K = Kern;
L = Lipidtropfen; M = Mitochondrium; P = Plastide; S = Stärkekorn; V = Vacuole;
W = Zellwand; X = Körper mit einfacher Umgrenzungsmembran. Auf den Unterschied
zwischen den undifferenzierten Plastiden der sekundären Zentralzelle und den zu Chloroplasten entwickelten Plastiden der Wandzellen sei hingewiesen. In den Mitochondrien befinden sich kleinere schwarze Partikel. Vergr. ~20000mal

nur noch wenige, meistens kurze Thylakoide sowie mehrere kleinere
Vesikel (Abb. 2, 3). Eine Vacuolisierung und eine Degeneration der

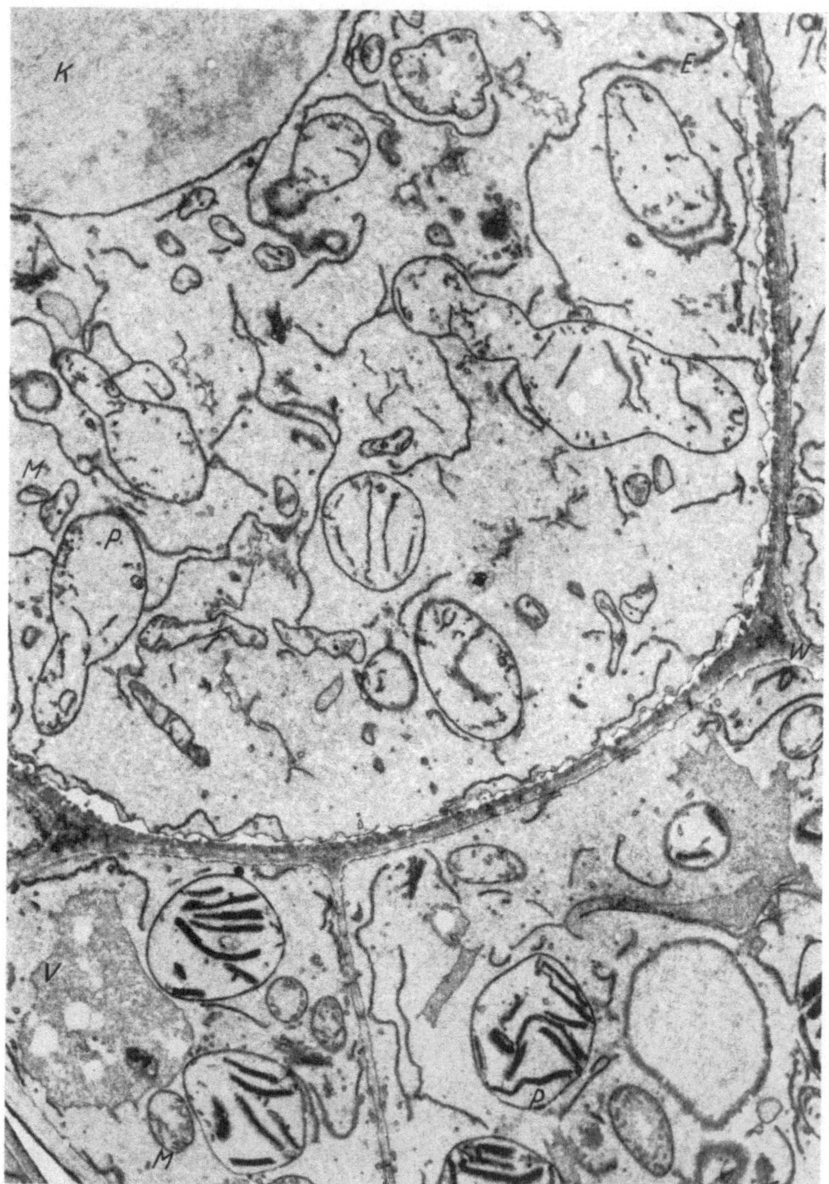

Abb. 3. Unterer Teil einer jungen Eizelle, die sich von den angrenzenden Wandzellen abzu-
lösen beginnt. E = endoplasmatisches Reticulum; K = Kern; M = Mitochondrium; P =
Plastide; V = Vacuole; W = Zellwand. Die Plastiden in den Archegonwandzellen sind zu
Chloroplasten mit stark ausgebildetem Thylakoidsystem differenziert. Vergr. ∼10000 mal

Plastiden und Mitochondrien, wie sie in den völlig entsprechenden Stadien bei *Pteridium aquilinum* nach BELL und MÜHLETHALER [1, 2] auftritt, ist hier nicht festzustellen. Bisher erkannte man auf den Schnitten nur gelegentlich einzelne kleinere Stärkekörner, die von nun an in der

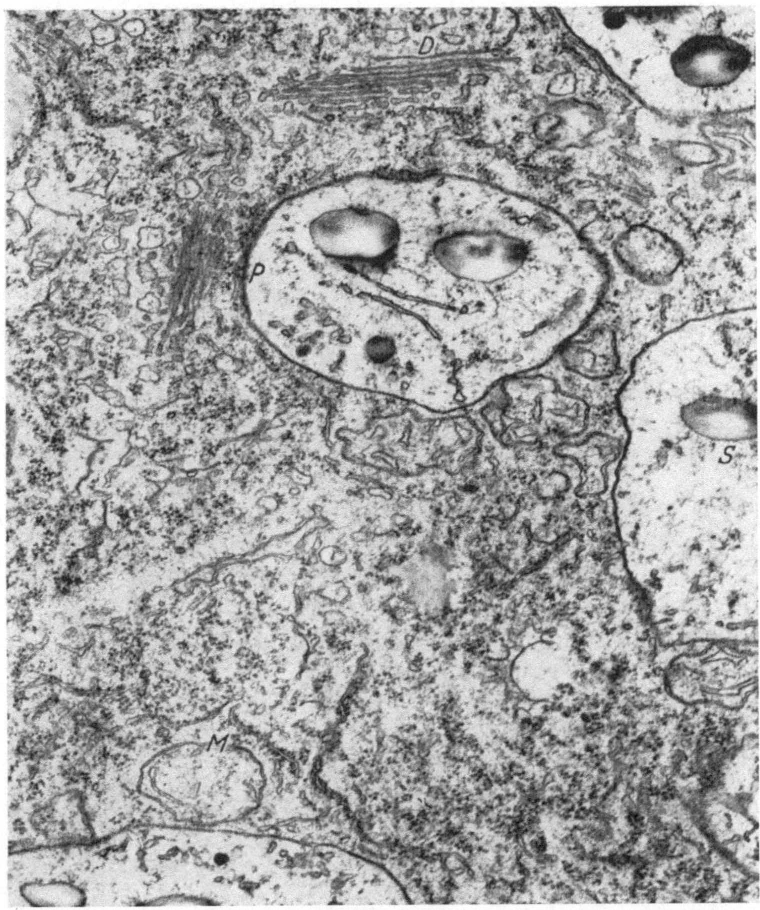

Abb. 4. Ausschnitt aus einer heranwachsenden Eizelle. D = Dictyosom; M = Mitochondrium; P = Plastiden mit Stärkekörnern (S), kleinen Vesikeln und kurzen Thylakoiden. Nach OsO$_4$-Fixierung werden die zahlreichen Ribosomen im Cytoplasma sichtbar. Vergr. $\sim$30000 mal

heranwachsenden Eizelle immer häufiger erscheinen (Abb. 4). Etwa von dem gleichen Entwicklungsstadium an zeigt der Eikern eine unregelmäßige Oberfläche. Er bildet Fortsätze aus, die oft tief in das Plasma hineinreichen. Auf Einzelschnitten erscheinen dann diese Evaginationen gelegentlich wie vom Kern abgelöste Teile. Schnittserien zeigen, daß in fast allen geprüften Fällen diese Teile doch noch in irgendeiner Schnitt-

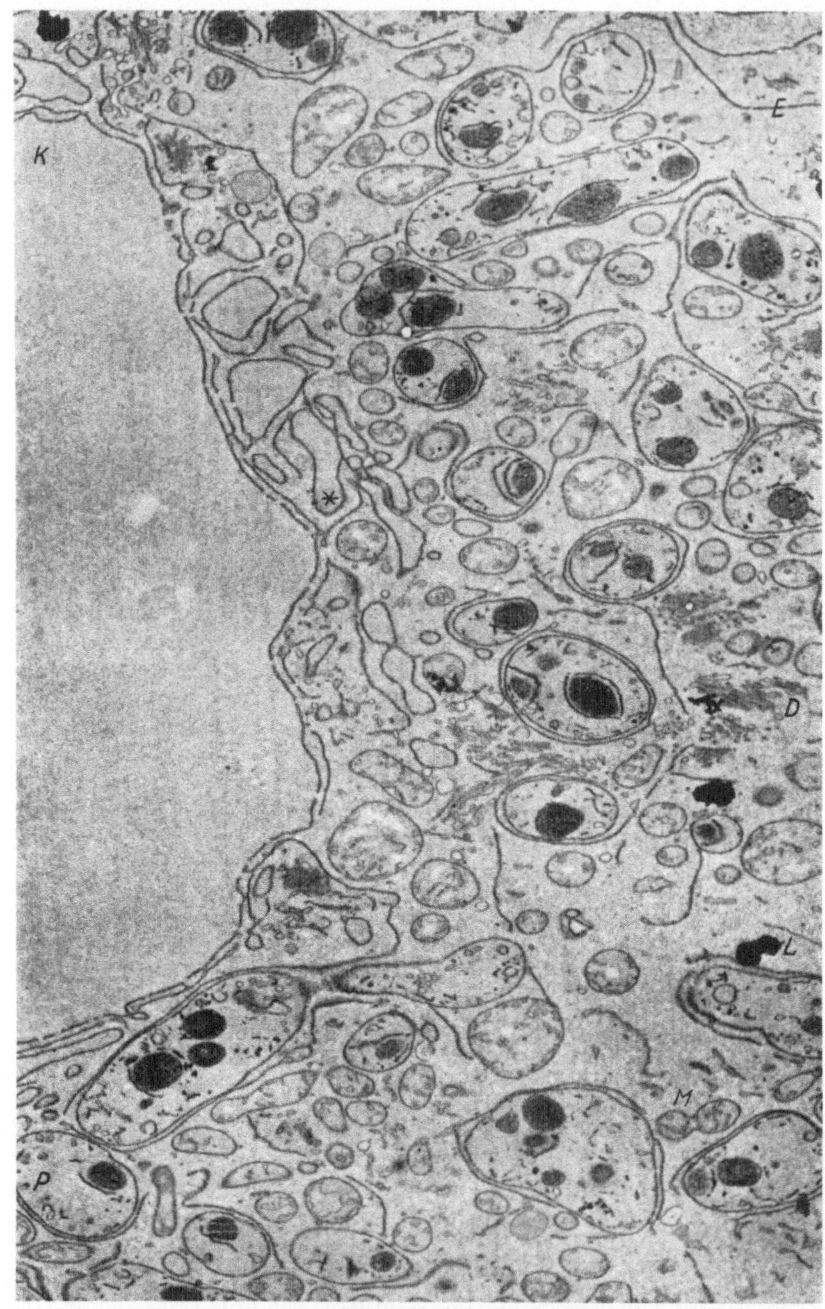

Abb. 5. Ausschnitt aus einer reifenden Eizelle. Vor dem Kern (K) liegen mehrere Abschnitte, die vom Nucleus abgelöst erscheinen. Hingewiesen sei auf den mit * bezeichneten Kernteil (vgl. Abb. 6). In den Plastiden (P) dunkel kontrastierte Stärkekörner. D = Dictyosom; E = endoplasmatisches Reticulum; L = Lipidtropfen; M = Mitochondrium.
Vergr. ~10000 mal

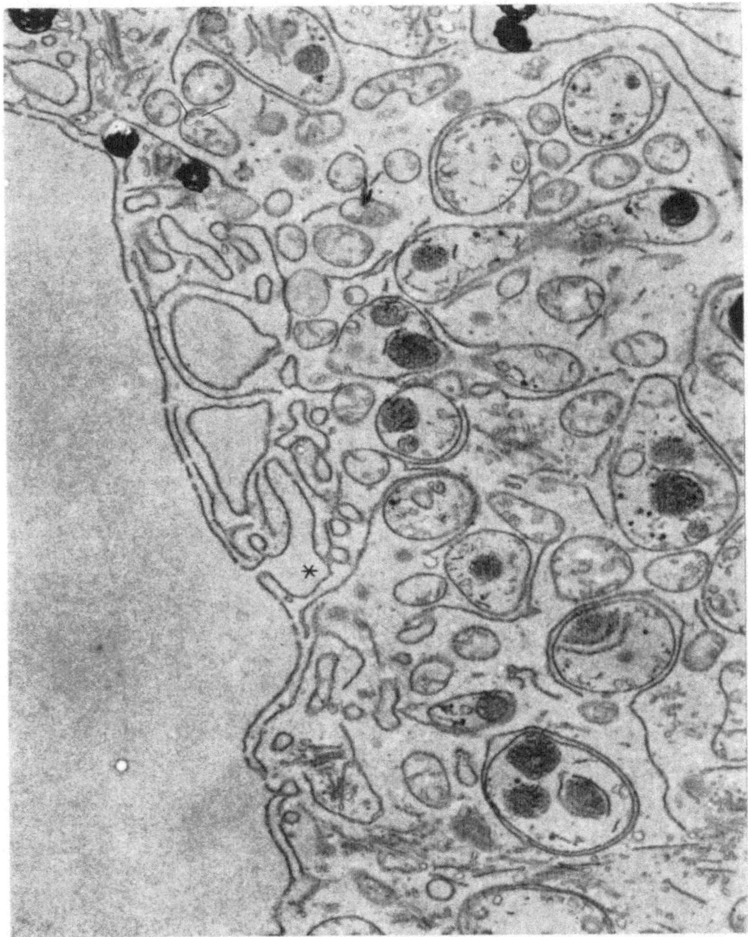

Abb. 6. Dieselbe Zellregion wie bei Abb. 5 etwa 250 nm (Dicke von 5 Schnitten) tiefer geschnitten. Man erkennt die Verbindung zwischen einem anscheinend abgelösten Kernteil *
(vgl. Abb. 5) und dem Eikern. Vergr. ~11000 mal

ebene eine Verbindung mit dem Kern besitzen (Abb. 5, 6). Es sprechen
jedoch einige Beobachtungen dafür, daß sich solche Abschnitte vom
Kern ablösen können. Anzeichen für eine Weiterentwicklung dieser abgetrennten Kernstücke zu Plastiden oder Mitochondrien konnten nicht
festgestellt werden.

Die elektronenmikroskopischen Bilder erlauben zusammenfassend
zunächst die Feststellung: In der Eizelle und ihren vorhergehenden Zellen
sind stets Plastiden und Mitochondrien zu erkennen. Vacuolisierte und
degenerierte Organellen sind nicht zu beobachten. Die Plastiden besitzen
in allen Wachstumsphasen die Fähigkeit, Stärke zu deponieren. Diese
Stärkeablagerung ist besonders in der Eizelle während ihrer Reifung aus-

geprägt, also gerade dann, wenn bei *Pteridium* nach BELL und MÜHLE
THALER [1] nur degenerierte oder keine Plastiden mehr erkennbar sind.
Ferner sind keine Anzeichen festzustellen, daß sich bei *Sphaerocarpus* die
Kernausstülpungen zu Plastiden oder Mitochondrien entwickeln können.

Da alle Entwicklungsstadien erfaßt wurden, kann man nun versuchen,
auch zahlenmäßig das Verhalten der Zellbestandteile von der Primärzelle
der axialen Zellreihe über die primäre und sekundäre Zentralzelle bis
schließlich zur reifen Eizelle hin zu verfolgen. Eine solche quantitative
Betrachtung ist hier durchführbar, weil die auszuwertenden Zellen keine
oder nur sehr kleine Vacuolen besitzen, und die fraglichen Plasmabestandteile zufallsgemäß verteilt sind; allenfalls in der reifenden Eizelle
findet man gelegentlich kleine schmale Randbereiche, in denen nicht so
viele Organellen wie in den übrigen Zellbezirken zu liegen scheinen. Diese
Abweichungen in der zufallsgemäßen Verteilung sind nach rechnerischer
Überprüfung so gering, daß sie unberücksichtigt bleiben können. Zellen
unmittelbar vor oder während der Teilung schieden bei der Auswertung
aus, da dann die Organellen nicht mehr zufallsgemäß verteilt sind. Aus
Gründen der strengen Vergleichbarkeit wurden nur solche Archegonien
berücksichtigt, die mit $KMnO_4$ fixiert und in Araldit eingebettet waren.
Von jedem Entwicklungsstadium wurden möglichst viele mediane Längschnitte, die die Eizelle oder ihre vorhergehenden Zellen ganz oder zum
größten Teil zeigten, ausgewählt. Die Zell- und Kernflächen wurden planimetriert und alle angeschnittenen Mitochondrien, Plastiden und Stärkekörner ausgezählt. Um für alle Wachstumsphasen und Schnitte eine einheitliche Bezugsfläche zu erhalten, wurden bei jedem Schnitt die erhaltenen Werte für die Zellbestandteile auf 100 μ^2 Protoplasmafläche umgerechnet und aus diesen Einzelwerten für jedes Entwicklungsstadium der
Mittelwert bestimmt.

Die nach diesem Verfahren erhaltenen größten Werte für die Zell-
und Kernflächen jedes Entwicklungsstadiums sind in Diagramm 1 zusammengestellt. Ferner ist die Kernfläche als Prozentwert der größten
Zellfläche angegeben. Es ist ersichtlich, daß die Zellfläche und damit
auch die Zellgröße sowie die Kernfläche zunehmen.

In Diagramm 2 sind für die Plastiden, Mitochondrien und Stärkekörner die Werte je 100 μ^2 Plasmafläche zusammengefaßt.

Mitochondrien: Die Zahlen für die Mitochondrien schwanken in
drei Bereichen: In den Entwicklungsstadien I bis III zwischen 13,5 und
15, in den sich anschließenden Wachstumsphasen IV bis IX zwischen
23,5 und 26,5 sowie in den Stadien X bis XII zwischen 32 und 36 pro
100 μ^2 Plasmafläche. Die Unterschiede zwischen je zwei benachbarten
Zahlenbereichen lassen sich gut statistisch absichern. So errechnet sich t
beim Vergleich der beiden Mittelwerte von Entwicklungsstadium III und
IV mit $t_{(13)} = 12,23$, während für eine statistisch gesicherte Verschiedenheit schon $t_{(13)} = 4,22$ bei einer Wahrscheinlichkeit $p = 0,001$ genügt.
Ähnliches gilt für die Mittelwerte der Wachstumsstadien IX und X. Hier
beträgt $t_{(9)} = 8,62$, während schon $t_{(9)} = 4,78$ bei $p = 0,001$ ausreichen
würde. Das Diagramm zeigt, daß die Zahl der Mitochondrien pro 100 μ^2

Plasmafläche deutlich ansteigt. Dieser Anstieg ist besonders beträchtlich, da die Zellgröße nach Diagramm 1 während der Eientwicklung zunimmt.

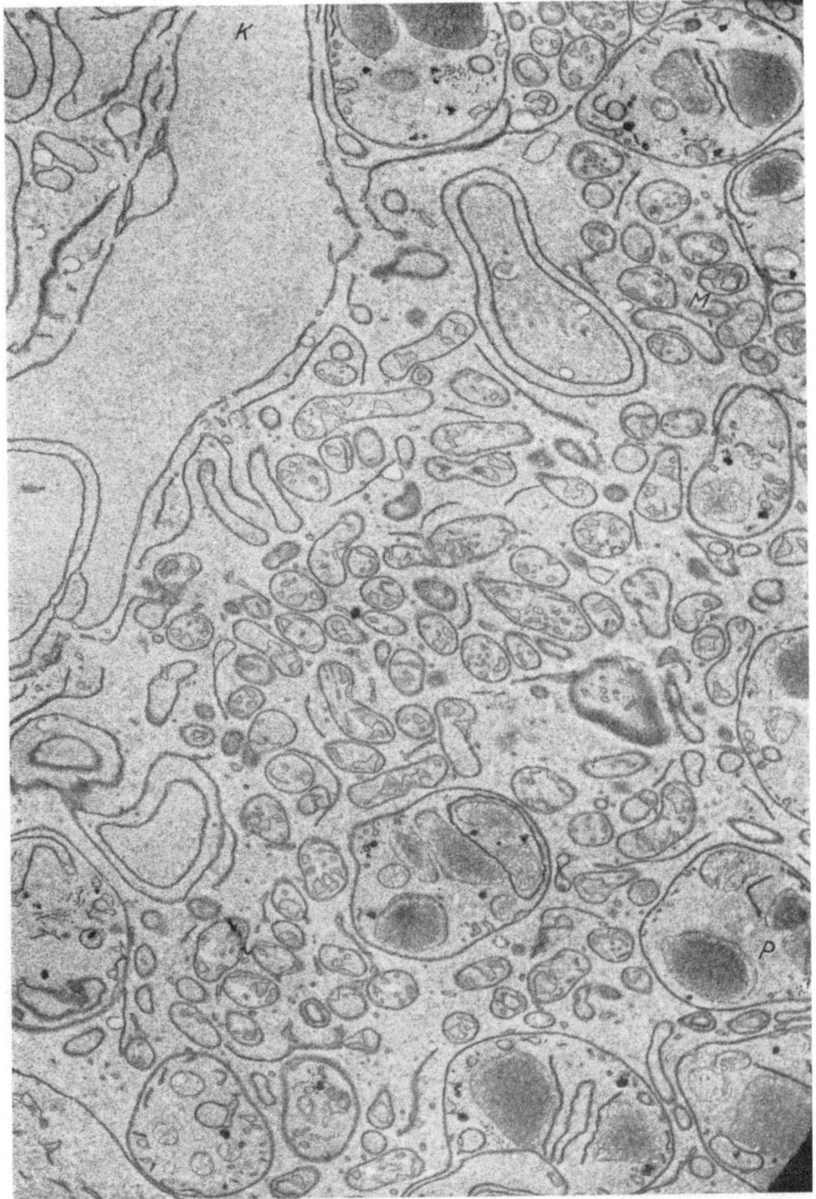

Abb. 7. Ausschnitt aus einer reifen Eizelle. K = Kern mit scheinbar abgeschnürten, frei im Plasma liegenden Teilen; M = Mitochondrien; P = Plastiden mit z. T. großen Stärkekörnern. Vergr. ~14500mal

Plastiden: Im Gegensatz zu den Mitochondrien bleibt die Zahl der Plastiden pro 100 μ^2 Plasmafläche annähernd konstant. Die Werte liegen fast alle zwischen 6 und 9. In der sekundären Zentralzelle (IV) und in der sehr jungen Eizelle (V) unterschreiten die ermittelten Zahlen den Bereich. Für die Mittelwerte von III und IV beträgt nach der statistischen Berechnung $t = 2,73$; es wäre $t_{(:\varepsilon)} = 4,22$ für einen gesicherten Unterschied

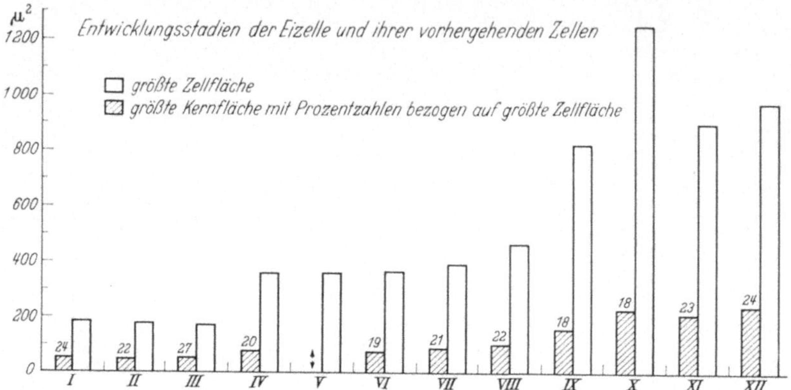

Diagramm 1. Die Zunahme der Zell- und Kernflächen auf medianen Längsschnitten während der Oogenese

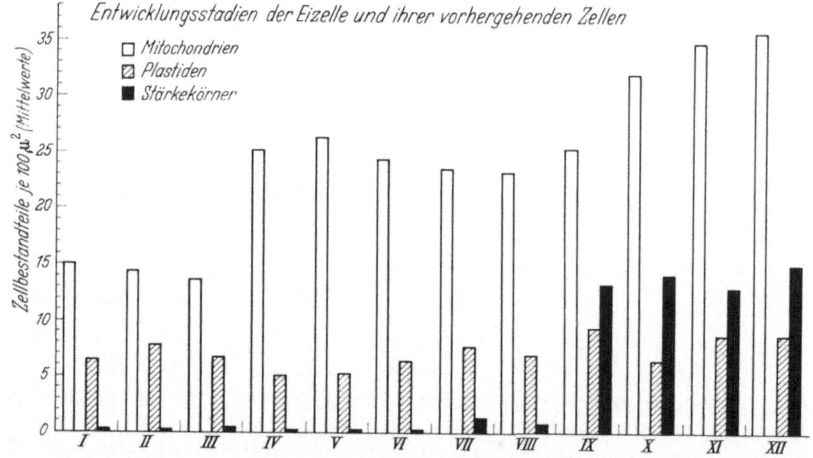

Diagramm 2. Das zahlenmäßige Verhalten von Mitochondrien, Plastiden und Stärkekörnern während der Entwicklung der Eizelle und ihrer vorhergehenden Zellen. Die Ziffern I bis XII bezeichnen die aufeinanderfolgenden Entwicklungsstadien: I Primärzelle der axialen Zellreihe; II primäre Zentralzelle; III sekundäre Zentralzelle in einem Archegon mit Halskanalmutterzelle; IV sekundäre Zentralzelle in einem Archegon mit zwei Halskanalzellen; V sehr junge Eizelle, späte Telophase der sich teilenden sekundären Zentralzelle; VI junge Eizelle in einem ~68 μ großen Archegon; VII junge Eizelle in einem ~77 μ großen Archegon; VIII junge Eizelle in einem ~96 μ großen Archegon; IX Eizelle abgelöst von Bauchkanalzelle; X Eizelle in einem Archegon mit degenerierender Bauchkanalzelle; XI fast reife Eizelle in einem Archegon mit vollständig aufgelöster Bauchkanalzelle, Halsspitze noch geschlossen; XII reife Eizelle, Spitze des Archegons geöffnet

bei $p = 0,001$ erforderlich; also sind die Mittelwerte nicht verschieden. Bei dem Vergleich der Werte von Stadium V und VI ergibt sich $t_{(17)} = 4,16$, während für $p = 0,001$ schon $t_{(17)} = 3,97$ ausreicht. Damit ist formal schon ein statistisch gesicherter Unterschied, nämlich eine Zunahme der Plastiden, gegeben. Die beiden Werte (4,16 und 3,97) weichen jedoch so wenig voneinander ab, daß man wohl kaum eine reale Verschiedenheit damit begründen kann. — Ferner übersteigt der Wert im Stadium IX mit 9,4 geringfügig den Bereich von 6—9 Plastiden pro 100 μ^2 Plasmafläche. Die statistische Berechnung der Mittelwerte von VIII und IX mit $t_{(9)} = 1,95$ sowie von IX und X mit $t_{(9)} = 2,39$ ergibt klar keinen Unterschied, der erst mit $t_{(9)} = 4,78$ bei $p = 0,001$ erreicht würde. — Schließlich sei noch auf den entsprechenden Vergleich der Mittelwerte von Stadium X und XI hingewiesen. Hier errechnet sich eine Verschiedenheit mit $t_{(15)} = 4,65$, weil für $p = 0,001$ bereits $t_{(15)} = 4,07$ genügt. Der Unterschied, der eine Plastidenzunahme bedeuten würde, ist formal gesichert. Da die Werte (4,65 und 4,07) so nahe beieinanderliegen, läßt sich jedoch aus dieser Berechnung schwerlich ein wirklicher Unterschied der Plastidenzahlen pro 100 μ^2 in den beiden Wachstumsphasen folgern. — Alle übrigen entsprechenden Mittelwerte benachbarter Entwicklungsstadien sind nach der statistischen Berechnung nicht voneinander verschieden. Zusammenfassend folgt aus dieser quantitativen Auswertung, daß mit fortschreitender Entwicklung nie ein statistisch gut gesichertes Absinken der Plastidenzahlen je 100 μ^2 zwischen zwei Wachstumsstadien eintritt. In fast allen Wachstumsphasen schwanken die gefundenen Werte zwischen 6 und 9 je 100 μ^2. Die beobachteten Unterschiede sind so gering, daß man die Plastidenzahlen pro 100 μ^2 Plasmafläche als annähernd gleichbleibend anzusehen hat. Da die Plasmafläche nach Diagramm 1 während der Entwicklung zunimmt, muß die Gesamtzahl der Plastiden in der Zelle ansteigen. Die Plastiden vermehren sich also offensichtlich.

Stärkekörner: Die Zahl der Stärkekörner je 100 μ^2 Plasmafläche bleibt bis nach der Teilung der sekundären Zentralzelle (V und VI) sehr niedrig. Erst mit dem Heranwachsen der Eizelle (VII, VIII) erhöht sich der Wert und steigt verhältnismäßig schnell auf ein Maximum, das schon etwa in dem Stadium (IX) erreicht ist, in dem sich die Bauchkanalzelle völlig von der Eizelle abgelöst hat, jedoch noch keine Anzeichen einer Degeneration aufweist. Dieser Wert von etwa 13—15 Stärkekörnern je 100 μ^2 Plasmafläche bleibt während der weiteren Reifung der Eizelle nahezu unverändert.

Die medianen Längsschnitte erlauben nun weiterhin, das Zell- und Kernvolumen sowie die Gesamtzahl der Plastiden und Mitochondrien in der ganzen Zelle abzuschätzen. Die Zellen wurden als Quader, Zylinder oder Kegelstümpfe betrachtet und die zur Inhaltsberechnung erforderlichen Größen jeweils dem medianen Längsschnitt entnommen, der die größte Zellfläche zeigte. Der Kern wurde als Kugel aufgefaßt; der Radius läßt sich aus der größten Kernfläche, die als Kreisfläche angenommen wurde, berechnen.

Zur Bestimmung der Plastiden -und Mitochondrienzahlen dienten die schon bekannten Mittelwerte der Organellen je 100 μ^2 Plasmafläche sowie Längenmessungen der Plastiden und Mitochondrien. Auf den Schnitten jedes Entwicklungsstadiums wurden die längsten Plastiden und Mitochondrien ausgemessen und die festgestellten Zahlen erheblich aufgerundet. Diese Werte betragen für die Plastiden 3—5 μ und für die Mitochondrien 1,5—3 μ. Die selten auftretenden, außergewöhnlich langen Organellen wurden bei diesen Messungen nicht berücksichtigt. Es wird nun angenommen, daß die Organellen, die in den Mittelwerten je 100 μ^2 Plasmafläche angegeben sind, nur solche 3—5 μ großen Plastiden oder 1,5—3 μ langen Mitochondrien darstellen, die genau senkrecht zur Schnittebene liegen. Dann befinden sich in einem Plasmaraum von 300—500 μ^3 bzw. 150—300 μ^3 so viele Plastiden bzw. Mitochondrien, wie die Mittelwerte je 100 μ^2 Plasmafläche angeben. Diese Werte werden auf die Differenz: Zellvolumen minus Kernvolumen bezogen und ergeben angenähert die Gesamtzahl der Plastiden bzw. Mitochondrien in der Zelle (Tab. I). Da nach den Schnittserien zahlreiche Plastiden und Mitochondrien kleiner sind als 3—5 μ bzw. 1,5—3 μ, stellen die so ermittelten Werte nur Mindestzahlen dar; wenn man diese mit 1,5 oder 2 multipliziert, würde man (niedrig gerechnet) überschlagsmäßig etwa die Maximalwerte erhalten.

Tabelle 1: *Die Zell- und Kernvolumina und die Gesamtzahlen der Mitochondrien und Plastiden je Zelle in einigen charakteristischen Entwicklungsstadien*

	Entwicklungsstadien	Zell-volumen	Kern-volumen	Mitochondrien	Plastiden
II	primäre Zentralzelle	1300 μ^3	100 μ^3	60—90	20—35
IV	sekundäre Zentralzelle (2 Halskanalzellen)	3200 μ^3	300 μ^3	250—400	30— 50
VI	Eizelle (Archegongröße ~ 68 μ)	3300 μ^3	300 μ^3	250—400	40— 65
VIII	Eizelle (Archegongröße ~ 96 μ)	6000 μ^3	450 μ^3	450—650	80—120
XI	Eizelle (Ei fast reif)	10 300 μ^3	1200 μ^3	1000—1600	160—270
XII	Eizelle (Ei reif)	9800 μ^3	1500 μ^3	800—1300	150—260

Nach Tab. 1 enthält die primäre Zentralzelle (II), die in ihrem Feinbau und ihrer Größe einer meristematischen Zelle des wachsenden Thallusscheitels entspricht, etwa dreimal mehr Mitochondrien als Plastiden. Von diesem Stadium an bis zur reifen (XII) oder fast reifen Eizelle (XI) nimmt das Zellvolumen etwa um das 6—8fache, das Kernvolumen etwa um das 12—15fache zu. Gleichzeitig vermehrt sich die Zahl der Mitochondrien etwa um das 13—17fache, die Zahl der Plastiden um das 8fache. In der Eizelle befinden sich schließlich etwa 5—7 mal mehr Mitochondrien als Plastiden. Es besteht also während der Oogenese eine klare gleichsinnige Beziehung zwischen der Vergrößerung des Plasma- und Kernraums und der Erhöhung der Plastiden- und Mitochondrienzahl.

Die quantitative Auswertung zeigt eindeutig, daß eine Vermehrung der Plastiden und Mitochondrien stattfinden muß. Können nun Angaben über die Vermehrungsweise gemacht werden? In keinem der Entwicklungsstadien wurden Anhaltspunkte dafür gefunden, daß die Organellen etwa entsprechend den Angaben von MÜHLETHALER und FREY-WYSSLING [11] bei *Elodea* aus sehr kleinen Anlagen ohne Innenstrukturen hervorgehen können. Auch waren keine Anzeichen für eine Herkunft aus anderen Membranen des Protoplasten festzustellen.

Mit Ausnahme der letzten Wachstumsphase der reifenden und reifen Eizelle (X—XII) beobachtet man bei allen vorhergehenden Entwicklungsstadien auf Einzelschnitten Plastiden und Mitochondrien mit engen Einschnürungen. So geformte Organellen werden oft als Teilungsstadien angesehen. Trifft diese Interpretation zu, dann müßte die enge Einschnürung allseitig sein. Schnittserien zeigten jedoch, daß eine solche Einschnürung häufig vorgetäuscht wird, wenn der Schnitt durch nur teilweise verschmälerte Organellenteile geführt ist, und sich die auf einem Einzelschnitt eng erscheinende Einschnürung in anderen Schnittebenen zur ganzen Breite eines Mitochondriums oder einer Plastide vergrößert. Aus dieser Beobachtung folgt, daß stark eingeschnürte Plastiden oder Mitochondrien auf einzelnen elektronenmikroskopischen Dünnschnitten keinen sicheren Hinweis auf eine Organellenteilung geben können. — Schnittserien beweisen, daß auch allseitige enge Einschnürungen bei Plastiden und Mitochondrien auftreten. In einem solchen Fall ist die Einschnürung nur auf wenigen Schnitten zu beobachten, während in den höheren und tieferen Schnittebenen die beiden Teile derselben Organelle wie zwei selbständige nebeneinanderliegende Plastiden bzw. Mitochondrien erscheinen. Solche Organellenformen lassen sich nun nicht nur als Teilungsstadien — wie es meistens geschieht — interpretieren. Man kann sie ebenso als Zeichen einer Vereinigung zweier Plastiden oder Mitochondrien betrachten. Zum Beispiel vermögen nach EPSTEIN und SCHIFF [7] Plastiden bei *Euglena* zu verschmelzen. Ferner wurden Fusionen von Mitochondrien häufig beschrieben (DRAWERT und MIX [6]; NOVIKOFF [13]). Schließlich muß noch auf eine dritte Deutungsmöglichkeit hingewiesen werden. Mitochondrien und undifferenzierte Plastiden meristematischer Zellen besitzen eine große Plastizität. Es ist daher vorstellbar, daß sich durch Formveränderungen auch enge Einschnürungen bei Plastiden und Mitochondrien ausbilden, die jedoch reversibel sind und unabhängig von Organellen-Teilungen oder -Verschmelzungen erscheinen und verschwinden können.

Wie bei fast allen Untersuchungen an fixiertem Material ist es zunächst nicht möglich, allein von einzelnen erfaßten Zuständen in den Schnitten auf einen ablaufenden Vorgang im lebenden Archegonium zweifelsfrei zurückzuschließen. Berücksichtigt man jedoch die Ergebnisse der quantitativen Auswertung, so läßt sich entscheiden, welche der drei Interpretationen hier wohl zutrifft. Die Deutung der engen Einschnürungen als Organellfusionen oder als reversible Formveränderungen müßte bei gleichzeitiger Zellvergrößerung (Diagramm 1 u. Tab. 1) zu sinkenden Zahlen pro 100 μ^2 Plasmafläche führen. Lediglich die Auffassung einer Organellteilung steht nicht im Widerspruch zu den quantitativen Auswertungen. Daher ist aus diesen Befunden zu folgern, daß sich während der Oogenese bei *Sphaerocarpus donnellii* Plastiden und Mitochondrien durch Teilung vermehren.

Ein wesentlicher Einwand läßt sich allerdings gegen diese Schlußfolgerung vorbringen. Die ganze Archegoniumentwicklung könnte sich so lange hinziehen, daß man trotz vieler untersuchter Wachstumsstadien die vollständige Oogenese nur mit beträchtlichen zeitlichen Lücken erfaßt hat. Gerade solche Stadien in denen die Plastiden bzw. Mitochondrien degenerieren und sich aus dem Eikern neubilden, wären dann vielleicht ausgelassen worden.

Bei den hier eingehaltenen Kulturbedingungen, die ein optimales Wachstum der Archegonien ermöglichen, dauert nach Lebendbeobachtungen die Entwicklung des Archegoniums von den ersten sichtbaren Erhebungen auf der Thallusoberseite bis zur reifen Eizelle 5—6 Tage, von der Teilung der sekundären Zentralzelle bis zur reifen Eizelle längstens 4 Tage. Insgesamt wurden bisher über 80 Archegonien elektonenmikroskopisch genauer geprüft, so daß alle wesentlichen Entwicklungsstadien in zeitlich enger Folge untersucht sein dürften. Hierbei wurden besonders eingehend die Wachstumsphasen der sekundären Zentralzelle und der jungen Eizelle berücksichtigt, in denen bei *Pteridium aquilinum* nach Bell und Mühlethaler eine Mitochondrien- und Plastidendegeneration zu beobachten ist. Bei *Sphaerocarpus* konnten in den entsprechenden Stadien degenerierte Organellen *nicht* festgestellt werden. In diesem Zusammenhang ist darauf hinzuweisen, daß die Degeneration und Elimination so großer Organellen wie der Plastiden und Mitochondrien nicht sehr schnell vor sich gehen dürfte; zumindest ihre Reste müßten wohl noch längere Zeit im Plasma wahrnehmbar sein.

Wie bereits erwähnt, sind bei *Sphaerocarpus* keine Anzeichen festzustellen, daß sich aus den Evaginationen des Eikerns Plastiden bzw. Mitochondrien entwickeln. Zum ersten Mal erscheinen Kernvorstülpungen in einer Wachstumsphase, die etwa zwischen Stadium VIII und IX liegt. Bevor sie in der Eizelle auftreten, findet schon eine Vermehrung der Mitochondrien und Plastiden statt (Diagramm 2 u. Tab. 1). Diese neu gebildeten Organellen können also sicher nicht von Kernprotuberanzen stammen. Sie entstehen nach den bereits dargelegten Überlegungen durch Teilung schon vorhandener Plastiden und Mitochondrien. Dieselbe Vermehrungsart dürfte auch in den Reifungsstadien der Eizelle verwirklicht sein, in denen Kernvorstülpungen und eine Zunahme von Organellen zu beobachten ist. Denn auch in diesen Wachstumsphasen weisen Plastiden und Mitochondrien gelegentlich enge Einschnürungen auf, die nach der quantitativen Auswertung als Teilungsformen anzusehen sind.

In den bisher veröffentlichten elektronenmikroskopischen Untersuchungen über die Eientwicklung bei Pflanzen finden sich einige Beobachtungen über das Verhalten der Plastiden und Mitochondrien. So berichtet Camefort [3] von einer deutlichen Degeneration der Chondriosomen und einer ausgeprägten Umgestaltung der Plastiden in der Eizelle von *Pinus laricio*. Rodkiewicz und Mikulska [14] sprechen ebenfalls von einer Degeneration vieler Mitochondrien während der Eientwicklung bei

Lilium candidum. Zu diesen beiden Arbeiten ist zu bemerken, daß die Beobachtungen nach OsO_4-Fixierung gemacht wurden. Nach eigenen Untersuchungen sowie nach den Erfahrungen von BELL und MÜHLE-THALER [2] bleiben jedoch bei OsO_4-Fixation die Membranen nicht immer gut erhalten. — Im Gegensatz zu diesen Befunden bei Spermatophyten ist nach Untersuchungen von MENKE und FRICKE [9] ein Abbau von Plastiden und Mitochondrien sowie die Neubildung dieser Organellen aus dem Eikern bei dem Farn *Dryopteris filix-mas* nicht festzustellen.

Faßt man alle Befunde zusammen, so lassen die Ergebnisse zumindest den Schluß zu, daß eine Rekonstitution des Plasmas in dem erwähnten Sinne während der Oogenese bei Pflanzen nicht allgemein zutrifft.

Die Untersuchungen wurden mit Unterstützung der Deutschen Forschungsgemeinschaft durchgeführt.

Summary

All cells taking part in oogenesis of the liverwort *Sphaerocarpus donnellii Aust.* were investigated under the electron microscope after fixation in $KMnO_4$, OsO_4 and glutardialdehyde. In the egg and its preceding cells plastids and mitochondria are well recognizable. A degeneration and elimination of these organelles was not observed. In all developmental stages the plastids may deposit starch. There are no indications that nuclear evaginations which appear in the maturing egg cell, may develop into mitochondria and plastids.

In a quantitative investigation the relative proportions of mitochondria, plastids and starch grains were followed from the primary cell of the axial row, over the mother cell of the axial row (= primäre Zentralzelle) and the various growth stages of the central cell (= sekundäre Zentralzelle) up to the mature egg (diagram 1,2). According to this analysis the total number of mitochondria and plastids considerably increases during oogenesis (Tab. 1). With all probability these organelles multiply by division. This view is strongly supported by the fact that in all developmental stages of the egg cell, except the mature egg, mitochondria and plastids possess narrow, according to serial sections, three-dimensional constrictions. From the results being published on the oogenesis of other plants, one has to conclude that there is no general validity of the degeneration and elimination of mitochondria and plastids during the development of the egg as MÜHLETHALER and BELL (1962) had suggested interpreting their investigations on the fern *Pteridium aquilinum*.

Literatur

[1] BELL, P. R., and K. MÜHLETHALER: J. Ultrastruct. Res. **7**, 452 (1962).
[2] —— J. Cell Biology **20**, 235 (1964).
[3] CAMEFORT, H.: Ann. Sc. nat. Bot. Biol. vég., 12e Sér. **3**, 265 (1962).
[4] DIERS, L.: Planta (Berl.) **66**, 165 (1965).
[5] — Z. Naturforsch. 20b, 795 (1965).
[6] DRAWERT, H., u. M. MIX: Flora (Jena), **151**, 487 (1961).
[7] EPSTEIN, H. T., and J. A. SCHIFF: J. Protozool. **8**, 427 (1961).
[8] HAUSTEIN, E.: Z. Vererbungsl. **93**, 531 (1962).
[9] MENKE, W., u. B. FRICKE: Z. Naturforsch. 19b, 520 (1964).
[10] MEYER, A.: Das Chlorophyllkorn in chemischer, morphologischer und biologischer Beziehung. Leipzig: A. Felix Verlag 1883.
[11] MÜHLETHALER, K., and A. FREY-WYSSLING: J. biophys. biochem. Cytol. **6**, 507 (1959).
[12] —, u. P. R. BELL: Naturwissenschaften **49**, 63 (1962).
[13] NOVIKOFF, A. B.: Mitochondria (Chondriosomes) in: The Cell, Vol. II. Edit. BRACHET, J. and A. E. MIRSKY. New York and London: Academic Press 1961.

[14] RODKIEWICZ, B., u. E. MIKULSKA: Flora (Jena) 154, 383 (1963).
[15] SCHIMPER,A. F. W.: Jb. wiss. Bot. 16, 1 (1885).
[16] SCHMITZ, F.: Verh. naturw. Verein Rheinl. u. Westfal. 40, 1 (1882).
[17] SCHÖTZ, F.: Planta (Berl.) 58, 333 (1962).
[18] STUBBE, W.: Z. Vererbungsl. 93, 175 (1962).

Diskussion

Vorsitz: HAGEMANN

P. Sitte: Bevor wir in die Diskussion eintreten, möchte ich — wohl in unser aller Namen — mein Bedauern darüber ausdrücken, daß es Herrn Kollegen MÜHLETHALER nicht möglich war, die an ihn ergangene Einladung zu unserem Symposium anzunehmen.

Schötz: Degeneration und Restitution der Chloroplasten und Mitochondrien sollen, wie Herr MÜHLETHALER kürzlich sagte, in einem Zeitraum von wenigen Stunden erfolgen. Ich möchte nun Herrn DIERS fragen, ob er auf Grund seiner Beobachtungen derartiges für möglich hält, und ob er Kriterien dafür hat, daß er — falls es einen solchen schnellen Ablauf gibt — bei seinen Untersuchungen die fragliche Zeitspanne sicher erfaßt hat.

Diers: Nach BELL und MÜHLETHALER [2] geht die Degeneration und Elimination der Mitochondrien innerhalb weniger Stunden vor sich und ist nur in der Zentralzelle und in der jungen Eizelle festzustellen. Der Abbau und das Verschwinden der Plastiden dauert wesentlich länger und läßt sich vom Entwicklungsstadium der Zentralzelle bis zur fast reifen Eizelle verfolgen [1]. Für die reife Eizelle schreiben die Autoren [1]: "No plastids can be identified with confidence." Nimmt man an, daß bei *Sphaerocarpus* die Degeneration und Elimination dieser Organellen in den gleichen Entwicklungsstadien wie bei *Pteridium* erfolgt, dann hätte man sicher solche Degenerationsformen oder zumindest Restkörper der abgebauten Organellen finden müssen. So etwas wurde bei *Sphaerocarpus* nicht gefunden, übrigens auch nicht von MENKE und FRICKE [9] bei *Dryopteris filix-mas*.

Sphaerocarpus unterscheidet sich von den Farnen auch dadurch, daß hier keine verdickte Membran um die Eizelle gebildet wird. Damit entfällt bei unserem Objekt die Möglichkeit der Elimination der Restkörper in die Wand, die von BELL und MÜHLETHALER [Nature 195, 198 (1962)] für *Pteridium* postuliert worden war.

P. Sitte: Ich glaube die zuletzt erwähnte Behauptung von BELL und MÜHLETHALER ist mehr nebenher aufgestellt worden und ohne Bedeutung für das Grundsätzliche.

Schötz: MÜHLETHALER hat vor vier Wochen auf einem Symposium in Belgien über das Auftreten von Thylakoiden in Kernprotuberanzen von *Dryopteris*-Eizellen berichtet.

Diers: Der schlüssige Beweis dafür, daß es sich in diesem Fall um Thylakoide handelt, ist wohl noch nicht erbracht worden. Daß Membranen in Kernen auftreten können, weiß man ja von elektronenmikroskopischen Untersuchungen, zumal tierischer Eizellen [z. B. KESSEL, R. G., J. Cell. Biol. 24, 471 (1965)].

Schnepf: Bei Permanganat-Fixierung ist es möglich, daß Membranen verschiedener Organellen miteinander verschmelzen, beispielsweise Mitochondrien- mit Plastidenmembranen (von HONGLADAROM kürzlich gezeigt; Intern. Botanikerkongreß, Edinburgh 1964). Ich halte es für zumindest nicht ausgeschlossen, daß in dem von MÜHLETHALER gezeigten, gerade besprochenen Einzelfall etwas Ähnliches sich ereignet haben könnte.

Bajer: What is the evidence for your assumption that the "division stages" you have shown really represent divisions and not fusions? In living cells the shape of the organelles is different during fusion and during division. Judging from your EM pictures I would be inclined to interpret what you have shown as typical fusion and not division, as you do.

Diers: According to my quantitative results there is an increase in number of mitochondria and plastids. That is why I interpreted these stages as plastids or mitochondria in division.

Schnepf: Wenn die Möglichkeit besteht, daß Partikel sich nicht nur teilen, sondern auch fusionieren, dann sollte man weniger die Zahl als vielmehr das Gesamtvolumen aller Partikel, z. B. von Mitochondrien pro Zelle bestimmen. Das ist auch bei elektronenmikroskopischen Aufnahmen rechnerisch ohne weiteres möglich.

Klima: Wenn eine Organellenart mit einheitlicher Gestalt vorliegt, ist es erlaubt, lediglich Anschnittflächen zu zählen und daraus eine Vermehrung zu erschließen. Ist dagegen ein Gestaltwechsel anzunehmen, läßt sich eine Vermehrung tatsächlich allenfalls durch Messung des Gesamtvolumens mit Hilfe der Treffermethode beweisen. Dabei sind gewisse Korrekturen erforderlich, die sich aus der Gestalt, räumlichen Verteilung und Formvariabilität der Organellen ergeben.

Diers: Obwohl das Gesamtvolumen der Mitochondrien pro Zelle nicht bestimmt wurde, ist nach den ausgewerteten Bildern anzunehmen, daß es während der Eizellentwicklung zunimmt. Man erkennt nämlich auf Schnitten durch die reifende und reife Eizelle neben vielen kleinen auch zahlreiche große Mitochondrien, die etwa die gleiche Größe wie die Mitochondrien jüngerer Entwicklungsstadien besitzen. Da in den späten Entwicklungsphasen die Mitochondrienzahl erheblich zunimmt (Tab. 1), folgt, daß sich auch das Gesamtvolumen der Mitochondrien vergrößert.

Hagemann: Ist die Verschmelzung von Mitochondrien ein normaler und regelmäßig zu beobachtender Vorgang ? In welcher Zeit läuft er ab ?

Bajer: I was not studying this problem in detail but I observed hundreds of cases in vitro. Fusion is very fast; it will take less than a minute, sometimes only 20 seconds. You often will get simply breaks because mitochondria are very sensitive to strong illumination and I suppose you do not consider this the way they multiply.

Schnepf: Wurden bei der Untersuchung der Oogonentwicklung von *Sphaerocarpus* eigentlich auch während der Nacht Fixierungen durchgeführt ?

Diers: Es wurde morgens, vormittags, nachmittags und abends fixiert. Cytologische Besonderheiten, die mit einer bestimmten Fixierungszeit verbunden sein könnten, wurden nicht festgestellt. Außerdem zeigen Befruchtungsversuche, daß Eizellen sowohl tagsüber als auch nachts befruchtet werden können. Nach unseren Beobachtungen vermag ferner das reife Archegonium die Spermatozoiden nur während einer Zeit von längstens $^3/_4$—1 Std anzulocken. Auf Grund dieser Befunde möchten wir nicht annehmen, daß die Eireifung nach einer engbegrenzten zeitlichen Rhythmik abläuft, in dem Sinne, daß etwa nur nachts wesentliche Entwicklungsvorgänge erfolgen.

Falk: Sind an diesen Eizellen schon histochemische Untersuchungen gemacht worden, z. B. in bezug auf die Lokalisation lytischer Enzyme ? Wenn Organellen eliminiert werden, müßte man zumindest eine zeitlang einen Anstieg solcher Enzymaktivitäten erwarten, wie dies für die bei tierischen Zellen beschriebenen Cytolysomen gefunden wurde.

Diers: An pflanzlichen Eizellen sind solche Untersuchungen noch nicht durchgeführt worden.

Döbel: Lassen sich nach Glutaraldehyd- oder Osmiumtetroxid-Fixierung Unterschiede in der Matrixstruktur zwischen Kern und Plastiden feststellen ? Die Kernmatrix läßt sich dabei ja relativ gut charakterisieren.

Diers: Die Matrix von Gebilden, die sich möglicherweise vom Kern abgelöst haben könnten, erscheint bei diesen Fixierungen genau wie die Kernmatrix und unterscheidet sich somit recht deutlich von derjenigen der Plastiden. Eingehende Untersuchungen hierüber sind bei uns im Gange.

Wohlfarth-Bottermann: Kann zwischen Mitochondrien und Plastiden in jedem Fall sicher unterschieden werden ?

Diers: In Einzelschnitten nicht immer, in Serienschnitten aber stets absolut sicher.

Stoeckenius: Was wird aus den Kernausstülpungen, die sich ganz abschnüren ?

Diers: Die Untersuchungen darüber sind im Gange. Ich darf vielleicht schon sagen, daß sich in ihnen keine Membranen bilden.

Nucleinsäuren und Proteinsynthese in Plastiden

Von

B. Parthier und R. Wollgiehn, Halle/Saale

Mit 5 Abbildungen

Seitdem bekannt ist, daß subcelluläre Strukturen wie Plastiden in der Regel nicht *de novo* entstehen, sondern aus bereits vorhandenen Organellen durch Teilung hervorgehen, und nachdem aus zahlreichen Befunden die Plastiden als genetisch autonom erkannt wurden, ist es wichtig zu wissen, ob das genetische System der Plastiden dem des Zellkerns prinzipiell vergleichbar ist und sich in seiner Gesamtheit vom Gehalt an aktiver DNA bis zur Proteinsynthese in das bekannte, vor allem aus Befunden an Bakterien und einigen tierischen Systemen entwickelte Schema einordnen läßt. Es soll versucht werden, den heutigen Stand des Wissens darüber aufzuzeigen.

A. DNA in Plastiden

Erst in jüngster Zeit konnte das Vorkommen von DNA in Plastiden mit letzter Sicherheit nachgewiesen werden, nachdem sich methodischer Unzulänglichkeiten wegen über lange Zeit positive und negative Befunde gegenübergestanden haben.

Der Nachweis von DNA in Plastiden konnte auf mehreren Wegen erbracht werden.

I. Cytochemischer Nachweis

Der sichere cytochemische Nachweis von DNA in Plastiden ist mit größeren Schwierigkeiten verbunden. Die mangelhafte Spezifität einzelner Farbreaktionen und der sehr geringe DNA-Gehalt der Plastiden ließen in zahlreichen Fällen keine eindeutigen Aussagen zu.

Die Befunde von Metzner [77] an Chloroplasten von *Agapanthus* deuteten darauf hin, daß im Stroma nur RNA, in den Grana aber RNA und DNA lokalisiert sind, wobei jedoch auf den DNA-Gehalt nur auf Grund der wenig spezifischen Methylgrün-Pyronin-Färbung bei negativer *Feulgen*reaktion geschlossen wurde. Dagegen beobachtete Chiba [25] eine positive *Feulgen*reaktion an Chloroplasten von *Selaginella sativa, Tradescantia fluminescens* und *Rhoeo discolor*, während Spiekermann [112] ebenfalls mit der *Feulgen*reaktion und Ausbleiben der Färbung nach Desoxyribonuclease-Behandlung DNA im Primärgranum der Plastiden meristematischer Zellen von *Chlorophytum* und *Helianthus tuberosus* nachweisen konnte.

Besonders überzeugend waren die cytochemischen und elektronenmikroskopischen Befunde von Ris und Plaut [94] an Chloroplasten von *Chlamydomonas*. Von der sicher ausreichenden Charakterisierung ausgehend, daß DNA eine säurelösliche Substanz ist, die nach Hydrolyse mit der *Feulgen*reaktion gefärbt wird, nach Acridin-Orange-Färbung charakteristisch gelb-grün fluoresciert und gegen DNase, nicht aber gegen RNase empfindlich ist, fanden sie in Nachbarschaft der Pyrenoide deutlich lokalisierbare DNA-haltige Regionen zwischen den Chloroplastenlamellen. Wie die elektronenmikroskopischen Aufnahmen erkennen lassen (Abb. 1a und b), enthalten diese Regionen Granula, die offenbar Chloroplasten-Ribosomen darstellen. Dazwischen liegen Bezirke geringerer Dichte mit Fibrillen von 25—30 Å Dicke, die durch DNase-Behandlung verschwinden, also offenbar DNA darstellen.

Ähnliche Befunde ließen sich später auch an Chloroplasten junger Blätter von *Beta vulgaris* erbringen [52]. In der Matrix der interlamellaren Bereiche finden sich neben ribosomenartigen Partikeln fibrilläre Komponenten, die mit DNase extrahiert werden können unter solchen Bedingungen, unter denen auch die Kern-DNA entfernt wird. Die entsprechenden Bezirke sind vor der Behandlung mit DNase feulgenpositiv und färbbar mit Uranylacetat (Abb. 2).

II. Nachweis durch Einbau von ³H-Thymidin

Für den endgültigen Nachweis des Vorkommens von DNA in Plastiden war der Befund von besonderer Bedeutung, daß ³H-Thymidin, ein spezifischer DNA-Precursor, in die Chloroplasten eingebaut werden kann.

Nach den ersten Beobachtungen von Brachet [15] über den Einbau von Thymidin in die Chloroplasten kernloser Teile von *Acetabularia* und in die Chloroplasten von *Spirogyra* durch Stocking und Gifford [111] wiesen Scher und Sagan [98] nach, daß der Thymidin-Einbau in die Chloroplasten von *Euglena* tatsächlich in eine mit DNase extrahierbare Fraktion, also in die DNA erfolgt. Wir konnten dann diese Befunde an Chloroplasten einer höheren Pflanze *(Nicotiana rustica)* bestätigen und ebenfalls mit Sicherheit den Einbau des ³H-Thymidins in die Chloroplasten-DNA nachweisen [118, 119] (Abb. 3). An Chloroplasten von *Clivia miniata* und *Bilbergia spec.* [81], von *Beta vulgaris* [62] und *Euglena* [96] und an Proplastiden der Eizellen von *Pteridium aquilinum* [6] und von Wurzelspitzen von *Allium cepa* [23] wurden später gleiche Ergebnisse erzielt.

III. Biochemische Untersuchungen an isolierten Chloroplasten

1. DNA-Gehalt der Chloroplasten

Bereits vor 30 Jahren unternahm Menke [74] die ersten Versuche, Chloroplasten zu isolieren und ihren Nucleinsäure-Gehalt zu bestimmen. Trotz ständiger methodischer Weiterentwicklungen sind jedoch bis heute keine in jeder Weise befriedigenden Verfahren zur Chloroplasten-Isolierung gefunden worden.

Um zuverlässige Aussagen aus Befunden an isolierten Plastiden machen zu können, ist vorauszusetzen, daß die Chloroplasten beim Homogenisieren des Gewebes weder mechanisch zerstört werden noch osmotisch bedingte Veränderungen erfahren, um ein Auswaschen chloroplasteneigener Substanzen weitgehend zu verhindern. Außerdem müssen die Präparate möglichst frei sein von anderen Zellbestandteilen, vor allem von Zellkernbruchstücken und von Mikroorganismen. Diese Anforderungen können heute weitgehend erfüllt werden, wenn die Chloroplasten in nicht-wäßrigen Medien isoliert, oder aber nach Isolierung in wäßrigen Medien im Dichtegradienten gereinigt werden.

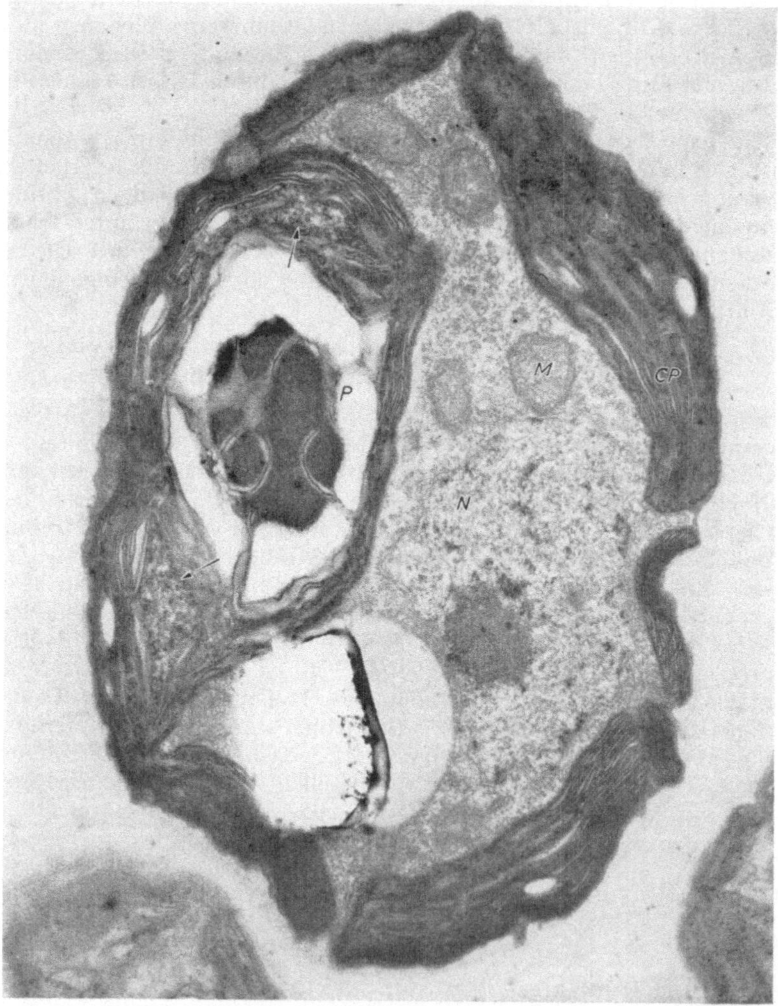

Abb. 1 a. Schnitt durch eine Zelle von *Chlamydomonas moewusii.* Fixierung im Kellenberg-Medium. *CP* = Chloroplast, *M* = Mitochondrien, *N* = Nucleus, *P* = Pyrenoid. Die Pfeile weisen auf die DNA-haltigen Strukturen in Chloroplasten. Vergr. 24 000 : 1

Mit ständig verbesserten Methoden wurden in den letzten Jahren von verschiedenen Autoren immer wieder Chloroplasten isoliert und auf

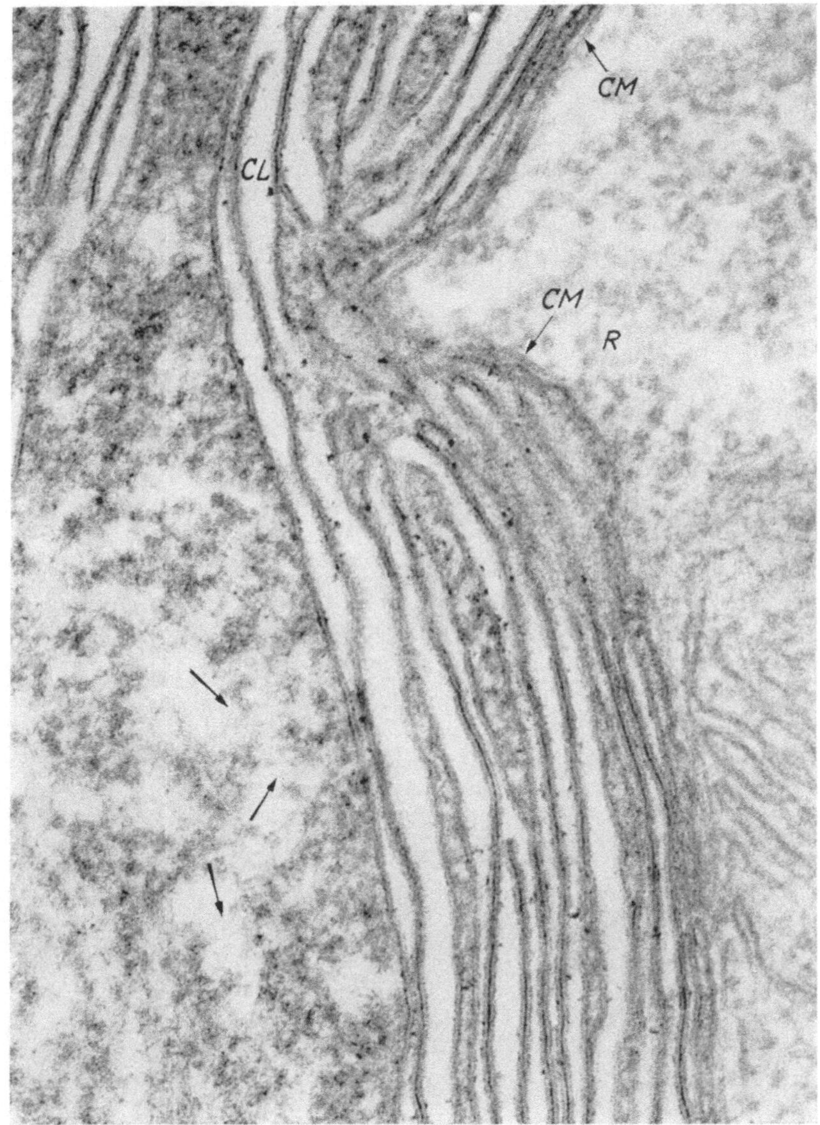

Abb. 1 b. Feulgen-positive Region aus einem Chloroplasten von *Chlamydomonas moewusii*. Der erweiterte Raum zwischen den Chloroplasten-Membranen enthält dichtere Partikel [Chloroplasten-Ribosomen (*R*)] und 25 Å dicke Fibrillen (Pfeile), die vermutlich DNA darstellen. *CM* = äußere Chloroplasten-Membran, *CL* = Chloroplastenlamellen. Vergr. 110000 : 1. (Mit freundlicher Genehmigung aus Ris und Plaut [94])

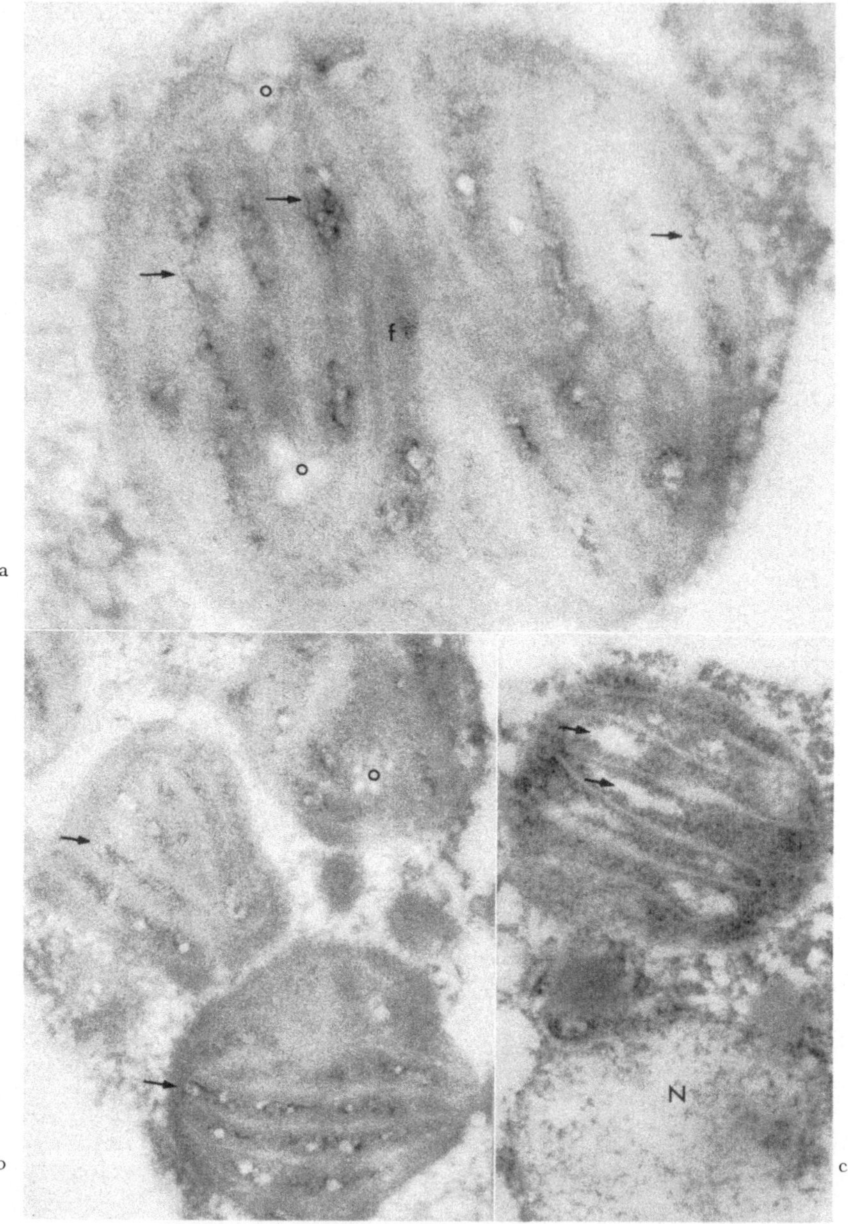

Abb. 2 a—c. Chloroplasten aus Blättern von *Beta vulgaris*. a. Nach RNase-Behandlung. Die Pfeile zeigen auf Uranyl-färbbare Filamente. f = Ferritin-Granula, o = helle Hohlräume die durch Extraktion osmiophiler Tropfen entstanden sind. Vergr. 71000 : 1. b. Wie a, jedoch Vergr. 49000 : 1. c. Wie a, aber nach Extraktion mit DNase. Die Färbung im Nucleus (N) und in den fibrillären Bezirken (Pfeile) ist reduziert, während die Ribosomen und die entsprechenden Partikel in der Chloroplasten-Matrix noch gefärbt sind. Vergr. 49000 : 1.
(Mit freundlicher Genehmigung aus Kislev, Swift und Bogorad [62])

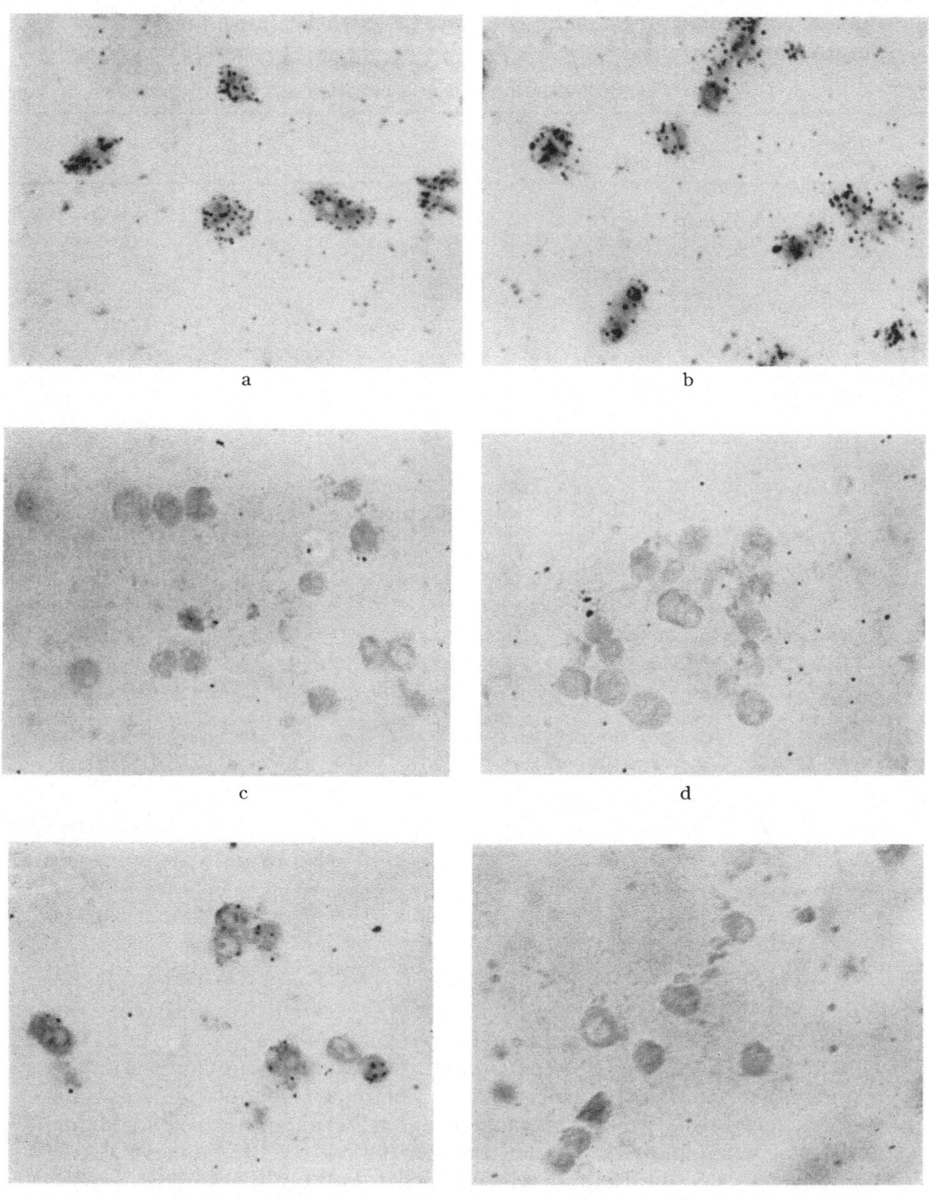

Abb. 3 a—f. Autoradiogramme isolierter Chloroplasten von *Nicotiana rustica*. a, b. Einbau von ³H-Thymidin in die Chloroplasten junger Blattstecklinge (*a*) und junger Blätter (*b*). c, d. Kein Einbau von ³H-Thymidin in die Chloroplasten alter Blattstecklinge (*c*) und alter Blätter (*d*). e, f. Chloroplasten junger Blattstecklinge (wie a), aber nach Behandlung mit DNase (*e*) und 1,6 n Perchlorsäure, 20 min 70° (*f*) (aus [*119*])

ihren DNA-Gehalt untersucht. Die wesentlichen Befunde sind in Tab. 1 zusammengefaßt.

Tabelle 1: *DNA-Gehalt der Plastiden*

Objekt	in mg pro Chloroplast	in μg pro mg Protein	in % des Trockengewichts	in mg pro mg Chlorophyll	Autor
Euglena	1×10^{-11}	—	—	—	Brawerman u. Eisenstadt (1964) [19]
Chlamydomonas * .	$1,2 \times 10^{-11}$	1,56	—	—	Sager u. Ishida (1963) [97]
	$6,2 \times 10^{-11}$	2,7	—	—	
Acetabularia . . .	1×10^{-13}	0,2	—	—	Gibor u. Izawa (1963) [39]
Nicotiana tabacum .	—	—	0,5	—	Holden (1952) [48]
Nicotiana tabacum .	—	10	—	—	Jagendorf u. Wildman (1954) [56]
Nicotiana tabacum .	$8,0 \times 10^{-11}$	—	0,88	122	Chiba u. Sugahara (1957) [26]
Spinacia	$4—7 \times 10^{-11}$	—	0,58	66	
Nicotiana tabacum .	—	—	0,7	—	Cooper u. Loring (1958) [30]
Nicotiana rustica .	$6—10 \times 10^{-11}$	—	0,2—0,6	80	Böttger u. Wollgiehn (1958) [12] Wollgiehn (1958) [116]
*Nicotiana rustica** .	—	—	0,25	40	Wollgiehn, unveröffentlicht
*Zea mays**	—	—	0,17	17	Orth u. Cornwell (1963) [82]
*Phaseolus vulgaris**	—	—	0,15	11	Kirk (1963) [59]
*Spinacea**	—	3—4	—	—	Pollard (1964) [88]
*Antirrhinum*** . .	$5,6 \times 10^{-12}$	—	0,025	—	Ruppel (1965) [95a]

In den mit * versehenen Versuchen wurden die Chloroplastenpräparate durch Zentrifugation im Dichtegradienten gereinigt, in den anderen Fällen durch verschiedene Waschungen. ** Isolation der Chloroplasten im nichtwäßrigen Medium.

Die beobachteten Unterschiede im DNA-Gehalt der Chloroplasten haben gewiß zum Teil methodische Ursachen. Daneben muß aber mit größeren biologischen Schwankungen gerechnet werden, denn der DNA-Gehalt ist offenbar von Objekt zu Objekt verschieden und darüber hinaus vom Entwicklungszustand der Pflanze abhängig. Dennoch gibt es auch bemerkenswerte Übereinstimmungen in den Befunden. Die DNA-Menge pro Chloroplast beträgt bei den meisten Objekten 10^{-11} bis 10^{-12} mg. Sicher lassen sich die Befunde verschiedener Autoren nicht ohne weiteres miteinander vergleichen. Das Trockengewicht der Plastiden könnte sehr stark durch ihren Stärkegehalt beeinflußt werden. Dennoch sind übereinstimmend DNA-Mengen von 0.1—0,5% des Chloroplasten-Trockengewichts gefunden worden. Der Anteil der Chloroplasten-DNA an der Gesamt-DNA der Zelle soll nach übereinstimmenden Angaben bei *Spinacia, Beta vulgaris* [27], *Chlamydomonas* [27, 97], *Chlorella* [27] und

Euglena [*19*] nur 1—5% betragen. Bei der Annahme von 100 Plastiden pro Zellkern wird der DNA-Gehalt in einem Chloroplasten mit 0.01% der Kern-DNA angegeben [*41*], das entspricht 1% der Gesamt-DNA der Zelle in Plastiden. Allerdings wurde für *Chlorella* auch ein Wert von 10% angegeben [*53*]. Wir hatten dagegen bei gut wachsenden Blattstecklingen von *Nicotiana rustica* Werte bis zu 25% errechnet, die offenbar auf die extrem starke Vermehrung der Chloroplasten in den sich nicht teilenden Zellen zurückzuführen sind [*12, 116*].

Besonders zuverlässig erscheinen die Befunde an Chloroplasten kernloser Teile steril kultivierter *Acetabularien*, wo Verunreinigungen durch Kernbruchstücke mit Sicherheit ausgeschlossen werden können [*4, 39*]. Die von GIBOR und IZAWA [*39*] ermittelten, verglichen an anderen Chloroplasten sehr niedrigen Werte von 10^{-13} mg DNA pro Chloroplast entsprechen etwa dem DNA-Gehalt von *Vaccinia*-Viren [*1*]. Geht man von einem Triplet-Code und einer Doppelstrang-DNA aus, so könnte diese DNA-Menge $3 \cdot 10^4$ Aminosäuren codieren, das entspräche einigen hundert verschiedenen Proteinen mit einem durchschnittlichen Molekulargewicht von 20000 [*39, 41*]. Der DNA-Gehalt anderer Plastiden ist aber wesentlich höher und liegt in der Größenordnung des DNA-Gehaltes von Bakterien (*Aerobacter aerogenes* $2 \cdot 10^{-12}$ mg/Zelle [*24*], *E. Coli* 10^{-11} mg/Zelle [*5*]), so daß die gefundenen Werte tatsächlich biologisch sinnvoll erscheinen.

2. Charakterisierung der Plastiden-DNA

Versuche zur Charakterisierung der Chloroplasten-DNA haben deutliche Unterschiede zur Kern-DNA der gleichen Zelle ergeben. IWAMURA [*51*] hatte nach differenziertem Zentrifugieren von *Chlorella*-Homogenaten sowohl aus der Kernfraktion als auch aus der Chloroplasten-Fraktion DNA extrahiert, die sich in ihrer Basenzusammensetzung und ihrer Stoffwechselaktivität voneinander unterschieden. Später gelang es dann mehreren Arbeitsgruppen, diese Unterschiede an verschiedenen Objekten näher zu charakterisieren.

Beim Zentrifugieren von DNA-Präparaten aus Chloroplasten-haltigem Gewebe im CsCl-Dichtegradienten treten neben der Hauptfraktion (Zellkern-DNA), die meist 94—98% der Gesamt-DNA der Zelle ausmacht, eine oder zwei kleine Nebenfraktionen auf (*,,Satelliten"-DNA*) [*17, 19, 27, 33, 34, 62, 63, 90, 91*]. Eine dieser Fraktionen fehlt in der DNA gereinigter Kernpräparate, während umgekehrt in gereinigten Chloroplasten diese ,,Satelliten"-DNA sehr stark angereichert ist, also Chloroplasten-eigene DNA darstellt (Abb. 4). Die von einigen Autoren gefundene zweite ,,Satelliten"-DNA-Fraktion ist offenbar Mitochondrien-DNA [*19, 27, 34, 91*].

Die im CsCl-Dichtegradienten bisher gewonnenen Werte für die Dichte (*"buoyant density"*) der Chloroplasten-DNA verschiedener Organismen im Vergleich zur jeweiligen Kern-DNA sind in Tab. 2 dargestellt.

Von verschiedenen Autoren ermittelte Werte für den gleichen Organismus — *Euglena* wurde am intensivsten untersucht — stimmen

Tabelle 2: *Die im CsCl-Dichtegradienten ermittelte Dichte ("buoyant density") und der Guanin + Cytosin-Gehalt der Chloroplasten-DNA im Vergleich zur Kern-DNA*

	Kern-DNA		Chloroplasten-DNA		2. „Satelliten"-DNA (Mitochondrien-DNA) Dichte	Literatur
	Dichte	G + C-Gehalt in Mol-%	Dichte	G + C-Gehalt in Mol-%		
Euglena	1,707	51	1,685	24	1,690	Ray u. Hanewalt (1964) [90]
Euglena	1,707	47	1,684	25	1,692	Brawerman u. Eisenstadt (1964) [19]
Euglena	1,707	28*	1,686	26*	1,691	Edelman, Schiff u. Epstein (1965) [34]
Chlamydomonas . .	1,728	62	1,702	39	—	Sager u. Ishida (1963) [97]
Chlamydomonas . .	1,723	64*	1,695	36*	—	Chun, Vaugham u. Rich (1963) [27]
Chlorella	1,716	57*	1,695	36*	—	Chun, Vaugham u. Rich (1963) [27]
Chlorella	1,72	57	1,69	44	—	Iwamura (1960) [51], Iwamura u. Kuwashima (1964) [53]
Spinacia oleracea . .	1,695	36*	1,705**	60*	1,719**	Chun, Vaugham u. Rich (1963) [27]
Beta vulgaris	1,695	36*	1,705**	60*	1,719**	
Beta vulgaris	1,689	31*	1,700	42*	—	Kislev, Swift u. Bogorad (1965) [62]
Nicotiana tab. . . .	1,690	—	1,703	—	—	Shipp et al. (1965) [100a]

* Berechnet aus der im CsCl-Dichtegradienten ermittelten Dichte der DNA.

** Diese Fraktionen wurden nicht genauer zugeordnet. Beide wurden in der angereicherten Chloroplastenfraktion gefunden. Es ist wahrscheinlich, daß eine der Fraktionen ebenfalls Mitochondrien-DNA ist.

gut überein. Beträchtliche Unterschiede ergeben sich dagegen beim Vergleich der bisher untersuchten Algen mit den höheren Pflanzen. Die Chloroplasten-DNA von *Spinacia* und *Beta vulgaris* hat im Gegensatz zu den Algen eine höhere Dichte als die Kern-DNA.

Die Dichte der DNA steht in direkter Beziehung zu ihrer Basenzusammensetzung. Es ist möglich, aus der ermittelten Dichte den Gehalt an Guanin + Cytosin bzw. Adenin + Thymin zu berechnen. Einige Autoren haben die Basenzusammensetzung der Chloroplasten-DNA auch analytisch bestimmt und erhielten die gleichen Werte, wie sie aus der Dichte berechnet wurden. Zur Charakterisierung der Basen-

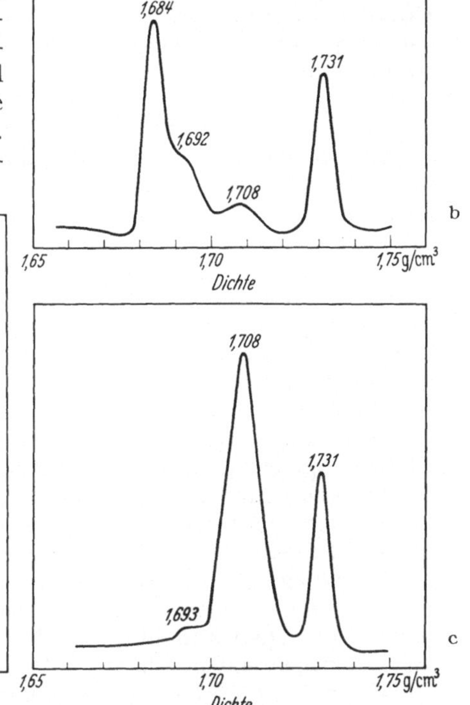

Abb. 4 a—c. CsCl-Dichtegradienten-Zentrifugation von DNA aus *Euglena gracilis*. Zur Markierung wurde in b und c DNA aus *Micrococcus lysodeikticus* mit bekannter Dichte (1,731) zugesetzt. a. DNA-Präparat aus ganzen Chloroplasten, noch stark verunreinigt mit Kern-DNA (1,708). b. Angereicherte Chloroplasten-DNA (1,684), extrahiert aus einem 23000 g-Sediment der in einem 0,005 M MgCl₂ + 0,5% Na-Desoxycholat-haltigen Tris-Puffer aufgebrochenen Chloroplasten. c. DNA aus einer Zellkern-Präparation (1,708). (Mit freundlicher Genehmigung aus BRAWERMAN und EISENSTADT [*19*])

zusammensetzung der Chloroplasten-DNA sind einige G + C-Werte in Tab. 2 zusammengestellt. Auch KIRK [*59*] bestimmte die Zusammensetzung der DNA von *Phaseolus*-Chloroplasten und fand ein molares Verhältnis von Adenin : Guanin von 1,67 gegenüber einem Wert von 1,54 für Kern-DNA.

Die Kern-DNA von *Euglena* zeichnet sich noch durch ihren Gehalt an 2,3% 5-Methylcytosin aus, während diese Base in der Chloroplasten-DNA nicht nachgewiesen werden konnte [*19, 90*].

Ebenso wie die Dichte wird auch die thermische Denaturierung der DNA (T_m) durch ihre Basenzusammensetzung bestimmt. Die geringere Dichte im CsCl-Gradienten und der niedrigere G + C-Gehalt der Chloroplasten-DNA von *Euglena* ließen auch eine niedrigere Denaturierungstemperatur gegenüber der Kern-DNA erwarten. Das hat sich experimentell bestätigt. Die T_m-Werte betragen für Chloroplasten-DNA 78–80° und für Kern-DNA 89–91° [*19*].

Das Verhalten der Chloroplasten-DNA im Dichtegradienten nach Hitzedenaturierung, das Schmelzprofil und der Gehalt an gleichen Mengen der komplementären Basenpaare sprechen für das Vorliegen von Doppelstrang-Helix-Molekülen.

IV. DNA-Synthese in Plastiden

Aus verschiedenartigen Experimenten konnten – teilweise nur indirekte – Beweise dafür erbracht werden, daß die Chloroplasten-DNA auch tatsächlich in den Plastiden selbst gebildet wird und nicht aus dem Zellkern stammt.

Iwamura [*51*, *52*] isolierte aus *Chlorella* 2 DNA-Fraktionen, die sich in ihrem Stoffwechsel stark voneinander unterschieden. Die Stoffwechselaktivität der den Chloroplasten zugeordneten DNA war im Licht wesentlich höher als die der Kern-DNA. Im Dunkeln war die Synthese der Chloroplasten-DNA sehr stark herabgesetzt bei einer wesentlich größeren Unabhängigkeit der Kern-DNA von der Belichtung.

Auch die autoradiographischen Untersuchungen (^{3}H-Thymidin-Einbau) sprechen für eine unabhängige Chloroplasten-DNA-Synthese. Bei unseren Untersuchungen gingen wir von der Beobachtungv aus, daß beim Wachstum bewurzelter Blattstecklinge von *Nicotiana rustica* nicht nur die Protein- und RNA-Menge, sondern auch der DNA-Gehalt sehr stark anstieg. Da wir Zellteilungen nicht beobachteten und das sehr geringe endomitotisch bedingte Zellkernwachstum in keinem Verhältnis zur DNA-Zunahme stand, vermuteten wir, daß die sich sehr stark vermehrenden Chloroplasten den DNA-Gehalt der Zelle beeinflussen [*12*, *117*]. Es zeigte sich dann [*118*, *119*], daß ^{3}H-Thymidin tatsächlich nur in die DNA (Kontrolle durch Säurehydrolyse, DNase- und RNase-Behandlung) sich teilender Chloroplasten gut wachsender Blattstecklinge oder sehr junger Blätter eingebaut wird, während die Chloroplasten ausgewachsener Blätter und Blattstecklinge keine DNA-Synthese mehr erkennen lassen (Abb. 3). Swift et al. [*62*] konnten auch an *Beta vulgaris* zeigen, daß nur die Chloroplasten junger Blätter Thymidin inkorporieren. Darüber hinaus konnten Budd et al. [*23*] in kurzfristigen Versuchen mit *Allium cepa*-Wurzelspitzen einen Einbau in die Proplastiden und Mitochondrien auch dann feststellen, wenn keine DNA-Synthese im Zellkern ablief. Diese weitgehende Trennung des DNA-Stoffwechsels der Kerne von dem der Chloroplasten spricht sehr stark für eine *Autonomie des Stoffwechsels der Chloroplasten-DNA*.

Besonders überzeugende Befunde konnten jedoch wiederum an *Euglena* erbracht werden. Die Fähigkeit der *Euglena*-Zellen zur Bildung voll entwickelter, grüner Chloroplasten kann mit chemischen oder physikalischen Mitteln zerstört werden. Durch milde UV-Bestrahlung entstehen Mutanten, die über Generationen hin nicht wieder ergrünen [66]. Das Maximum des Wirkungsspektrums bei 260 mμ macht einen UV-Einfluß auf die Nucleinsäuren sehr wahrscheinlich.

Mit spezieller Methodik konnten GIBOR und GRANICK [40] zeigen, daß farblose Mutanten bereits dann entstehen, wenn nur das Cytoplasma mit UV-Licht bestrahlt wird, während eine Bestrahlung des Kerns allein keine Mutanten entstehen läßt. Der Zellkern der Mutanten bleibt nach der Bestrahlung voll funktionsfähig und müßte, wenn er für die Bildung der Chloroplasten-DNA verantwortlich wäre, das Ausbleichen verhindern.

Die größere Empfindlichkeit der Chloroplasten-DNA gegen UV hängt offenbar mit ihrem höheren Thymin-Gehalt zusammen. Die Dimerisierung des Thymins könnte eine Hemmung der Replikationsfähigkeit der DNA verursachen.

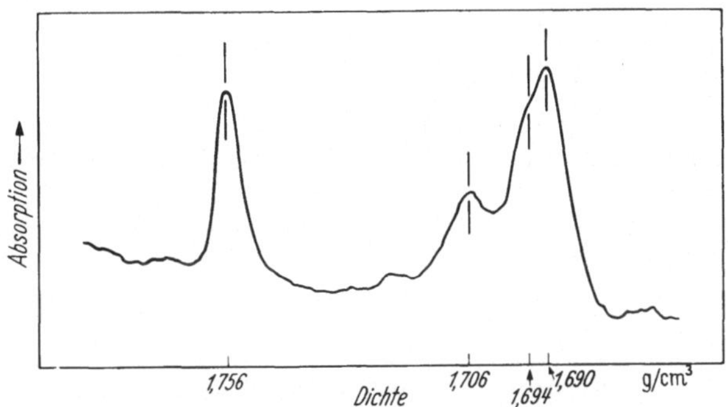

Abb. 5. CsCl-Dichtegradienten-Zentrifugation des angereicherten DNA-Bereiches mit niedriger Dichte aus einem DNA-Präparat einer farblosen UV-Mutante von *Euglena gracilis*. 1,756 DNA aus E. coli als Markierung. 1,706 restliche Kern-DNA. 1,690-Bereich vermutlich Mitochondrien-DNA. Chloroplasten-DNA (1,685) fehlt vollständig. (Mit freundlicher Genehmigung aus RAY und HANEWALT [91])

Analysen der *Euglena*-DNA haben ergeben, daß alle jene Mutanten, die keine Chloroplasten bilden, auch keine Chloroplasten-DNA enthalten [34, 91]. Abb. 5 zeigt das Ergebnis eines entsprechenden Versuchs. Aus einer durch UV-Bestrahlung gewonnenen farblosen *Euglena*-Mutante wurde in einer Vorfraktionierung die gesamte „Satelliten"-Region der DNA isoliert und angereichert. Eine erneute Zentrifugation im CsCl-Dichtegradienten ergab, daß neben einem Rest Kern-DNA (Dichte 1,700 g/cm³) nur noch DNA vom Typ 1,690 (offenbar Mitochondrien-DNA), aber keine Chloroplasten-DNA (1,685) nachzuweisen ist [91]. Aus allen Versuchen ergibt sich überzeugend, daß die Chloroplasten-DNA nicht im Zellkern, sondern in den Chloroplasten selbst gebildet wird.

B. RNA in Plastiden

I. RNA-Nachweis und RNA-Gehalt

Bereits in älteren Untersuchungen wurde mit cytochemischen Methoden neben DNA auch RNA in Plastiden nachgewiesen [77, 112]. Nach elektronenmikroskopischen Befunden an *Zea mays* [55] und *Spinacia* [79] ist die RNA mit 150—170 Å großen Partikeln innerhalb des Stroma verbunden (Chloroplasten-Ribosomen), die den Cytoplasma-Ribosomen vergleichbar sind und mit RNase entfernt werden können. Auch durch Einbau von ^{3}H-Cytidin in die Plastiden von *Zea mays* [55] und ^{3}H-Uridin in die Chloroplasten von *Clivia miniata* [81] konnte das Vorkommen von RNA in Plastiden nachgewiesen werden.

Bei quantitativen Bestimmungen des RNA-Gehaltes der Plastiden ergaben sich größere Schwankungen als bei DNA-Bestimmungen. Zweifellos wird der RNA-Gehalt in viel stärkerem Maße vom physiologischen Zustand der Chloroplasten und von der Isolierungstechnik bestimmt. So ist es nicht verwunderlich, daß die Angaben verschiedener Autoren zwischen 0,1 und etwa 7% des Trockengewichts schwanken [8, 12, 26, 30, 48, 56, 58, 71, 82, 95, 106, 116] oder aber in der Größenordnung von 10—50 μg RNA pro mg Chloroplasten-Protein liegen [12, 30, 47, 106, 116]. Der Anteil der Chloroplasten-RNA an der Gesamt-RNA der Zelle wurde übereinstimmend für Tabakblätter mit 15—20% [116], für Bohnen, Tabak und Spinat mit 25—35% [47], für junge Gerstenblätter mit 20—30% [62] und für *Euglena* im Normalfalle mit 10—20% [22] berechnet.

II. RNA-Synthese in Chloroplasten

Wenn die Chloroplasten ein autonomes, durch die Chloroplasten-DNA gesteuertes, Protein-synthetisierendes System enthalten, dann muß erwartet werden, daß sie ihre RNA selbst synthetisieren. Sie müssen alle RNA-Typen enthalten, die für die Proteinsynthese erforderlich sind: Messenger-RNA, Transfer-RNA und ribosomale RNA bzw. Ribosomen.

In vivo-Versuche können über eine chloroplasteneigene RNA-Synthese nur indirekt Aufschluß geben. Aus Befunden wie dem Einbau von markierten Vorstufen in die Chloroplasten-RNA [55, 81] oder dem Anstieg des RNA-Gehaltes der Plastiden beim Ergrünen farbloser Euglenen [22] und der Hemmung dieser RNA-Synthese durch verschiedene Inhibitoren der Nucleinsäure-Synthese (5-Fluoruracil [107], Hadacidin [72], Actinomycin D [13, 70, 87]) kann noch nicht auf die Synthese der RNA in den Plastiden selbst geschlossen werden.

Den exakten Beweis konnten erst Versuche mit isolierten Chloroplasten erbringen. Semal et al. [100] zeigten, daß der Einbau von ^{32}P-ATP in die RNA einer angereicherten Chloroplastenfraktion wesentlich höher ist als in die RNA der angereicherten Kern-Fraktion. Kirk [60, 61] gelang es ebenfalls, in hochgereinigten Chloroplastenpräparaten aus Bohnenblättern einen Einbau von ^{14}C-ATP und ^{14}C-GTP zu erzielen

und Unterschiede zum Einbau in eine isolierte Kernfraktion heraus-
zuarbeiten (vgl. auch B, III, 2). Auch isolierte Chloroplasten aus kernlosen Teilen von *Acetabularia* sind in der Lage, ^{14}C-Uracil in die RNA
einzubauen [*99*].

III. Analyse der Chloroplasten-RNA

1. Allgemeine Befunde

Es liegen bisher nur wenige analytische Angaben zur Charakterisierung der
Chloroplasten-RNA vor. Die ersten Versuche waren mit unzureichenden Mitteln
durchgeführt worden und konnten nur Hinweise auf die tatsächlichen Verhältnisse
in intakten Chloroplasten geben.

Nach ersten Andeutungen über das Vorkommen mehrerer, wenn auch nicht
genauer definierter RNA-Fraktionen in den Chloroplasten von *Allium cepa* durch
MENKE [*75*], fraktionierten SZARKOWSKI et al. [*113*] RNA-Präparate aus Chloroplasten von *Secale cereale* an Ecteola-Säulen und konnten die RNA grüner Chloroplasten in 4 Fraktionen, die RNA etiolierter Chloroplasten dagegen nur in 2 Fraktionen trennen. Auch durch differenzierte Zentrifugation aufgebrochener Chloroplasten von Spinat [*47*] konnten mehrere RNA-Fraktionen gefunden werden,
die entweder im Lamellen-System oder im Stroma lokalisiert sind. Eine Extraktion
der RNA aus Chloroplasten von *Allium* in einem Phenol-Puffer-System zeigte
2 RNA-Fraktionen unterschiedlicher Basenzusammensetzung [*95*]. Neuere Untersuchungen ergaben darüber hinaus zahlreiche direkte oder indirekte Beweise für die
Existenz der funktionell unterschiedlichen RNA-Typen in Chloroplasten.

2. DNA-abhängige RNA (Messenger- oder Template-RNA)

Die Annahme von JACOB und MONOD [*54*], daß die Proteinsynthese in Bakterien
durch eine kurzlebige, die genetische Information der DNA tragende RNA gesteuert
wird, ist inzwischen an vielen Objekten — auch an Pflanzen — auf ihre Allgemein-
gültigkeit untersucht worden. Alle vier Forderungen, die nach Befunden an Bakterien an eine Informations-RNA gestellt werden müssen: a) Basenzusammensetzung der homologen DNA entsprechend, b) schnelle Markierung bei kurzer
Lebensdauer, c) Hetrogenität in der molaren Größe, d) die Fähigkeit, die Proteinsynthese in zellfreien Ribosomensystemen zu stimulieren, sind bisher in keinem
Falle an Chloroplasten oder anderem Pflanzenmaterial gleichzeitig untersucht
worden. Obwohl in den wenigen vorliegenden Experimenten bisher nur einzelne
Charakteristika geprüft worden sind, sprechen sie doch für das Vorkommen einer
Messenger-RNA in Pflanzen [*11, 50, 65, 93*]. Es muß jedoch angenommen werden,
daß die Informations-RNA in höheren Pflanzen im Vergleich zur Messenger-RNA in
Bakterien eine relativ lange Lebensdauer von mindestens einigen Stunden hat
[*32, 50*].

Der erste experimentelle Nachweis einer DNA-abhängigen RNA-
Synthese in Chloroplasten konnte von KIRK [*60, 61*] erbracht werden.
Der Einbau von ^{14}C-ATP und von ^{14}C-GTP in die RNA sorgfältig gereinigter Chloroplasten von *Phaseolus* wird sowohl durch DNase als auch
durch Actinomycin D, einem spezifischen Inhibitor der DNA-abhängigen
RNA-Polymerase gehemmt. Die RNA-Synthese in isolierten Chloroplasten ist intensiver als in isolierten Kernen; sie ist auch wesentlich
empfindlicher gegen Actinomycin und DNase und stärker abhängig vom
Zusatz der 4 Nucleosidtriphosphate als die Kern-DNA-Synthese. Es
konnte erwartet werden, daß die neu gebildete RNA der Chloroplasten-
DNA ähnlicher ist als der Kern-DNA, wenn die RNA-Synthese durch
eine RNA-Polymerase mit Chloroplasten-DNA als Matrize erfolgt. Das
hat sich weitgehend bestätigt. Beim Einbau von ^{14}C-ATP und ^{14}C-GTP

entsprach zwar das Verhältnis von A zu G (2.02) nicht völlig dem A : G-Verhältnis der Chloroplasten-DNA (1,67), was bei der *in vitro*-Synthese von Kern-DNA der Fall ist (in Kern-DNA und neu gebildeter RNA ist A : G = 1,56), aber es ist möglich, daß nicht die gesamte DNA der Chloroplasten bei der RNA-Synthese aktiv war.

Die Befunde wurden an isolierten Chloroplasten von *Acetabularia* bestätigt, wo von vornherein Verunreinigungen durch Kernmaterial ausgeschlossen werden können. Der Einbau von ^{14}C-Adenin in die Chloroplasten-RNA, der im Dunkeln geringer ist als im Licht, und der durch Streptomycin und Penicillin nicht beeinflußt wird, also nicht auf Verunreinigungen durch Mikroorganismen zurückzuführen ist, wird ebenfalls durch Actinomycin D und DNase deutlich gehemmt [99]. Schließlich konnte aus *Euglena*-Chloroplasten eine RNA-Fraktion isoliert werden, die in starkem Maße die Proteinsyntheseaktivität von Ribosomen stimuliert, also eine Messenger-RNA darstellen müßte; allerdings ist sie bisher analytisch nicht auf ihre Zusammensetzung geprüft worden [20].

3. Transfer-RNA („lösliche" RNA)

Über das Vorkommen „löslicher" oder Transfer-RNA in Chloroplasten liegen kaum Befunde vor. Ein Grund für fehlende Experimente mag sein, daß T-RNA logischerweise angenommen werden muß, wenn eine aktive Proteinsynthese stattfindet. Andererseits ist ein analytischer Nachweis sicher besonders schwierig, weil die Gefahr des Auswaschens beim Isolieren der Chloroplasten gerade für lösliche RNA besonders groß ist. Beim differenzierten Zentrifugieren osmotisch aufgebrochener Chloroplasten findet man eine RNA-Fraktion, die bei 100000 g nicht sedimentiert, also eine „lösliche" RNA zu sein scheint [47]. Aus dem 100000 g-Überstand aufgebrochener Chloroplasten von Erbsenkeimlingen haben Sissakian et al. [109] eine RNA isoliert und sie mit einem Gemisch von ^{14}C-Aminosäuren auf ihre Acceptor-Aktivität überprüft. Auch nach Versuchen anderer Autoren zur Proteinsynthese isolierter Chloroplasten kann angenommen werden, daß die Chloroplasten selbst T-RNA enthalten [109].

4. Ribosomale RNA

Siehe folgenden Abschnitt!

C. Ribosomen in Plastiden

Die Ribosomen als Reaktionsorte der Proteinbiosynthese sind aus dem Cytoplasma von Bakterien, tierischer und pflanzlicher Zellen seit mehr als einem Jahrzehnt bekannt. Sie sind reine Nucleoproteinkörper mit einem RNA-Gehalt von 40—50%. (Weitere Literatur über Pflanzen-Ribosomen s. [114].) 1962 hat Lyttleton [67] zum ersten Mal Ribosomen aus isolierten, aufgebrochenen Chloroplasten beschrieben. Seitdem ist diese Beobachtung mehrfach bestätigt worden, indem Ribonucleoprotein-Partikeln aus Chloroplasten verschiedener Species isoliert und charakterisiert werden konnten (Tab. 3). Nach elektronenmikroskopischen Untersuchungen liegen die basophilen Körperchen von etwa 170 Å Durchmesser in der Matrix der Chloroplasten vor [55, 62, 79].

Besonders reich an Ribosomen sind junge Chloroplasten [55]. Ferner
sollen die Chloroplasten aus belichteten Blättern mehr Ribosomen enthalten als aus verdunkelten [23]. Schließlich fehlen sie weitgehend in
Chloroplasten farbloser, bei Dunkelheit angezogener *Euglena*-Zellen
[17, 36]. Erst nach Beleuchtung der Kulturen, mit dem Ausbilden der

Tabelle 3: *Sedimentationskonstanten in Svedberg-Einheiten für Ribosomen aus iso-
lierten Chloroplasten im Vergleich zu entsprechenden Cytoplasma-Ribosomen.* In
Klammern: Untereinheiten, durch Herabsetzung des Mg^{++}-Gehaltes entstanden.
(Die Werte sind nicht in allen Fällen auf unendliche Verdünnung bezogen)

Pflanzenspecies	Werte für Sedimentationskonstanten		Literatur
	Chloroplasten	Cytoplasma	
Spinacia oleracea . . .	66 (47, 33)	—	LYTTLETON 1962 [67]
Clivia, Chenopodium .	67 (48, 30)	70	MIKULSKA et al. 1962 [78]
Spinacia oleracea . . .	80 (71, 50)	—	APP and JAGENDORF 1963 [2]
Euglena gracilis . . .	43 (30, 24)	49 (36, 27)	BRAWERMAN 1963 [17, 18]
Euglena gracilis . . .	60 (36, 30)	70	EISENSTADT and BRA-WERMAN 1964 [37]
Brassica pekinensis . .	68	83 (55, 40, 26)	CLARK et al. 1964 [28, 29]
verschiedene Species .	67—74 (42, 50)	67—73 (49, 18)	ODINTSOVA et al. 1964 [80]
Nicotiana tabacum . .	70	80	BOARDMAN et al. 1965 [10]
Pisum sativum	62 (46, 32)	76 (54, 38)	SISSAKIAN et al. 1965 [105]
Spinacia oleracea . . .	70	80	SPENCER 1965 [108a]

typischen Chloroplastenstrukturen, treten Ribosomen von unterschiedlichem Sedimentationsverhalten und anderer Basenzusammensetzung
der RNA im Vergleich zu den Cytoplasmaribosomen auf.

Obwohl die Angaben der einzelnen Autoren für die Sedimentationskonstanten unterschiedlich ausfallen, dürfte feststehen, daß sich die
Ribosomen aus Chloroplasten von denen des Cytoplasma eindeutig
unterscheiden (Tab. 3). Während die Cytoplasma-Ribosomen dem
80 S-Typ angehören und damit den Ribosomen aus tierischen Zellen
nahestehen, weisen die Ribosomen aus Chloroplasten Sedimentationskonstanten um 70 S auf, die auch für Bakterienribosomen charakteristisch sind. Ob dieser Sedimentationsunterschied einem Struktur-
unterschied im molekularen Bereich entspricht und ob sich daraus
auch funktionelle Differenzen ergeben, ist noch nicht endgültig geklärt.

Einige Angaben über die Basenzusammensetzung der RNA von Chloroplastenribosomen im Vergleich zur RNA der entsprechenden Cytoplasma-
ribosomen sind in Tab. 4 zusammengestellt. Offensichtlich bestehen bei
höheren Pflanzen keine Unterschiede in der Basenzusammensetzung der
RNA beider Ribosomenarten. Auffallend ist die bemerkenswerte Konstanz in der Zusammensetzung der Ribosomen-RNA verschiedener

Pflanzenarten. (Vor allem beim Vergleich einer großen Zahl von Angaben, die in der Tabelle nicht berücksichtigt wurden, weil in den betreffenden Arbeiten eine direkte Gegenüberstellung von Ribosomen des Cytoplasma und der Chloroplasten fehlt [*29, 49, 67, 89*]). Typisch ist ein stets hoher Guaningehalt von meist über 30 Mol-%. Bei *Euglena* bestehen dagegen signifikante Unterschiede in der RNA-Zusammensetzung von beiden Ribosomenarten [*18, 20, 37*].

Tabelle 4: *Basenzusammensetzung der RNA von Ribosomen der Chloroplasten und des Cytoplasma verschiedener Pflanzenarten* (⁺ RNA ganzer Chloroplasten).
Angaben in Mol-%

Objekt	Ribosomenart	A	G	C	U	Literatur
Euglena	Cytoplasma	23	29	27	21	Eisenstadt und
	Chloroplasten	31	27	17	25	Brawerman
						(1964) [*37*]
Allium porrum⁺	Cytoplasma	24	31	21	24	Ehring (1963)
	Chloroplasten	23	32	22	24	[*35*], Ruppel
						(1964) [*95*]
Chenopodium	Cytoplasma	25	30	24	21	Odintsova,
album	Chloroplasten	25	29	25	21	Golubeva u.
Phaseolus vul-	Cytoplasma	25	29	25	21	Sissakian
garis	Chloroplasten	26	29	24	21	(1964) [*80*]

Ribosomen enthalten normalerweise zwei RNA-Typen von 28 S und 18 S. Das ist auch in Cytoplasma-Ribosomen von pflanzlichen Geweben (Keimlinge, Wurzeln, etiolierte Sprosse, Blütenstände) der Fall, wobei sich beide RNA-Typen in ihrer Basenzusammensetzung unterscheiden können [*89*]. Für Chloroplastenribosomen höherer Pflanzen fehlt dieser Nachweis noch völlig. In *Euglena* enthalten die Chloroplastenribosomen zwei RNA-Komponenten von 14 S und 19 S; Cytoplasmaribosomen stellen eine einheitliche Fraktion dar mit einer mittleren Sedimentationskonstanten von 19 S. Die Basenzusammensetzung der beiden RNA-Typen aus Chloroplasten soll gleich sein, aber von der RNA der Cytoplasmaribosomen deutlich verschieden [*20*].

Die Ribosomen aus *Euglena*-Chloroplasten unterscheiden sich auch in ihrem Bindungsverhalten zur Template-RNA von den Ribosomen des Cytoplasmabereichs [*37*]. Darauf wird später (S. 266) noch näher einzugehen sein.

Wie Cytoplasma- und Bakterienribosomen zerfallen Ribonucleoproteinpartikel aus Chloroplasten bei herabgesetzter Magnesiumkonzentration in Untereinheiten. Aus den bisher verfügbaren experimentellen Daten geht hervor, daß die Zerfallprodukte sehr wahrscheinlich 30 S- und 50 S-Untereinheiten sind. Gelegentlich beobachtete 18 S-Banden im Schlierenbild der Ultrazentrifuge sind wohl als adsorbierte lösliche Proteine anzusehen.

Polyribosomen (Polysomen) sind in Bakterien und tierischen Zellen mehrfach als Strukturen von 5—40 Einzelribosomen beschrieben worden, die durch einen Faden von Template-(oder Messenger-)RNA untereinander verbunden sind. Die Polyribosomen gelten als die aktiven Formen bei der Biosynthese eines Polypeptids. Auf den Mechanismus der Reaktion kann hier nicht eingegangen werden. — *Brassica*-Blattzellen enthalten im Cytoplasma [*29*] und auch in Chloroplasten [*28*]

Polysomenstrukturen, wie aus Untersuchungen der Sedimentations-
profile in der Ultrazentrifuge hervorgeht. Nach diesen Beobachtungen ist
die Aggregation von Einzelribosomen zu polymeren Strukturen im Licht
gefördert; bei Dunkelheit herrscht Zerfall vor. Dieses Wechselspiel
ist bei den Chloroplastenribosomen stärker ausgeprägt als bei den
80 S-Ribosomen im Cytoplasma. Die Autoren [28, 29] vermuten deshalb
einen engen Zusammenhang zwischen lichtstimulierter Neubildung einer
hochmolekularen RNA, die als Template für die gesteigerte Proteinbiosynthese fungiert. Der Versuch, Polyribosomen auch in Tabak-
Chloroplasten nachzuweisen, war jedoch erfolglos [10]. Zur endgültigen
Sicherung des Polysomenvorkommens in Chloroplasten fehlen bisher
weitere experimentelle Befunde.

D. Proteinsynthese in Plastiden

In den Plastiden sind somit alle Voraussetzungen für die Biosynthese
ihrer eigenen Proteine gegeben: Sie enthalten DNA, die verschiedenen
RNA-Typen und Ribosomen bzw. Polyribosomen als spezifische Strukturen der Proteinsynthese. Es gilt nun zu prüfen, ob auch die Chloroplastenproteine nach dem gleichen Mechanismus gebildet werden, den
wir von anderen Systemen her kennen. Unsere Kenntnisse darüber sind
jedoch noch sehr lückenhaft, wie aus der folgenden Darstellung hervorgeht.

I. Proteingehalt der Chloroplasten

Der Gesamtproteingehalt der Plastiden variiert nicht nur von Art
zu Art, sondern er wird auch vom Entwicklungszustand bestimmt und
hängt darüber hinaus in starkem Maße von der gewählten Methode zur
Isolation der Organellen ab.

Von den beiden hauptsächlich benutzten Verfahren zur Isolierung, nämlich
a) in wäßrigen, salz- oder zuckerhaltigen, annähernd isotonischen Medien und b) in
organischen Lösungsmitteln nach Lyophilisierung der Blattgewebe, ist dem zweiten
der Vorrang einzuräumen, wenn es gilt, Gehaltsbestimmungen an pufferlöslichen
Substanzen in Chloroplasten durchzuführen. Wäßrig isolierte Chloroplasten verlieren in der Regel einen großen Teil ihrer Matrixproteine während des Homo-
genisierens und Waschens. Für die meisten biochemischen Experimente sind nicht-
wäßrig isolierte Chloroplasten jedoch nicht oder wenig aktiv, und auch der Einwand
einer Adsorption von Cytoplasmaproteinen an den Chloroplasten ist nicht leicht zu
widerlegen, auch wenn der prozentuale Anteil gering sein mag.

Diese Schwierigkeiten in der Isolationsmethodik mögen denn auch der
Grund dafür sein, daß erstaunlich wenige Bestimmungen über den
Proteingehalt sauberer Chloroplasten vorliegen. Mit organischen Lösungsmitteln aus *Vicia*- und Getreideblättern gewonnene Chloroplasten
enthalten mehr als 50% der gesamten Zellproteine [45]. Während aus
Tabakblättern mit NaCl-Phosphatpuffer isolierte Chloroplasten nur
noch 25—30% des Gesamtproteingehaltes aufweisen [84], können mit
Saccharose-Boratpuffer gewonnene *Oenothera*-Chloroplasten bis zu 75%
der Blattproteine enthalten [120]. 45—55% der Chloroplastenproteine
liegen als Lipoproteinkomplexe vor [64, 76] und gehören zum lamellaren
Bereich der Organellen; etwa 50% sind als pufferlösliche Matrixproteine
anzusehen [45].

Diese löslichen Proteine stellen das Sammelbecken für die Enzyme der im Chloroplasten ablaufenden Dunkelreaktionen dar. Nicht nur Nucleinsäure-, Protein- und Lipidstoffwechsel finden hier statt, sondern auch die ohne Licht katalysierten Prozesse der Photosynthese bis zur Stärkebildung. Die Enzyme der Lichtreaktionen hingegen dürften in oder an den Granalamellen bzw. deren submikroskopischen Untereinheiten zu lokalisieren sein [76, 83]. Ausschließlich auf Chloroplasten beschränkt erscheinen Carboxydismutase, Ribulokinase und die Enzyme des photosynthetischen Elektronentransports, die NADP-abhängige 3-Phosphoglycerat-Dehydrogenase und eine alkalische Fructose-Diphosphatase, während die übrigen Enzyme des intermediären Kohlenhydratstoffwechsels teilweise oder ganz im Cytoplasma lokalisiert sind [107].

II. Proteinsynthese *in vivo*

Die Chloroplasten als spezifische Träger der Photosyntheseprozesse verdanken als einzige Organellen der Blattzellen auch ihre Entwicklung dem Licht. Sie gewinnen dadurch gegenüber allen anderen subcellulären Strukturen eine einzigartige Stellung: die starke Förderung der Substanzsynthesen, besonders auch der Proteine, bei Belichtung. Es ist noch nicht entschieden, ob die zugeführte elektromagnetische Energie in ihren ersten Umwandlungsprodukten zu chemischer Energie oder das Überangebot von Bausteinen für die Synthesesteigerungen der Makromoleküle verantwortlich sind, vielleicht beides.

Die erhöhte Synthese der Chloroplastenproteine im Licht ist mehrfach experimentell untersucht worden. Besonders wenn etiolierte Gewebe oder Zellen nach Belichtung ergrünen, steigt der Proteingehalt der Plastidenfraktion parallel zum Chlorophyllgehalt in den ersten Stunden linear an [22, 31, 73, 92]. Das ist leicht verständlich, da die wachsenden Plastiden zur Ausbildung ihrer Membranen Nettosynthesen benötigen. Es ist sehr wahrscheinlich, daß dieser Massensynthese von Proteinen lichtinduzierte Neubildungen von Nucleinsäuren und in begrenztem Maße auch von Proteinen vorausgehen, da Ribosomen in Proplastiden vorhanden sind [107]. Wie und wo diese Induktionen durch Belichtung ausgelöst werden, ist noch unbekannt, obwohl Vermutungen darüber nicht fehlen [115]. — Auch die Synthesen einer Reihe von Photosynthese-Enzymen wie NADP-abhängige Glycerinaldehyd-Dehydrogenase, Ribulose-1,5-Diphosphat-Carboxylase oder Ferredoxin werden im Licht induziert oder verstärkt, wie an den Aktivitätszunahmen gemessen werden

Tabelle 5: *Hemmung der ^{35}S-Methionin-Inkorporation in die Proteine verschiedener subcellulärer Fraktionen in Licht und Dunkelheit durch Chloramphenicol.* Die spezifischen Aktivitäten der Proteine (Imp $\times$ 10^2/min/mg Protein) werden nach Homogenisieren der Blattgewebe von *Nicotiana rustica* und Auftrennung der Fraktionen durch Differentialzentrifugation bestimmt (nach [85])

Fraktion	Spezifische Aktivitäten der Proteine				% Hemmung zu den Wasserkontrollen	
	Licht		Dunkelheit			
	Wasser	Chloramph.	Wasser	Chloramph.	Licht	Dunkelheit
7 min 1000 g (Chloroplasten)	44,5	12,4	24,9	15,9	72	36
20 min 20 000 g (Mitochondrien)	70,0	36,4	59,1	39,5	48	33
60 min 120 000 g (Ribosomen)	34.3	20,9	32,9	21,8	39	32
Überstand (lösliche Proteine)	24,0	14,4	22,2	14,6	40	34

kann [*21, 47, 44, 57, 69, 107*]. Durch Einsatz von spezifischen Inhibitoren der Proteinsynthese kann auch meist entschieden werden, daß die Erhöhung des Substratumsatzes nicht auf der Aktivierung eines vorhandenen Enzymproteins, sondern auf dessen *de novo*-Synthese beruht.

In den Chloroplasten ausgewachsener Blätter von *Nicotiana rustica* findet im Licht gleichfalls verstärkter Aminosäure-Einbau in die Proteine statt. Mit Hilfe eines Hemmstoffs der Proteinsynthese, des Chloramphenicols, können wir nachweisen, daß der ^{35}S-Methionin-Einbau in jenen Proteinfraktionen der Zelle nahezu vollkommen gehemmt wird, die im Licht gegenüber Dunkelheit vermehrt synthetisiert erscheinen [*85*]. Das sind vor allen Dingen die Proteine aus der Chloroplastenfraktion (Tab. 5).

Mehrere Beobachtungen deuten darauf hin, daß die Membranproteine im Licht gegenüber den Matrix- oder Stromaproteinen schneller markiert werden, wenn die Chloroplasten $^{14}CO_2$ photosynthetisch verwerten [*46, 84, 89*]. Die Radioaktivität tritt in kurzzeitigen Experimenten nur in den Aminosäuren Alanin, Glycin, Serin und Asparaginsäure der Proteine auf, in Aminosäuren also, die unmittelbar vom CO_2-Reduktions-

Tabelle 6: *Spezifische Aktivitäten* (Imp/min/mg Protein) *verschiedener Proteinfraktionen aus isolierten Chloroplasten von Nicotiana rustica nach 5 min* $^{14}CO_2$*-Photosynthese der Blätter* (nach [*84*])

Fraktion	Sedimentation bei		mg Protein pro mg Chlorophyll	Spez. Aktivität der Proteine
	(min)	(g)		
ganze Chloroplasten (einmal gewaschen)	8	1 000	5,48	920
Fragmente I . . .	8	1 000	6,38	1900
Fragmente II . . .	12	3 000	4,56	900
Fragmente III . .	15	6 000	4,44	690
Fragmente IV . .	30	20 000	4,12	700
Chloroplasten-Ribosomen . . .	60	120 000	26,8	290
lösliche Chloroplasten-Proteine	Überstand		∞	240
Cytoplasma-Prot. .	Überstand nach 60 min 120 000 g des Homogenats		—	430

cyclus abgeleitet werden können. In Tab. 6 sind die spezifischen Aktivitäten (Imp./min./mg Protein) einiger Fraktionen aus osmotisch aufgebrochenen Chloroplasten angegeben, die wir aus Tabakblättern nach 5 min $^{14}CO_2$-Photosynthese isolierten. Es ist deutlich zu erkennen, daß die Membranfragmente (Granafraktionen) höhere Werte aufweisen als die Fraktionen der löslichen Chloroplastenproteine oder der löslichen Cytoplasmaproteine [*84*].

Aus den vorliegenden Beobachtungen dieser *in vivo*-Experimente ergibt sich die noch ungeklärte Frage nach der Synthese von Strukturproteinen in Chloroplasten. Wie wir oben erwähnt haben, sind nach den elektronenmikroskopischen Bildern die Ribosomen im Matrix-Bereich der Chloroplasten lokalisiert. Hier allein sollte nach den heutigen Vorstellungen die Synthese der Chloroplastenproteine stattfinden. Da die Strukturproteine im Licht jedoch weit schneller markiert erscheinen als die Matrixproteine, kann zunächst nicht ausgeschlossen werden, daß die Markierung der ersten auch über Wege möglich ist, die mit den geltenden Auffassungen über die Synthesereaktionen an den Ribosomen nicht identisch sind. Nach [7, 101, 108] sollen bei der Aminosäure-Inkorporation in die Chloroplastenproteine Lipidverbindungen eine wichtige Rolle spielen. Lipoproteine oder -peptide erscheinen am schnellsten markiert und geben die Radioaktivität auch wieder schnell an Proteine ab.

III. Proteinsynthese *in vitro*

Von den einzelnen Reaktionen der Polypeptidbiosynthese im zellfreien System ist der erste Schritt, die anhydridartige Verknüpfung eines Aminosäurerestes mit dem Adenylrest des ATP unter Pyrophosphat-Abspaltung (die sog. Aminosäure-Aktivierung), in Chloroplasten mehrfach untersucht worden, am eingehendsten von Bové und Raacke [14], sowie von Marcus [68]. Danach scheinen Unterschiede in der Enzymaktivität für bestimmte Aminosäuren in Chloroplasten gegenüber dem Cytoplasma zu bestehen. Die Aktivität ist nur im löslichen Chloroplastenextrakt vorhanden, sie ist nicht strukturgebunden [14].

Die nächsten Schritte, Kopplung der „aktivierten" Aminosäuren an eine Transfer-RNA und die Übertragung auf die Chloroplastenribosomen, sind bisher noch nicht näher studiert worden, von einer Ausnahme abgesehen [105]. Die Aussagekraft solcher Experimente ist besonders stark von der Sauberkeit der benutzten Chloroplastenfraktion abhängig, da bereits geringe Cytoplasma-Verunreinigungen zu Fehlinterpretationen führen können.

Schon vor den ersten Experimenten zur Aminosäure-Aktivierung ist die Inkorporation von radioaktiv markierten Aminosäuren in die Proteine isolierter Chloroplasten als Maß für deren Synthesekapazität erkannt worden [102, 110]. Diese Versuche wurden später von den verschiedensten Arbeitsgruppen erfolgreich wiederholt [2, 3, 18, 36, 37, 38, 42, 84, 86, 103, 104, 105, 108, 109]. Wie Tab. 7 zeigt, sind die Chloro-

Tabelle 7: *In vitro-Einbau von Valin-1-¹⁴C in die Proteine verschiedener subcellulärer Fraktionen von jungen Pisum-Blättern, ohne Zusatz von Cofaktoren, Enzymen oder Substraten.* Inkubation 60 min bei 25° C in Dunkelheit (nach [86])

Bezeichnung der Fraktion	Spez. Aktivität der Proteine (Imp/min/mg)
Chloroplasten, einmal gewaschen	3600
Chloroplastenfragmente	3400
Chloroplastenribosomen	940
Mitochondrien (+ Chloroplastenfragmente).	1400
Cytoplasmaribosomen	150
Cytoplasmatischer Überstand.	0
Chloroplasten + Überstand zugesetzt . . .	120
Chloroplastenribosomen + Überstand zuges.	0

plasten allen anderen subcellulären Fraktionen überlegen, wenn die Experimente ohne zugesetzte Cofaktoren, Enzyme und Substrate der Proteinbiosynthese durchgeführt werden. Im partikelfreien Überstand befindet sich sogar ein Inhibitor der Aminosäure-Inkorporation [84, 105]. Die Tatsache des starken Einbaus in die Proteine der Chloroplastenfraktionen kann am einfachsten damit erklärt werden, daß alle notwendigen Substanzen dafür in den Plastiden vorhanden sind, einschließlich Transfer- und Template-RNA. Aber auch Chloroplastenfragmente sind noch relativ aktiv, im Gegensatz zur Fraktion der Chloroplastenribosomen, wenn ohne Zusätze inkubiert wird [37, 86].

Allerdings ist die Beteiligung von Bakterien bei den Chloroplasten-Inkorporationsversuchen wegen der ähnlichen Sedimentationseigenschaften nicht immer auszuschließen [3]: Darauf deuten ein gelegentlich über Stunden linear ansteigender Einbau und die fehlende Hemmung durch RNase im Chloroplastensystem hin (Bakterienmembranen sind für RNase impermeabel). Dagegen sprechen jedoch die Einbau-Stimulierung durch Licht bei den für Bakterienwachstum relativ ungünstigen Inkubationstemperaturen von 20—25° C und auch die Unwirksamkeit bestimmter stark bactericider Antibiotica wie Streptomycin [47].

Eine Übersicht über die Erfordernisse bzw. über die Wirkungen von Hemmstoffen bei der Aminosäure-Inkorporation in die Proteine ganzer Chloroplasten *in vitro* gibt Tab. 8, zusammengestellt aus den Experimenten verschiedener Autoren. Bei der Heterogenität von Ausgangsmaterial, Methodik und eingesetzter Konzentrationen ist eine völlige Übereinstimmung nicht zu erwarten. Es wird aber deutlich, daß spezifische Hemmstoffe wie Chloramphenicol und Puromycin stets wirkungsvolle Inhibitoren sind, daß der Zusatz eines nicht markierten Gemisches von proteinogenen Aminosäuren in der Regel unwirksam ist, Magnesium notwendig erscheint und zwischen ATP-Erfordernis und Lichtwirkung offensichtlich eine gegenseitige Ersetzbarkeit besteht (vgl. [37, 108a]). Die Frage nach der RNA-Beteiligung ist noch zu vage und unregelmäßig gestellt worden, als daß man hieraus schon bindende Schlüsse ziehen kann. Werden synthetische Polynucleotide wie Poly-U zu ganzen Chloroplasten oder Chloroplastenribosomen zugesetzt [9, 36, 37, 85, 109], so wird zwar in den meisten Fällen eine leichte, 2—3 fache Stimulierung der Phenylalanin-Inkorporation erreicht, sie steht jedoch in keinem Verhältnis zu den an Bakterien-Ribosomen beobachteten Effekten. Nach BISWAS [9] tritt diese schwache Stimulierung erst auf, wenn die Chloroplastenribosomen mit RNase präinkubiert werden, wobei die endogene Template-RNA von den Ribosomen abgespalten wird, und die Bindungsorte an den Ribosomen für die Anheftung des Poly-U frei werden.

Unklar ist auch noch die Wirkung von Actinomycin auf die Proteinsynthese im isolierten Chloroplasten. Während Plastiden aus Grünalgen einen deutlichen Abfall der Aminosäure-Inkorporation nach Hemmstoffzusatz zeigen, weisen die Chloroplasten aus höheren Pflanzen keine kurzfristigen Einbauhemmungen auf. Möglicherweise widerspiegelt sich hier ein Unterschied in der Synthese bestimmter RNA zwischen Chloroplasten aus aktiv wachsenden, sich vermehrenden und bereits ausgewachsenen Zellen, bzw. zwischen der Lebensdauer des Messengers.

Ein direkter Nachweis dafür, daß der Chloroplasten-DNA eine genetische Rolle bei der Proteinsynthese der Chloroplasten zukommt, liegt bisher noch nicht vor. Der Aminosäure-Einbau in die Chloroplastenproteine aus längere Zeit kernlos gehaltener *Acetabularia* [42] ist zwar ein starker Hinweis gegen eine Kern-DNA-abhängige Proteinsynthese in solchen Chloroplasten, es ist aber noch kein Beweis für eine „Steuerung" durch Chloroplasten-DNA.

Sind die Versuchsergebnisse mit ganzen Chloroplasten aus den oben erwähnten Gründen nicht immer eindeutig als chloroplasteneigene

Tabelle 8: *Zusammenstellung über Erfordernisse und Hemmstoffe der Aminosäure-Inkorporation in isolierten Chloroplasten.* + = wirksam (stimulierender bzw. hemmender Effekt), — = ohne Wirkung

Pflanzenspecies	Licht	ATP	Mg^{++}	Amino-säure-Gemisch	RNA[1] oder Poly-U[2]	Einbauhemmung durch				Literatur
						Chloram-phenicol	Puro-mycin	Actino-mycin	RNase	
Nicotiana tabacum	+	—	+	—					—	[110]
Spinacia oleracea	+	—	+	—		+	+		—	[2, 3]
Nicotiana tabacum	—	+	+	+	—[1] +[2]	+	+	—	+	[109]
Acetabularia	—	—	+			+	+	+	—	[42]
Euglena gracilis	—(+)	+(—)	+		+[1] +[2]	+	+	+	+	[36, 37]
verschiedene Species	+	—	+	—			+			[101, 104]
Pisum sativum						+	+	—	+	[105]
Nicotiana, Spinacia	+	—	+	—	—[2]	+	+	—		[84, 85]
Spinacia oleracea		+	+	+	+[2]	+	+	—	+	[108a]

Proteinsynthese zu interpretieren, so beweisen erfolgreiche Experimente mit Ribosomenfraktionen aus Chloroplasten die Spezifität der Proteinsynthese in Plastiden recht deutlich. Besonders aus *Euglena*-Chloroplasten hat Brawerman [17, 18, 36, 37] aktive Ribosomen gewonnen, obwohl sie nicht die hohen Inkorporationswerte der ganzen Chloroplasten oder der Cytoplasmaribosomen erreichen. Die Proteinsynthese-Eigenschaften der Chloroplastenribosomen unterscheiden sich in mancher Hinsicht von denen ganzer Chloroplasten oder Cytoplasmaribosomen. Der wichtigste Unterschied besteht darin, daß ein Zusatz von Template-RNA die Aminosäure-Inkorporation nur im System mit Chloroplastenribosomen stimuliert. Diese Beobachtung kann nur dadurch verständlich werden, daß ganze Chloroplasten genügend Template-RNA (*„Messenger"*) für die Polypeptidbildung besitzen, die beim Prozeß der Ribosomen-Isolierung ins Medium verloren geht. Der Zusatz von Template-RNA stimuliert denn auch den Aminosäure-Einbau mit Chloroplastenribosomen um das Zehnfache [37]; zugegebene Ribosomen-RNA zeigt diese Wirkung nicht. Eine leichte Stimulierung der Proteinsynthese bei Zusatz von Triphosphaten und die entsprechende Hemmung durch Actinomycin weisen darauf hin, daß ganze *Euglena*-Chloroplasten in gewissem Ausmaße zur Neusynthese von Template-RNA fähig sind.

Der Unterschied zwischen Chloroplasten- und Cytoplasmaribosomen aus *Euglena* bezüglich ihres Verhaltens zur Template-RNA wird auf die ungleichen Bindungseigenschaften der beiden Ribosomenarten gegenüber dem RNA-Molekül zurückgeführt. Mit Detergentienbehandlung

(0,5% Desoxycholat) kann die Template-RNA bedeutend leichter von den Chloroplastenribosomen freigesetzt werden als von Cytoplasmaribosomen [37].

Die Beweiskette vom Vorkommen funktionsfähiger DNA bis zur Synthese spezifischer Enzymproteine in Plastiden ist experimentell noch nicht lückenlos geschlossen. Von biochemischer Seite scheinen jedoch alle Voraussetzungen für eine *genetische Autonomie der Plastiden* gegeben zu sein. Das bedeutet aber nicht, daß die cytoplasmatischen Organellen vom Zellkern völlig unabhängig sein müssen. Noch bleibt zu untersuchen, wie weit neben den plastideneigenen Genen auch noch Gene des Zellkerns die Vorgänge innerhalb der Plastiden steuern oder beeinflussen. Nach einer Hypothese von GIBOR und GRANICK [41] könnte es in den Chloroplasten konstitutive und induzierbare Gene geben. Die konstitutiven Gene wären verantwortlich für die Reduplikation der Organellen und für einige biochemische Prozesse, während andere Vorgänge wie die Differenzierung induzierbar sind und dementsprechend von Genen des Regulator-Typs einer Operon-Einheit kontrolliert werden. Es wäre denkbar, daß auf die induzierbaren Gene der Plastiden auch solche Faktoren wie einige notwendige Metabolite, bestimmte Aktivatoren oder Inhibitoren einwirken, die von den Genen des Zellkerns kontrolliert werden.

Größere Schwierigkeiten experimenteller und materialbedingter Art müssen noch überwunden werden, bis unsere Kenntnisse über die Plastiden so umfangreich sind wie über Bakterien und einige tierische Systeme.

Summary

Nucleic acids and protein synthesis in plastids

Since it is known that plastids are genetically autonomous structures, which are formed by division of existing plastids, it seems valuable also from the biochemical point of view, to relate the existing results on nucleic acids and protein synthesis in plastids with the corresponding facts and theories known from other biological material.

DNA in plastids has been detected by cytochemical, biochemical and ^{3}H-thymidine incorporation studies. The content has been estimated as 10^{-11} to 10^{-12} mg DNA per chloroplast (see Table 1). These values are within the range of magnitude of the DNA content in bacteria, and the amount is sufficient to code the amino acids of at least several hundred proteins with a molecular weight of 20,000. The DNA of chloroplasts seems to have a double helical structure. Its buoyant density has been determined as 1.684 to 1.695 g/cm³ from algae chloroplasts and as 1.700 to 1.705 g/cm³ from isolated chloroplasts of higher plants. These data are significantly different from the corresponding nuclear DNAs (1.707 to 1.725 and 1.690 to 1.685, respectively). The differences in the buoyant densities have a parallel in the differences of the base compositions betweeen the DNA of chloroplasts and the corresponding nuclei (see Table 2). A difference exists also in the T_m values (78—80° and 89 to 91°C, respectively). Furthermore, ^{3}H-thymidine incorporation experiments *in vivo* suggest an autonomous metabolism of the chloroplast DNA.

RNA has been localized in plastids by various methods; however, great differences have been reported from quantitative analysis due to the method choosen and due to the loss of soluble RNA in aqueous media. RNA synthesis in chloroplasts has been confirmed with precursor studies both *in vivo* and *in vitro*. Although there is until now no definite evidence of a messenger RNA in chloroplasts (in the sense of a bacterial messenger), DNA-dependent RNA synthesis and the existence of a RNA with template activity have been demonstrated.

Chloroplasts contain ribosomes in their matrix areas. The ribonucleoprotein particles show sedimentation values of 65—70 S, in contrast to the 75—83 S values of the corresponding ribosomes in the cytoplasm. The base composition of ribosomal RNA from chloroplasts seems to be the same as in the RNA of cytoplasmic ribosomes, at least in higher plants (see Table 4). There are also indications that polyribosomes occur in plastids.

About 50% of the total protein content of a green cell is located in the chloroplasts and ca. 50% of the chloroplast protein is buffer-soluble protein. The *in vivo* protein synthesis in plastids is strongly stimulated by light, as has been shown by quantitative analysis, incorporation of radioactive precursors, and inhibitor studies. *In vitro* experiments with isolated chloroplasts demonstrate amino acid activation, aminoacyl binding to a soluble RNA fraction, and the incorporation of radioactive amino acids into chloroplast proteins (see Table 8). Whole chloroplasts seem to contain most of the cofactors necessary for a normal protein synthesis, but isolated ribosomes from *Euglena* chloroplasts are unable to incorporate amino acids into proteins without addition of cofactors. This system is also sensitive to RNase and actinomycin.

Although many of the facts mentioned in this paper suggest a scheme of protein synthesis similar to that known from bacteria or animal systems, there is no direct proof until now that the identical mechanism exists in plastids. There are still some important missing links in the chain from plastid DNA (gene) function to the synthesis of a specific chloroplast protein. Furthermore, it must be kept in mind that, besides the genetic autonomy of the plastids, certain relevant correlations between the plastids and the nucleus or cytoplasm, respectively, should be suggested.

Literatur

[1] Allison, A. C., and D. C. Burke: J. gen. Microbiol. 27, 181 (1962).
[2] App, A. A., and A. T. Jagendorf: Biochim. biophys. Acta (Amst.) 76, 286 (1963).
[3] — — Plant Physiol. 39, 772 (1964).
[4] Baltus, E., and J. Brachet: Biochim. biophys. Acta (Amst.) 76, 490 (1963).
[5] Barner, H., and S. S. Cohen: J. Bact. 72, 115 (1956).
[6] Bell, P. R., and K. Mühlethaler: J. molec. Biol. 8, 853 (1964).
[7] Besinger, E. N., M. J. Molchanov, and N. M. Sissakian: Biochimija 29, 749 (1964).
[8] Biggins, J., and R. B. Park: Nature (Lond.) 203, 425 (1964).
[9] Biswas, S., and B. B. Biswas: Experientia (Basel) 21, 251 (1965).
[10] Boardman, N. K., R. B. Francki, and S. G. Wildman: Biochemistry 4, 872, (1965).
[11] Böttger, I., u. I. Lüdemann: Flora 155, 331 (1964).
[12] —, u. R. Wollgiehn: Flora 146, 302 (1958).
[13] Bogorad, L., and A. B. Jacobson: Biochem. biophys. Res. Commun. 14, 113 (1964).
[14] Bové, J., and I. D. Raacke: Arch. Biochem. 85, 521 (1959).
[15] Brachet, J.: Exp. Cell Res. 6, 78 (1959).
[16] — in: Progr. Biophys. molec. Biol. 15, 99 (1965).
[17] Brawerman, G.: Biochim. biophys. Acta (Amst.) 61, 313 (1962).
[18] — Biochim. biophys. Acta (Amst.) 72, 317 (1963).
[19] —, and J. M. Eisenstadt: Biochim. biophys. Acta (Amst.) 91, 477 (1964).
[20] — — J. molec. Biol. 10, 403 (1964).
[21] —, and N. Konigsberg: Biochim. biophys. Acta (Amst.) 43, 374 (1960).
[22] —, A. O. Pogo, and E. Chargaff: Biochim. biophys. Acta (Amst.) 55, 326 (1959).
[23] Budd, G. C., and G. M. Mills: Nature (Lond.) 205, 524 (1965).
[24] Caldwell, P. C., and C. Hinshelwood: J. chem. Soc. 1950, 1115.
[25] Chiba, Y.: Cytologia (Tokyo), 16, 259 (1951).
[26] —, and K. Sugahara: Arch. Biochem. 71, 367 (1957).

[27] Chun, E. H., M. H. Vaugham, and A. Rich: J. molec. Biol. 7, 130 (1963).
[28] Clark, M. F.: Biochim. biophys. Acta (Amst.) 91, 671 (1964).
[29] — R. E. F. Matthews, and R. K. Ralph: Biochim. biophys. Acta (Amst.) 91, 289 (1964).
[30] Cooper, W. D., and H. S. Loring: J. biol. Chem. 228, 813 (1957).
[31] de Deken-Grenson, M.: Biochim. biophys. Acta (Amst.) 14, 203 (1954).
[32] Dure, L., and L. Waters: Science 174, 410 (1965).
[33] Edelman, M., H. T. Epstein, and J. A. Schiff: Proc. nat. Acad. Sci (Wash.) 52, 1214 (1964).
[34] — J. A. Schiff, and H. T. Epstein: J. molec. Biol. 11, 767 (1965).
[35] Ehring, R.: Z. Naturforsch. 17 b, 837 (1963).
[36] Eisenstadt, J., and G. Brawerman: Biochim. biophys. Acta (Amst.) 76, 319 (1963).
[37] — — J. molec. Biol. 10, 392 (1964).
[38] Francki, R. B. I., N. V. Boardman, and S. G. Wildman: Biochemistry 4, 865 (1965).
[39] Gibor, A., and M. Izawa: Proc. nat. Acad. Sci. (Wash.) 50, 1164 (1963).
[40] —, and S. Granick: J. Cell Biol. 15, 599 (1962).
[41] — — Science 145, 890 (1964).
[42] Goffeau, A., and J. Brachet: Biochim. biophys. Acta (Amst.) 95, 302 (1965).
[43] Hageman, R. H., and D. I. Arnon: Arch. Biochem. 57, 421 (1955).
[44] Hall, D. O., R. C. Huffaker, L. M. Shannon, and A. Wallace: Biochim. biophys. Acta (Amst.) 35, 540 (1959).
[45] Heber, U.: Z. Naturforsch. 15 b, 95 (1960).
[46] — Nature (Lond.) 195, 91 (1962).
[47] — Planta 59, 600 (1963).
[48] Holden, M.: Biochem. J. 51, 433 (1952).
[49] Hsiao, T. C.: Biochim. biophys. Acta (Amst.) 91, 598 (1964).
[50] Ingle, J., J. L. Kay, and R. E. Holm: J. molec. Biol. 11, 730 (1965).
[51] Iwamura, T.: Biochim. biochem. Acta (Amst.) 42, 161 (1960).
[52] — Biochim. biophys. Acta (Amst.) 61, 427 (1962).
[53] —, and S. Kuwashima: Biochim. biophys. Acta (Amst.) 80, 678 (1964).
[54] Jacob, F., and J. Monod: J. molec. Biol. 3, 318 (1961).
[55] Jacobson, A. B., H. Swift, and L. Bogorod: J. Cell Biol. 17, 557 (1963).
[56] Jagendorf, A. T., and S. G. Wildman: Plant Physiol. 29, 270 (1954).
[57] Keister, D. L., A. T. Jagendorf, and A. San Pietro: Biochim. biophys. Acta (Amst.) 62, 332 (1962).
[58] Kern, H.: Protoplasma (Wien) 50, 505 (1959).
[59] Kirk, J. T. O.: Biochim. biophys. Acta (Amst.) 76, 417 (1963).
[60] — Biochem. biophys. Res. Commun. 14, 393 (1964).
[61] — Biochem. biophys. Res. Commun. 16, 233 (1964).
[62] Kislev, N., H. Swift, and L. Bogorad: J. Cell Biol. 25, 327 (1965).
[63] Leff, I., M. Mandel, H. T. Epstein, and J. A. Schiff: Biochem. biophys. Res. Commun. 13, 126 (1963).
[64] Lichtenthaler, H. K., and R. B. Park: Nature (Lond.) 198, 1070 (1963).
[65] Loening, U. E.: Nature (Lond.) 195, 467 (1962).
[66] Lyman, H., H. T. Epstein, and J. A. Schiff: Biochim. biophys. Acta (Amst.) 50, 301 (1961).
[67] Lyttleton, J. W.: Exp. Cell Res. 26, 312 (1962).
[68] Marcus, A.: J. biol. Chem. 234, 1238 (1959).
[69] Margulies, M. M.: Plant Physiol. 39, 579 (1964).
[70] McCalla, D. R., and R. K. Allan: Nature (Lond.) 201, 504 (1964).
[71] McClendon, J. A.: Amer. J. Bot. 39, 275 (1952).
[72] Mego, J. L.: Biochim. biophys. Acta (Amst.) 79, 221 (1964).
[73] —, and A. T. Jagendorf: Biochim. biophys. Acta (Amst.) 53, 237 (1961).
[74] Menke, W.: Z. physiol. Chem. 252, 43 (1938).
[75] — Z. Vererbungsl. 91, 152 (1960).
[76] — Ann. Rev. Plant Physiol. 13, 27 (1962).
[77] Metzner, H.: Biol. Zbl. 71, 257 (1952).

[78] Mikulska, E. J., M. S. Odintsova, and N. M. Sissakian: Biochimija (Moskau) 27, 1061 (1962).
[79] Murakami, S.: Exp. Cell Res. 32, 398 (1963).
[80] Odintsova, M. S., E. V. Golubeva, and N. S. Sissakian: Nature (Lond.) 204, 1090 (1964).
[81] Olszewska, M. J., and E. Mikulska: Experientia (Basel) 20, 267 (1964).
[82] Orth, G. M., and D. G. Cornwell: Biochim. biophys. Acta (Amst.) 71, 734 (1963).
[83] Park, R. B., and N. G. Pon: J. molec. Biol. 3, 1 (1961).
[84] Parthier, B.: Z. Naturforsch. 19 b, 235 (1964).
[85] — Z. Naturforsch. (im Druck).
[86] —, u. R. Wollgiehn: Naturwissenschaften 50, 598 (1963).
[87] Pogo, G. T., and A. O. Pogo: J. Cell Biol. 22, 296 (1964).
[88] Pollard, C. G.: Arch. Biochem. 105, 114 (1964).
[89] — Biochem. biophys. Res. Commun. 17, 171 (1964).
[90] Ray, D. S., and P. C. Hanewalt: J. molec. Biol. 9, 812 (1964).
[91] — — J. molec. Biol. 11, 760 (1964).
[92] Rhodes, M. J. C., and E. W. Yemm: Nature (Lond.) 200, 1077 (1963).
[93] Richter, G., u. H. Senger: Biochim. biophys. Acta (Amst.) 95, 362 (1965).
[94] Ris, H., and W. Plaut: J. Cell Biol. 13, 383 (1962).
[95] Ruppel, H. G.: Biochim. biophys. Acta (Amst.) 80, 63 (1964).
[95a] — Z. Pflanzenphysiol. 53, 32 (1965).
[96] Sagan, L., Y. Ben-Shaul, J. Schiff, and H. T. Epstein: J. Cell Biol. 23, 81 A (1964).
[97] Sager, R., and M. R. Ishida: Proc. nat. Acad. Sci. (Wash.) 50, 725 (1963).
[98] Scher, S., and L. Sagan: J. Protozool. 2, 13 p (1962).
[99] Schweiger, H. G., u. S. Berger: Biochim. biophys. Acta (Amst.) 87, 533 (1964).
[100] Semal, J., D. Spencer, K. Y. Kim, and S. G. Wildman: Biochim. biophys. Acta (Amst.) 91, 205 (1964).
[100a] Shipp, W. S., F. J. Kieras, and R. Haselkorn: Proc. Nat. Acad. Sci. 54, 207 (1965).
[101] Sissakian, N. M., E. N. Besinger, and A. S. Marchukaitis: Dokl. Akad. Nauk SSSR, Odt. 147, 1493 (1962).
[102] —, and I. I. Filippovich: Dokl. Akad. Nauk SSSR, Odt. 102, 579 (1955).
[103] — — Biochimija (Moskau) 22, 375 (1957).
[104] — —, and E. N. Svetailo: Dokl. Akad. Nauk SSSR, Odt. 147, 488 (1962).
[105] — — — ,and K. A. Aliyev: Biochim. biophys. Acta (Amst.) 95, 474 (1965).
[106] —, and M. S. Odintsova: Biochimija (Moskau) 21, 577 (1956).
[107] Smillie, R. M.: Canad. J. Bot. 41, 123 (1963).
[108] Smirnov, B. P., and M. A. Rodionova: Biochimija (Moskau) 29, 386 (1964).
[108a] Spencer, D.: Arch. Biochem. 111, 381 (1965).
[109] —, and S. G. Wildman: Biochemistry 3, 954 (1965).
[110] Stephenson, M. L., K. V. Thimann, and P. C. Zamecnik: Arch. Biochem. 65, 194 (1956).
[111] Stocking, C. R., and E. M. Gifford: Biochem. biophys. Res. Commun. 1, 159 (1959).
[112] Spiekermann, R.: Protoplasma (Wien) 48, 303 (1957).
[113] Szarkowski, W., u. T. Golaszewski: Naturwissenschaften 48, 622 (1961).
[114] Ts'o, P. O. P.: Ann. Rev. Plant Physiol. 13, 45 (1962).
[115] Virgin, H. J., A. Kahn, and D. v. Wettstein: Photochem. Photobiol. 2, 83 (1963).
[116] Wollgiehn, R.: Diplomarbeit Halle 1958.
[117] — Flora 150, 117 (1961).
[118] —, u. K. Mothes: Naturwissenschaften 50, 65 (1963).
[119] — — Exp. Cell Res. 35, 52 (1964).
[120] Zucker, M., and H. T. Stinson: Arch. Biochem. 96, 637 (1962).

Diskussion

Vorsitz: *Hagemann*

Beermann: Sind diese 10^{-14} g DNS in einem Stück, oder sind das mehrere Stücke?

Wollgiehn: Man weiß das nicht genau, nimmt aber an, daß es sich um ein Molekül handelt. Das Molekulargewicht müßte dann etwa 10^9 betragen. Wenn man wesentlich niedrigere Werte findet, so ist an eine Degradation während der Präparation zu denken (für die DNS von E. coli gilt bekanntlich dasselbe).

Abel: Daß Sie in älteren Blättern keinen Einbau von ^{3}H-Thymidin bekommen, ist im Hinblick auf Endomitosen und die damit verbundene Chloroplastenvermehrung überraschend.

Wollgiehn: Gewiß, aber der Thymidin-Einbau ist bei älteren Blättern praktisch Null, auch wenn wir die Versuche über mehrere Tage ausdehnen. Erstaunlich ist auch, daß bei jungen Blättern Thymidin wahrscheinlich in *alle* Chloroplasten eingebaut wird, und nicht etwa nur in solche, die gerade vor einer Teilung stehen.

Tschermak-Woess: Endomitotische Polyploidisierung erfolgt ja nur in jüngeren Blättern im Anschluß an die Mitosetätigkeit.

Abel: BUTTERFASS hat gewisse Indizien für das Vorliegen von Endomitosen auch in älteren Blättern.

Tschermak-Woess: Im allgemeinen schließt das endomitotische Wachstum sofort an das mitotische an und kann sich je nach Gewebe noch verschieden lang hinziehen; es erfolgt z. B. in der Stengelepidermis von *Vicia faba* noch in gestreckten Internodien. Dennoch ist es praktisch ausgeschlossen, daß in alten Blättern noch Endomitosen stattfinden.

Hagemann: Selbst wenn noch Endomitosen stattfinden, müßte man nicht unbedingt einen Einbau bekommen. Wenn die Endomitosen in synchronen Schüben erfolgen, hinge der Erfolg stark von der Applikationszeit ab.

Karlson: Darf man annehmen, daß alle wichtigen Proteine (Enzymproteine der Photosynthese, Strukturproteine usw.) vollständig in der Chloroplasten-DNS codiert sind? Oder gibt es Hinweise dafür, daß die Information für bestimmte Enzyme im Kern liegt?

Wollgiehn: Da bestimmte Enzyme ausschließlich in Chloroplasten vorkommen, könnte man annehmen, daß sie dort gebildet werden. Experimentelle Hinweise gibt es m. W. keine.

Stubbe: Viele genetische Befunde zeigen, daß die normale Entwicklung der Chloroplasten von einem Zusammenspiel zwischen Genom und Plastom abhängt. Nur können wir nicht sagen, um welche Enzyme es sich dabei im einzelnen handelt.

Duspiva: Besteht hinsichtlich der Chloroplastenvermehrung eine Synchronisation? Vermehren sich die Chloroplasten zu einem bestimmten Zeitpunkt im Zellcyclus?

Wollgiehn: Spezielle Versuche dazu stehen noch aus. Sicher werden aber in Zellen, die sich nicht weiter teilen, noch viele Chloroplasten gebildet.

Bier: In diesem Zusammenhang ist von Interesse, daß PARSONS bei *Tetrahymena* nachgewiesen hat, daß Thymidin in eine Cytoplasmafraktion — vermutlich mitochondriale DNS — eingebaut wird, wenn im Kern keine DNS-Synthese ist und umgekehrt [J. Cell Biol. **25**, 641 (1965)].

Wollgiehn: Dem entspräche der im Vortrag erwähnte Befund von BUDD an Proplastiden der Zwiebelwurzelspitze.

Neupert: Von Untersuchungen über die Proteinsynthese in isolierten Mitochondrien ist bekannt, daß RN-ase unwirksam ist, weil sie in intakte Mitochondrien nicht eindringen kann. Es wäre also möglich, daß die Unempfindlichkeit des Systems gegen RN-ase auf der Unversehrtheit der Chloroplasten beruht und nicht auf Verunreinigung mit Bakterien zurückgeht. Vielleicht erklärt das die unterschiedlichen Befunde bei Chloroplasten aus dem gleichen Objekt.

Hagemann: Zur vorhin diskutierten Frage, ob alle Plastiden-Enzyme ausschließlich von der Plastiden-DNS codiert werden, möchte ich meinen, daß das sehr unwahrscheinlich ist. Für Plastiden-Enzyme liegen zwar entsprechende Ergebnisse noch nicht vor. Aber mir scheint zum Vergleich der an Mitochondrien erhobene

Befund wichtig, daß bestimmte Cytochrome nicht nur als Folge von plasmatischen Mutationen, sondern auch durch Genmutationen ausfallen können.

Klima: Wenn man den Einfluß der Kerngene einerseits und der Plastiden-DNS andererseits aufklären will, ist genau zu prüfen, ob der Einfluß der Kerngene nicht auf die Bereitstellung gewisser Zwischenprodukte beschränkt bleibt, die dann in spezifische Enzyme eingebaut werden (z. B. die Haemgruppe bei Cytochromen).

Beermann: Beruht der Unterschied zwischen Chloroplasten-Ribosomen und Ribosomen des Plasmas auf der RNS-Komponente oder auf dem ribosomalen Protein? Ist schon geprüft, ob sich ribosomale RNS von Chloroplasten-Ribosomen mit Chloroplasten-DNS hybridisieren läßt? Wenn man feststellen würde, daß z. B. Transfer- und ribosomale RNS von der Zelle zur Verfügung gestellt werden müßten, wäre eine zwangsläufige Symbiose zwischen beiden Elementen gegeben.

Wollgiehn: Über Unterschiede in der Proteinkomponente liegen keine Befunde vor. Zur Frage der RNS sind — außer den Untersuchungen an *Euglena*, keine weiteren Befunde in der Literatur mitgeteilt. Unsere ersten analytischen Versuche an höheren Pflanzen lassen aber erkennen, daß in den Chloroplasten neben niedermolekularer RNS noch zwei Fraktionen ribosomaler RNS vorliegen. Über ihre Herkunft ist damit noch nichts ausgesagt.

Parthier: Chloroplasten-DNS von *Euglena* hat eine kleinere Dichte als die *Euglena*-Kern-DNS; bei höheren Pflanzen ist es umgekehrt. Während bei *Euglena* die Basenzusammensetzung von Cytoplasmaribosomen-RNS und Chloroplastenribosomen-RNS verschieden ist, konnte entsprechendes bei höheren Pflanzen nicht nachgewiesen werden. Schließlich läßt sich die RNS-Synthese in den *Euglena*-Chloroplasten durch Actinomycin hemmen, nicht aber in den Chloroplasten höherer Pflanzen.

Schaller: If an incorporation of amino acids in plastids depends on ribo-nucleoside triphosphates in contrast to the incorporation due to isolated ribosomes then you could argue that there are not involved bacteria because the triphosphates could not penetrate the membrane.

Beermann: You have mentioned the studies which apparently have shown the existence of polysomes in plastids. But I have not seen any one in electron microscopic pictures demonstrated here today. Then you mentioned that the main incorporation occurs not in the ribosomal fraction, but in the fragmented membrane fraction. Could it be, that the polysomes are attached to the membranes coming down in centrifugal field?

Parthier: I think, Gunning has shown polyribosomes in plastids by electron microscopy. During homogenizing the tissue endogenous nucleases become active, rapidly attacking unprotected RNA, particularly the sensitive messenger RNA of polysomes. Therefore, a lack of polyribosomes in our fractions does not imply their non-existence.

P. Sitte: Polyribosomes are not easily seen in the electron microscope unless they are scattered on a membrane which is seen in surface view.

Miller: But you can easily see them in erythroblasts where membranes of the endoplasmic reticulum are absent.

Stoeckenius: Der Terminus „Strukturprotein" ist mehrfach gebraucht worden. Wie läßt er sich in diesem Falle definieren?

P. Sitte: „Strukturprotein" bedeutet auch hier wohl jenes Protein, das beim Aufbrechen der Chloroplasten nicht in Lösung geht. Dieses Protein ist von Menke u. Mitarb. als „lamellares Strukturproteid" näher beschrieben worden. In neuerer Zeit ist mehrfach kontraktiles Protein nachgewiesen worden.

Stoeckenius: Zahlreiche Enzyme sind an die Membranen gebunden. Das „lamellare Strukturproteid" hat aber keine Enzymaktivität. Vielleicht handelt es sich wenigstens z. T. um denaturierte Enzyme.

Meyer: Kontraktiles Protein ist mehrfach auch bei Mitochondrien gefunden worden. Insofern besteht eine interessante Parallele zwischen Plastiden und Mitochondrien.

P. Sitte: Diese Parallele geht sogar so weit, daß in beiden Fällen ein Zusammenhang zwischen Kontraktion und Intensität des Elektronentransports nachgewiesen werden konnte.

Die Plastiden als Erbträger

Von

WILFRIED STUBBE, Köln

Mit 1 Abbildung

1934 veröffentlichte OTTO RENNER in den Berichten der Sächsischen Akademie der Wissenschaften seine wohlbegründeten Ansichten über „Die pflanzlichen Plastiden als selbständige Elemente der genetischen Konstitution" [*33*]. Diejenigen, die damals angesichts der großen Erfolge der Chromosomengenetik der Lehre vom Kernmonopol der Vererbung huldigten, konnten daraus entnehmen, daß ihre Auffassung von dem genetischen System der Zelle zu eng gefaßt war. Heute, nach mehr als 30 Jahren, dürfte es weniger Zweifler am Bestehen einer plasmatischen Vererbung geben als damals; dennoch kann man nicht sagen, daß die Lehre von den extrachromosomalen Erbanlagen schon vorbehaltlos allgemein anerkannt sei. Das liegt wohl zum Teil daran, daß sich das Vorhandensein plasmatischer Erbanlagen weniger gut demonstrieren läßt als das der chromosomalen Gene. Zum anderen herrscht vielfach keine Klarheit darüber, mit welchen Methoden man die Erbträgernatur bestimmter Plasmaorganellen beweisen kann, oder wie man bestimmte Merkmalsunterschiede mit Plasmabestandteilen in Verbindung bringt. Im folgenden Referat sollen die Fakten erörtert werden, auf die sich heute unsere Ansichten von der genetischen Selbständigkeit der Plastiden gründen. Es scheint geboten, dabei methodische Fragen mit Vorrang zu behandeln.

Wenn die Plastiden Träger eigener Erbanlagen sind, deren Summe von RENNER [*33*] als Plastom bezeichnet wurde, so muß es in den Plastiden ein Kontinuum geben, welches seine Replikation selbst katalysiert und deshalb bei Verlust von der übrigen Zelle nicht neugebildet werden kann. Die morphologische *Kontinuität* der Plastiden wird dadurch sichergestellt, daß sie sich nur durch Teilung vermehren, wie schon früh von SCHMITZ, MEYER und SCHIMPER [*36, 23, 34, 35*] erkannt wurde. Wie von jeder genetischen Information, so darf man auch von einer in den Plastiden verankerten eine strenge *Spezifität* erwarten, deren Konstanz nur selten durch einen Fehler bei der Replikation (Mutation) durchbrochen wird. Dank dieser Spezifität entfaltet das Plastom seine *genetische Aktivität* immer in ganz bestimmter Weise. Als Teil des übergeordneten genetischen Systems der gesamten Zelle muß das Plastom mit allen Komponenten dieses Systems harmonieren, das heißt, angepaßt sein. Deshalb kann man es nicht beliebig gegen ein anderes Plastom austauschen. Selbstverständlich hängt auch die Replikation des Plastoms von dem harmonischen Zusammenwirken aller Komponenten des Idioplasma ab.

Will man prüfen, ob die Plastiden auch *de novo* entstehen können, somit nur zufälligerweise Kontinuität aufweisen, so muß man Zellen experimentell plastidenfrei (apoplastidisch) machen und sie unter geeigneten Bedingungen weiterhin vermehrungsfähig erhalten. Wegen der mannigfaltigen, zum Teil noch unbekannten Funktionen, welche die Plastiden in der Zelle erfüllen, ist für derartige Experimente jedoch mit Schwierigkeiten zu rechnen. Es wird nicht leicht sein, die ausfallenden Plastidenfunktionen durch künstliche Bedingungen zu kompensieren. Am geringsten dürften die Schwierigkeiten für Apoplastidie-Experimente bei Protisten sein, weil in diesem Bereich des Organismenreichs die Evolution denselben Weg schon einmal beschritten hat, als das Tierreich sich vom Pflanzenreich sonderte. Phytoflagellaten, welche sich wahlweise autotroph (nach Art der Pflanzen) oder heterotroph (nach Art der Tiere) ernähren können, wie zum Beispiel *Euglena gracilis*, stellen deshalb ein bevorzugtes Objekt für derartige Versuche dar . Abgesehen von den reversiblen Modifikationen des Plastidenphänotyps, welche sich im Zusammenhang mit der verschiedenen Lebensweise in Anpassung an die jeweils herrschenden Umweltbedingungen einstellen, entstehen spontan mit einer Häufigkeit von 1—2% irreversibel farblose *Euglena*-Kolonien aus zuvor grünen Zellen [*16*]. Diese spontane Mutationsrate läßt sich durch Einwirkung von erhöhter Temperatur, UV-Strahlen und einer Reihe von chemischen Agentien beträchtlich erhöhen, in gewissen Fällen bis zu 100% [*16, 21*]. Es wurde vielfach angenommen, daß die bleichen Zellen apoplastidisch geworden seien, wobei man sich entweder vorstellte, daß eine Verlangsamung der Plastidenvermehrung bei unverminderter Zellteilungsrate zu einem Ausverdünnen der Plastiden geführt habe [*49, 12*], oder daß die Plastiden zerfallen und aufgelöst worden sind [*31*]. Zweifel an der Apoplastidie dieser bleichen *Euglena*-Stämme tauchten auf, als man durch elektronenmikroskopische Untersuchungen fand, daß in bestimmten Fällen anstelle der Plastiden andersartige Gebilde zu finden sind, die wegen ihrer konzentrischen Lamellenstruktur CL-Körper genannt werden [*42*]. Seitdem ist die Frage nach der Apoplastidie der bleichen Algen wieder offen [*16, 32*]. Verstärkt werden die Zweifel noch dadurch, daß man schon festgestellt hat, daß plastidenfreie Zellen sich nicht mehr teilen können [*20, 8*]. Hinweise darauf, daß Plastiden bei Verlust in irgendeiner Weise neugebildet werden könnten, haben alle diese Untersuchungen jedenfalls nicht ergeben. Man darf aber annehmen, daß sich die Plastiden mit zunehmender phylogenetischer Differenzierung als immer unentbehrlicher erweisen. L. BAUER stellte nach umfangreichen Versuchen, bei Moosen Apoplastidie und Plastidenmutationen zu erzielen, fest: ,,Es scheint nach diesen Erfahrungen, daß bei den Moosen die Teilung von Kern und Zelle an das Vorhandensein intakter Plastiden geknüpft ist'' [*8*]. Diesen Eindruck erhält man auch, wenn man bei *Oenothera* extrem disharmonische Genom-Plastom-Kombinationen herstellt [*48*]. Man kann auch darin einen Hinweis dafür sehen, daß Plastiden nicht de novo entstehen können.

Um so überraschender war es deshalb, daß in jüngster Zeit auf Grund elektronenmikroskopischer Untersuchungen an dem Farn *Pteridium aquilinum* von MÜHLETHALER und BELL [*30*] behauptet wurde, daß die Plastiden und Mitochondrien vom Zellkern neugebildet werden könnten. Sie fanden, daß bei der Eireifung die alte Garnitur der Plastiden und Mitochondrien aufgelöst wird. Aus Protuberanzen des Zellkerns soll dann noch vor der Befruchtung der Eizelle die Neubildung der eliminierten Organellen vor sich gehen. Die Autoren halten es für möglich, daß diese Rekonstituierung des Plasma aus dem Eikern für alle niederen und höheren Pflanzen sowie Tiere Gültigkeit habe, und vertreten schließlich noch die Ansicht, daß eine Neubildung der Plastiden aus dem Zellkern in bester Übereinstimmung mit den genetischen Befunden von E. BAUR, CORRENS, RENNER, v. WETTSTEIN und MICHAELIS stehe, nach denen die Plastiden einen dominant mütterlichen Erbgang aufweisen. Auf die Zurückweisung dieser Behauptung durch SCHÖTZ und STUBBE [*40*] werden wir noch zurückkommen. Inzwischen konnten MENKE und FRICKE sowie DIERS [*22, 13, 14*] zeigen, daß sich die Auflösung der Plastiden und Mitochondrien für andere Archegoniaten nicht bestätigen läßt. Der Fall von *Pteridium aquilinum*, den MÜHLETHALER und BELL beschreiben, scheint offenbar eine Ausnahme zu sein. Die sich in allen Fällen zeigenden Fortsätze des Zellkerns der reifen Eizelle brauchen deshalb nicht im Sinne von MÜHLETHALER und BELL interpretiert zu werden. Die anderen Autoren weisen darauf hin, daß MÜHLETHALER und BELL uns den Beweis dafür, daß die blasenförmigen Ausstülpungen des Eikerns sich tatsächlich zu Plastiden und Mitochondrien entwickeln, schuldig geblieben sind. Es sei hier auch erwähnt, daß selbst dann, wenn die Plastiden der Eizelle sich auflösen, ihre Kontinuität für die Nachkommenschaft ausnahmsweise durch väterlichen Erbgang aufrechterhalten werden kann, wie man es z. B. bei bestimmten *Oenothera*-Bastarden beobachtet hat [*48*].

Besser als durch den Nachweis der Kontinuität und Autoreduplikation der Plastiden läßt sich ihre Erbträgernatur durch den *Nachweis spezifischer Plastomdifferenzen* begründen, denn diese schließen das Vorhandensein eines genetischen Kontinuums mit ein. Solange eine physicochemische Charakterisierung des Plastoms noch aussteht, müssen wir allerdings die Erbanlagen in den Plastiden auf indirekte Weise, nämlich auf Grund ihrer phänotypischen Wirkung kennzeichnen (also ebenso, wie es auch für die Gene der Chromosomen üblich ist). Dazu benötigen wir phänotypische Unterschiede, von denen wir nachgewiesen haben, daß sie auf Erbanlagen in den Plastiden zurückgehen. Einem nichtmendelnden extrachromosomal vererbten Phän ist es jedoch nicht ohne weiteres anzusehen, ob seine Erbanlage in den Plastiden oder außerhalb derselben ihren Sitz hat. Damit stellt sich für den Nachweis spezifischer Plastomdifferenzen zunächst einmal das *Lokalisationsproblem* im engeren Sinne.

Nicht immer liegen die Verhältnisse so günstig, daß der Erbträger zugleich auch Merkmalsträger ist. Wir wissen, daß die Plastiden Träger von Merkmalen sein können, deren Gene in den Chromosomen liegen (z. B. Pigmentierungsgene).

Andererseits kennen wir auch Plastomwirkungen, die über den Bereich des eigenen Erbträgers hinausgehen und die gesamte Zelle (z. B. die Keimfähigkeit eines Pollenkorns) oder Gewebe und Organe (z. B. die Gestalt von Blättern) beeinflussen. Eine ausreichend befriedigende Antwort auf die Lokalisationsfrage läßt sich zunächst nur geben, wenn die Plastomwirkung sich auf den eigenen Erbträger beschränkt. Das bedeutet also, daß auch eine gegenseitige Beeinflussung von verschiedenen homologen Erbträgern unterbleiben muß, sofern sich diese zusammen in derselben Zelle, der „Mischzelle", befinden. Wenn bei einem Objekt erst einmal die Lokalisation einer bestimmten Erbanlage in den Plastiden gelungen ist, so kann man im Anschluß daran weitere Plastomdifferenzen ermitteln, wie noch zu zeigen sein wird.

Für eine zuverlässige Lösung des Lokalisationsproblems bietet sich die cytogenetische Methode an, die auf Boveri [*11*] zurückgeht. Wir können für die plasmatischen Erbträger nach denselben Prinzipien verfahren, die seinerzeit angewendet wurden, um die Mendelfaktoren in den Chromosomen zu lokalisieren. Das heißt, wir verknüpfen die von der Cytologie erarbeiteten Befunde über Entwicklungsgeschichte, Vermehrung, Häufigkeit und Weitergabe der Zellbestandteile bei der Zellteilung mit den Gesetzmäßigkeiten, welche die Genetik für die Segregation der phänotypischen Unterschiede nach Hybridisierung beobachtet. Bei den Mendelfaktoren wird durch den besonderen Verteilungsmechanismus der Chromosomen dafür gesorgt, daß das Mischungsverhältnis der Gene während der Mitose und anschließenden Zellteilung erhalten bleibt. Für die plasmatischen Erbträger fehlt ein solcher Mechanismus. Deshalb kann bei Vorliegen eines Gemisches das Mischungsverhältnis der homologen plasmatischen Erbträger mit verschiedenen Erbanlagen durch *jede* Zellteilung zufallsgemäß abgeändert werden. Diese Überlegung liegt auch der Entmischungshypothese zugrunde, mit welcher E. Baur [*9*] die Scheckungsmuster bei *Pelargonium zonale* durch eine zufallsgemäße Verteilung grüner und weißer Plastiden erklärte. Die statistischen Gesetzmäßigkeiten, welche sich bei derartigen Entmischungsvorgängen ergeben, sind in den letzten 10 Jahren sehr gründlich von Michaelis [*24—28*] ausgearbeitet und an Beispielen demonstriert worden. Es wird dabei deutlich, daß sich für die in unterschiedlicher Anzahl in der Zelle vorhandenen Erbträger verschiedener Kategorien im Verlauf der Ontogenese der Pflanzen charakteristisch verschiedene Verteilungsmuster ergeben. Deshalb ist es möglich, durch Vergleich der phänotypischen Muster mit den berechneten oder im Modellversuch ermittelten schließlich zur Beantwortung der Lokalisationsfrage zu gelangen.

Wir wollen an dieser Stelle kurz auf einige praktische Fragen eingehen, die sich im Zusammenhang mit der Anwendung der cytogenetischen Methode auf unsere plastidengenetischen Probleme ergeben:

a) *Die Herstellung der Mischzellen* kann auf verschiedene Weise erfolgen: 1. Der natürlichste Weg ist die Gametenkopulation. Sie führt jedoch nur dann zum Ziel, wenn sowohl das Plasma des ♀ wie das des ♂ Gameten erhalten bleiben und an die Teilungsprodukte der Zygote weitergegeben werden. Nicht selten werden die Plastiden des ♂ Gameten eliminiert (z. B. bei *Spirogyra*). Als hervorragendes Beispiel für biparentale Plastidenübertragung sei *Oenothera* erwähnt. 2. Bei asexuellen Organismen, wie z. B. *Euglena*, müßte man versuchen, die Zellverschmelzung auf dem Wege der Pfropfung zu erreichen, oder nach dem

Verfahren, wie es z. B. von HAWKINS und COLE [18] bei *Amoeba proteus* erfolgreich angewandt wurde, indem eine kleinere Plasmaportion in eine ganze Zelle injiziert wird. 3. Durch Mutation entstandene Mischzellen eignen sich ebenfalls für das Studium der somatischen Segregation, wie es z. B. MICHAELIS bei Epilobium wiederholt gezeigt hat [24–28]. Für die statistische Auswertung der Entmischungsmuster ist es wichtig, daß sich die Herstellung der Ausgangsmischzelle möglichst oft reproduzieren läßt. Bei allen Organismen, die nur einen einzigen Chloroplasten pro Zelle enthalten, kann sich die Untersuchung somatischer Segregationsvorgänge nur auf die außerplastidischen Erbträger erstrecken, was unter Umständen bei negativem Befund nützliche Hinweise auf eine Plastidenvererbung ergeben kann.

b) *Die Erfassung der Entmischungsmuster* zum Zwecke der Rekonstruktion der Entmischungsvorgänge verlangt eine genaue Ermittlung der Zelldeszendenzen. DE DEKEN-GRENSON und MESSIN [12] haben deshalb für *Euglena*-Kulturen ein halbfestes Medium verwendet, um die genealogischen Zusammenhänge feststellen zu können. Vorzüglich dürften sich Pflanzen eignen, bei denen die Zelldeszendenzen alle in einer Ebene angeordnet sind, und wo die Zellteilungen mit großer Regelmäßigkeit erfolgen, beispielsweise die Rotalge *Caloglossa leprieurii* [43]. Bei höheren Pflanzen mit ihrem vielschichtigen Bau und der besonderen Wachstumsweise der Meristeme und Organe ist die Ermittlung der Zelldeszendenzen erfahrungsgemäß sehr aufwendig. Man vergleiche hierzu die beispielhaften Untersuchungen von BARTELS an Epilobium [3–6]. Durch Mutation markierte Zelldeszendenzen erleichtern jedoch auch wiederum die Rekonstruktion der entwicklungsgeschichtlichen Zusammenhänge, wie schon von RENNER [33] betont wurde. Die direkte Beobachtung der Entmischungsvorgänge ist bei höheren Pflanzen meist nicht möglich, da sich das Merkmal oft spät in der Entwicklung manifestiert (wie z. B. die Pigmentierung der Plastiden). Wenn die Zygote die Ausgangsmischzelle ist, die grün und weiß determinierte Plastiden enthält, so bleibt der größte Teil des Entmischungsmusters unsichtbar, denn erst die Assimilationsgewebe der Blätter zeigen das Scheckungsmuster als Resultat des Entmischungsvorgangs. Die Entwicklungsgeschichte eines echten Entmischungsmusters für verschiedenfarbige Plastiden sieht so aus, daß zunächst in der Nähe der Ausgangsmischzelle ein kleinfleckiges Mosaik vorliegt, welches nach und nach großfleckiger wird (weil jede reinerbig gewordene Zelle nur wieder reinerbige Zellen produziert). Schließlich findet man die beiden komplementären („allelen") Phänotypen in sauber getrennten Sektoren vor. Von diesem Zustand aus kann das Muster ganz in den einen oder anderen Phänotyp übergehen, wenn die über den Vegetationskegel hinweglaufende Sektorengrenze durch ungleiches Wachstum der verschiedenen Sektoren oder durch Neuanlage von weiteren Vegetationspunkten außerhalb des Meristems zu liegen kommt. Bei dem Schichtenbau des Angiospermenvegetationskegels ist es dabei zudem möglich, daß eine Schicht diesen, die andere Schicht jenen Phänotyp repräsentiert. Wir haben dann eine Periklinalchimäre vor uns, welche gewöhnlich stabil ist.

c) Für das Lokalisationsproblem ist es nun wichtig, eine *statistische Auswertung* vieler homologer Entmischungsvorgänge vorzunehmen, indem jeweils der Anteil der beiden reinerbigen und der gemischten Zelltypen in einer bestimmten Zellgeneration ermittelt wird, um danach die Entmischungsgeschwindigkeit beurteilen zu können. MICHAELIS [*24, 25, 26*] hat darauf hingewiesen, daß die Entmischungsgeschwindigkeit im wesentlichen von der Anzahl der plasmatischen Erbträger je Zelle bestimmt wird, so daß aus der Art des Musters geschlossen werden kann, in welcher annähernden Zahl der dem Muster zugrundeliegende Erbträger in jeder Zelle vorhanden sein muß. Für das Erreichen des statistischen Entmischungsendes, ausgedrückt durch die Zellteilungsfolge, in der Mischzellen nur noch einen unbedeutenden Anteil an der Gesamtzahl der Zellen ausmachen, spielt nach MICHAELIS das Ausgangsmischungsverhältnis keine Rolle mehr. Nach einer Faustregel liegt nach n · 10 Zellteilungsfolgen der Anteil der Mischzellen bereits so niedrig, daß man vom Ende der Entmischung sprechen kann (n = Zahl der Erbträger unmittelbar nach der Zellteilung).

Diesen Berechnungen liegt ein Modell zugrunde, bei dem die Zelle durch eine Urne repräsentiert wird, welche eine bestimmte Anzahl von Kugeln enthält, die z. B. die Plastiden vor einer Zellteilung repräsentieren. Die verschiedenen Phänotypen der Plastiden werden durch verschiedene Farben gekennzeichnet. Durch Herausgreifen der Hälfte der Kugeln und nachfolgende Verdoppelung der Kugeln entsprechend ihrem Mischungsverhältnis wird eine Zellteilungsfolge dargestellt. Mit Hilfe dieses Modells kann man sowohl die Muster praktisch ermitteln, wie auch die geeigneten mathematischen Formeln für die Berechnung der Gesetzmäßigkeiten der Entmischung ableiten (vgl. [*24, 25*]). Für die Übertragung dieses Modells auf das biologische Objekt werden aber Korrekturen notwendig sein, wie man sich leicht klarmachen kann, wenn man die Voraussetzungen dieses Modells an der lebenden Zelle überprüft. Folgendes müßte sichergestellt sein, wenn das Urnenmodell gültig sein soll: 1. Die Wirkungsweise der Erbanlagen muß auf den Erbträger beschränkt bleiben. 2. Die homologen Erbträger müssen gleiche Vermehrungsgeschwindigkeit besitzen. 3. Sie replizieren sich in Anlehnung an den Rhythmus des Zellteilungsgeschehens, damit die Zahl der Erbträger je Zelle innerhalb bestimmter Zahlenbereiche konstant bleibt. 4. Nach jeder Verdoppelung der Erbträger bzw. vor jeder Zellteilung muß eine gute Durchmischung stattfinden. 5. Die Zellteilungen sorgen für äquale Verteilung der Erbträger. 6. Die Zellteilungsfolge muß bekannt sein. Dieses Modell liefert uns den langsamsten Fall der Entmischung. Bei Unterschieden in der Vermehrungsrate der allelen Erbträger, bei mangelnder Durchmischung und bei inäqualer Zellteilung wird der Entmischungsvorgang beschleunigt.

Ob man die von MICHAELIS ausgearbeitete Methode der Entmischungsmusteranalyse anwenden kann, hängt also davon ab, wie weit die oft recht mühsamen Untersuchungen gediehen sind, welche sich mit den Voraussetzungen des zugrundeliegenden Modells befassen. Ein wertvoller Beitrag, der strenggenommen allerdings nur für dieses Objekt gilt, ist die Bearbeitung der Entwicklungsgeschichte von *Epilobium hirsutum* durch BARTELS [3—6]. Das Ziel dieser Untersuchungen war es, die Zellteilungsschritte bzw. die Zelldeszendenzen von der Zygote oder den jeweiligen Initialen der Gewebe bis zu dem vom Genetiker analysierten fertigen Zellmuster zu ermitteln. Folgende Ergebnisse seien für unsere Überlegungen herausgegriffen: Die Mesophyllinitialen des ersten Blattwirtels werden in der 10. Teilungsfolge angelegt, die der folgenden Wirtel entstehen jeweils 3 Teilungsfolgen später. In den einzelnen Blättern laufen noch etwa 20—25 Zellteilungen ab. Die ersten Blüten werden nach etwa 20 Laubblattwirteln angelegt, so daß zwischen Zygote und Blütenanlage etwa 67 Zellteilungen erfolgt sind. Man ist damit bei *Epilobium* in der Lage, die Zellteilungsfolge bzw. Zellgeneration als Bezugsgröße für die statistische Auswertung der Entmischungsmuster zu benutzen. Ist die Lage der Ausgangsmischzelle bekannt, so kann man z. B. für bestimmte Entmischungsmuster angeben, nach wieviel Zellteilungen das Entmischungsende erreicht wird. An diesen Mustern kann man das Entmischungsende daran erkennen, daß keine verschachtelten Flecken mehr, sondern nur noch klar begrenzte Sektoren gebildet werden. Ist es also möglich anzugeben, an welchem Blattwirtel dieser Zustand erreicht wurde, so hat man im Falle von *Epilobium* auch die Möglichkeit, das Entmischungsende durch das sehr korrekte Maß der Zellteilungsfolge auszudrücken, falls die Lage der Ausgangsmischzelle bekannt war. Leider ist es notwendig, entsprechende entwicklungsgeschichtliche Untersuchungen für jedes Objekt gesondert durchzuführen. Für *Oenothera* stehen sie noch aus. So mußte sich SCHÖTZ [38] z. B., als er Unterschiede in der Entmischungsgeschwindigkeit bei verschiedenen Plastidenkombinationen nachwies, noch mit zeitlichen Angaben begnügen, statt sich auf die Zellteilungsfolge zu beziehen. Für die Beantwortung der Lokalisationsfrage müßten außerdem noch ausgedehnte Erhebungen über die Erbträgerzahlen in meristematischen Zellen angestellt werden. Es bleibt zu hoffen, daß wenigstens für einige bevorzugte Objekte der extrachromosomalen Vererbungsforschung diese entwicklungsgeschichtlichen und cytologischen Untersuchungen noch durchgeführt werden. Bis dahin wird man versuchen, die Lokalisationsfrage mit Hilfe von Indizien zu beantworten.

Als Musterbeispiel für einen solchen Indizienbeweis sei mir erlaubt, mein eigenes Objekt *Oenothera* heranzuziehen. Von den besonderen Vorzügen, welche dieses Objekt für das Studium der plasmatischen Vererbung bietet (vgl. [46]), seien für das hier zur Debatte stehende Lokalisationsproblem nur die wichtigsten erwähnt: Bei den kultivierten Arten findet man mit ausreichender Häufigkeit spontane Mutationen, welche extrachromosomal vererbte Plastidenmerkmale betreffen (z. B. Pigment- und Strukturunterschiede). Ferner: Eine Mischung der verschiedenen Plastidensorten ist auf dem Wege der Kreuzung möglich und sicher reproduzierbar, da auch der Pollen Plastiden und Plasma auf die Nachkommen überträgt. Kreuzt man z. B. normalgrüne und weiße Zweige einer Pflanze in beiden Richtungen miteinander, so zeigt sich in der Nachkommenschaft ein quantitativer Reziprokenunterschied: Die Nachkommen besitzen zum überwiegenden Teil Plastiden der Mutter, sind jedoch teilweise von Plastiden des Vaters gescheckt. Dieser Reziprokenunterschied läßt sich durch die unterschiedlichen Plasmamengen erklären, welche Eizelle und Spermazelle zur Zygote beisteuern. Die Lokalisationsfrage stellt sich folgendermaßen: Handelt es sich bei den beiden Plastidenphänotypen, wie sie sich in den ausdifferenzierten Mesophyllzellen manifestieren, nun um solche, die von erblichen Unterschieden im Plastom hervorgerufen werden, oder sind dafür erbliche Differenzen im Cytoplasma verantwortlich?

Die schnelle Entmischung, die meist schon im Bereich der ersten Rosettenblätter abgeschlossen ist, läßt vermuten, daß der verantwortliche Erbträger in geringer Anzahl in der Zelle vorhanden ist. Legt man die bei *Epilobium* gewonnenen Anhaltspunkte zugrunde, so dauert der Entmischungsvorgang durchschnittlich nicht länger als 20 bis 30 Zellteilungsfolgen. Möglicherweise wird die Entmischung durch mangelhafte Durchmischung beschleunigt. Pflanzen mit ausschließlich mütterlichen Plastiden gehen teilweise sicherlich auf die inäquale erste Teilung der Zygote zurück, wobei die väterliche Plasmaportion manchmal ganz dem Suspensor zugeteilt werden kann. Die charakteristische Entwicklung des Scheckungsmusters und die relativ schnelle Entmischung weisen auf echte Plastidenentmischung hin, wenn man annimmt, daß alle anderen in Betracht kommenden Zellpartikel in wesentlich größerer Zahl vorkommen. Die für *Epilobium* und *Oenothera* ermittelten Plastidenzahlen liegen für junge Palisadenzellen bei einem Wert von n = 12 [*29, 41, 7*]. Ob diese Zahl auch im Meristem vorliegt, ist ungeklärt. Schröder vermutet höhere Zahlen [*41*].

Das Hauptindiz liefert jedoch der Nachweis echter Mischzellen, in denen die beiden Plastidenphänotypen ebenso klar getrennt nebeneinander vorkommen, wie man sie sonst in rein weißen und rein grünen Geweben antrifft. Wenn Übergangsformen zwischen den Plastidentypen fehlen, darf man diese Tatsache als Argument dafür ansehen, daß der phänotypische Unterschied nicht cytoplasmatisch bedingt ist; denn wäre letzteres der Fall, so sollte man erwarten, daß Plastiden im Grenzbereich der beiden Plasmotypen sich intermediär ausbilden würden, oder daß Mischzellen überhaupt fehlen. Die völlig freie Mischbarkeit des mütterlichen und väterlichen Plasma wird hierbei vorausgesetzt, da beide Gameten von derselben Pflanze stammen. Es sei nochmals darauf hingewiesen, daß Mischzellen in der Regel erwartungsgemäß nur an den unteren Teilen der Pflanzen vorkommen. In der Blütenregion findet man sie nur in seltenen Ausnahmefällen [*44*]. Mischzellen wurden bei *Oenothera* erstmals von Krumbholz nachgewiesen, später auch von Schötz und von Stubbe [*37, 44*]. Der Nachweis echter Mischzellen in Verbindung mit einer dazupassenden Entmischungsmusterentwicklung wird auch bei anderen Objekten als beweiskräftig für eine Plastidenvererbung angesehen. Eine Zusammenstellung der beobachteten Fälle findet man z. B. bei Hagemann [*17*]. Wir brauchen darauf nicht näher einzugehen.

Es sei nur noch erwähnt, daß Schötz und Stubbe [*39, 47*] in ihrer Erwiderung an Mühlethaler und Bell auch mit dem Nachweis gemischter Eizellen argumentierten. STUBBE [44] hat bei Pflanzen, die in der Blütenregion noch ein verschachteltes grün/weißes Scheckungsmuster zeigten, durch Kreuzung mit einem gelbgrünen Vater nachgewiesen, daß tatsächlich noch gemischte Eizellen mit grün und weiß determinierten Plastiden vorhanden waren. Wenn die Annahme von MÜHLETHALER und BELL [30] zuträfe, daß die Plastiden vom Eikern neugebildet würden, nachdem die vorherige Plastidengarnitur eliminiert worden ist, so müßte der Zellkern eine Art von Gedächtnis besitzen, das ihn dazu befähigt, genau diejenigen Plastidensorten neuzubilden, mit denen er vorher kombiniert war. Diese Vorstellung ist aber wenig wahrscheinlich. Bei *Oenothera* kann man ferner die verschiedensten Plastiden-Wildtypen jeweils mit dem gleichen Genotyp kombinieren, wobei die Plastiden ihre Eigenart für unbegrenzte Generationenfolgen beibehalten. SCHÖTZ [39] stellt mit Recht die Frage: „Was soll bei vorhergehender Zerstörung und Ausschaltung der Plastiden die genetisch gleich konstituierten Kerne der Eizellen veranlassen, verschiedene Plastidentypen nachzubilden?" Die Annahme eines Kontinuums, welches in den Plastiden selbst weitergegeben wird und die genetische Information für bestimmte Eigenschaften bewahrt, bleibt also die plausibelste Erklärung für das Vererbungsverhalten der Plastiden.

Neuerdings hat die mit biochemischen Methoden arbeitende Genetik diesem Kontinuum nachgespürt. Daß dabei in erster Linie der Nachweis von Nucleinsäuren in den Plastiden eine Rolle spielt, ist nur zu natürlich. Wie GIBOR und GRANICK [15] zusammenfassend dargelegt haben, ist die Ansicht, daß die Plastiden eine autonome DNS enthalten, gut gestützt. Die gleichen Autoren haben auch bereits darauf hingewiesen, daß es nicht allein genügt, den Nachweis für das bloße Vorhandensein von DNS und RNS in den Plastiden zu erbringen. Es muß vor allem gezeigt werden, daß die DNS sich in den Plastiden repliziert, daß sie die Funktion hat, Boten-RNS zu machen, und daß diese Boten-RNS spezifisch für gewisse Proteine oder Enzyme der Organelle kodiert. Mit anderen Worten, ihre genetische Spezifität und Aktivität müssen bewiesen werden. An diesem Punkt treffen sich die klassische und die moderne molekulare Genetik, indem sie für den Beweis der Spezifität der genetischen Information der Plastiden auf vergleichende Untersuchungen angewiesen sind an einem Material, welches sich in einem oder wenigen Faktoren unterscheidet. Während die klassische Genetik die Erbanlage mit Hilfe der phänotypischen Wirkung, die unter Umständen in komplizierte Reaktionsfolgen eingreift, kennzeichnet, ist die molekulare Genetik vielleicht in der Lage, den Unterschied im genetischen Kode direkt zu beschreiben. Derartige Ergebnisse, welche chemische Unterschiede in der DNS mit bestimmten Plastidendifferenzen in Verbindung bringen, stehen noch aus. Man ist aber geneigt, GIBOR und GRANICK zuzustimmen, wenn sie in der mutativen Veränderung der Plastiden durch direkte UV-Bestrahlung einen indirekten Beweis dafür sehen, daß hierdurch eine DNS geschädigt wurde, die nicht vom Kern ersetzbar ist. Die Untersuchung der Plastiden-DNS in zellfreien Systemen könnte Ergebnisse zeitigen, welche die Kenntnis der genetischen Infor-

mation der Plastiden beträchtlich erweitern werden. Auch die Mutationsforschung, vor allem, wenn sie gezielt die DNS angreift [21], wird ihre
Rolle dabei spielen.

Bei allen Erfolgen, die wir dem Forschungszweig verdanken, der die
Nucleinsäuren zum Gegenstand seiner Untersuchungen macht, sollten
wir jedoch nicht übersehen, daß es noch andere Möglichkeiten der
Speicherung von genetischer Information geben kann. Als Beispiel
möchte ich auf die Vererbung einer bestimmten Cortex-Struktur bei
Paramecium hinweisen, der die Sonneborn-Schule ihre Aufmerksamkeit
widmete [10]. Hier wird eine ganz bestimmte Ordnung weitergegeben,
welche auf keinerlei Differenzen im Erbgut des Zellkerns oder des
Cytoplasma zurückgeführt werden kann. Man sollte mit entsprechenden Möglichkeiten wenigstens rechnen, wenn man die Plastiden genetisch analysiert.

Abschließend wollen wir noch einen Blick auf das Spektrum der
plastomabhängigen Merkmale werfen, wie es sich aus den Erfahrungen
an *Oenothera* ergibt. Wenn man das Lokalisationsproblem für bestimmte
chlorophylldefiziente Plastidenmutationen als geklärt ansieht, kann man
in Kreuzungen jeweils den einen Kreuzungspartner mit diesen mutierten
Plastiden markieren und damit die Plastidensorte des anderen Kreuzungspartners in den Vergleich einbeziehen. Bei *Oenothera* lassen sich
auf diesem Wege unter vorteilhafter Ausnutzung der Komplexheterozygotie in großer Zahl Genom-Plastom-Kombinationen herstellen, an
denen man die phänotypische Wirkung der in den Wildarten vorkommenden natürlichen Plastomdifferenzen studieren kann. Ein umfassendes
Beispiel für derartige Untersuchungen hat STUBBE [45, 46] mit der Untergattung *Euoenothera* gegeben. Hier lassen sich 5 Wildtyp-Plastome und
3 haploide Grundgenotypen unterscheiden, aus denen man also in der
Diplophase 30 verschiedene Genom-Plastom-Kombinationen herstellen
kann (s. Abb. 1). Die Plastomdifferenzen zeigen sich, wenn man alle
5 Plastome jeweils in Gegenwart desselben Genotyps vergleicht. Das
Zusammenwirken von Genom und Plastom bei der Merkmalsbildung
ergibt dann unterschiedliche Phänotypen. Harmonische Beziehungen
zwischen Genom und Plastom führen zu lebensfähigen Pflanzen, disharmonische (Inkompatibilität) ergeben die verschiedensten Entwicklungsstörungen. Auf diese Weise wird uns dann ein Einblick gegeben,
in welche physiologischen Prozesse das Plastom in irgendeiner Weise
eingreift: Eine große Rolle spielen naturgemäß die verschiedensten
Abwandlungen der normalen Plastidenentwicklung. Die Störungen sind
in Abb. 1 vereinfachend durch nur wenige Symbole dargestellt. Nicht nur
die Differenzierung der Plastiden zu Chloroplasten, sondern auch ihre
Teilungsfähigkeit kann beeinträchtigt sein. Da zwischen Plastidenvermehrung und Zellteilung offenbar enge Korrelationen bestehen, kann es
im Extremfall dazu kommen, daß die Zellen mit gehemmten Plastiden
sich nicht mehr weiterentwickeln [19, 48]. Die Plastomdifferenz wirkt
dann als Letalfaktor. Daneben gibt es unabhängig von derartigen Unverträglichkeitsfolgen auch bei harmonischen Genom-Plastom-Kombinationen nachweisbare Unterschiede in der Vermehrungsgeschwindig-

keit der Plastiden [37], welche plastomabhängig sind. Bestimmte haplontische Genom-Plastom-Kombinationen unterscheiden sich dadurch, daß die Form der Stärkekörner im Pollenschlauch in Abhängigkeit von Plastomdifferenzen verschieden ist (z. B. rund oder spitzspindelförmig) [46]. Außerdem kann die Keimfähigkeit des Pollens aufgehoben sein, ohne daß irgendein Zusammenhang mit anderen Defekten gegeben wäre (z. B. Chlorophylldefizienz in der Diplophase).

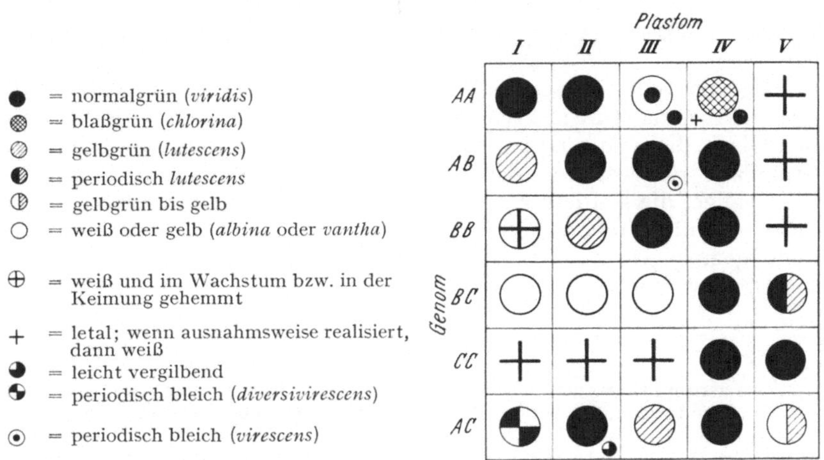

Abb. 1. Kombinationsrechteck, welches eine Übersicht über die phänotypische Wirkung der einzelnen Genom-Plastom-Kombinationen gibt. (Nach STUBBE 1959)

Daß die Keimungsunfähigkeit des Pollens von Plastomdifferenzen abhängt, wird an grün-weiß-gescheckten Pflanzen ersichtlich: Weiße Sektoren produzieren keimfähigen und grüne Sektoren keimungsunfähigen Pollen. Die Lokalisation der Erbanlage für die Pollenunterschiede basiert also darauf, daß die Grenze für die Grün-weiß-Scheckung mit der für die Keimungs- oder Stärkeformunterschiede des Pollens zusammenfällt.

Nicht zuletzt werden auch Wachstumsvorgänge und damit morphogenetische Prozesse von den Stoffwechselvorgängen der Plastiden beeinflußt. Das kann dazu führen, daß bei verschiedenen Plastomen in Verbindung mit demselben Genotyp Form- und Größendifferenzen auftreten, die besonders bei den Laubblättern auffallen. Gewöhnlich ist dabei auch die Ergrünungsfähigkeit der betreffenden Kombinationen unterschiedlich. Bei *Oenothera lutescens* (= *flavens · flavens*) z. B. ist die Kombination mit Plastom II im Laub bleich und hat eine normale Blattform, während die Kombination mit Plastom III dunkelgrüne, oberseits eingerollte Blätter entwickelt. Weitere plastomabhängige Merkmale wurden in der Untergattung *Raimannia* durch die Schwemmle-Schule beschrieben [1].

Infolge des Fehlens von Austauschvorgängen zwischen verschiedenen Plastomen ist es leider schwierig, das Plastom in einzelne Erbfaktoren aufzugliedern. Ansatzpunkte hierfür ergeben sich bisher allein aus der Beobachtung des Mutationsgeschehens. Falls nämlich von den Wirkungen eines bestimmten Plastidentyps einzelne mutativ verändert werden, ohne daß dabei die übrigen in Mitleidenschaft gezogen werden, so darf man wohl daraus schließen, daß hier verschiedene funktionelle Untereinheiten des Plastoms unabhängig voneinander mutieren. Mit spontanen Mutanten ließen sich in dieser Hinsicht einige Ergebnisse erzielen [45], während die experimentelle Auslösung von Plastommutationen bei höheren Pflanzen, speziell bei *Oenothera*, noch keine Erfolge gebracht hat [2].

RENNER schrieb 1934 [33]: „Die Evidenz für die Autonomie der Plastiden ist nach meiner Überzeugung heute so weit gefördert, daß es sich lohnt, diese Autonomie zur These zu machen, und diese These zur Grundlage für weitere, auch deduktive Behandlung des Problems." Die Erfolge, die seitdem auf dem Gebiet der Plastidenvererbung erzielt wurden, sind von Forschern erarbeitet worden, die sich RENNERs Einstellung zu eigen gemacht haben. Sie konnten die damals geäußerten Ansichten weitgehend bestätigen.

Summary

If we assume that plastids are bearers of genetic information, they necessarily should be endowed with continuity, specificity and genetic activity. In order to demonstrate that plastids arise only from preexisting plastids, experiments have been undertaken to destroy or eliminate the plastids of the cell (for example in *Euglena gracilis*, a phytoflagellate which then can be cultivated under heterotrophic conditions). In most of these experiments it remains still an open question, whether the resulting bleached strains are free from plastids or whether they contain tiny proplastids which, in consequence of a mutation, cannot differentiate into normal chloroplasts. In no case, however, has it been possible to demonstrate convincingly a *de novo* synthesis of plastids.

Observations which prove specific heritable differences of plastids also argue for a continuum, however, it is difficult to demonstrate that this continuum is localized within the plastid itself. By combining plastids of different origin with the same nucleus (genome) and cytoplasm (plasmone) it should become clear that resulting phenotypic differences must be due to plastome differences. One must exclude, though, that the hereditary factor depends on contaminating cytoplasmic particles outside of the plastids. So long as no characterization of plastome differences in physico-chemical terms has been reached, the cytogenetic method provides the only way to solve this localization problem. This method links the cytological data on frequency, multiplication, development and distribution of the organelles during cell division with the rules of somatic segregation observed after hybridizing different phenotypes.

As shown by MICHAELIS, the probability of sorting out of a mixture of cell constituents is a function of the number of independent elements per cell. Therefore a statistical treatment of the process of sorting out and the patterns of somatic segregation, respectively, enable us to distinguish between different kinds of plasmic inheritance (e.g. between plastid and mitochondrial inheritance). Difficulties for the applicability of this method result from insufficient knowledge of the different factors which influence the segregation process of the cell constituents. In addition, in higher plants, a great deal of work is required to determine the details of ontogenesis. It is necessary, to know the cell lineages and the number of cell divisions between the initial "mixed cell" and the tissue in which the sorting out has practically finished in a statistical sense. This situation will be reached the earlier the smaller the number of cell constituents concerned. Furthermore the speed of sorting out accelerates if there are differences in multiplication rate of the segregating hereditary units, if the mingling is incomplete, or (and) if the cell division is inequal.

In view of the many requirements investigators in most cases resort to circumstantial proof. In *Oenothera*, e. g., plastome dependence of certain chlorophyll deficiencies is regarded proven, because of 1) the biparental inheritance is coupled with reciprocal differences and 2) at the same time somatic segregation takes place rapidly. Of main importance is 3) the occurence of mixed cells (with two well discernible plastid phenotypes) which can be found regularly in patterns of the *albomaculata status*. Beyond that in *Oenothera* we succeeded in producing plants with three different plastid types. This was achieved by pollinating flowers with mixed egg cells (containing green and white determined plastids) by pollen with a third plastid type (yellowish green determined).

The opinion of MÜHLETHALER and BELL, that the nucleus of the egg cell is apt to produce plastids and mitochondria *de novo*, and that this procedure is common in all plants and in best agreement with the genetic observations on maternal inheritance, is not reconcilable with the concept of plastid inheritance which is well established.

In quest of the chemical substance representing the continuum (plastome), DNA, of course, has been suspected. Other possibilities of conserving genetic information, however, should not be overlooked beside this. In every case it will be necessary to demonstrate the connexion between phenotypic differences and differences in the genetic substance.

Investigations in the genus *Oenothera* have revealed by crossing experiments a series of plastome dependent characters. On account of many cases of incompatibility between genome and plastome it was shown that the plastome controls: 1. the normal development of chloroplasts or different kinds of variation, 2. the multiplication of plastids in meristematic tissues or the inhibition of this multiplication, 3. the differences in multiplication rate of plastids, 4. the differences of shape of the starch grains in the pollen tube, 5. the germination or inhibition of germination of pollen, 6. morphogenetic processes (e. g. shape of leaves). Since by plastome mutation several single characters may be altered without influencing the others it seems justified to regard the plastome composed of different functional subunits.

Literatur

[1] ARNOLD, C.-G.: Ber. dtsch. bot. Ges. **76**, 3 (1963).
[2] —, u. M. KRESSEL: Z. Vererbungsl. **97**, 213 (1965).
[3] BARTELS, F.: Flora **144** , 105 (1956).
[4] — Flora **149**, 206 (1960).
[5] — Flora **149**, 225 (1960).
[6] — Flora **150**, 552 (1961).
[7] — Planta (Berl.) **60**, 434 (1964).
[8] BAUER, L.: Flora **136**, 30 (1943).
[9] BAUR, E.: Z. indukt. Abstamm.- u. Vererb.-L. **1**, 330 (1909).
[10] BEISSON, J., and T. M. SONNEBORN: Genetics **53**, 275 (1965).
[11] BOVERI, T.: Ergebnisse über die Konstitution der chromatischen Substanz des Zellkerns. Jena: G. Fischer 1904.
[12] DE DEKEN-GRENSON, M., et S. MESSIN: Biochim. biophys. Acta (Amst.) **27**, 145 (1958).
[13] DIERS, L.: Ber. dtsch. bot. Ges. **77**, 369 (1964).
[14] — In: Probleme der biologischen Reduplikation (P. SITTE, Hsg.), Berlin-Heidelberg-New York: Springer 1966.
[15] GIBOR, A., and S. GRANICK: Science **145**, 890 (1964).
[16] GRENSON, M.: Int. Rev. Cytol. **16**, 37 (1964).
[17] HAGEMANN, R.: Plasmatische Vererbung. Beitr. 4 d. Reihe: Genetik. Grundlagen, Ergebnisse u. Probleme in Einzeldarstellungen. Hrsg. H. STUBBE. Jena: VEB G. Fischer Verlag 1964.
[18] HAWKINS, S. E., and R. J. COLE: Exp. Cell Res. **37**, 26 (1965).
[19] KISTNER, G.: Z. indukt. Abstamm.- u. Vererb.-L. **86**, 521 (1955).

[20] LWOFF, A., et H. DUSI: C. R. Soc. biol. (Paris) 119, 1092 (1935).
[21] McCALLA, D. R.: Science 148, 497 (1965).
[22] MENKE, W., u. B. FRICKE: Z. Naturforsch. 19b, 520 (1964).
[23] MEYER, A.: Das Chlorophyllkorn in chemischer, morphologischer und biologischer Beziehung. Leipzig: A. Felix 1883.
[24] MICHAELIS, P.: Cytologia 20, 315 (1955).
[25] — Züchter 25, 209 (1955).
[26] — Planta (Berl.) 50, 60 (1957).
[27] — Planta (Berl.) 51, 600; 51, 722 (1958).
[28] — Flora 151, 162 (1961).
[29] — Protoplasma (Wien) 55, 177 (1962).
[30] MÜHLETHALER, K., u. P. R. BELL: Naturwissenschaften 49, 63 (1962).
[31] PRINGSHEIM, E. G.: Zur Cytologie der Apochlorose. Vortr. Gesamtgeb. Bot. Hrsg. Dtsch. bot. Ges. NF. Nr. 1, 85 (1962).
[32] — Ber. dtsch. bot. Ges. 75, 335 (1963).
[33] RENNER, O.: Ber. Verh. Sächs. Akad. Wiss. Leipzig. Math.-phys. Kl. 86, 241 (1934).
[34] SCHIMPER, A. F. W.: Bot. Ztg. 41, 105, 121, 137, 153, 809 (1883).
[35] — Jb. wiss. Bot. 16, 1 (1885).
[36] SCHMITZ, F.: Verh. naturw. Verein Rheinl. u. Westfalen 40, 1 (1882).
[37] SCHÖTZ, F.: Planta (Berl.) 43, 182 (1954).
[38] — Planta (Berl.) 51, 173 (1958).
[39] — Planta (Berl.) 58, 333 (1962).
[40] —, u. W. STUBBE: Naturwissenschaften 49, 211 (1962).
[41] SCHRÖDER, K. H.: Z. Bot. 50, 348 (1962).
[42] SIEGESMUND, K. A., W. G. ROSEN, and S. R. GAWLIK: Amer. J. Bot. 49, 137 (1962).
[43] STOCKER, O.: Grundriß der Botanik. Berlin-Göttingen-Heidelberg: Springer 1952. Vgl. Abb. 49 B.
[44] STUBBE, W.: Ber. dtsch. bot. Ges. 70, 221 (1957).
[45] — Z. Vererbungsl. 90, 288 (1959).
[46] — Z. Bot. 48, 191 (1960).
[47] — Z. Vererbungsl. 93, 175 (1962).
[48] — Z. Vererbungsl. 94, 392 (1963).
[49] TERNETZ, CH.: Jb. wiss. Bot. 51, 435 (1912).

Diskussion

Vorsitz: *Hagemann*

Klima: Könnte es so sein, daß Plastiden, die sich langsam vermehren (Plastom IV), genügend Zeit bleibt, allenfalls gesetzte Schäden zu reparieren, während die sich schnell vermehrenden Plastiden nicht mehr regulieren können und deshalb viel unverträglicher und empfindlicher sind?

Stubbe: Das ist durchaus möglich, aber noch nicht nachgewiesen. Wir haben von dem phylogenetisch relativ ursprünglichen Plastom IV bisher noch keine Mutanten gefunden, im Gegensatz zu allen anderen Plastidensorten. Das ließe sich im Sinne Ihrer Annahme deuten.

Hagemann: Die Feststellung, daß die verschiedenen Plastidensorten eine verschiedene Vermehrungsgeschwindigkeit haben, bezieht sich zunächst einmal auf die Konkurrenz zweier Plastidensorten. Beim Vorhandensein von nur *einer* Plastidensorte in der Zelle entfällt naturgemäß eine solche Konkurrenz.

Stubbe: Wie extrem disharmonische Genom-Plastom-Kombinationen im Vergleich mit den harmonischen zeigen, besteht eine Korrelation zwischen Plastidenvermehrung und Zellvermehrung. Es ist aber noch ungeklärt, ob auch geringfügige Unterschiede in der Vermehrungsgeschwindigkeit der Plastiden die Teilungsrate der Meristemzellen beeinflussen oder nicht. Aussagen zu diesem Problem darf man erhoffen, wenn die Untersuchungen von SCHROEDER [41] unter Verwendung von verschiedenen Plastidensorten vergleichend weitergeführt würden.

Karlson: Wenn für die Plastidenentwicklung gewisse Produkte des Zellkerns notwendig sind und deshalb nur der langsam wachsende Typ genügend von diesem Material vorfindet, dann sollten diese Plastiden auch langsamer wachsen. Das müßte sich entweder im Wuchs oder in der Zahl der Plastiden pro Zelle klar zeigen.

Stubbe: Ich habe Wuchsunterschiede tatsächlich festgestellt. Doch wird die Deutung in Ihrem Sinne dadurch erschwert, daß zugleich leichte Chlorophylldefekte auftreten.

Schötz: Ich schätze, daß sich Sorte I höchstens anderthalbmal schneller vermehrt als Sorte IV; ob das ausreicht, um zu postulieren, daß die „schnellere" Sorte nicht genügend Stoffe aus dem Cytoplasma nachziehen kann, erscheint zweifelhaft.

Hagemann: Besteht nicht auch die Möglichkeit, daß eine absolut schnellere Vermehrung der einen oder anderen Plastidensorte dadurch vorgetäuscht wird, daß sie sich zu verschiedenen Zeitpunkten innerhalb der Interphase vermehren?

Stubbe: Haben Sie denn den Verdacht, daß die beigegebene weiße Plastidensorte, die als Bezug für das Messen der Vermehrungsgeschwindigkeit dient, die andere Sorte beeinflußt?

Hagemann: Wenn die Plastidenvermehrung gestoppt wird, sobald die für die Zelle typische Plastidenzahl erreicht ist, wird sich jene Plastidensorte durchsetzen, die — genetisch bedingt — etwas früher mit der Teilung anfängt. Absolut brauchen sich beide Sorten in ihrer Vermehrungsgeschwindigkeit dann überhaupt nicht zu unterscheiden.

Stubbe: Das ist eine sehr interessante, neue Deutungsmöglichkeit.

Duspiva: Gibt es für die Plastiden bestimmte Phasen im Zellcyclus, in denen ihre DNS repliziert wird, bzw. in denen sie sich teilen?

Bartels: Aus Regenerationsversuchen an der somatischen, unteren Blattepidermis von *Peperomia metallica* [BARTELS, Zeit. Bot. **52**, 572—599 (1965)] wissen wir, daß hier die Plastidenzahl der verschiedenartigen Zellen — gewöhnliche Epidermiszelle und Stoma-Nebenzelle mit je 4,65 ($\pm$ 0,8), Drüsen-Nebenzelle mit je 7,53 ($\pm$ 1,5) Plastiden — sich zunächst verdoppelt, dann erst setzen Kern- und Zellteilungen ein. Die experimentelle Überprüfung, wieweit diese Ergebnisse auch für die Zellteilungen in primären Meristemen zutreffen, steht noch aus.

Karlson: Can you give some more details about the kind of interaction between nucleus and plastids in terms of development of the whole cell? What is the kind of these interactions which you have designated as „harmonische Wechselwirkung"?

Stubbe: This is a term introduced by RENNER. It means cooperation or interaction. It is quite clear that this interaction is an interaction between a plastom and a genom. But I see no possibility to make any speculations about it in biochemical terms.

Karlson: In my opinion this phenomenon has very much in common with virusinfection. There are mutants (or strains) of many plants who are not infected by certain viruses, while other of the same species are easely infected. In this case also the nucleus must have some influence on virus reduplication.

Hagemann: There are two kinds of interaction between the nucleus and the plastids. Firstly a physiological one: The plastids need some compounds produced under the influence of the nucleus, and the whole cell needs some compounds produced under the influence of the plastids. On the other hand, there is a genetic interaction: certain mutated genes can induce a plastid mutation which, being induced, is inherited extrachromosomally. Moreover, there exist data which indicate a genetic influence of the plastids on the nucleus.

Karlson: And the way to interpret this?

Stubbe: If certain gene products are necessary for the multiplication of the genetic information in the plastids it is possible to influence the reproduction of the genetic information in the plastids, if the gene products are not readily produced or if they are altered.

Karlson: This would make an incompatibility, but not a true mutation of the plastids. This would be — in my knowledge — the first case of a directed mutation of genetic material.

Hagemann: That a gene-mutation is able to induce a mutation of an extrachromosomal carrier of genetic information is well established in several instances (e.g. in *Arabidopsis,, Epilobium, Hordeum, Nepeta, Oryza, Zea* and — somewhat different—in *Paramecium*). The reciprocal fact (a certain type of plastids inactivates some functions of the nucleus) is known from the work by J. SCHWEMMLE.

Schaller: Has it ever been noticed that only the cells having a mixed population of plastid-mutants would survive? There could be perhaps a kind of complementation (in terms of phage genetics).

Stubbe: We don't have such mutants. It is very difficult to induce plastid-mutants by chemical or physical agents.

Schaller: What I mean is: Only those cells should survive, which have the right combination of plastids, if there is any interaction from plastid to plastid or nucleus/plastid.

Stubbe: We are looking for such types of plastid mutations. We often tried to cross two chlorophylldeficient plastid-types, but we never got green plastids by combination.

Beermann: I would like to know how much work has been done in the direction of in vitro systems of plastids, to find out their autonomy with respect to DNA-, RNA- and protein-metabolism? Is it possible perhaps to breed plastids in a complete in-vitro-system as an organism on its own, for they are semi-autonomous organelles which have a long evolution, may be from a symbiont origin.

Parthier: Probably it has been tried. But it was completely unsuccessful until now to grow or to multiply plastids in vitro.

Hagemann: A possible way for further investigations would be transplantation of plastids.

Taylor: I wondered if anyone has studied at the cellular level the kinds of changes that occur in these so-called mutations.

Hagemann: If a plant is homozygous for a certain recessive allele (e.g. *as* in barley, *ij* in maize), in many cells some of the plastids mutate from green to white. The result is a plant which is striped, because normal and mutated plastids are sorted out at random during vegetative divisions. There are no plastid mutations in the heterozygous state of this nuclear gene, and no mutations in plants with homozygous normal nuclei. By backcrosses we get plants containing normal nuclei and both green and white plastids in their cells, which are sorted out. But even in this case the white plastids never become green, and there are no back-mutations from white to green.

Karlson: Behind the phenotypes so far investigated there can be hidden a large number of genetic defects, which are not easily seen.

Parthier: HEBER and GOTTSCHALK have detected a gene-mutation where the hydrogen transfer in photosynthesis is affected.

Karlson: This information from the nucleus must come somehow to the chloroplast to yield this enzyme. Either the enzyme-protein must be transported from the cytoplasm to the chloroplast, or messenger-RNA must be transported into the chloroplast.

Morphologische Aspekte der Mitochondrien-Vermehrung[1]

Von

K. E. Wohlfarth-Bottermann, Bonn

Mit 10 Abbildungen

Während Struktur und Funktion der schon Anfang dieses Jahrhunderts [9] als Zellorganellen erkannten Mitochondrien heute als weitgehend geklärt gelten können, ist das alte Problem, wie es (anläßlich der Zellteilung oder einer Schädigung des Chondrioms) zu einer Vermehrung bzw. erneuten Komplettierung dieser wichtigen Zellstrukturen kommt, noch nicht in einer Weise gelöst, die allgemeine Zustimmung gefunden hat. Auch in der neueren Literatur findet man zahlreiche, einander widersprechende Theorien, die sich allerdings zum Teil auf sehr verschiedene Zellen beziehen. Es besteht also die Möglichkeit, daß in verschiedenen Zellarten die Vermehrung der Mitochondrien auf unterschiedliche Weise erfolgt.

Der folgende Beitrag berichtet über morphologische Untersuchungen an drei Ciliaten-Arten (*Tetrahymena pyriformis, Paramecium aurelia* und *Paramecium caudatum*) und behandelt zur Einführung und zur späteren Diskussion der eigenen Ergebnisse einige der wichtigsten Auffassungen, die sich auf Untersuchungen an Bakterien, Hefen und mehrzelligen Organismen stützen.

Die phasenkontrastmikroskopische Beobachtung lebender tierischer Zellen läßt in Verbindung mit einer kinematographischen Auswertung zumindest *einen* Weg der Mitochondrienvermehrung erkennen: *Die Entstehung von Mitochondrien aus ihresgleichen*. Wie Abb. 1 zeigt, kommt es bei langen fadenförmigen Mitochondrien von Gewebekulturzellen häufig quer zur Längsachse des Mitochondriums zu einer Fragmentierung in zwei oder mehrere Teilstücke, die daraufhin unter Längenzunahme wieder heranwachsen können. Über den genauen Modus dieses „Fragmentierungsvorganges" lassen sich phasenkontrastmikroskopisch keine Aufschlüsse gewinnen. Sicher ist aber, daß dabei in den meisten Fällen keine Teilung der Mitochondrien in zwei gleich große Tochtermitochondrien erfolgt, sondern daß diese sehr verschiedene Länge besitzen können. Oft werden von einem Mitochondrium nur kleine Stücke abgeschnürt, so daß man eher von einer „Knospung" als von einer „Mitochondrienteilung" sprechen sollte.

---

[1] Der Herr Kultusminister des Landes Nordrhein-Westfalen unterstützte die Untersuchungen durch eine Sachbeihilfe aus Überschußmitteln des Westdeutschen Rundfunks.

Ein anscheinend grundsätzlich verschiedener Vermehrungsmodus, *die Entstehung von Mitochondrien aus andersartigen Zellstrukturen*, ist zwar häufig beschrieben worden, aber immer noch umstritten. Fast alle Zellbestandteile sind für eine Neubildung von Mitochondrien verantwortlich gemacht worden, so die Zellmembran [*19, 47, 48*], der Kern [*5, 6, 7, 12, 24, 38, 39, 55*], endoplasmatische Membranen [*3, 34*], der Golgi-Apparat [*33*], das perinucleare Cytoplasma [*10*], Pinocytosevesikel [*20*] und kleinere membranöse oder granuläre Plasmadifferenzierungen [*8, 25, 49, 50, 53, 58, 59, 60, 62, 68*]. Keine der oben genannten Annahmen hat jedoch bisher allgemeine Anerkennung gefunden.

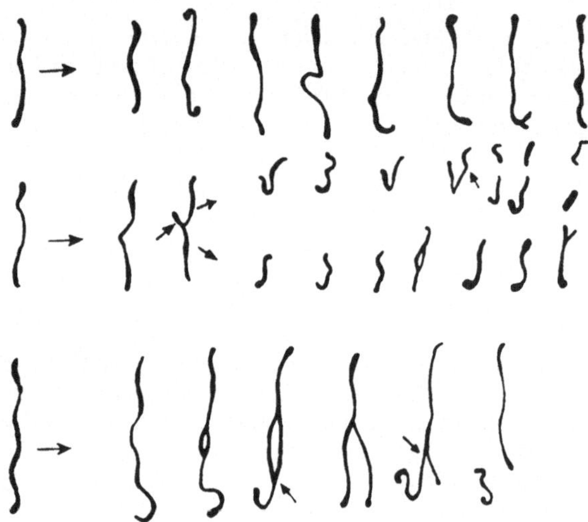

Abb. 1. Darstellung des Formwechsels von Mitochondrien nach kinematographischen Lebendbeobachtungen von Frederic und Chèvremont [*18*]. Beachte die Fragmentierung des Mitochondriums der 2. Reihe

Die Schwierigkeiten einer Analyse der Mitochondrienvermehrung liegen zum Teil in der Fähigkeit dieser Organellen zu einem schnellen Strukturwandel [vgl. *1, 18, 68*]. Der hierdurch bedingte Polymorphismus erschwert bei der morphologischen Analyse von einzelnen Zustandsbildern die Entscheidung, ob bestimmte Formen wirklich als Entwicklungs- bzw. als Vermehrungsstadien anzusprechen sind.

Das hohe Interesse an der Frage der Mitochondrienvermehrung ergibt sich aus der zentralen Bedeutung dieser Organellen für den Stoffwechsel der Zelle [vgl. *28, 31, 43*] und der oft geäußerten Vermutung, daß sie Träger plasmatischer Vererbungsfaktoren darstellen. Das Vorkommen von Desoxyribonucleinsäure zumindest in bestimmten Mitochondrien [*41, 42*] kann nach neueren morphologischen und cytochemischen Untersuchungen wohl nicht mehr bezweifelt werden. Bisher waren entsprechende Befunde aus biochemischen Untersuchungen an Zellfraktionen wegen der naheliegenden Gefahr von Verunreinigungen mit großer Vorsicht aufgenommen worden.

Ergebnisse

Das begrenzte Auflösungsvermögen des für Vitaluntersuchungen an Mitochondrien besonders geeigneten Phasenkontrastmikroskopes hat über ihre Feinstruktur keine wesentlichen Aussagen zugelassen. Dagegen wurden schon bald nachdem geeignete Schneideverfahren für die elektronenoptische Untersuchung von Zellen und Geweben zur Verfügung standen, zwei verschiedene Bautypen von Mitochondrien entdeckt, die

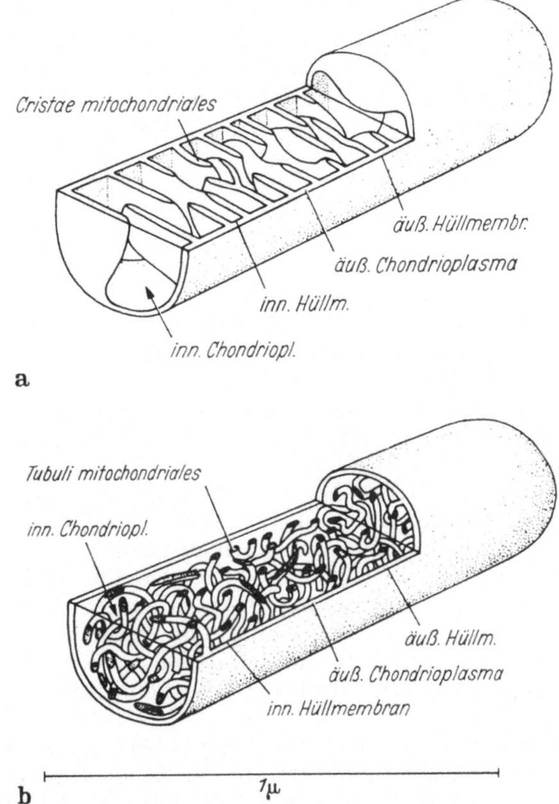

Abb. 2. Schematische Darstellung der Feinstruktur der Mitochondrien vom Crista-Typ (*a*) und vom Tubulus-Typ (*b*) nach elektronenmikroskopischen Untersuchungen [*63*]

durch Schemata in Abb. 2 charakterisiert werden. Der „Crista-Typ" kommt in den meisten vielzelligen tierischen Organismen vor, während der „Tubulus-Typ" vorwiegend bei Einzellern und pflanzlichen Organismen verbreitet ist, bei tierischen Geweben aber nur selten vorkommt. Bei den Ciliaten ist der Tubulus-Typ in besonders klarer Form verwirklicht und auch erstmals beschrieben worden [*61*].

Um der Frage nach dem Vermehrungsmodus der Mitochondrien von Ciliaten näherzukommen, wurden an drei Arten der Ciliaten-Gattungen

Tetrahymena und *Paramecium* folgende elektronenmikroskopische Untersuchungen[1] vorgenommen:

1. Vergleichende Untersuchungen von Zellen aus Kulturen mit hoher und niedriger Teilungsrate (Kultur bei 25° C bzw. 10° C). Bei 25° C erfolgen bei *Paramecium* durchschnittlich etwa vier Zellteilungen innerhalb von 24 Std, bei 10° C in der gleichen Zeit nur eine Zellteilung.

2. Untersuchung von Zellen, die bei 10° C kultiviert worden waren, vor der Fixation aber mehrere Stunden im Kulturmedium bei 25° C gehalten wurden (und umgekehrt).

3. Untersuchung definierter Stadien bezogen auf den Teilungsrhythmus der Zelle. Es wurden insbesondere Zellen $^1/_2$, 1 und 2 Std vor Beginn der Zellteilung, während der Zellteilung und $^1/_2-2$ Std nach erfolgter Zellteilung isoliert und in bezug auf das Vorkommen von Vermehrungsstadien einzeln geprüft.

4. In Versuchen mit 0,0025 M Na-Azid, das $^1/_2-1$ Std auf die Zellen einwirkte, wurde die Atmungsfunktion der Mitochondrien blockiert. (Nach Keilin (vgl. [46], p. 510) wird die Oxydation von Cytochrom C durch NaN_3 und KCN durch eine Cytochromoxydase-Hemmung gestört.) Zur Untersuchung gelangten sowohl Zellen unmittelbar nach Einwirkung des Atmungsgiftes als auch während und nach einer bestimmten Regenerationszeit in giftfreier Chalkley-Lösung.

Die Temperaturversuche und die Analyse definierter Teilungsstadien bezweckten die bessere Erfassung von hier zu vermutenden Vermehrungsstadien von Mitochondrien unter den natürlichen Gegebenheiten eines schnellen Zellwachstums. Die Vergiftungsversuche wurden in der Hoffnung unternommen, durch eine Unterbrechung der Atmungsfunktion der Mitochondrien die Zellen zu veranlassen, das Chondriom zu vergrößern und evtl. diesen Vorgang sichtbar machen zu können.

Anhand der Abb. 3—10 sollen im folgenden die Ergebnisse dieser Versuche verdeutlicht werden. Abb. 3—5 geben die morphologischen Aspekte wieder, die für eine Beteiligung von Vorstadien bei der Mitochondrienvermehrung sprechen, während die Abb. 6—10 diejenigen Befunde zeigen, die eine Vermehrung der Mitochondrien aus ihresgleichen nahelegen.

Wie schon früher beschrieben wurde, treten bei Ciliaten unter Kulturbedingungen, die eine hohe Teilungsrate der Zellen bewirken [62] und besonders in bestimmten Stadien der Zellteilung [15] in vermehrtem Umfange blasige Gebilde auf, die im Innern Anschnitte von Mikrotubuli aufweisen (Abb. 3b, 4a—c). Nach Größe und Gestalt dieser Vesikel liegt der Verdacht nahe, daß es sich hier um Vorstadien von Mitochondrien handeln könne, zumal in ihnen wie bei normalen Mitochondrien Anschnitte röhrenförmiger Strukturen gefunden wurden. Diese ,,Mikrotubuli'' sind allerdings in den meisten Fällen kleiner als die Tubuli mitochondriales. Sie besitzen nur etwa die Hälfte des Durchmessers normaler Tubuli und kommen zudem in geringer Zahl vor. Weiterhin unter-

---

[1] Für die Mitarbeit beim Ansatz und bei der Auswertung der Versuche danke ich Frau B. Koeppen-Lesche.

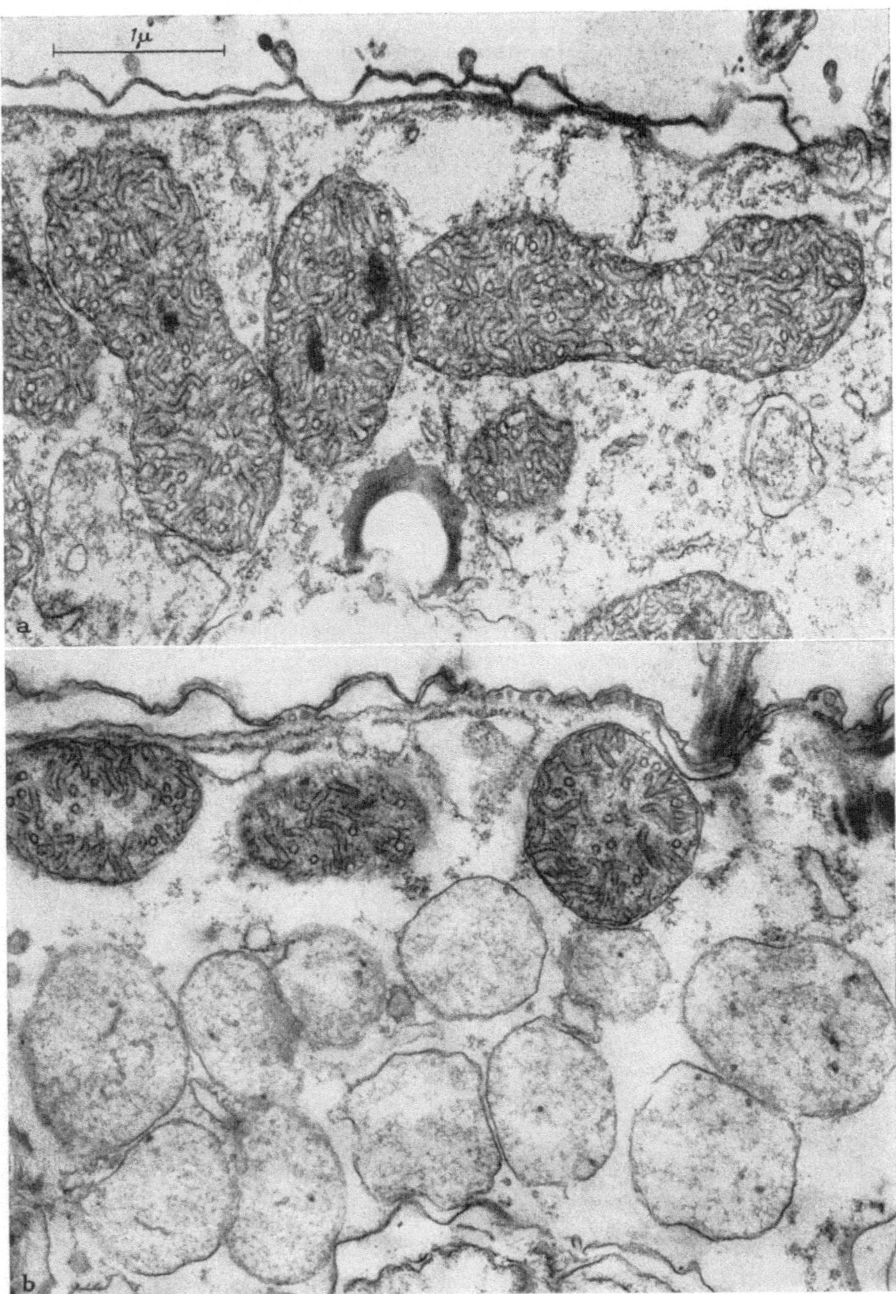

Abb. 3. *Tetrahymena pyriformis*. a Corticaler Bereich einer Zelle aus einer Kultur bei 10° C,
Fix. 2% OsO_4 + 2% $K_2Cr_2O_7$, pH 7,0. b Kultur bei 25° C, Fix. 1% OsO_4, pH 8,0. In Kul-
turen mit rascher Teilungsfolge der Zellen (25° C) treten vermehrt blasenförmige Körper auf
(Bild b, unten), die Anschnitte röhrenförmiger Strukturen („Mikrotubuli") aufweisen.
Vergrößerung 22 000 : 1

scheiden sich diese blasenförmigen Körper von normalen Mitochondrien
in der Mehrzahl aller Fälle dadurch, daß sie nur *eine* begrenzende Membran

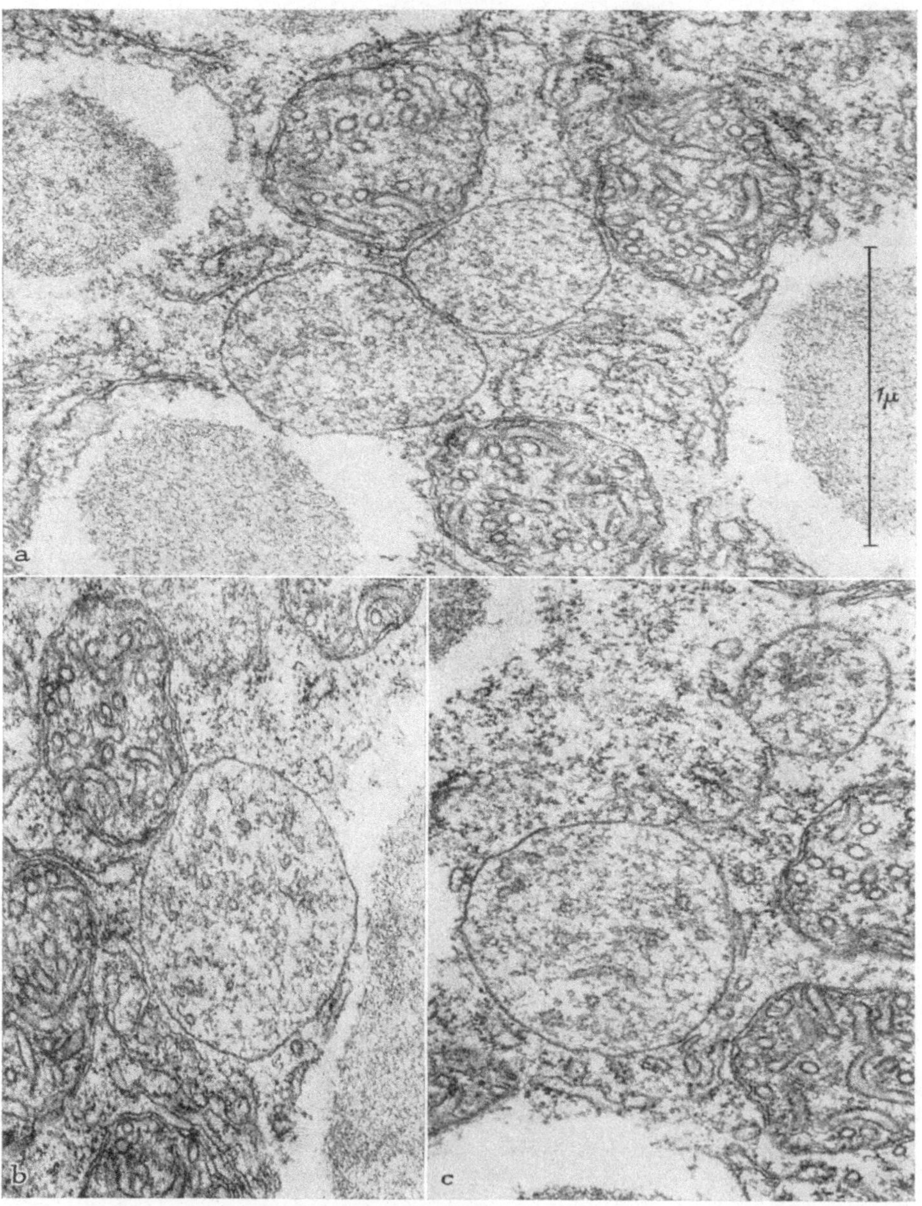

Abb. 4. *Paramecium caudatum*, kultiviert bei 25° C, Fix. 2% OsO_4 + 2% $K_2Cr_2O_7$, pH 7,0.
a—c Ausschnitte aus Zellen, die sich zur Teilung vorbereiten („Teilungsstadien"). Neben
normalen Mitochondrien Vorkommen von blasenförmigen Körpern mit Anschnitten von
Mikrotubuli. Vergrößerung 40 000 : 1

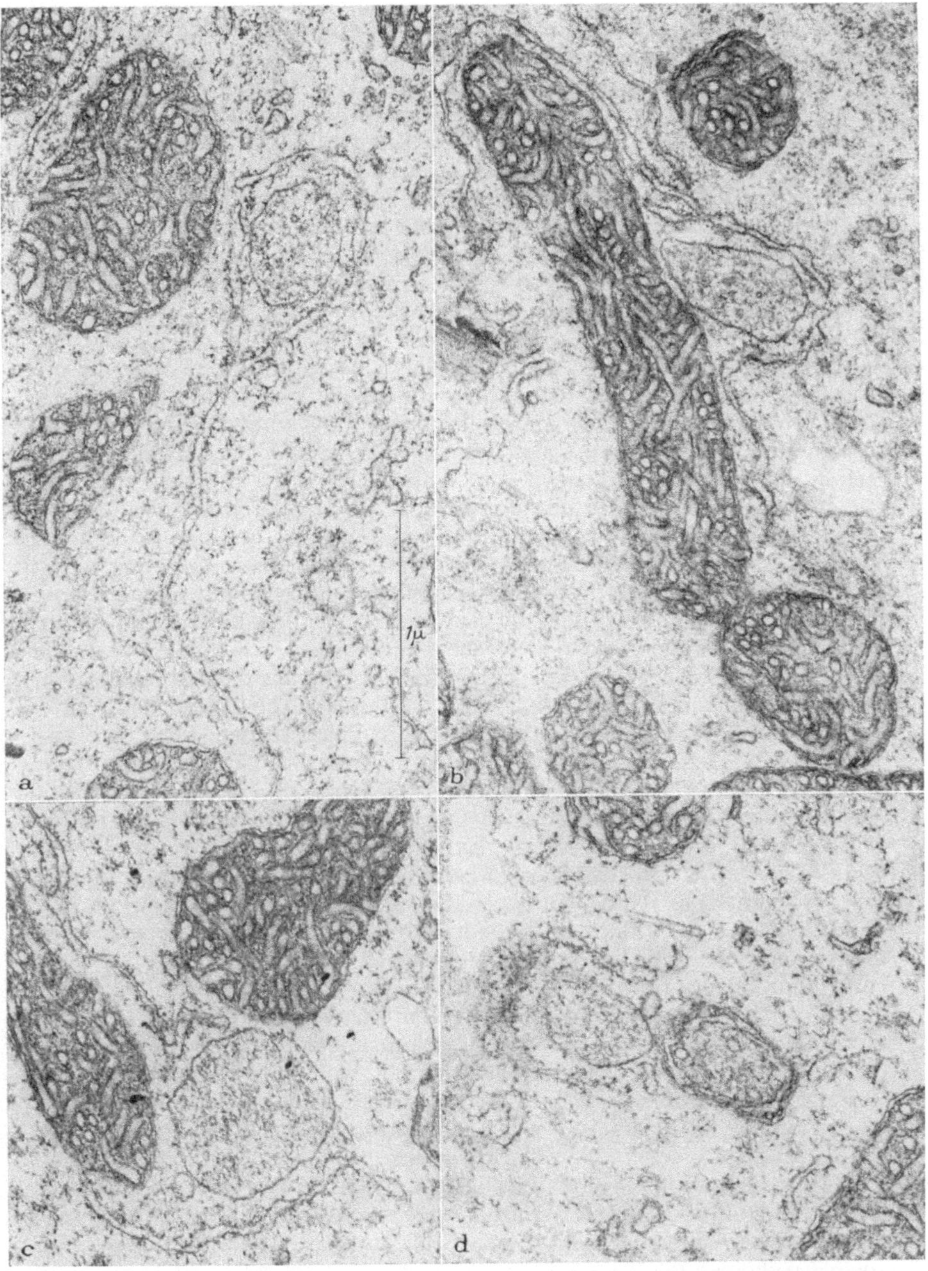

Abb. 5. *Paramecium caudatum*. Ausschnitte aus Zellen, die 60 min der Wirkung von 0,0025 M Na-azid ausgesetzt waren und sich vor der Fixation 90 min in Chalkley-Lösung von der Azid-Wirkung erholten. Fix. 2% OsO_4 + 2% $K_2Cr_2O_7$, pH 7,0. a—c Blasenförmige Körper mit Mikrotubuli. Beachte die enge räumliche Lagebeziehung dieser Gebilde mit schlauchförmigen Membranstrukturen des Cytoplasmas. d mögliche Übergangsformen der blasenförmigen Körper zu Mitochondrien: Auftreten von Tubuli mitochondriales. Vergrößerung 33000 : 1

aufweisen, während normale Mitochondrien eine äußere und eine innere Hüllmembran besitzen (vgl. Abb. 2 und 3a).

Auch nach Einwirkung von Na-Azid auf die Zellen treten diese blasenförmigen Körper vermehrt auf, sofern den vergifteten Zellen Gelegenheit geboten wurde, sich $1^1/_2$ Std von der Giftwirkung in einem giftfreien Medium zu erholen (Abb. 5a—d). Nach solchen Vergiftungsversuchen ist sehr häufig eine auffällig enge Lagebeziehung zwischen den fraglichen Körpern und endoplasmatischen Membranen zu erkennen (Abb. 5a—c).

Für die Entscheidung der Frage, ob es sich bei den blasenförmigen Körpern (die unter allen geprüften Bedingungen einer gesteigerten Vermehrung des Chondrioms in vermehrtem Umfange auftreten) wirklich um Vorstadien von Mitochondrien handelt, ist es vom morphologischen Standpunkt wichtig, ob sich eindeutige Übergangsformen von diesen Körpern zu normalen Mitochondrien finden lassen. Trotz Auswertung eines großen Untersuchungsmaterials (etwa 1200 Aufnahmen von 270 verschiedenen Zellen) war uns dies in unseren jetzigen Untersuchungen nur selten und dann auch nicht in ganz eindeutiger Weise möglich. Abb. 5d zeigt neben einem blasenförmigen Körper ein Gebilde mit Tubuli mitochondriales, das als Übergangsstadium gedeutet werden könnte. Solche Stadien mit doppelter Hüllmembran und Anschnitten röhrenförmiger Gebilde, die in ihrem Durchmesser den Tubuli mitochondriales gleichen, kommen jedoch zu selten vor, um sie mit Sicherheit als Übergangsstadien zu normalen Mitochondrien ansprechen zu können.

Dagegen finden sich Hinweise für eine Vermehrung der Mitochondrien aus ihresgleichen im Vergleich dazu reichlich: Abb. 6a—c zeigt drei Mitochondrien aus Zellen, die bei 10° C kultiviert wurden, vor der Fixation aber in ihrem ursprünglichen Kulturmedium 4 Std auf 25° C erwärmt wurden. Hierdurch sollte eine kurzfristige Steigerung des Stoffwechsels bewirkt werden. Bei den abgebildeten Mitochondrien handelt es sich in allen Fällen um hantelförmige Einschnürungen relativ großer Mitochondrien. Das Vorliegen einer eindeutigen hantelförmigen Einschnürung ist aus Dünnschnitten nur in den Fällen als sicher anzusehen, wo an der Einschnürungsstelle die begrenzenden Hüllmembranen des betreffenden Mitochondriums deutlich abgebildet sind. In solchen Fällen muß es sich um mehr oder weniger exakte Querschnitte der Membranen handeln. Dies aber beweist, daß das betreffende Mitochondrium an der fraglichen Einschnürungsstelle wirklich eine Einschnürung der gezeigten Dimension aufweist und daß eine Einschnürung nicht durch einen tangentialen Anschnitt einer geringeren Einschnürung vorgetäuscht wird. Eine absolute Sicherheit in der Erkennung von Einschnürungen bringt die Analyse von Serienschnitten.

Abb. 7a—c zeigt ebenfalls stark taillierte Mitochondrien, die aus frühen Teilungsstadien von *Paramecium* stammen. Zellen, die sich zur Teilung anschicken, können relativ früh an ihrer Größe und später an dem Erscheinen der ersten Andeutung einer Teilungsfurche erkannt und isoliert werden. Nach physiologischen Untersuchungen [*36*] stoppt bei *Tetrahymena* das Wachstum der respiratorischen Aktivität („das spekulativ mit der Synthese respiratorischer Enzyme in Verbindung gebracht werden kann") vor dem sichtbaren Eintritt der Zellteilung. Die Aktivität beginnt mit verdoppelter Rate zu wachsen, bevor die Zellteilung beendet ist. Entsprechend findet man in Zellen, die soeben die Anlage einer Teilungsfurche erkennen lassen, also in dem Stadium, das der Zellteilung unmittelbar vorausgeht, taillierte Mitochondrien der in Abb. 7 gezeigten

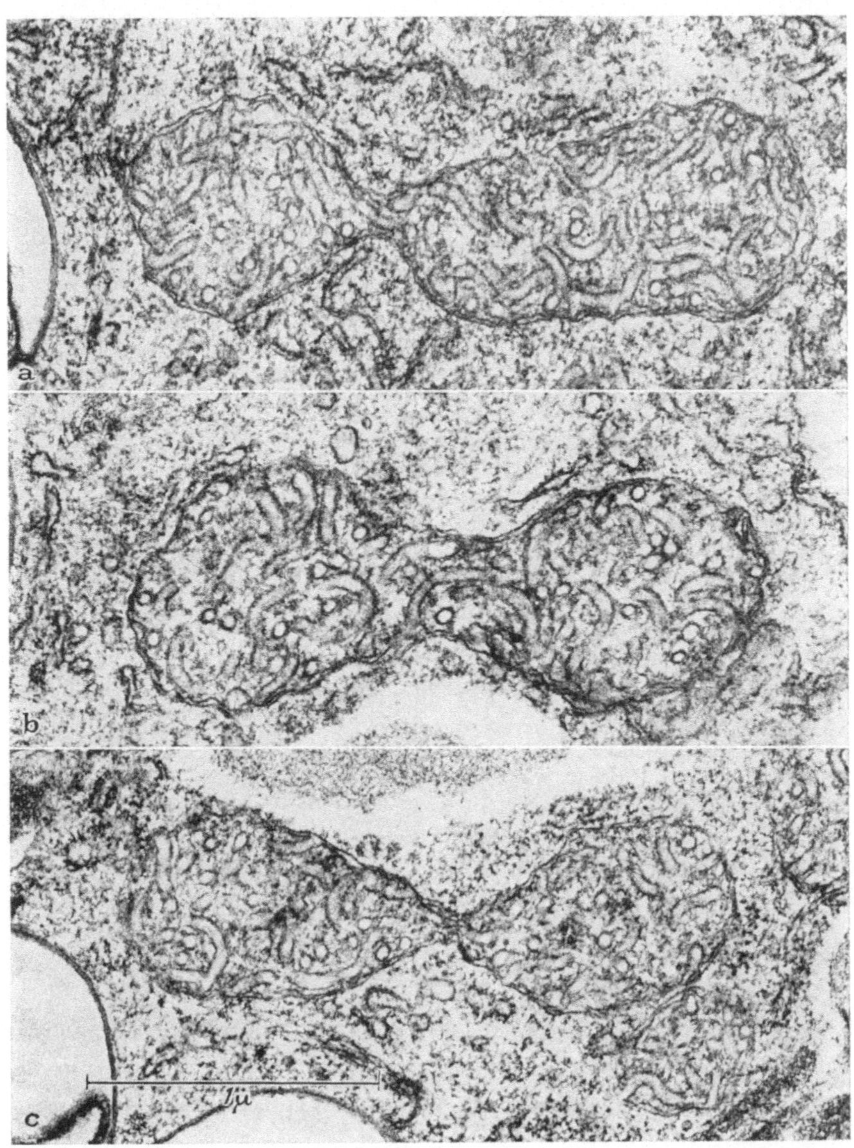

Abb. 6. *Paramecium aurelia.* Mitochondrien aus Zellen, die bei 10° C kultiviert wurden, vor der Fixation aber im Kulturmedium 4 Std bei 25° C gehalten worden waren. Fix. 2% OsO_4 + 2% $K_2Cr_2O_7$, pH 7,0. Mitochondrien mit verschieden stark ausgeprägten Taillen. Vergrößerung 38000 : 1

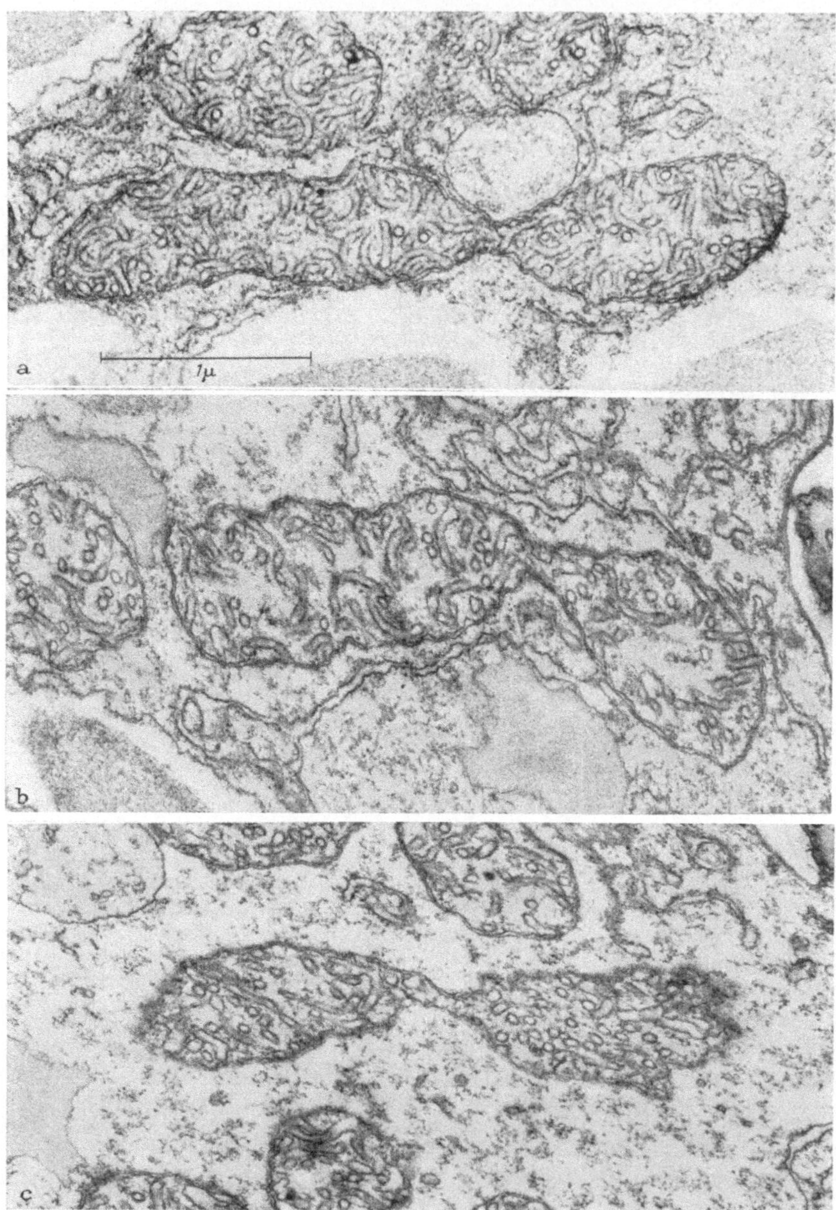

Abb. 7. *Paramecium caudatum*. Mitochondrien aus Zellen, die bei 25° C kultiviert wurden und kurz vor der Teilung standen (soeben erkennbare Anlage der Teilungsfurche bei den betreffenden Zellen). Fix. 2% OsO_4 + 2% $K_2Cr_2O_7$, pH 7,0. Hantelartige Verformung der Mitochondrien. Vergrößerung 28000 : 1

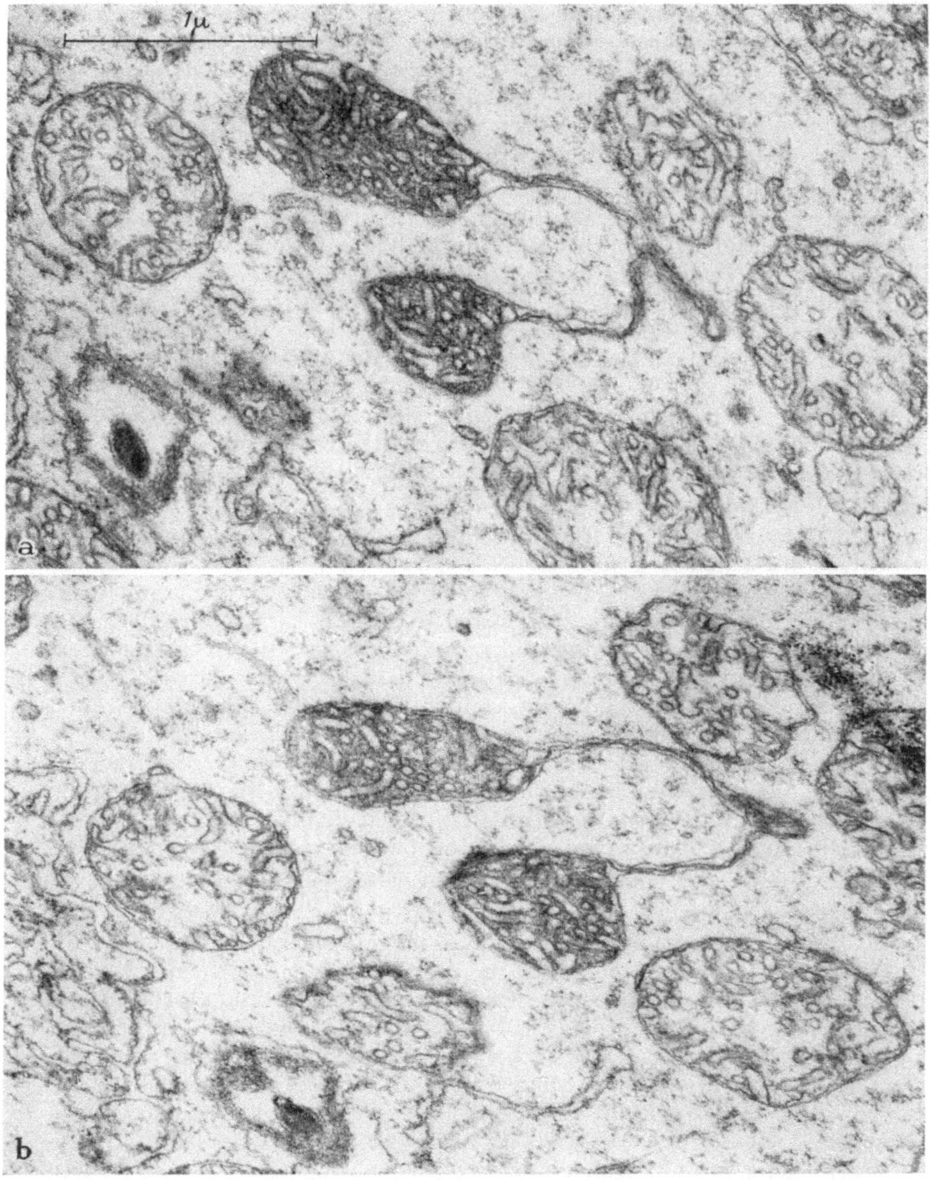

Abb. 8. *Paramecium caudatum*. Wie Abb. 7. a und b Serienschnitte durch ein Mitochondrium, dessen beide Hauptteile nur durch die äußere und innere Mitochondrien-Hüllmembran miteinander in Verbindung stehen. Beachte die größere Elektronendichte der Matrix dieses Mitochondriums im Vergleich zur Matrix der benachbarten Mitochondrien. Vergrößerung 33 000 : 1

Art. Weder Zellen mit hoher Teilungsrate noch Paramecien, die bei 10° C kultiviert und mehrere Stunden durch Erwärmung auf 25° C zu einem erhöhten Stoffwechsel angeregt wurden, zeigen (mit Ausnahme der bereits diskutierten Vesikel) am Chondriom andere morphologische Aspekte, die sich auf eine Vermehrung der Mitochondrien beziehen lassen, als die in Abb. 6 und 7 beschriebenen Taillen-Formen.

Nur selten werden Erscheinungen beobachtet, wie sie in Abb. 8 charakterisiert sind. Die abgebildeten Serienschnitte (8a u. b) zeigen ein Mitochondrium, dessen beide Hauptstücke nur durch einen dünnen Schlauch in Verbindung stehen; er besteht aus der äußeren und inneren Hüllmembran des Mitochondriums. Von besonderem Interesse ist hier der große Elektronenkontrast der Matrix des aus zwei Teilstücken bestehenden Mitochondriums im Vergleich zu der Matrix der benachbarten Mitochondrien. In jüngster Zeit ist der Nachweis geführt worden [2], daß die Matrix, d. h. das innere Chondrioplasma der Mitochondrien (vgl. Abb. 2), für ihre Kontraktilität verantwortlich ist und daß sich eine größere Elektronendichte der Matrix auf einen Zustand stärkerer Kontraktion beziehen läßt.

Unterbricht man die Atmungsfunktion der Mitochondrien durch Einsetzen der Zellen in Lösungen von 0,0025 M Na-Azid, so ist das Ausmaß der Vergiftung der Zellen an ihrer Volumenzunahme recht gut abzuschätzen. In unseren Versuchen waren Zellen von *P. caudatum*, die *länger* als 1 Std dem Gift in der Konzentration von 0,0025 M ausgesetzt waren, in der Mehrzahl nicht mehr regenerationsfähig. Überträgt man jedoch Zellen nach nur einstündiger Gifteinwirkung in eine Chalkley-Lösung, so nehmen die Zellen innerhalb von 30—60 min wieder ihre normale Gestalt an und erweisen sich als teilungsfähig. Offenbar werden nur sehr wenige Mitochondrien irreversibel geschädigt: Ganz selten findet man in regenerierenden Zellen Degenerationsformen von Mitochondrien. Diese ähneln dann Abbaustadien von Mitochondrien, wie sie nach Strahlenschäden auftreten [65] und daher leicht zu erkennen sind.

Die elektronenmikroskopische Untersuchung der morphologischen Folgen der Giftwirkung zeigt nach 30 min Regeneration erheblich vergrößerte Mitochondrien (Abb. 9 u. 10a—c) und zugleich ein scheinbar stark zerstörtes Cytoplasma. Nach 90 min Regeneration (Abb. 5 und 10d) findet man das Cytoplasma bereits wieder normal strukturiert, stark vergrößerte fadenförmige Mitochondrien bestimmen aber immer noch das Bild des Chondrioms. Das Längen/Breiten-Verhältnis der Mitochondrien von *Paramecium* geht im allgemeinen über 3 : 1 nicht hinaus. Ausgesprochene Fadenformen (Abb. 5b, 9) sind eine Folge der Natrium-Azid-Einwirkung. Sowohl nach kürzerer als auch nach längerer Regenerationszeit (30 bzw. 90 min) erweisen sich die meisten fadenförmigen Mitochondrien, die offenbar längsgeschnitten sind, als stark tailliert (Abb. 9 u. 10). Abb. 9a u. b stellen Serienschnitte durch ein solches Mitochondrium dar, aber auch Formen, wie sie in Abb. 10 gezeigt werden, sind typisch für Na-Azid-vergiftete Zellen, die sich in der Erholungsphase befinden (Abb. 9, 10a—c) oder sich schon erholt haben (Abb. 5, Abb. 10d).

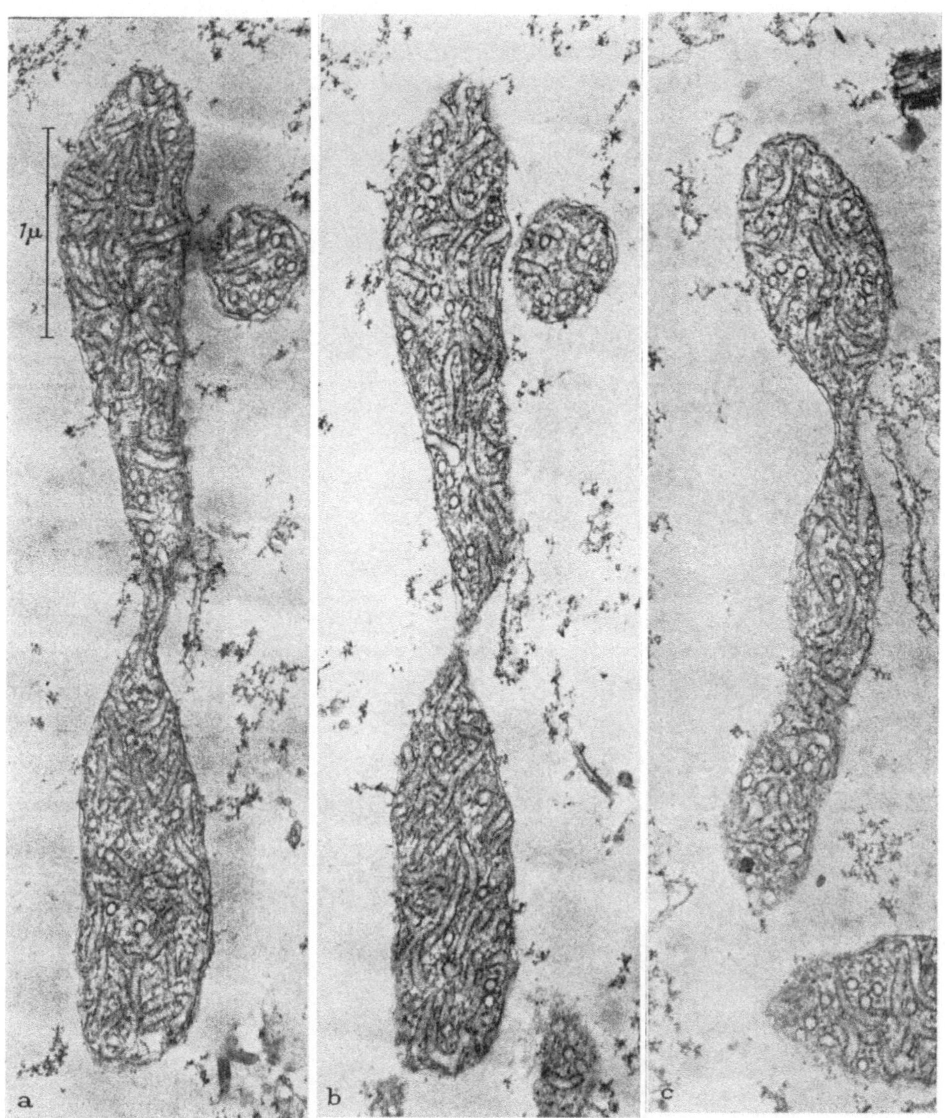

Abb. 9. *Paramecium caudatum*. Mitochondrien aus Zellen, die 60 min der Wirkung von 0,0025 M Na-azid ausgesetzt waren und sich vor der Fixation 30 min in Chalkley-Lösung von der Azid-Wirkung erholten. Fix. 2% OsO_4 + 2% $K_2Cr_2O_7$, pH 7,0. Stark vergrößerte Mitochondrien mit extremer hantelartiger Einschnürung. a und b Serienschnitte. Die schlechte Strukturerhaltung des Cytoplasmas ist eine Azid-Wirkung, die nach 30 minütiger Erholung der Zellen noch nicht kompensiert ist, sondern erst nach 90 minütiger Erholung (vgl. Abb. 5 und Abb. 10 d) zurückgeht. Vergrößerung 28 000 : 1

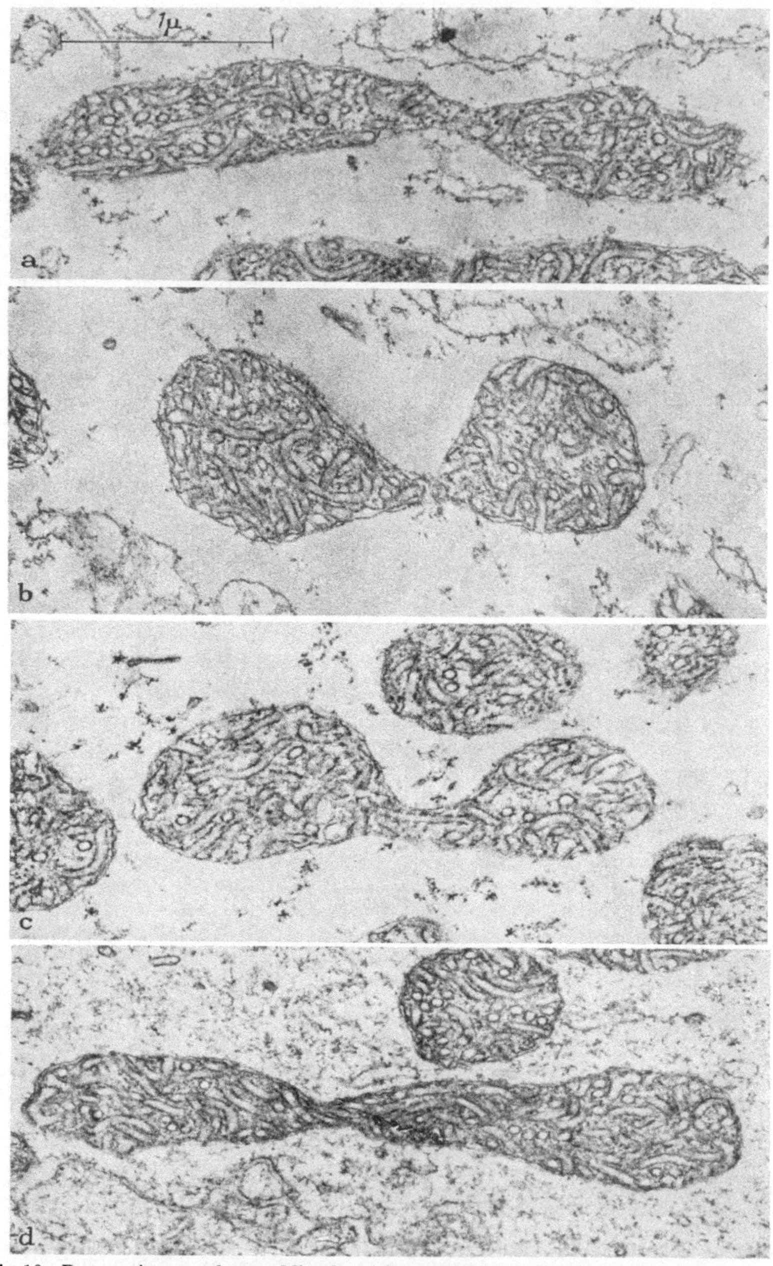

Abb. 10. *Paramecium caudatum.* Mitochondrien aus Zellen, die 60 min der Wirkung von
0,0025 M Natriumazid ausgesetzt waren und sich vor der Fixation 30 min (a—c) bzw.
90 min (d) in Chalkley-Lösung von der Azid-Wirkung erholten. Fix. 2% OsO_4 + 2%
$K_2Cr_2O_7$, pH 7,0. Hantelartige Einschnürung der Mitochondrien. Beachte die nach 90 minütiger Regeneration wieder normale Struktur des Cytoplasmas in 10 d im Vergleich zu
10 a—c, wo nach nur 30 minütiger Erholung das Cytoplasma kaum dargestellt werden kann.
Vergrößerung 28 000 : 1

Diskussion und Deutung

In früheren Untersuchungen zum Problem der Mitochondrienvermehrung [62] waren die in Abb. 3—5 dargestellten Vesikel mit Mikrotubuli als „Promitochondrien", also als Entwicklungsstadien von Mitochondrien gedeutet worden. Mit Hilfe der inzwischen verbesserten Präparationstechnik (Polyester statt Methacrylat als Einbettungsmittel) lassen sich jedoch eindeutige Übergangsstadien zwischen den mit Mikrotubuli besetzten Vesikeln und normalen Mitochondrien nicht in einem genügend großen Umfange sichtbar machen, um die frühere Deutung noch mit Sicherheit aufrechterhalten zu können. Sollten die Vesikel mit Mikrotubuli Vorstadien von Mitochondrien sein, so müßte man annehmen, daß ihre Umwandlung zu normalen Mitochondrien so schnell erfolgt, daß Übergangsstadien nur selten erfaßt werden. Obgleich diese Möglichkeit nicht auszuschließen ist, müssen wir aus den vorliegenden morphologischen Untersuchungen zunächst den Schluß ziehen, daß bei Ciliaten eine Entstehung von Mitochondrien aus Vorstadien, den sog. Promitochondrien, zwar nicht ausgeschlossen ist, aber auch nicht als bewiesen gelten kann. Zur Klärung der Natur der hier diskutierten Körper, die zu Zeiten einer Vergrößerung des Chondrioms vermehrt auftreten, sind weitere Untersuchungen erforderlich, vor allem wohl ihre Isolierung nach fraktionierender Zentrifugation der Zellbestandteile mit nachfolgender Analyse der enzymatischen Aktivitäten.

Dagegen erscheint die Vermehrung des Chondrioms durch Einschnürung langer Mitochondrien und nachfolgender Abschnürung von zwei mehr oder weniger gleich großen oder ungleich großen Teilstücken, also eine Vermehrung der Mitochondrien aus ihresgleichen, nach den Abb. 6 bis 10 kaum zu bezweifeln. In keinem Falle konnte in unseren Versuchen eine Teilung der Mitochondrien in Längsrichtung beobachtet werden. Zu allen Zeiten einer natürlichen starken Vermehrung des Chondrioms (hohe Teilungsrate, Zellen unmittelbar vor und nach der Zellteilung) traten in großem Umfange mehr oder weniger hantelförmig eingeschnürte Mitochondrien auf, die auch schon früher [62] als morphologisch sichtbarer Aspekt der Mitochondrien-Vermehrung gedeutet wurden. Man muß wohl annehmen, daß es bei der weiteren Zusammenschnürung des Mitochondriums zu einer Fusion der Hüllmembranen kommt, wodurch der Aufteilungsprozeß vollzogen wäre. Endstadien eines solchen Vorganges sind ebenfalls seinerzeit abgebildet worden [62], so daß sich ihre Darstellung an dieser Stelle erübrigt.

Weder früher noch in den jetzigen Untersuchungen ergaben sich Hinweise dafür, daß die Aufteilung eines Mitochondriums in zwei Teilstücke zu zwei quantitativ gleichen Tochtermitochondrien führt, vielmehr können diese recht unterschiedliche Größe besitzen.

Die Ergebnisse unserer Vergiftungsversuche stimmen hiermit überein: Nach Blockade der Atmungsfunktion kommt es zu einer Vergrößerung der einzelnen Mitochondrien, insbesondere durch Längenzunahme (Abb. 5, 9 u. 10), wie sie auch für pflanzliche Gewebe nach entsprechenden Versuchen beschrieben wurde [66]. Es erscheint plausibel, daß die Zelle nach teilweiser Blockierung ihrer Atmung diesen Zustand durch eine Vergrößerung ihrer Mitochondrien zu kompensieren versucht. Von besonderem Interesse in diesem Zusammenhang ist es, daß dabei gleichzeitig wieder in vermehrtem Umfange hantelförmige Mitochondrien auftreten, wie sie auch unter natürlichen Verhältnissen während einer Vergrößerung des Chondrioms gefunden werden. Es erscheint daher nicht abwegig, diese unter pathologischen Verhältnissen vorkommenden Hantelformen (Abb. 9 und 10) mit denen unter natürlichen Gegebenheiten auftretenden (Abb. 6 u. 7) zu vergleichen und beide im Sinne einer Mitochondrienvermehrung zu deuten.

Wie stimmen die an Ciliaten gewonnenen Ergebnisse mit Befunden überein, die an anderen Zellen gewonnen wurden? Es kann nicht Sinn dieses Beitrages sein, die vorliegende Literatur auch nur annähernd vollständig zu diskutieren, es soll aber versucht werden, übersichtsmäßig Aufschluß darüber zu gewinnen, ob bei anderen Zellen gleiche Verhältnisse wie bei Ciliaten vermutet werden dürfen.

Für eine Entstehung von Mitochondrien aus morphologisch nicht mit Mitochondrien identischen Vorstadien sprechen eine Reihe von Arbeiten, die bereits z. T. in der Einleitung zitiert wurden. Hier können diejenigen Annahmen außer acht gelassen werden, die offensichtlich nicht mit den Verhältnissen bei Ciliaten in Übereinstimmung stehen, so eine Entstehung neuer Mitochondrien aus der Zellmembran, aus dem Kern, dem Golgi-Apparat oder aus Pinocytose-Vesikeln. Alle diese Annahmen erscheinen auch entweder nicht genügend belegt, oder sind in Nachuntersuchungen bezweifelt worden (vgl. [14, 37]). Für die Entstehung von Mitochondrien aus granulären oder membranösen Cytoplasmadifferenzierungen, die man mit den fraglichen Vesikeln bei Ciliaten vergleichen könnte, sprechen Untersuchungen an *Paramecium* [15, 62], an den "microbodies" der Leberzellen [50] und an Gewebekulturzellen [58, 59, 60]. Jedoch gilt für diese Arbeiten der gleiche Einwand, der bei der Deutung der Vesikel mit Mikrotubuli in der vorliegenden Studie berücksichtigt werden muß: Eine lückenlose Erfassung zu fordernder Zwischenstadien, d. h. Übergangsstadien von den fraglichen Vorstadien bis zu normalen Mitochondrien sind auch bei den vorgenannten Arbeiten nicht in der wünschenswerten Deutlichkeit nachgewiesen worden. Zusammenfassend läßt sich also sagen, daß die Möglichkeit der Entstehung von Mitochondrien aus Vorstadien oder sonstigen Zellstrukturen bei verschiedenen anderen Zellarten als Ciliaten ebenfalls nicht ausgeschlossen werden kann, aber auch nicht bewiesen ist.

Eine Ausnahme bilden die Bakterien. Hier kann die Entstehung von mitochondrienähnlichen Körpern aus der die Bakterienzelle umhüllenden Cytoplasmamembran (Plasmalemma) kaum noch bezweifelt werden: Die sog. Membrankörper der Bakterien ("Chondrioide", Bakterien-Mitochondrien, "Mesosomen") [17, 21, 54] bestehen aus einer Hüllmembran, die zahlreiche tubuläre und vesikuläre Innenstrukturen umschließt. Morphologisch besitzen diese Körper also starke Ähnlichkeit mit den Mitochondrien vom Tubulus-Typ [21, 22]. Biochemische Untersuchungen der isolierten Membrankörper ergaben übereinstimmend die gleichen

Enzyme (Dehydrogenasen, Phosphorylasen, Cytochrome u. a.) wie in Mitochondrien höherer Organismen [*29, 40, 57, 67*]. Man sieht daher diese Membrankörper nach biochemischen, cytochemischen und morphologischen Kriterien als die Mitochondrien der Bakterien an (vgl. [*21, 27*]). Die Mitochondrien der Bakterien scheinen allerdings insofern eine Sonderstellung zu haben, als sie neben einer „normalen" Mitochondrienfunktion auch noch Aufgaben bei der Zellwand-Bildung, bei der Verdoppelung von Kernstrukturen und bei der Bildung der Endosporen übernehmen können [*44*]. Die Bakterien-Mitochondrien entstehen *de novo* durch Invagination der Cytoplasmamembran, mit der sie ein kontinuierliches Membransystem bilden [*17, 21, 26*]. Von besonderem Interesse in unserem Zusammenhang ist es, daß die Bakterien-Mitochondrien neben einer nachweisbaren *de novo*-Entstehung angeblich gleichzeitig die Fähigkeit besitzen sollen, sich durch Teilung zu vermehren [*21, 23, 51*].

Auch bei der Hefe *Turolopsis utilis* ist eine Neuentstehung von Mitochondrien aus plasmatischen Membransystemen sehr wahrscheinlich: Bei Zucht unter aeroben Bedingungen enthält diese Hefe Mitochondrien, während unter anaeroben Zuchtbedingungen keine Mitochondrien und dementsprechend auch keine Cytochrome nachweisbar sind. Statt dessen findet sich ein charakteristisches „Membransystem", das 2 Dehydrogenasen der Atmungskette enthält. Bei Belüftung der anaerob wachsenden Zellen werden Cytochrome synthetisiert, während gleichzeitig aus plasmatischen Membranen neue Mitochondrien entstehen sollen [*34*].

Elektronenmikroskopisch sind in Teilung befindliche Mitochondrien vom Crista-Typ schon verschiedentlich abgebildet worden [*16, 56, 62*]. Eindeutige Hinweise, wie eine Teilung der Mitochondrien aus Leber-Zellen vor sich zu gehen scheint, enthält eine Arbeit von LAFONTAINE und ALLARD [*30*]. Daraus geht hervor, daß die Vorbereitung der Abschnürung der beiden Teilstücke bei diesem Mitochondrien-Typ offenbar etwas anders verläuft als bei Mitochondrien vom Tubulus-Typ: Unter dem Einfluß von 2-Me-DAB (2-Methyl-Dimethylaminoazobenzol, einer dem Buttergelb ähnlichen, jedoch nicht cancerogenen Verbindung) wird die Zahl der Lebermitochondrien erheblich vergrößert. Gleichzeitig mit dieser Vergrößerung des Chondrioms finden sich häufig Mitochondrien, die an einer Einschnürungsstelle durch zwei Membranen quer in zwei Stücke geteilt sind, wobei diese Doppelmembranstruktur mit der inneren Hüllmembran des Mitochondriums in Verbindung steht. Schließlich finden die oben genannten Autoren Stadien, bei denen die beiden Teilstücke eines Mitochondriums etwas auseinandergerückt und nur noch durch die äußere Hüllmembran miteinander verbunden sind.

Der Unterschied im Vorgang der Aufspaltung eines Mitochondriums in zwei Teilstücke besteht beim Crista-Typ im Vergleich zum Tubulus-Typ vielleicht darin, daß beim Crista-Typ die Membranen der Cristae, kommunizierend mit der inneren Hüllmembran des Mitochondriums (vgl. Abb. 2), zunächst eine vollständige Querseptierung zwischen den beiden Teilstücken anlegen, bevor es zur vollständigen Trennung der Teilstücke kommt. Beim Tubulus-Typ kann dies nicht beobachtet werden, vielmehr durchziehen hier auch an den Stellen starker Durchschnürung (vgl. Abb. 9a) oft noch 1 oder mehrere Tubuli *in Längsrichtung* das Mitochondrium und verbinden so noch die beiden präsumptiven Teilstücke miteinander. Eine stärkere Abgrenzung der Teilstücke wie bei Mitochondrien vom Crista-Typ konnte jedenfalls bei den Mitochondrien von *Paramecium* nicht beobachtet werden.

Wie für Tubulus-Mitochondrien finden sich also in der Literatur Anhaltspunkte, daß auch der Vorgang der Teilung von Crista-Mitochondrien (vgl. auch [*13, 32, 45*]) morphologisch analysierbar ist und wenn auch nicht gleich, so doch ähnlich verläuft wie bei den Tubulus-Mitochondrien.

Das morphologische Ergebnis, wonach Mitochondrien beider Bautypen sich durch Abschnürungsvorgänge in zwei mehr oder weniger verschieden große Hälften teilen können, stimmt sowohl mit Erfahrungen phasenkontrastmikroskopischer Lebendbeobachtungen als auch mit biochemischen und quantitativen Untersuchungen überein, wonach eine Vermehrung des Chondrioms durch Einlagerung neuen Materials [*1, 11, 35*] in bereits existierende Mitochondrien (Wachstum) und ihre nachfolgende Vermehrung durch Teilung [*4*] erfolgen kann. Zweifellos kann man auch unter bestimmten Bedingungen und bei bestimmten Objekten durch geeignete Untersuchungen eine Vermehrung von Mitochondrien aus Vorstufen ausschließen [*35*]. Jedoch erscheint es verfrüht, hieraus die generelle Schlußfolgerung zu ziehen, daß eine „Neuentstehung" von Mitochondrien nicht möglich wäre. Wir wissen, daß eine Neuentstehung von Mitochondrien zumindest bei Bakterien vorkommt und anscheinend zugleich eine Vermehrung durch Teilung [*21, 23, 26, 51*]. Eine Neubildung von Mitochondrien bei Hefen ist zumindest sehr wahrscheinlich [*34*]. Diese Tatsachen sollten in Verbindung mit den morphologischen Hinweisen auf mögliche Vorstufen auch bei höheren Organismen manche Autoren zur Vorsicht mahnen, Ergebnisse an bestimmten Objekten nicht zu generalisieren und auf Grund von Einzeluntersuchungen eine Neubildung von Mitochondrien in den Bereich der reinen Spekulation zu verweisen. Zusammenfassend läßt sich sagen, daß eine Entstehung von Mitochondrien aus ihresgleichen heute als *ein* Weg der Vermehrung des Chondrioms wohl als bewiesen gelten kann. Ob es bei den Eukaryonten der einzig mögliche ist, erscheint noch fraglich.

Abschließend noch einige Gedanken zu der Frage, inwieweit wir berechtigt sind, die Mitochondrien als Reduplikanten anzusehen. Hierzu ist zunächst zu berücksichtigen, daß wir trotz zahlreicher Bemühungen keinen Beweis dafür haben, daß die Mitochondrien bei der Zellteilung stets auf eine bestimmte Weise auf die beiden Tochter-Zellen verteilt werden [*13*]. Die elektronenmikroskopischen Bilder bestätigen nun die phasenkontrastmikroskopische Beobachtung, daß bei Teilung von Mitochondrien keine quantitativ identischen Tochtermitochondrien entstehen, ein Befund, der allerdings qualitativ identische Tochtermitochondrien nicht ausschließt. Da bei der Teilung aber jedenfalls keine quantitativ identischen Mitochondrien entstehen, leuchtet es ein, daß es bei der Zellteilung auch anscheinend nicht wichtig ist, genau quantitativ gleiche Hälften des Chondrioms auf beide Tochterzellen zu verteilen.

Der in jüngerer Zeit geführte Nachweis von Desoxyribonucleinsäure in Mitochondrien [*41, 42*] beweist ebenfalls keineswegs zwingend eine morphologische Kontinuität dieser Zellorganellen, wenn man sich unsere heutigen Kenntnisse über die Vermehrung der auch DNS-haltigen Basalkörper vergegenwärtigt.[1]

Immerhin muß man Mitochondrien aber wohl als den *Ort der Reduplikation* von charakteristischen Struktur- und Enzym-Mustern bezeichnen. Das identisch zu reproduzierende Grundmuster der Struktur liegt in der elektronenmikroskopischen Größenordnung der Cristae und Tubuli, das Grundmuster der Funktion, d. h. der den Cristae und Tubuli aufsitzenden und der in der Matrix befindlichen Enzyme, in der Größenordnung der Makromoleküle. Von entscheidender Wichtigkeit ist wohl vor allem die Reproduktion des Funktionsmusters, also eines Musters makromolekularer Größenordnung. Um ein Muster makromolekularer Größenordnung identisch zu reproduzieren, erscheint das komplexe elektronenmikroskopische Gefüge eines ganzen Mitochondriums nicht unbedingt notwendig; jedenfalls haben wir für eine solche Notwendigkeit bisher keinen Beweis. Bei Bakterienmitochondrien persistiert offenbar *nicht* das Strukturmuster, sondern nur das Funktionsmuster, und zwar wohl in der Cytoplasmamembran, aus der neue Bakterienmitochondrien entstehen können. Ob das Strukturmuster bei Mitochondrien höherer Organismen persistiert (also immer nur sich selbst redupliziert), oder unter Mitwirkung von Nucleinsäuren aus „einfacheren" Strukturen (z. B. Membranen, die nicht als Cristae oder Tubuli differenziert sind) neu aufgebaut werden kann, ist eine Frage, die heute noch nicht sicher zu beantworten ist. In Anbetracht der Dynamik und Konvertibilität cytoplasmatischer Membranen [*28, 52, 63, 64*] scheint der Zelle eine Neubildung von Membranen jedenfalls relativ leicht möglich zu sein.

Da das Funktionsmuster der Mitochondrien, d. h. die räumliche Anordnung der Enzyme (die zu charakteristischen Reaktionsketten führt) offensichtlich persistiert und redupliziert wird, erscheint vielleicht die Beantwortung der Frage, ob die vergleichsweise „gröbere" Gestalt des Mitochondriums immer erhalten bleibt oder evtl. aus Vorstadien bzw. Teilstücken neu aufgebaut werden kann, von geringerer Bedeutung, als diesem Problem heute gemeinhin zugemessen wird. Dagegen werden morphologische und cytochemische Hinweise auf das Vorkommen von Desoxyribonucleinsäure in Mitochondrien [*41, 42*; vgl. auch den Beitrag Tuppy u. Wintersberger, p. 325] im Zusammenhang mit der Frage der

---

[1] Die Basalkörper der Ciliaten, die mit den Centriolen zu homologisieren sind, enthalten DNS [J. Randall et al.: Proc. roy. Soc., B, **162**, 473 (1965)], vermehren sich aber entgegen früheren Annahmen (Lwoff) nicht durch Teilung, sondern entstehen neu — allerdings in enger räumlicher Beziehung zu bereits vorhandenen Basalkörpern [J. Randall et al.: Proc. Linn. Soc. Lond. **174**, 31 (1963)]. Bei der Amöbe *Naegleria gruberi* werden Basalkörper beim Übergang der Amöbenform zur Flagellatenform sogar neugebildet, ohne daß die Amöbenform vorher Basalkörper enthält [F. Schuster: J. Protozool. **10**, 297 u. 313 (1963)]. Dabei besteht im Falle der Basalkörper der Ciliaten kein Zweifel, daß sie entscheidende morphogenetische Funktionen bei der Entstehung der komplexen corticalen Zelldifferenzierungen der Ciliaten besitzen; insbesondere sind sie für die Ausbildung der Cilien unabdingbar.

Mitochondrienvermehrung in Zukunft sicher größere Aufmerksamkeit verdienen, wenn es darum geht, die molekularen Mechanismen bei der Reduplikation des mitochondrialen Funktionsmusters zu analysieren.

Summary

Electron microscopic studies on *Tetrahymena pyriformis*, *Paramecium caudatum*, and *P. aurelia* were made to reveal the mode of multiplication of mitochondria in ciliates. Morphological evidence from cells with a high multiplication rate of the chondriome (under natural and experimental conditions) is in favour of the view, that mitochondria arise from preexisting mitochondria by division into two daughter-mitochondria of different or equal size. This result is in agreement with experiences from living tissue culture-cells and with biochemical and quantitative observations, pointing to a multiplication of mitochondria by incorporation of new material into preexisting mitochondria, which subsequently grow and divide. The way of division of the ciliate mitochondria (tubulus-type) is illustrated by electron micrographs.

On the other hand, the possibility, that mitochondria may also arise from non identical precursor-structures ("promitochondria"), cannot be excluded. In ciliates, vesicles bounded only by one unit-membrane and containing microtubules appear under all investigated conditions of a high mitochondrial multiplication rate, but transitional forms to morphologically normal mitochondria are rarely found. In contrast to this situation in ciliates, the generation of new mitochondria from other cell structures has been unequivocally demonstrated in Bacteria and yeast cells.

The problem, whether mitochondria can be nominated autoduplicating organelles is discussed. During cell division, no definite mode of distribution of the chondriome in equal halves into the daughter-cells is known as yet. This finding is consistent with the morphological experience indicating to an unequal division of these cell organelles: The partition of mitochondria does not deliver two quantitative identical parts; therefore it seems to be irrelevant for the cell to distribute quantitative identical parts of the chondriome to its descendants.

It is quite reasonable to suppose, that the mitochondria are the scene of the multiplication of characteristic patterns of structure and of functionally related enzyme chains as well. The pattern of structure lies in the order of magnitude of electron microscopic resolution, while the pattern of function in contrast to this is on the macromolecular level. There is no definite proof as yet that the cell needs the complex mitochondrial structure as a whole to reduplicate the macromolecular pattern of enzymes. In Bacteria and yeast cells, the pattern of structure is not perpetual; whether in cells of higher organims the situation is different (i. e. whether the multiplication of mitochondria by division must be considered as the *only* way to multiplicate the chondriome,) is a matter of conjecture.

Literatur

[1] André, J.: J. Ultrastruct. Res. Suppl. 3, 1—185 (1962).
[2] Aoki, A., M. H. Burgos, and M. T. Tellez de Inon: J. Microscopie 4, 217 (1965).
[3] Bade, E. G.: Z. Zellforsch. 61, 754 (1964).
[4] Bahr, G. R., and E. Zeitler: J. Cell Biol. 15, 489 (1962).
[5] Barton, A. A., and G. Causey: J. Anat. (Lond.) 92, 399 (1958).
[6] Bell, P. R., and K. Mühlethaler: J. Ultrastruct. Res. 7, 452 (1962).
[7] — — J. Cell Biol. 20, 235 (1964).
[8] Belt, W. D.: J. biophys. biochem. Cytol. 4, 337 (1958).
[9] Benda, C.: Ergebn. Anat. u. Entwickl-Gesch. 12, 743 (1902).
[10] Berger, E. R.: J. Ultrastruct. Res. 11, 90 (1964).
[11] Brosemer, R. W., W. Vogell u. T. Bücher: Biochem. Z. 338, 854 (1963).
[12] Causey, G., and H. Hoffman: Brit. J. Cancer 9, 666 (1955).
[13] Danneel, R., u. E. Güttes: Naturwissenschaften 38, 117 (1951).
[14] Diers, L.: Ber. dtsch. Bot. Ges., Tagung München 1964, p. 369 (1965).

[15] EHRET, C. F., and G. DE HALLER: J. Ultrastruct. Res. Suppl. 6, 1 (1963).
[16] FAWCETT, D. W.: J. nat. Cancer Inst. Suppl. 15, 1475 (1955).
[17] FITZ-JAMES, P. C.: J. biophys. biochem. Cytol. 8, 507 (1960).
[18] FREDERIC, J., u. M. CHÈVREMONT: Film W₂, Inst. f. d. wiss. Film, Göttingen.
[19] GEREN, B. B., and F. O. SCHMITT: Proc. nat. Acad. Sci. (Wash.) 40, 863 (1954).
[20] GEY, G. O., P. SHAPRAS, and E. BORYSKO: Ann. N. Y. Acad. Sci. 58, 1089 (1954).
[21] GIESBRECHT, P.: Zbl. Bakt., 1. Abt. Orig. 179, 538 (1960).
[22] — Zbl. Bakt., I. Abt. Orig. 187, 452 (1962).
[23] HAGEDORN, H.: Zbl. Bakt., II. Abt. 113, 134 (1960).
[24] HOFFMAN, H., and G. W. GRIGG: Exp. Cell Res. 15, 118 (1958).
[25] HUDSON, G., and J. F. HARTMANN: Z. Zellforsch. 54, 147 (1961).
[26] IMAEDA, T., and M. OGURA: J. Bact. 85, 150 (1963).
[27] KELLENBERGER, E., and A. RYTER: Selected applications of the electron microscope in bacteriology. In: Modern Development in Electron Microscopy, Ed. B. M. SIEGEL, p. 335. New York 1964.
[28] KOMNICK, H., u. K. E. WOHLFARTH-BOTTERMANN: Fortschr. Zool. 17, 2 (1964).
[29] KUSUNOSE, M., and E. KUSUNOSE: Ann. Rep. Jap. Soc. Tuberc. 4, 17 (1959).
[30] LAFONTAINE, J. G., and C. ALLARD: J. Cell Biol. 22, 143 (1964).
[31] LEHNINGER, A.: Pediatrics 26, 466 (1960).
[32] LEON, H. A., and S. F. COOK: Science 124, 123 (1956).
[33] LEVER, J. D.: J. biophys. biochem. Cytol. Suppl. 2, 313 (1956).
[34] LINNANE, A. W., E. VITOLS, and P. G. NOWLAND: J. Cell Biol. 13, 345 (1962).
[35] LUCK, D. J. L.: J. Cell Biol. 24, 461 (1965).
[36] MAZIA, D.: Mitosis and the Physiology of Cell Division. In: The Cell. Vol. 3, 77. Ed. J. BRACHET and A. E. MIRSKY. New York: Academic Press 1961.
[37] MENKE, W., u. B. FRICKE: Z. Naturforsch. 19 b, 520 (1964).
[38] MÜHLETHALER, K., u. P. R. BELL: Naturwissenschaften 49, 63 (1962).
[39] MÜHLPFORDT, H.: Z. Tropenmed. Parasit. 14, 357 (1963).
[40] MUDD, S., T. KAWATA, J. I. PAYNE, T. SALL, and A. TAKAGI: Nature (Lond.) 189, 79 (1961).
[41] NASS, M. M. K., and S. NASS: J. Cell Biol. 19, 593 (1963).
[42] — — J. Cell Biol. 19, 613 (1963).
[43] NOVIKOFF, A. B.: Mitochondria (Chondriosomes). In: The Cell. Vol. 2, 299. Ed. J. BRACHET and A. E. MIRSKY. New York: Academic Press 1961.
[44] OHYE, D. F., and W. G. MURRELL: J. Cell Biol. 14, 111 (1962).
[45] PAYNE, F.: J. Morph. 91, 555 (1952).
[46] PEARSE, A. G. E.: Histochemistry theoretical and applied. 2nd ed. London: Churchill 1960.
[47] ROBERTIS, E. D. P. DE, and H. BLEICHMAR: Z. Zellforsch. 57, 572 (1962).
[48] ROBERTSON, J. D.: Biochem. Soc. Symp. 16, 3 (1959).
[49] ROTH, L. E.: J. biophys. biochem. Cytol. 3, 985 (1957).
[50] ROUILLER, C., and W. BERNHARD: J. biophys. biochem. Cytol. Suppl. 2, 355 (1965).
[51] RYTER, A., et F. JAKOB: C. R. Acad. Sci. (Paris) 257, 3060 (1963).
[52] SCHNEIDER, L.: Verh. dtsch. Zool. Ges. Kiel 1964, Zool. Anz. Suppl. 28, p. 243 (1965).
[53] SCHULZ, H.: Beitr. path. Anat. 119, 71 (1958).
[54] SHINOHARA, C., K. FUKUSHI, J. SUZUKI, and K. SATO: J. Electronmicroscopy 6, 47 (1958).
[55] STEINERT, M.: J. biophys. biochem. Cytol. 8, 542 (1960).
[56] TAHMISIAN, T. N., E. L. POWERS, and R. L. DEVINE: J. biophys. biochem. Cytol. Suppl. 2, 325 (1956).
[57] VANDERWINKEL, E., and R. G. E. MURRAY: J. Ultrastruct. Res. 7, 185 (1962).
[58] WEISSENFELS, N.: Z. Naturforsch. 12 b, 168 (1957).
[59] — Z. Naturforsch. 13 b, 182 (1958 a).
[60] — Z. Naturforsch. 13 b, 203 (1958 b).

[61] Wohlfarth-Bottermann, K. E.: Z. Naturforsch. **11 b**, 578 (1956).
[62] — Z. Naturforsch. **12 b**, 164 (1957).
[63] — Verh. dtsch. Zool. Ges. Münster 1959, Zool. Anz. Suppl. **23**, 393 (1960).
[64] — Naturwissenschaften **50**, 237 (1963).
[65] —, u. L. Schneider: Strahlentherapie **116**, 26 (1961).
[66] Wrischer, M., u. Z. Devidé: Z. Naturforsch. **20 b**, 260 (1965).
[67] Yamaguchi, J.: Sci. Rep. Res. Inst. Tohoku Univ. **9**, 125 (1960).
[68] Zollinger, H. U.: Rev. Hémat. **5**, 696 (1950).

Diskussion

Vorsitz: *Bruns*

David: Eine besondere Form der Vermehrung von Mitochondrien fanden wir in Leberzellen von *Salamandra maculata* (Abb. I a).

Die schematische Abbildung (I b) zeigt, wie wir uns den Vorgang denken: Zuerst entsteht eine ringförmige Struktur in der Matrix, die sich dann der Wand anlagert; nach Ausbildung einer doppelten Membran erfolgt schließlich die Ausstoßung.

P. Sitte: Have you any striking reasons for reading your last picture (cf. Abb. I b) from the left to the right and not from the right to the left?

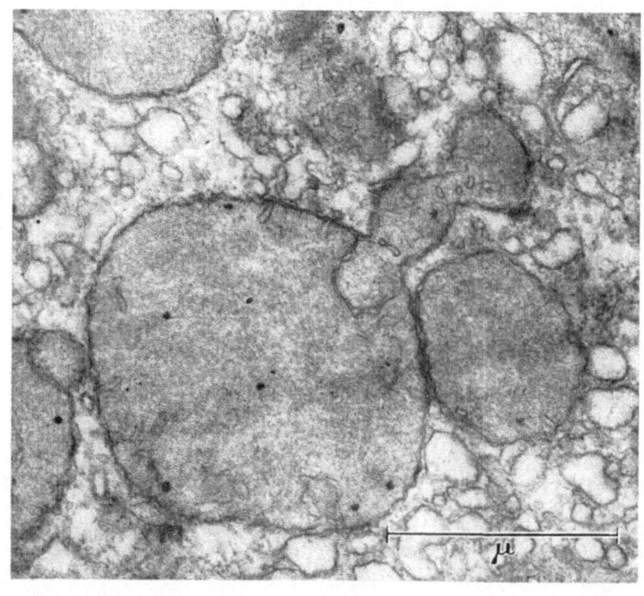

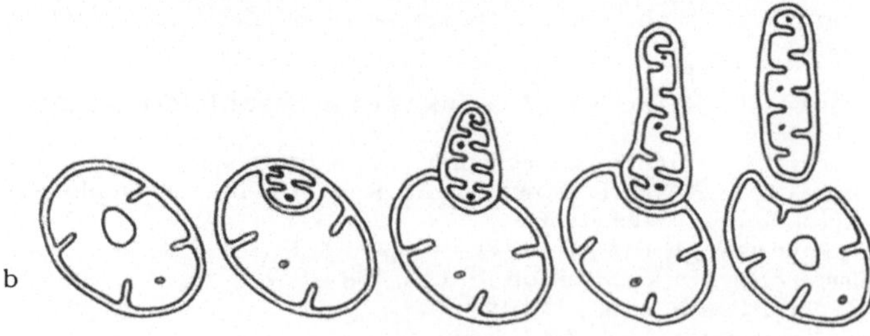

Abb. I a, b

David: Man sieht sehr viele Vorstufen in der Matrix, die sich meiner Meinung nach nicht erklären lassen, wenn man es umgekehrt deutet.

H. Sitte: If you find different stages of development at the same object, then, I believe, it is impossible to recognize the direction of this process. This could be possible only by introducing a time marker.

Miller: It should be mentioned in this connection that KARNOVSKY (Exper. Molec. Pathol. *2*, 347, 1963) has shown rather striking morphological changes of mitochondria at different times of the year, for instance in summer and winter frogs.

Bajer: From observations in the living state, it is well established that mitochondria are connected in some way with nuclei. In the light-microscope we cannot

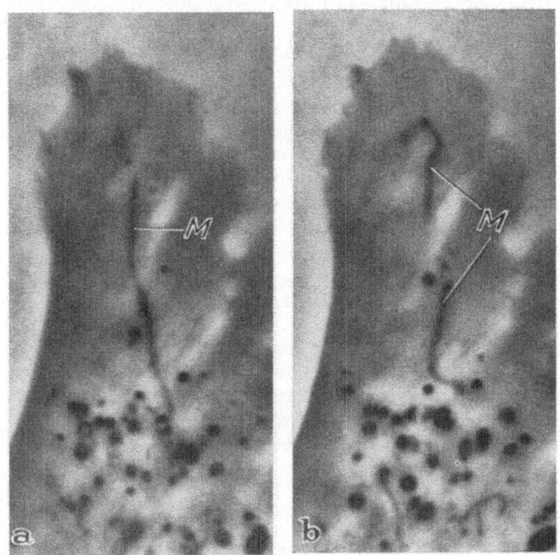

Abb. II a und b. Phasenkontrastmikroskopische Lebendaufnahmen (Aufnahmeabstand 12 sec) eines Zellausläufers von einem gezüchteten Hühnerherzmyoblasten
Das fadenförmige Mitochondrium (M in Abb. a) hat sich durch Fragmentation (M in Abb. b) vermehrt. Vergr.: 2150:1

see, whether they are really attached to or only touch the nucleus. However, these findings support in some way the conception that mitochondria may be formed by the nuclear membrane.

Wohlfarth-Bottermann: This, indeed, is often seen in amebas, too. But these connections, I think, have to be interpreted in a physiological manner. There is no clear evidence from electron microscopy, that there occurs a generation of mitochondria from the nuclear membrane. But, of course, one cannot exclude it with certainty.

Grundmann: We also saw the attachment of mitochondria to the nucleus often in fibroblast cultures. The mitochondria go to the nuclear membrane, and they move away again. So we think, this movement is due to metabolic activity.

Weissenfels: Wie sich Mitochondrien in der lebenden Zelle bilden und verändern, kann besonders gut bei Gewebekulturen beobachtet werden (Abb. II a, b). Man sieht da, wie Mitochondrien aus gerade noch sichtbaren Körperchen von etwa 0,2 μ Durchmesser rasch heranwachsen, sich verzweigen und fragmentieren, wobei die Teilstücke wieder zu normalen Mitochondrien heranwachsen. Auch Fusionen kommen vor.

Bei der elektronenmikroskopischen Untersuchung interessieren in erster Linie die kleinen Körperchen, die zu Mitochondrien auswachsen. Man findet hier eiförmige

Gebilde mit einfacher Hüllmembran und relativ kontrastreichen Innenstrukturen, die jedoch von Ansatz zu Ansatz bei diesen gezüchteten Zellen recht verschieden aussehen können. Eine eindeutige Aussage darüber, welche von diesen Körpern tatsächlich Vorstufen von Mitochondrien sind, ist naturgemäß sehr schwierig. Vielleicht helfen hier Untersuchungen weiter, die den Umstand ausnützen, daß durch subletale Röntgenbestrahlung die in einer Zelle vorhandenen Mitochondrien zerstört werden können (wahrscheinlich werden sie aufgelöst und resorbiert). Zwei Stunden später erkennt man phasenoptisch das Auftreten kleiner Körnchen, die erneut zu Mitochondrien heranwachsen.

Wohlfarth-Bottermann: Herr Schneider hat Strahlenschäden bei *Paramecium* untersucht. Große Anteile des Chondrioms degenerieren in Nahrungsvacuolen oder ähnlichen Gebilden (Protoplasma *53*, 530 und 554 [1961]; Strahlentherapie *116*, 25 [1961]). Die Zahl der Promitochondrien war jedoch in der Regenerationsphase nicht vergrößert.

Grundmann: I should be very interested, what pictures of the mitochondria are to be seen during mitotic division, when respiratory activity of the cell is decreased.

Wohlfarth-Bottermann: Wir haben auch Zellteilungsstadien untersucht, die man ja leicht isolieren und identifizieren kann. Da war das Chondriom völlig unauffällig.

Zur Frage, ob es sich bei den Hantelformen nicht um Fusionen handeln könnte, möchte ich sagen, daß man das bei Gewebekulturzellen nur schlecht entscheiden kann, bei Ciliaten aber verhältnismäßig gut. Einerseits haben wir ja gerade das Stadium untersucht, in dem die Chondriomvermehrung anzunehmen ist. Andererseits kommen bei den Ciliaten außer nach Natriumacideinwirkung nie Fadenformen vor. So ist die Wahrscheinlichkeit doch ziemlich groß, daß hier wirklich Teilungsstadien vorliegen.

Miller: Before speaking of microbodies or dense bodies as mitochondrial precursors one should keep in mind that there are two different types of them: on the one hand, at least in uricase containing organs, microbodies seem to contain this enzyme together with catalase as shown by Baudhuin, Beaufay, and de Duve (J. Cell Biol. *26*, 219, 1965). On the other hand there are dense bodies of the type of secondary lysosomes (or residual bodies).

Weissenfels: Promitochondrien und ,,microbodies'' sind lichtmikroskopisch nicht zu unterscheiden. Promitochondrien lassen sich auf relativ frühen Entwicklungsstadien mit Janusgrün, ,,microbodies'' dagegen mit Trimethylthionin vital anfärben. Ich vermute, daß sich aus Mitochondrienvorstufen, besonders bei einem Überangebot von Proteinen im Kulturmedium, auch Lysosomen zu bilden vermögen.

Stoeckenius: I think, we have here a good example of the limitations of electron microscopy . We cannot solve these problems using only morphological observations.

Wintersberger: Concerning the decision, whether a microbody or a vesicle is a precursor of a mitochondrion: At least in the case of yeast it was shown, that inactive mitochondria (as they occur in the cytoplasmic ,,petite''-mutant or even in anaerobically grown yeast cells) do not have an electron-transport chain. However, they still have succinate dehydrogenase bound to the membrane. If it would be possible to stain for succinate dehydrogenase, one could perhaps decide, whether something is a precursor of a mitochondrion or not, because succinate dehydrogenase seems to be one of the very first enzymes that is bound to a mitochondrial membrane.

H. Sitte: Is it really possible that an organelle which has only one surrounding unit membrane (e. g. ,,microbody'') gets a second one , like a mitochondrion has ?

Wohlfarth-Bottermann: Da wir experimentell nachgewiesen haben (Zool. Anz., Suppl. *23*, 393, 1959), daß aus dem Grundplasma neue Elementarmembranen entstehen können, sehe ich theoretisch keine prinzipielle Schwierigkeit dafür. Das soll aber nicht bedeuten, daß es auch hier tatsächlich der Fall ist.

H. Sitte: Sind Fälle bekannt, in denen Membranen ,,offen'' sind, d. h. wo Elementarmembranen im Plasma frei endigen ?

Wohlfarth-Bottermann: Es scheint mir nicht notwendig, daß Membranen bei ihrer de-novo-Entstehung zunächst offen sind.

Stoeckenius: All cellular membranes enclose compartments which apparently have a composition different from that of the outside medium. There are rare cases, where you find open-ended unit membranes. However it is questionable, if these are real structures and not artifacts of fixation. The only exception I know of where open-ended unit membranes seem to occur, is in the formation of pox viruses in the cytoplasm. There must, of course, exist a mechanism, to form what we call an envelope. This could easily be explained by an invagination in a vesicle which subsequently is pinched off, or the fusion of a number of flattened vesicles along their edges. I don't think there is evidence that any cellular membranes are formed de novo in the absence of pre-existing membrane. Dr. WOHLFARTH, how can be sure that you are not observing growth of pre-existing membranes rather than formation of membranes from some other structures?

Wohlfarth-Bottermann: Seit den 20iger Jahren ist durch SPEK und HEILBRUNN bekannt, daß das Ektoplasma von Amöben durch quellende Ionen in Endoplasma transformiert wird (Lebendbeobachtung mit dem Ultramikroskop). Im optisch leeren Ektoplasma lassen sich auch elektronenoptisch — außer dem Plasmalemma — nie Membranen nachweisen. Solche treten jedoch sofort unter dem Einfluß von SCN-Ionen auf (Verhd. Dtsch. Zool. Ges. Münster 1959, Zool. Anzeiger Suppl. *23*, 393).

Stoeckenius: Das Ultramikroskop kann Strukturen nur dann zeigen, wenn sie nicht allzuklein sind und sich im Brechungsindex von der Umgebung hinreichend unterscheiden.

H. Sitte: I believe that the micropinocytotic vesicles, for example, are not visible in the ultramicroscope. It could be possible that the formation of membranes starts from such vesicles.

Wohlfarth-Bottermann: Wir haben aber auch mit dem Elektronenmikroskop im Ektoplasma von unbehandelten *Hyalodiscus* keine Membranen oder Vorstufen von solchen gesehen.

Brdiczka: Wie erklärt sich die Mitochondrienvermehrung bei Atmungshemmung, da doch die Bildung von mitochondrialen Proteinen an den energieliefernden Stoffwechsel der Mitochondrien geknüpft ist? Könnte es sich bei den vergrößerten Mitochondrien nicht vielmehr um Degenerationserscheinungen infolge Vergiftung handeln?

Wohlfarth-Bottermann: Wir vermuten, daß wir die Atmung nicht völlig blockiert haben.

Jakob: A relation of this abnormal form of mitochondria and respiration must be excluded in the case of yeast. If yeast is cultivated in a chemostate where cells are perfectly respiring (rate of division limited by glucose), the time of generation is threefold the normal one. Due to these conditions, the cells elongate, and the mitochondria become huge, as YOTSUYANAGI has shown. These structural changes are completely reversible.

Bruns: Wir können vielleicht doch als ein Ergebnis unserer Diskussion feststellen, daß vom Morphologischen her gesehen eine Vermehrung der Mitochondrien sui generis sichergestellt zu sein scheint, wie Sie es in Ihrem Vortrag herausgestellt haben. Für eine Entstehung de novo oder aus besonderen Vorstufen haben wir keine verläßlichen Beweise.

The Biogenesis of Mitochondria in Neurospora
A Summary of Present Findings[1]

By

DAVID J. L. LUCK, New York

With 5 Figures

In the 75 years since the publication of ALTMANN's "bioblast theory" the question of how mitochondria are formed by growing cells has been a continuing point of active speculation and discussion. During this period all possible mechanisms for mitochondrial biogenesis have been advanced, but in recent years greater attention has been given in speculations as well as in experimental work to the possibility that new mitochondria are derived through growth and division of pre-existing organelles. The observation that mitochondria contain a DNA of distinct base composition [6, 9] has provided an additional support for such a mechanism, and for some biologists it is now "established" that mitochondria are "self replicating" cell organs, just like ALTMANN's bioblasts. The term "self replication" does not properly describe our current understanding of the problem: it fails to acknowledge some important uncertainties concerning mitochondrial biogenesis, and if applied in its strictest sense, it is incorrect.

A review of some experimental observations made with *Neurospora crassa* can provide a framework for discussion of mitochondrial biogenesis. The organism has been found to be especially useful for studying this question because its biology and genetics have been well studied. When *Neurospora* is grown under exponential growth conditions its mitochondrial mass doubles at the same rate as does the total cell mass (doubling times 120—150 minutes) and these mitochondria can easily be isolated as highly purified fractions.

Mitochondrial Variability

The chemical composition and morphology of mitochondria produced by exponentially growing cultures of *Neurospora* have been shown to be variable within substantial limits [3]. In a choline requiring strain, for example, a change in the available level of choline in the culture medium can influence the phospholipid content of mitochondria. When shaken liquid cultures are started from conidia and grown at a choline chloride level just adequate to support exponential growth at maximal rates

---

[1] This work has been supported by a grant from the National Science Foundation.

(1 μg/ml), they produce mitochondria which have a phospholipid phosphorous to protein ratio of 8 μg/mg. An identical conidial inoculum grown at a choline level 10 times higher produces mitochondria in which the corresponding ratio is 16 μg/mg. This difference in chemical composition is reflected in the densities of the two types of mitochondria, which can be clearly demonstrated by observing the behavior of mitochondrial fractions in isopycnic sucrose gradient centrifugation. Figure 1 is a photograph of centrifuge tubes containing continuous sucrose gradients.

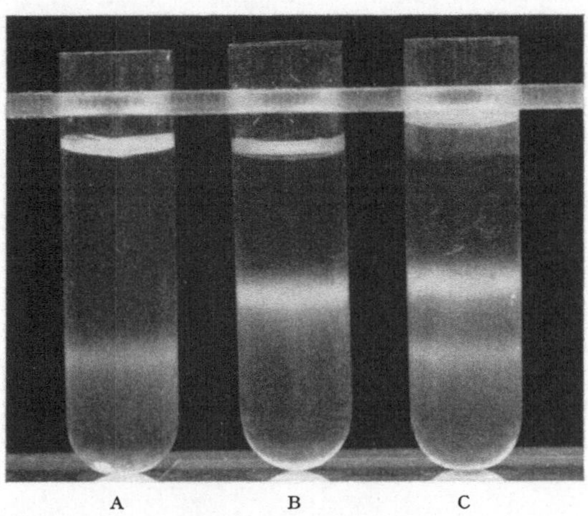

Fig. 1. Sedimentation in sucrose gradients of mitochondria derived from cultures of the *chol-1* strain of *N. crassa* growing at two levels of choline supplementation, shown in a photograph of the centrifuge tubes. Mitochondria derived from 15 hour cultures in 1 μg/ml choline chloride (Tube A) and 10 μg/ml (Tube B) were prepared as a crude fraction by differential centrifugation, resuspended in .44 M sucrose, and layered over a continuous sucrose gradient (1.9 to 0.9 M sucrose). The photograph shows the appearance of the tubes after 1.5 hours' centrifugation at 39,000 RPM in the SW 39 rotor of the Spinco preparative ultracentrifuge. Tube C was layered with an equal mixture of both types of mitochondria. The position of the bands was not changed in experiments where centrifugation was prolonged to 8 hours

Mitochondrial fractions layered over the gradients have been centrifuged to their equilibrium position. Mitochondria from low choline cultures form a band as shown in tube A, while those isolated from high choline cultures form a band at the lower density shown in tube B.

These differences in phospholipid composition resulting from different growth conditions are also associated with differences in mitochondrial structure. Figure 2 A shows a typical mitochondrial profile seen in a low choline cell. The diameters of these mitochondria range from 0.6 to 0.7 μ while those encountered in high choline cells (Fig. 2 B) are much smaller, with diameters ranging from 0.2 to 0.25 μ. These modifications of mitochondrial compositions and structure are not associated with differences in cytochrome content or in the specific activity (based on total mitochondrial protein) of malate or succinate dehydrogenase.

Similar variations in chemical compositions can be brought about by varying the leucine supplement for growth of a leucine requiring mutant or by providing amino acids or choline in the growth medium of a wild-

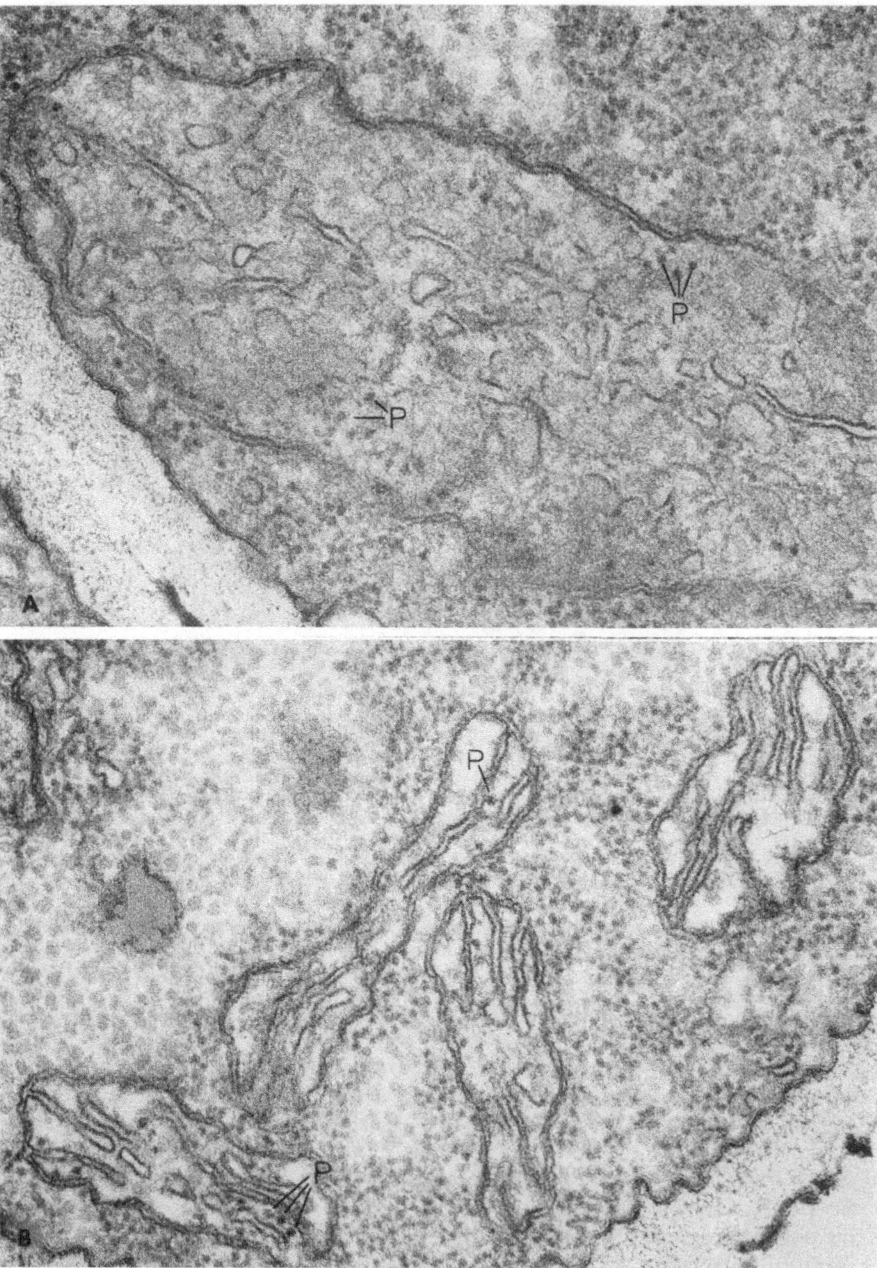

Fig. 2. Electron micrographs of mitochondria from cultures at two levels of choline suppli-
mentation. A 1 μg/ml choline chloride. B 10 μg/ml choline chloride. These profiles are typical
of the mitochondria encountered under the two culture conditions. The average diameter of
low choline mitochondria is 0.6—0.7 μ, while for high choline the value is 0.2—0.25 μ.
Intramitochondrial particles (P) which bear some resemblance to cytoplasmic ribosomes are
found in small numbers in both types of mitochondria. $\times$ 72,000

type strain. In all of these cases, the mitochondrial variability is a consequence of changes in the availability of precursors. Instances of modification of mitochondrial structure and composition are encountered in
glucose-repressed aerobic yeast [16] where the variability seems to reflect
physiological requirements. When the status of mitochondria in anaerobic
yeast is fully clarified [14, 12] it may come to represent an extreme case
of mitochondrial variation. It seems clear from the examples, however,
that mitochondria are not "replicated" in the sense that new mitochondria are not necessarily produced according to rigid, unchangeable
specifications, and that any mechanism proposed for mitochondrial
biogenesis must provide for the possibility of varying the structure and
chemical composition in response to physiological needs or availability of
precursors.

Growth and Division of Mitochondria

Experiments carried out with *Neurospora crassa* have provided
support for the hypothesis that mitochondria are formed by growth and
division of pre-existing organelles. Some years ago we studied a choline
requiring organism to find out how pre-existing membrane material was
distributed among individual mitochondria in a growing population [2].
Cultures labeled with tritiated choline were transferred to unlabeled
medium and permitted to grow for three doubling cycles. The distribution
of label among individual organelles in highly purified mitochondrial
fractions was determined by quantitative radioautography. The results
showed that throughout three doubling cycles of the mitochondrial mass,
the label was uniformly distributed among all mitochondria. Additional
short term labeling experiments indicated that, after exposure to tritiated
choline for only 1/12 of a division cycle, all mitochondria were labeled.
These results suggested that the entire mitochondrial mass grows
continuously adding new membranous material to the pre-existing
framework, and that the population of organelles increases by division.

Recently we have made use of the mitochondrial density shifts brought
about by changing the choline supplementation to a cholineless mutant
to test further this hypothesis [4]. If one carries out an experiment
in which a culture growing exponentially at low choline (producing the
mitochondria of density shown in Fig. 2, tube A) is transferred to a
high choline medium, a study of the density characteristics of mitochondria produced under the growth conditions can provide a clue to the
mechanism of their formation.

If our earlier observations were correct, and mitochondria grow and
divide, it is to be expected that in this experiment the entire population
of mitochondria would shift in density toward the low density position
characteristic of high choline cultures. There should be no residual heavy
population.

If mitochondria are produced *de-novo* from small molecular weight
precursors, two populations should be detectable: a stable band of heavy
mitochondria representing pre-existing organelles, and a growing band

at the low density position representing mitochondria formed in the high choline conditions.

If mitochondria are formed through the transformation of non-mitochondrial intracellular membranes, the results should be quite similar, though the number of mitochondria of intermediate density, and

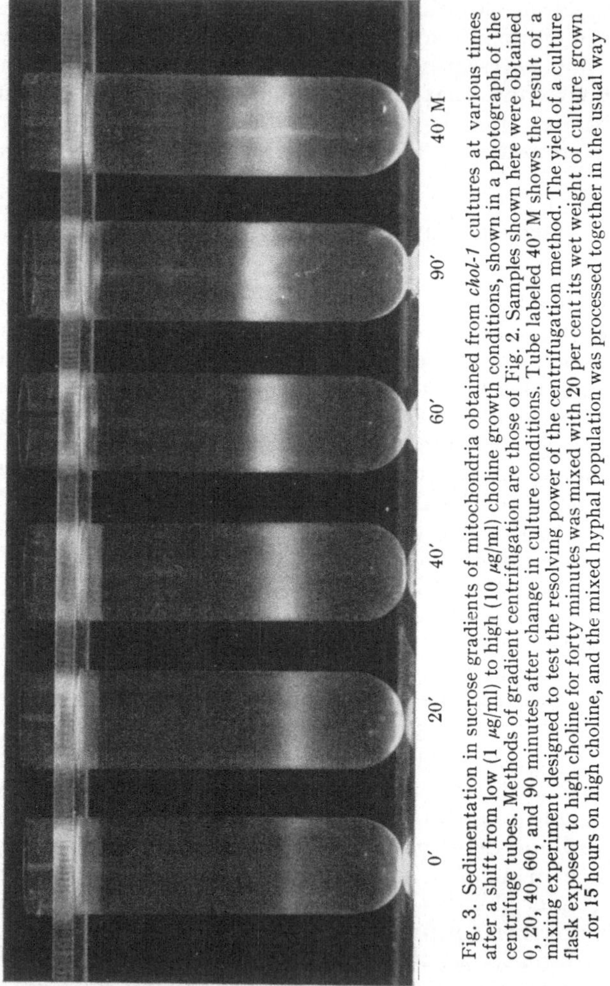

Fig. 3. Sedimentation in sucrose gradients of mitochondria obtained from *chol-1* cultures at various times after a shift from low (1 μg/ml) to high (10 μg/ml) choline growth conditions, shown in a photograph of the centrifuge tubes. Methods of gradient centrifugation are those of Fig. 2. Samples shown here were obtained 0, 20, 40, 60, and 90 minutes after change in culture conditions. Tube labeled 40' M shows the result of a mixing experiment designed to test the resolving power of the centrifugation method. The yield of a culture flask exposed to high choline for forty minutes was mixed with 20 per cent its wet weight of culture grown for 15 hours on high choline, and the mixed hyphal population was processed together in the usual way

the delay in the appearance of a distinct band at low density would be determined by the size of the precursor membrane pool. A study of the time course of choline incorporation into mitochondria [3] in this type of experiment indicates that such a pool is small or absent.

Figure 3 shows the results of a transfer experiment from low to high choline medium. The mitochondria were isolated and centrifuged to equilibrium at 0, 20, 40, 60, and 90 minutes after addition of high choline. There was a gradual shift toward lower density which involved the entire mitochondrial band. During these 90 minutes of observation the mitochondrial mass had increased by about 50%, yet there was no suggestion of a satellite band.

Tube 40′ M represents an attempt to test the resolving power of the method. A culture exposed to the new growth conditions for forty minutes was mixed with 20 percent its wet weight of cells grown continuously in high choline, and the mixture processed in the usual way. The proportions

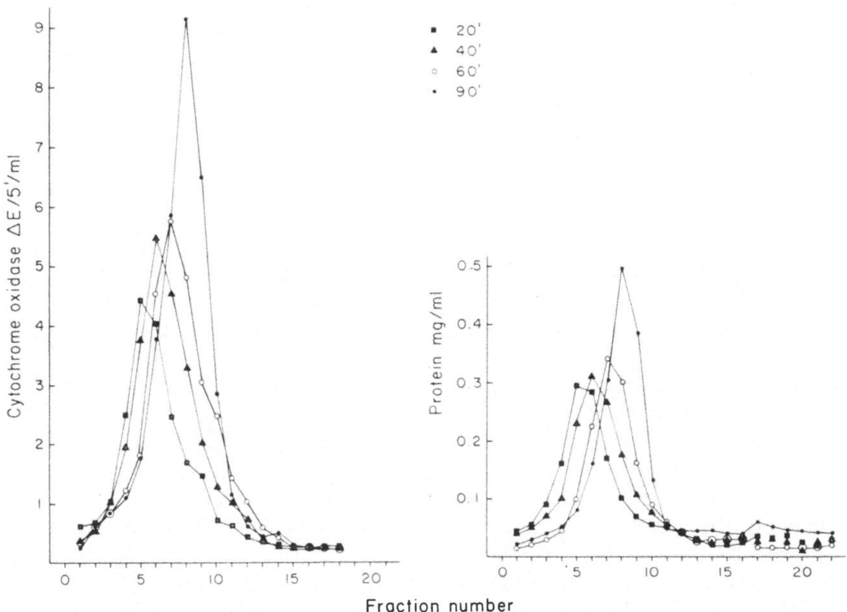

Fig. 4. Distribution of cytochrome oxidase and protein in sucrose gradients after isopycnic centrifugation of mitochondrial fractions obtained from *chol-1* cultures 20, 40, 60, and 90 minutes after a shift from low to high choline culture conditions. Serial fractions were collected through a pinhole in the bottom of centrifuge tubes similar to those shown in Fig. 3. Diluted fractions were assayed for cytochrome oxidase activity and for total protein. Fraction 1 represents material obtained from the bottom of the tube

of the mixture were chosen to simulate the expected ratio of old and new mitochondria after forty minutes' growth. It is apparent from the results that, when present, small satellite populations can be detected by the method, even in the presence of a shifting band of pre-existing mitochondria.

A study of the distribution of cytochrome oxidase activity and total protein was made by collecting serial fractions from tubes like those shown in Fig. 3. These results, summarized in Fig. 4, identify the bands as mitochondria, provide a more quantitative estimate of mitochondrial distribution, and substantiate the conclusion that a gradual shift in density which affects the entire mitochondrial population has taken place during the 90 minute observation. Table 1 gives data on the phospholipid to protein ratio of the mitochondrial bands at the various time points, and indicates that shifts in density can be correlated with change in phospholipid content.

Table 1: *Change in Mitochondrial Phospholipid to Protein Ratio in Cultures Converted from Low* (1 μg/ml) *to High* (10 μg/ml) *Choline*

Time after addition of supplemental choline	μg phospholipid phosphorus/mg protein	
	Exp. 1	Exp. 2
0	8.1	8.0
20′	8.6	8.2
40′	10.0	9.6
60′	11.5	11.1
90′	12.0	11.6

It has been possible to influence the density of mitochondria by growing cells on ^{15}N ammonium chloride as the sole nitrogen source. Although the density difference between ^{15}N mitochondria and ^{14}N mitochondria is small, an experiment like the choline experiment shown in Fig. 3 could be carried out, and analogous results were obtained [5]. In the case of the choline experiment the density changes reflected distributions of phospholipid among pre-existing and progeny mitochondria; in the ^{15}N experiment it reflects, for the most part, the distribution of protein.

In agreement with the earlier radioautographic data, these results support the hypothesis that in exponentially growing *Neurospora* the mitochondrial mass grows by addition of new components to the existing mitochondria and the population is increased by mitochondrial division. The results do not give information concerning the site of synthesis of the molecules and macromolecules involved in mitochondrial growth, nor do they specify the form in which these molecules are added to the existing structure. The fact that new components are randomly distributed in the mitochondrial division process does suggest that the accruing "pieces" are small and that they are incorporated at random into the pre-existing organelle.

These uncertainties concerning the site of synthesis of mitochondrial components indicate that the use of the term "self replication" obscures important aspects of the mitochondrial biogenetic process. Experimental data obtained in other cell types suggest that the mitochondrion may in fact depend on the cell for synthesis of many of its components. In the case of the liver cell, the synthesis of lecithin from CDP-choline may be carried out exclusively by microsomes [15]. When amino acid incorporation by isolated mitochondrial fractions has been studied, the rates of incorporation have always been very low. In experiments where attempts have been made to characterize the radioactive product [11] it has been found that soluble proteins are not labeled and that radioactivity is present exclusively in the insoluble lipo-protein residue. The possible significance of this finding, and other aspects of protein synthesis will be considered further in the discussion of mitochondrial DNA.

Mitochondrial DNA

DNA has been isolated from highly purified mitochondrial fractions of *Neurospora crassa* prepared by isopycnic gradient centrifugation [6]. The yield is small, amounting to less than 1 μg/mg mitochondrial protein,

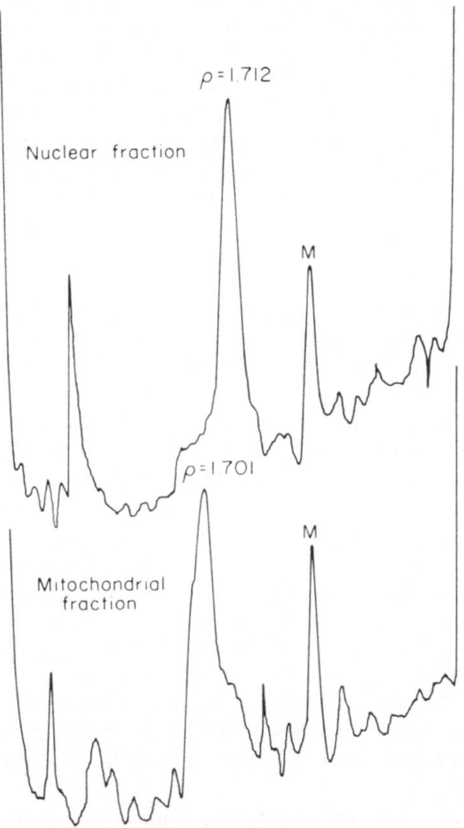

Fig. 5. Densitometer tracings of UV photographs made in the Spinco Model E ultracentrifuge. DNA's obtained from the fractions indicated had been centrifuged in cesium chloride for 20 hours at 44,770 RPM. The marker (*M*) was bacteriophage SP 8 DNA, density 1.742. The gradient increases in concentration from left to right

and accounting for less than 1% of the total cellular DNA. The DNA obtained from mitochondria has a buoyant density different from that of the nucleus. Figure 5 is a densitometer tracing of UV photographs made with the analytical ultracentrifuge after establishment of a cesium chloride gradient: DNA from the nuclear fraction has a buoyant density of 1.712 while the peak of mitochondrial DNA has an apparent density of 1.701. Careful examination of this and other photographs indicates that there are in fact two distinct mitochondrial components (density 1.699 and 1.702) present in equal amounts, but lying so close together that they are not resolved by the densitometer. In the mitochondrial DNA pattern shown, there is evidence of only small contamination with nuclear material, and this small trace can be eliminated by treating the intact mitochondria with DNase. The fact that the recovery of the bulk of the mitochondrial DNA is not affected by this treatment is further evidence for an intramitochondrial localization of this DNA.

Upon heat denaturation, there are the expected shifts in density for the mitochondrial DNAs indicating that they are present in the double stranded form. Estimates of the base composition of the nuclear and mitochondrial DNAs obtained from melting curves are in good agreement with those obtained from the buoyant density and indicate a GC content of 53% for the nuclear DNA and about 41% for the mitochondrial DNAs. The DNA isolated from *Neurospora* mitochondria had a molecular weight of $\sim 13 \cdot 10^6$ as estimated from electronmicrographs made by the Kleinschmidt technique [6]. An estimate based on the recovery of DNA per mg mitochondrial protein and the number of mitochondria per mg protein indicates that if the DNA is distributed uniformly, each mitochondrion would have a piece equivalent to a molecular weight $\sim 130 \cdot 10^6$.

While the conclusion that *Neurospora* mitochondria contain a DNA of unique buoyant density is well founded, and similar data have now been obtained for mitochondria of other cells [9], there is at present no direct evidence that this DNA has genetic function. Yet such a function is not excluded because in *Neurospora crassa* several mutations affecting purely mitochondrial functions have been described [8], and some of these fail to show linkage with other loci of the genetic map. Growth and division of mitochondria, proposed on the basis of experiments already described, would provide a structure, with continuity, for segregation of the mitochondrial DNA. It is therefore likely that this DNA will prove to embody those genetic determinants of mitochondrial function which are maternally inherited. Since in *Neurospora* there are known to be nuclear loci which affect mitochondrial function [8], it is clear that not all genetic determinants for the organelle are present in mitochondrial DNA.

We have made some observations, concerning the mechanism of replication and mode of hereditary transmission of mitochondrial DNA in *Neurospora* [7]. Using ^{15}N to label the DNA we have shown that during exponential growth in ^{14}N medium this DNA changes density in a manner consistent with semiconservative replication. In other experiments, still in preliminary form, we have obtained results which indicate that in a sexual cross, the character of the filial mitochondrial DNA is determined exclusively by the maternal parent. These properties of the DNA molecule are those which would be expected if it had genetic function.

At the present time there is little experimental evidence relating to the nature and function of the components controlled by mitochondrial

DNA. We have demonstrated that an RNA polymerase, resembling a DNA dependent enzyme, is present in mitochondrial fractions of *Neurospora* [6]. The product of such a polymerase could be active in protein synthesis by the mitochondrion. However, on the basis of the genetic evidence concerning the role of the nucleus in determining mitochondrial function cited above[2], and the results obtained with amino acid incorporation in isolated mitochondria referred to before, it seems unlikely that the mitochondrial DNA could direct synthesis of all mitochondrial proteins. A widely discussed hypothesis, consistent with many of the genetic and biochemical data, is that the mitochondrial DNA directs synthesis of a few proteins which function in the organization of the structure of the organelle, these proteins would have the property of binding, in a specific way, cytochromes and soluble enzymes as well as phospholipids. A change in the chemical structure of such a protein could affect its binding sites and thereby result in multiple changes in mitochondrial function. A protein fraction, which may have some of the properties of this hypothetical protein has been prepared from beef heart mitochondria. This preparation, described as "structural protein", is extremely insoluble and accounts for a substantial fraction of the total mitochondrial protein [1, 10]. It is of interest to recall that in experiments with amino acid incorporation by isolated mitochondria the radioactivity was present mainly in the insoluble lipoprotein fraction [11].

Concluding Remarks

Biological, genetic, and biochemical data obtained from *Neurospora crassa* have provided insight into the means by which mitochondria are formed in growing cells. While the results apply strictly to this organism, and in some cases strictly to the exponential growth phase, it is likely that the mechanisms outlined, and the existence of mitochondrial DNA, apply to other cells as well. It seems safe to assume that this DNA has a physiologic function, but information related to the specific components controlled by the molecule is not available at present and may be difficult to obtain. Further characterization of "structural protein" and study of this "protein" in cytoplasmic mutants affecting mitochondrial function in *Neurospora* and *Saccharomyces* may provide a reasonable approach to the problem. Questions relating to the exact means by which mitochondrial membranes are assembled and mitochondrial structure determined are more difficult to approach. Especially if, in terms of the "structural protein hypothesis", the assembly of membrane depends on the coming together of an insoluble protein synthesized within the mitochondrion and soluble proteins transported into the mitochondrion.

The mechanism of mitochondrial formation which has been proposed, and the presence in mitochondria of a unique DNA, raise questions with

---

[2] Clearer evidence of control of synthesis of a specific mitochondrial protein (cytochrome c) by a nuclear locus comes from a study of mutants of *Saccharomyces* [13].

evolutionary implications. While it is apparent that the mitochondrion does not embody all the properties of the "bioblast" it could be looked upon as an organism once capable of more or less independent existence and now transformed into an intracellular symbiont. If so, the "elementary organism" would seem to have transferred to the host cell, or otherwise lost, many of its synthetic requirements and genetic determinants. Whether or not the mitochondrion arose independently of its ultimate host cell, or developed as a derivative of the cell nucleus, its occurrance as a DNA-bearing structure separate from the nucleus and present in many copies per individual cell, confers great stability on the genetic determinants it bears. This stability is of course not absolute, and organisms like *Poky* and Mi-3 in *Neurospora*, and the non-segregating *"petit"* of yeast indicate that change may occur.

References

[1] Criddle, R. S., R. M. Bock, D. E. Green, and H. D. Tisdale: Biochem. biophys. Res. Commun. **5**, 75 (1961).
[2] Luck, D.: J. Cell Biol. **16**, 483 (1963).
[3] — J. Cell Biol. **24**, 445 (1965).
[4] — J. Cell Biol. **24**, 461 (1965).
[5] — Amer. Naturalist (in press).
[6] —, and E. Reich: Proc. nat. Acad. Sci. (Wash.) **52**, 931 (1964).
[7] — — Unpublished results.
[8] Mitchell, M. B., H. K. Mitchell, and A. Tissières: Proc. nat. Acad. Sci. (Wash.) **39**, 606 (1953).
[9] Rabinowitz, M., J. Sinclar, L. deSalle, R. Haselkorn, and H. H. Swift: Proc. nat. Acad. Sci. (Wash.) **53**, 1126 (1965).
[10] Richardson, S. H., H. O. Hultin, and S. Fleischer: Arch. Biochem. **105**, 254 (1964).
[11] Roodyn, D. B., J. W. Suttie, and T. S. Work: Biochem. J. **83**, 29 (1962).
[12] Schatz, G.: Biochim. biophys. Acta (Amst.) **96**, 342 (1965).
[13] Sherman, F.: Genetics **49**, 39 (1964).
[14] Wallace, D. G., and A. W. Linnane: Nature (Lond.) **201**, 1191 (1964).
[15] Wilgram, G. F., and E. P. Kennedy: J. biol. Chem. **238**, 2615 (1963).
[16] Yotsuyanagi, Y.: J. Ultrastruct. Res. **7**, 121 (1962).

Discussion

(cf. p. 335)

Mitochondrien als Träger genetischer Information

Von

Hans Tuppy und Erhard Wintersberger, Wien

Mit 1 Abbildung

Die chromosomalen Erbfaktoren umfassen nicht die gesamte genetische Information eines Organismus. Seit der ersten Beschreibung einer den Mendelschen Gesetzen nicht gehorchenden Vererbung durch Correns (1909) ist bei Mikroorganismen, bei höheren Pflanzen und auch bei Tieren eine größere Zahl von Erbeigenschaften aufgefunden worden, für deren Ausprägung und Weitergabe nicht-chromosomale genetische Faktoren verantwortlich sind (Zusammenfassende Darstellungen: Correns [9], Caspari [5], Ephrussi [15], Oehlkers [45], Nanney [37], Sager [52] u. a.). Zu den best-untersuchten Beispielen für nicht-chromosomale Vererbung gehören die von Ephrussi entdeckten „*Petite colonie*"-Mutanten von *Saccharomyces cerevisiae* [16, 17, 19], auf die im vorliegenden Referat näher eingegangen werden soll. Diese Kleinkolonie-Mutanten der Hefe sind — ähnlich wie auch die „*Poky*"-Mutanten von *Neurospora crassa* [35, 36] — durch einen Atmungsdefekt gekennzeichnet.

Mutativ veränderte Hefe-Mitochondrien

Wenn Einzelzellen von *S. cerevisiae* auf einen festen Nährboden gebracht werden, der nur wenig Glucose enthält, treten mit einer Häufigkeit von etwa 1% „*petites colonies*" auf, die sich durch ihre geringe Größe von den übrigen Kolonien unterscheiden. Die Häufigkeit der Mutation vom Wild- zum Kleinkolonie-Typ läßt sich durch Behandlung der Hefe mit Acriflavin und durch verschiedene andere chemische oder physikalische Einwirkungen kräftig steigern. Die Mutanten sind stabil; Rückmutationen zum Wildtyp treten bei noch so langdauernder vegetativer Vermehrung nicht auf [20]. Wie Kreuzungsversuche gezeigt haben, ist das Kleinkolonie-Merkmal ein nicht-mendelnder Charakter; seine Determinanten sind extrachromosomaler Art.

Die geringe Ausdehnung der Kleinkolonien erklärt sich dadurch, daß die Teilungsrate der mutierten Hefezellen herabgesetzt ist; sie können Glucose nicht mehr oxydativ abbauen, sondern nur noch vergären, wobei die Energieausbeute viel niedriger ist [59]. Biochemische Untersuchungen haben gezeigt, daß der Atmungsdefekt durch eine völlige Abwesenheit von Cytochromoxydase in den Zellen der Kleinkolonie-Mutante verursacht ist; die Wirksamkeit einiger anderer mit der Zellatmung zusammenhängender Fermente und Elektronenüberträger ist von

der Mutation ebenfalls betroffen und in den Mutanten zumindest stark
herabgesetzt [*61*, *62*]. Da es sich bei der Cytochromoxydase und den
anderen durch die „*Petite*"-Mutation in Mitleidenschaft gezogenen Fermenten um Komponenten der Mitochondrien handelt, stand ursprünglich die Möglichkeit eines mutativen Totalverlustes der Mitochondrien
zur Diskussion. Diese Annahme [*18*] stellte sich jedoch als unzutreffend
heraus, da der Nachweis geführt werden konnte, daß auch in den atmungsdefizienten Hefezellen Mitochondrien vorkommen; diese unterscheiden
sich von den normalen Hefe-Mitochondrien sowohl morphologisch [*73*,
75] als auch durch eine unvollständige enzymatische Ausstattung [*21*, *31*,
56], sind ihnen jedoch in vieler Hinsicht noch sehr ähnlich [*33*, *56*]. Die
extrachromosomale Mutation rührt demnach keineswegs an die Existenz
der Mitochondrien, sondern verursacht nur eine teilweise Modifikation,
die sich in einem strukturellen und funktionellen Defekt äußert.

Außer den durch eine nicht-chromosomale Mutation entstandenen
Kleinkolonie-Stämmen der Hefe kennt man andere, ebenfalls atmungs-
defiziente Stämme, die phänotypisch den ersteren weitgehend gleichen,
jedoch auf die Mutation chromosomaler, mendelnder Gene zurückgehen
[*16*, *57*, *58*]. Voraussetzung für die Ausbildung aktiver Atmungsfermente
und somit auch für den Aufbau funktionstüchtiger Mitochondrien ist
demnach ein Zusammenspiel chromosomaler und nicht-chromosomaler
Erbfaktoren.

Desoxyribonucleinsäure in Hefe-Mitochondrien

Die Desoxyribonucleinsäure-(DNA-)Natur jener Gene, die von den
Chromosomen aus wirksam werden, ist über jeden Zweifel erhaben. Da
bekannt war, daß das bei der „*Petite*"-Mutation als Mutagen bewährte
Acriflavin eine besondere Affinität zu DNA besitzt [*23*, *30*], lag es nahe
anzunehmen, daß als Träger der nicht-chromosomalen genetischen
Information, deren mutative Änderung die Ausbildung morphologisch
und funktionell intakter Mitochondrien blockiert, ebenfalls DNA fungiert. Es ließ sich weiter vermuten, daß die nicht-chromosomalen genetischen Determinanten, die im Zusammenwirken mit chromosomalen
Genen den Aufbau der Mitochondrien beherrschen, in diesen Organellen
selbst zu finden sein könnten [*16*]. Diese Überlegungen veranlaßten uns,
das Vorkommen und die Funktion von DNA in Hefe-Mitochondrien zu
untersuchen.

Ein eindeutiger chemischer Nachweis und eine verläßliche quantitative Bestimmung von DNA in Mitochondrien haben eine rigorose
Reinigung dieser Organellen, vor allem aber eine vollkommene Abtrennung DNA-haltigen Zellkern-Materials, zur Voraussetzung. Wenn die subcellulären Partikeln eines Hefe-Homogenates voneinander ausschließlich
durch differentielle Zentrifugation getrennt werden, bleibt die Fraktion,
in der die Mitochondrien enthalten sind, mit Kerntrümmern stark verunreinigt. Auch durch Sedimentieren oder Flotieren der Mitochondrienfraktion in einem Rohrzucker-Dichtegradienten ließ sich der erforderliche
Reinheitsgrad nicht erzielen. Eine saubere Abtrennung aller verunreinigenden DNA-reichen Anteile des Homogenates gelang schließlich durch

eine isopyknische Flotation der durch differentielles Zentrifugieren vorgereinigten Mitochondrien in einem Dichtegradienten, der mit Hilfe des Röntgenkontrastmittels „Urografin" hergestellt war [54]. Abb. 1 zeigt, wie sich die Hefe-Mitochondrien im Urografin-Gradienten verteilen; das Maximum der Verteilungskurve, deren Verlauf durch Messung der mitochondrialen Succinat-Cytochrom c-Reductase-Aktivität ermittelt wurde, liegt bei einer Schwebedichte von $1{,}15$ g $\cdot$ cm^{-3}. In den einzelnen Dichtezonen mit der Diphenylaminmethode von BURTON [4] oder mit einer für

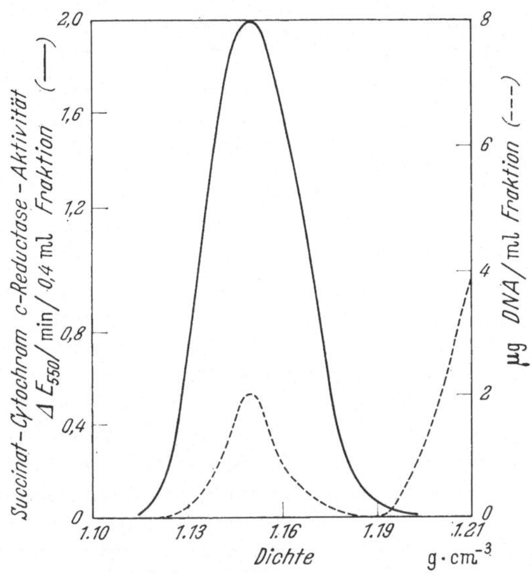

Abb. 1. Reinigung einer Mitochondrienfraktion aus aerob gezüchteter Hefe durch Flotation in einem „Urografin"-Gradienten: Verteilung von Succinat-Cytochrom c-Reductase-Aktivität (————) und DNA (— — —) [54]

diese Zwecke modifizierten Indolmethode von CERIOTTI [6] vorgenommene DNA-Bestimmungen ergaben kleine aber signifikante DNA-Werte, deren Verteilung im Gradienten jener der Mitochondrien genau entsprach [54]. Die somit gut gereinigten Hefe-Mitochondrien enthielten Protein und DNA in einem Gewichtsverhältnis von etwa $1:0{,}004$ [66].

Von besonderem Interesse war es zu prüfen, ob die atmungsdefizienten Mitochondrien der vegetativen Kleinkolonie-Mutante der Bäckerhefe ebenfalls DNA besitzen. Zahlreiche Versuche, auch diese „inkompletten" Mitochondrien durch Flotation in Urografin- oder anderen Gradienten zu reinigen, schlugen fehl. Der Mißerfolg dürfte ihrer im Vergleich zu normalen Mitochondrien viel größeren Fragilität [31, 56] zuzuschreiben sein. Eine andere Reinigungsmethode führte jedoch zum Ziel [66]: Die Mitochondrien wurden von der sie äußerlich verunreinigenden nuclearen DNA durch eine Behandlung mit Pankreas-Desoxyribonuclease bei 4° C befreit. Versuche hatten gezeigt, daß bei tiefer Temperatur nur die verunreinigende extramitochondriale DNA abgebaut wird, während intra-

mitochondriale DNA dem Zugriff des Enzyms entzogen bleibt [*32, 66*].
In den enzymatisch gereinigten Mitochondrien der Kleinkolonie-Mutante
wurde wie in jenen des Wildtyps DNA gefunden (Tab. 1), wenn auch in
etwas geringerer Menge (Protein : DNA = 1 : 0,002). Dieses Resultat
spricht dafür, daß die „*Petite*"-Mutation nicht mit einem vollständigen
Verlust der mitochondrialen DNA verbunden ist; ob es bei ihr zu einem
teilweisen Verlust oder einer Modifikation der in normalen Hefe-Mitochondrien vorhandenen DNA kommt, ist noch nicht entschieden.

Tabelle 1: DNA-Gehalt von Hefe-Mitochondrien

	μg DNA/mg Protein	
Normale Mitochondrien aerob gezüchteter Hefe	4,0	[*54*]
„Promitochondrien" anaerob gezüchteter Hefe	0,7	[*66*]
Mutierte Mitochondrien der „*Petite*"-Mutante	2,0	[*66*]

Wenn Zellen normaler, atmender Bäckerhefe unter anaeroben Bedingungen kultiviert werden, büßen sie ihre Atmungsfähigkeit und ihren
Besitz an Cytochromen ein [*60*]; elektronenoptische Untersuchungen
haben gezeigt, daß typische mitochondriale Strukturen nicht mehr vorhanden sind [*68*]. Im Gegensatz zur „*Petite*"-Mutation ist dieser Atmungs-
verlust jedoch reversibel: Belüftung in Gegenwart einer Energiequelle
wandelt die anaerob gezüchteten Zellen adaptiv wieder in atmende Zellen
um, die das klassische Cytochromsystem [*60*] und Mitochondrien besitzen
[*68*]. In den anaerob kultivierten, mitochondrien-freien Hefezellen kommen subcelluläre Teilchen vor, die einige für Mitochondrien charakteristische Enzyme wie z. B. Succinatdehydrogenase und Ferrochelatase
enthalten [*53*]. Diese Partikeln stehen anscheinend zu den voll ausgebildeten Mitochondrien der aeroben Hefe in Beziehung und sind von
Schatz [*53*] als deren Vorstufen, als „Promitochondrien", angesprochen
worden. Die Annahme eines adaptiven Überganges dieser Partikeln in
normale Mitochondrien ist durch den Nachweis, daß auch sie DNA enthalten (Tab. 1), gestützt worden.

Desoxyribonucleinsäure in den Mitochondrien anderer Organismen

Die Mitochondrien aerob gezüchteter Hefe in stationärer Wachstumsphase sind morphologisch und funktionell den Mitochondrien höherer
Organismen sehr ähnlich [*67, 74*]. Auch für die Reinigung der Mitochondrien verschiedener Säugetierorgane ließ sich die zunächst für Hefe-
Mitochondrien entwickelte Methode der Flotation in einem Urografin-
Dichtegradienten in der Ultrazentrifuge mit Erfolg anwenden. Die untersuchten gereinigten Säuger-Mitochondrien erwiesen sich ebenfalls als
DNA-haltig [*55*]; die ermittelten DNA-Werte lagen, auf Mitochondrien-
protein bezogen, um etwa eine Zehnerpotenz niedriger als für Hefe-
Mitochondrien. Über ähnliche analytische Ergebnisse berichteten auch
andere Arbeitsgruppen [*25, 29, 42*]; einige vergleichsweise sehr hohe

DNA-Werte sind vermutlich einer unzureichenden Reinigung der Mitochondrien zuzuschreiben (Tab. 2).

Durch den analytisch-chemischen Nachweis von DNA in isolierten, reinen Mitochondrien sind histoautoradiographische und elektronenmikroskopische Befunde, die auf das Vorkommen von DNA in Mitochondrien schließen lassen, erhärtet und in quantitativer Hinsicht ergänzt

Tabelle 2: *DNA-Gehalt isolierter Mitochondrien*

Herkunft der Mitochondrien	µg DNA/mg Protein*	Literatur
Saccharomyces cerevisiae. . . .	4,0	SCHATZ et al. [54]
Neurospora crassa	0,7	LUCK and REICH [32]
Rattenleber	0,46	SCHATZ et al. [55]
,, ...	0,65	NASS et al. [42]
Rattenniere	0,17	SCHATZ et al. [55]
Rinderherz	0,24	SCHATZ et al. [55]
,, ...	2,4	KROON [29]
Lammherz	4,5	KALF [25]

* Angegeben sind mittlere Werte.

worden. Die ersten Hinweise auf mitochondriale DNA hatten autoradiographische Untersuchungen gebracht, die CHÈVREMONT et al. [7, 8] schon vor mehreren Jahren mit experimentell modifizierten Gewebszellen anstellten; die belgischen Autoren beobachteten, daß in Gegenwart von Desoxyribonuclease *in vitro* kultivierte Fibroblasten radioaktiv markiertes Thymidin in die cytoplasmatischen Partikel einbauten und in diesen speicherten. Es blieb jedoch die Frage offen, ob das beobachtete Auftreten mitochondrialer DNA nicht eher als eine pathologische denn als eine normale Erscheinung zu werten sei. In letzter Zeit wurde jedoch mit der autoradiographischen Methode der Einbau tritierten Thymidins auch in Mitochondrien normaler Zellen nachgewiesen [1, 22, 26a]. NASS und NASS gelang die erste elektronenoptische Darstellung intramitochondrialer Fasern mit den Eigenschaften von DNA in tierischen und menschlichen Zellen verschiedener Art [38—40], darunter auch in Tumorzellen [41], während YOTSUYANAGI und GUERRIER [76] sowie KISLEV et al. [26a], ebenfalls im Elektronenmikroskop, die intramitochondriale DNA in Pflanzenzellen untersuchten.

Eine Isolierung von Mitochondrien-DNA glückte erstmals LUCK und REICH [32]. Die von ihnen aus sorgfältig gereinigten *Neurospora crassa*-Mitochondrien gewonnene DNA besaß, wie die Zentrifugation im Cäsiumchlorid-Dichtegradienten erwies, eine charakteristische Dichte (1,701 g · cm⁻³), die sich von jener der Kern-DNA (1,712 g · cm⁻³) unterschied. Da die durch isopyknische Zentrifugation ermittelte Dichte der DNA deren Basenzusammensetzung widerspiegelt, geht aus diesen Versuchen hervor, daß die mitochondriale DNA eine von der nuclearen DNA verschiedene chemische Struktur besitzt. Zu entsprechenden Ergebnissen gelangten neuerdings auch RABINOWITZ et al. [47], die aus Herz- und Leber-Mitochondrien von Hühnerembryonen isolierte DNA untersuchten.

Die nunmehr schon in größerer Zahl vorliegenden und übereinstimmenden Ergebnisse, die bei der Prüfung von Mitochondrien verschiedenster Herkunft erzielt worden sind, berechtigen zur Annahme einer *regelmäßigen* Anwesenheit von DNA in *allen* Mitochondrien. Als nächstes erhebt sich die Frage, ob diese DNA tatsächlich Träger einer genetischen Information ist. Wenn dem so wäre, müßte sie erstens zu einer Replikation und zweitens zu einer funktionellen Expression der Information fähig sein. Für eine replikative Vermehrung der Mitochondrien-DNA geben die schon erwähnten autoradiographischen Befunde einer Markierung durch den Einbau radioaktiven Thymidins [*1, 7, 8, 22*] sowie auch elektronenmikroskopische Aufnahmen, die für Verdopplungsvorgänge sprechen [*76*], gute Hinweise. Hinsichtlich der phänogenetischen Expression könnte man − in Analogie zu jener der chromosomalen Erbfaktoren − auch von cytoplasmatischen genetischen Faktoren erwarten, daß sie die Synthese spezifischer Proteine steuern, wobei RNA die Rolle des Informationsüberträgers zu spielen hätte. Wie eine einfacheRechnung zeigt, wäre die in Mitochondrien enthaltene DNA-Menge bei weitem groß genug, um dieser Aufgabe gerecht zu werden; so könnte die in einem Rinderherz-Mitochondrion vorhandene DNA die Information für eine Sequenz von etwa 10000 Aminosäureresten, d. h. für ungefähr 70 Polypeptidketten mit einem durchschnittlichen Molekulargewicht von 17000 abgeben [*55*] und die DNA-Menge in einem Rattenleber-Mitochondrion sogar die Struktur von 150 Proteinen mit einem Molekulargewicht von 20000 determinieren [*42*].

RNA- und Proteinsynthese in isolierten Mitochondrien

Die Fähigkeit isolierter Mitochondrien zum Einbau radioaktiv markierter Aminosäuren in Protein ist in vielen Laboratorien beobachtet und genau studiert worden [*13, 27, 34, 48, 50, 64, 65* u. a.]; sie kann heute als gesichert gelten. Dagegen blieb die Frage zu klären, ob diese Organellen auch imstande sind, RNA aufzubauen. In unserem Laboratorium gelang der Nachweis einer DNA-abhängigen RNA-Synthese zunächst in isolierten Rattenleber-Mitochondrien [*69*], sodann auch in hochgereinigten Hefe-Mitochondrien [*71, 72*]. In beiden Fällen wurden radioaktiv markierte RNA-Bausteine (Uridin bzw. ATP und UTP) einem säure-unlöslichen Material einverleibt, das durch Phenolextraktion gereinigt, mit heißer Säure, mit Alkali und mit Ribonuclease abgebaut und dadurch als RNA charakterisiert werden konnte. Der Einbau von ATP oder UTP zeigte sich von der gleichzeitigen Anwesenheit aller vier Ribonucleosid-Triphosphate abhängig, wie es für eine RNA-Synthese, die von einer DNA-abhängigen Polymerase katalysiert wird, zu erwarten war. Der Einbau wurde auch durch Actinomycin C, das als spezifischer Inhibitor der Transskription, der Informationsübertragung von DNA auf RNA, bewährt ist, stark gehemmt (Tab. 3). Luck und Reich [*32*] führten bei ihren Arbeiten über die Mitochondrien von *Neurospora crassa* ebenfalls den Nachweis einer DNA-abhängigen mitochondrialen RNA-Polymerase, während Neubert und Helge [*43*] und Kalf [*25*] zusätzliche Beobachtungen über die RNA-Synthese und ihre Actinomycin-Empfindlichkeit in Säuger-Mitochondrien anstellten.

Auf Grund dieser Ergebnisse ist erwiesen, daß der Mitochondrien-DNA eine wesentliche Funktion beim Aufbau der RNA zukommt. Da Versuche gezeigt haben [25, 28, 70], daß gleich der RNA-Synthese auch

Tabelle 3: *DNA-abhängiger Einbau von ³H-UTP in RNA in isolierten Hefe-Mitochondrien.* Das Reaktionsgemisch (komplettes System) enthielt im ml: 20 μMole Tris-HCl (pH 7,4); 10 μMole MgCl$_2$; je 50 mμMole ATP, GTP, CTP und UTP; 2 μC ³H-UTP (2,4 C/mMol); 1 mg Phosphoenolpyruvat; 100 μg Pyruvat-Kinase; und 1 mg Mitochondrien-Protein. Inkubiert wurde 10 min bei 37° C [71]

	$\mu\mu$-Mole eingebautes UTP/mg Protein
Komplettes System	106
— ohne GTP, ATP, CTP	0
— mit 5 μg Actinomycin C	25
— mit 10 μg Actinomycin C	18
— mit 20 μg Actinomycin C	10

die Proteinsynthese in isolierten Mitochondrien durch Actinomycin gehemmt wird (Tab. 4), scheint die biologische Information der mitochondrialen DNA nicht nur transskribiert, sondern auch für die Eiweiß-synthese herangezogen zu werden. Daraus folgt, daß wenigstens ein Teil

Tabelle 4. *DNA-abhängiger Einbau von ¹⁴C-Leucin in Protein in isolierten Hefe-Mitochondrien.* Das Reaktionsgemisch (komplettes System) enthielt im ml: 0,25 mMole Saccharose; 1 μMol EDTA; 20 μMole Tris-HCl (pH 7,4); 10 μMole MgCl$_2$; 40 μMole KCl; 0,2 μC ¹⁴C-Leucin (160 mC/mMol); und 1,5 mg Mitochondrien-Protein. Inkubiert wurde 60 min bei 37° C [70]

	$\mu\mu$-Mole eingebautes Leucin/mg Protein
Komplettes System	5,2
— mit 20 μg Actinomycin C	2,1

der mitochondrialen RNA der Informationsübertragung dienen sollte. In der Tat sprechen Ergebnisse, die bei einer Kurzzeit-Markierung der Hefe-Mitochondrien mit radioaktiven RNA-Bausteinen erhalten worden sind, für das mitochondriale Auftreten einer relativ kurzlebigen Messenger-RNA [71]. Auch andere Resultate legen die Annahme nahe, daß der Eiweißaufbau *in* den Mitochondrien nach einem ähnlichen Mechanismus abläuft wie die *extra*mitochondriale ribosomale Proteinsynthese. Auch in isolierten Mitochondrien wurden eine Hemmung des Aminosäure-einbaues in Protein durch die Antibiotica Chloramphenicol und Puro-mycin [24, 28, 70], das Vorkommen einer aminosäure-abhängigen ATP-Pyrophosphat-Austauschreaktion und Hydroxamatbildung [10, 12] sowie die Anwesenheit und Wirksamkeit von Aminoacyl-RNA-Syn-thetasen [70] festgestellt. In der aus hochgereinigten Hefe-Mitochondrien extrahierten RNA kommen, wie ihre Auftrennung in Rohrzucker-Dichtegradienten gezeigt hat, Fraktionen vor, die sich wie Transfer- und wie Ribosomen-RNA verhalten [70—72]. Schließlich isolierte Elaev [14] aus Rattenleber-Mitochondrien Ribonucleoprotein-Teilchen, derenEigenschaften jenen extramitochondrialer Ribosomen sehr ähnlich waren.

Trotz des intensiven Studiums der Eiweißsynthese in isolierten Mitochondrien ist es bisher nicht gelungen, den Einbau radioaktiv markierter Aminosäuren in ein wohldefiniertes mitochondriales Enzym nachzuweisen; am Beispiel der Rattenleber-Mitochondrien konnte klar erwiesen werden, daß ein solcher Einbau in die löslichen Enzyme Malatdehydrogenase und Katalase sowie auch in das Cytochrom c nicht stattfindet [51]. Übereinstimmend wurde hingegen bei Untersuchungen, die mit den Mitochondrien der Rattenleber [49], des Rinderherzens [63], der Hefe [70] und der Flugmuskulatur der Wanderheuschrecke [44] angestellt worden sind, eine sehr beträchtliche Inkorporation radioaktiver Aminosäuren in eine unlösliche, RNA-haltige Protein- oder Lipoproteinfraktion gefunden, die vermutlich zum „Strukturprotein" der Mitochondrien [11] in enger Beziehung steht. Nach Kalf und Grèce [26] sind Mitochondrien des Lammherzens imstande, Aminosäuren *in vitro* auch in ihr actomyosinähnliches „kontraktiles Protein" einzubauen; dieses ist gleich dem Strukturprotein eine Komponente der Mitochondrienmembran, läßt sich von dieser jedoch mit Hilfe von Salzlösungen abtrennen [46].

Da das Strukturprotein die Eigenschaft besitzt, einige von den Atmungsfermenten stöchiometrisch zu binden [11], wäre es denkbar, daß der Aufbau des Mitochondrions von dieser unlöslichen Proteinfraktion seinen Ausgang nimmt und die genetische Information der Mitochondrien in erster Linie für die Synthese eines funktionstüchtigen *Strukturproteins* erforderlich ist, das verschiedene Atmungsfermente in eine bestimmte räumliche Anordnung bringt und die Bildung der Elektronentransportpartikeln [2] ermöglicht. In den Atmungsmangelmutanten der Hefe wäre nach dieser Hypothese, die erst einer experimentellen Verifikation bedarf, die Ausbildung respiratorisch aktiver Elektronentransportpartikeln dadurch blockiert, daß das Strukturprotein eine Veränderung erfahren hat.

Es kann aber derzeit nicht ausgeschlossen werden, daß Mitochondrien *in situ* auch an der Synthese „löslicher" Proteine beteiligt sind. In diesem Zusammenhang ist erwähnenswert, daß die Syntheserate in isolierten Flugmuskel-Mitochondrien, wie Neupert [44] gefunden hat, nur etwa 10% von der *in vivo* gemessenen beträgt. Es ist also möglich, daß Mitochondrien im Zuge ihrer Reindarstellung die Fähigkeit zur Proteinsynthese teilweise einbüßen oder daß die Synthese bestimmter Fermente der Mitochondrien nur in Gegenwart von Induktoren abläuft, die bei den *in vitro*-Systemen fehlen [3].

Summary

Mitochondria as carriers of a genetic information

The demonstration by Ephrussi and others that there are extrachromosomal mutations affecting the mitochondrial respiration such, e.g., as the "petite" mutation in yeast, has led many authors to consider the mitochondria as being potential carriers of a cytoplasmic genetic information. This hypothesis has been strengthened by recent evidence for the presence and function of DNA in mitochondria.

Yeast mitochondria purified by density gradient flotation were shown to contain, per mg of mitochondrial protein, approximately 4 μg of DNA. In the mitochondria-like particles of the respiratory deficient "petite" mutant and in the mitochondrial precursor structures of anaerobically grown yeast cells significant though lower amounts of DNA were also found to be present. Examinations of mammalian mitochondria carried out in several laboratories likewise proved the occurence in such particles of a small quantity of DNA. These findings are in agreement with conclusions drawn from histochemical, autoradiographic and electronmicroscopic studies of animal and plant mitochondria. LUCK and REICH who were the first to isolate mitochondrial DNA demonstrated that mitochondrial DNA differed from nuclear DNA in its buoyant density and thus in its base composition, a finding which has since been confirmed by other authors.

Mitochondrial DNA appears to be capable of reduplication, as evidenced by the incorporation into mitochondria of tritiated thymidine, and it has been shown to be involved in mitochondrial RNA and protein synthesis. Isolated and highly purified mitochondria of rat liver, yeast, and other cells will incorporate labeled precursors into mitochondrial RNA, a process which proved to be inhibited by actinomycin and therefore is DNA dependent. So is the incorporation of radioactive amino acids into the protein of isolated mitochondria, which has been studied extensively in many laboratories. The results obtained indicate that the mechanism of mitochondrial protein synthesis is quite similar to that known from the study of other protein-synthesizing systems. The material most highly labeled in mitochondrial amino acid incorporation appears to be related to or identical with the "structural protein", whereas a number of soluble mitochondrial proteins failed to be labeled to any significant extent. The mitochondrial "structural protein" is known for its ability to bind components of the respiratory chain and, consequently, for its involvement in the formation of respiratory particles. If indeed the genetic information of the DNA occurring within the mitochondria is responsible, at least in part, for the synthesis of "structural protein", this might account for the impairment of respiration in mitochondria of cytoplasmic mutants such as in the "petite" mutant of *Saccharomyces cerevisiae*. *

Literatur

[1] BELL, P. R., and K. MÜHLETHALER: J. molec. Biol. **8**, 853 (1964).
[2] BLAIR, P. V., T. ODA, D. E. GREEN, and H. FERNÁNDEZ-MORÁN: Biochemistry **2**, 756 (1963).
[3] BORST, P.: In: The Regulation of Metabolic Processes in Mitochondria (Diskussionsbemerkung): Elsevier Publ. Co. (in press).
[4] BURTON, K.: Biochem. J. **62**, 315 (1956).
[5] CASPARI, E.: Advanc. Genet. **2**, 1 (1948).
[6] CERIOTTI, G.: J. biol. Chem. **198**, 297 (1952).
[7] CHÈVREMONT, M.: Biochem. J. **85**, 25 P (1962).
[8] — S. CHÈVREMONT-COMHAIRE et E. BAECKELAND: Arch. Biol. (Liège) **70**, 833 (1959).
[9] CORRENS, C.: In: Handbuch d. Vererbungswiss., Bd. II, S. 1 (1937).
[10] CRADDOCK, V. M., and M. V. SIMPSON: Biochem. J. **80**, 348 (1961).
[11] CRIDDLE, R. S., R. M. BOCK, D. E. GREEN, and H. TISDALE: Biochemistry **1**, 827 (1962) .
[12] DAS, H. K., S. K. CHATTERJEE, and S. C. ROY: Biochim. biophys. Acta (Amst.) **87**, 478 (1964).
[13] — — — J. biol. Chem. **239**, 1126 (1964).
[14] ELAEV, N. R.: Biokhimia **29**, 413 (1964).

* Die Untersuchungen wurden vom U. S. Public Health Service (Grant GM-11225) unterstützt.

[15] Ephrussi, B.: Nucleo-cytoplasmic Relations in Microorganisms. Oxford: Clarendon Press 1953.
[16] — Naturwissenschaften 43, 505 (1956).
[17] —, et A. M. Chimènes: Ann. Inst. Pasteur 76, 351 (1949).
[18] — P. L'Héritier, et H. Hottinguer: Ann. Inst. Pasteur 76, 64 (1949).
[19] —, and H. Hottinguer: Cold Spr. Harb. Symp. quant. Biol. 16, 75 (1951).
[20] — — et J. Tavlitzki: Ann. Inst. Pasteur 76, 419 (1949).
[21] —, and P. P. Slonimski: Nature (Lond.) 176, 1207 (1955).
[22] Guttes, E., and S. Guttes: Science 145, 1057 (1964).
[23] Heilweil, A. G., and Q. VanWinkle: J. phys. Chem. 59, 939 (1955).
[24] Kalf, G. F.: Arch. Biochem. 101, 350 (1963).
[25] — Biochemistry 3, 1702 (1964).
[26] —, and M. A. Grèce: Biochem. biophys. Res. Commun. 17, 674 (1964).
[26a] Kislev, N., H. Swift, and L. Bogorad: J. Cell Biol. 25, 327 (1965).
[27] Kroon, A. M.: Biochim. biophys. Acta (Amst.) 72, 391 (1963).
[28] — Biochim. biophys. Acta (Amst.) 76, 165 (1963).
[29] — In: The Regulation of Metabolic Processes in Mitochondria: Elsevier Publ. Co. (in press).
[30] Lerman, L. S.: Proc. nat. Acad. Sci. (Wash.) 49, 94 (1963).
[31] Linnane, A. W., and J. L. Still: Aust. J. Sci. 18, 165 (1956).
[32] Luck, D. J. L., and E. Reich: Proc. nat. Acad. Sci. (Wash.) 52, 931 (1964).
[33] Mahler, H. R., B. Mackler, S. Grandchamp, and P. P. Slonimski: Biochemistry 3, 668 (1964).
[34] McLean, J. R., G. L. Cohn, I. K. Brandt, and V. M. Simpson: J. biol. Chem. 233, 657 (1958).
[35] Mitchell, M. B., and H. K. Mitchell: Proc. nat. Acad. Sci. (Wash.) 38, 442 (1952).
[36] —, and A. Tissières: Proc. nat. Acad. Sci. (Wash.) 39, 606 (1953).
[37] Nanney, D. L.: In: The chemical basis of heredity, S. 134. Baltimore: John Hopkins Press 1957.
[38] Nass, M. M. K., and S. Nass: J. Cell Biol. 19, 593 (1963).
[39] — —, and B. A. Afzelius: Exp. Cell Res. 37, 516 (1965).
[40] Nass, S., and M. M. K. Nass: J. Cell Biol. 19, 613 (1963).
[41] — — J. nat. Cancer Inst. 33, 777 (1964).
[42] — —, and U. Hennix: Biochim. biophys. Acta (Amst.) 95, 426 (1965).
[43] Neubert, D., and H. Helge: Biochem. biophys. Res. Commun. 18, 600 (1965).
[44] Neupert, W.: In: The Regulation of Metabolic Processes in Mitochondria: Elsevier Publ. Co. (in press).
[45] Oehlkers, F.: Naturwissenschaften 40, 78 (1953).
[46] Ohnishi, T., and T. Ohnishi: J. Biochem. (Tokyo) 51, 380 (1962).
[47] Rabinowitz, M., J. Sinclair, L. de Salle, R. Haselkorn, and H. H. Swift: Proc. nat. Acad. Sci. (Wash.) 53, 1126 (1965).
[48] Reis, P. J., J. L. Coote, and T. S. Work: Nature (Lond.) 184, 165 (1959).
[49] Roodyn, D. B.: Biochem. J. 85, 177 (1962).
[50] — P. J. Reis, and T. S. Work: Biochem J. 80, 9 (1961).
[51] — J. W. Suttie, and T. S. Work: Biochem. J. 83, 29 (1962).
[52] Sager, R.: New Engl. J. Med. 271, 352 (1964).
[53] Schatz, G.: Biochim. biophys. Acta (Amst.) 96, 342 (1965).
[54] — E. Haslbrunner, and H. Tuppy: Biochem. biophys. Res. Commun. 15, 127 (1964).
[55] — — — Mh. Chem. 95, 1135 (1964).
[56] — H. Tuppy u. J. Klima: Z. Naturforsch. 18b, 145 (1963).
[57] Sherman, F.: Genetics 48, 375 (1963).
[58] — Genetics 49, 39 (1964).
[59] Slonimski, P. P.: Ann. Inst. Pasteur 76, 510 (1949).
[60] — La formation des enzymes respiratoires chez la levure. Paris: Masson 1953.
[61] —, et B. Ephrussi: Ann. Inst. Pasteur 77, 47 (1949).
[62] —, et H. M. Hirsch: C. R. Acad. Sci. (Paris) 235, 741 (1952).
[63] Truman, D. E. S.: Biochem. J. 91, 59 (1964).
[64] —, and A. Korner: Biochem. J. 83, 588 (1962).

[65] TRUMAN, D. E. S., and A. KORNER: Biochem. J. **85**, 154 (1962).
[66] TUPPY, H., E. HALSBRUNNER u. G. SCHATZ: Mh. Chem. (im Druck).
[67] VITOLS, E., E. J. NORTH, and A. W. LINNANE: J. biophys. biochem. Cytol. **9**, 701 (1961).
[68] WALLACE, P. G., and A. W. LINNANE: Nature (Lond.) **201**, 369 (1964).
[69] WINTERSBERGER, E.: Hoppe-Seylers Z. physiol. Chem. **336**, 285 (1964).
[70] — Biochem. Z. **341**, 409 (1965).
[71] — In: The Regulation of Metabolic Processes in Mitochondria: Elsevier Publ. Co. (in press).
[72] —, u. H. TUPPY: Biochem. Z. **341**, 399 (1965).
[73] YOTSUYANAGI, Y.: Nature (Lond.) **176**, 1209 (1955).
[74] — J. Ultrastruct. Res. **7**, 121 (1962).
[75] — J. Ultrastruct. Res. **7**, 141 (1962).
[76] —, et C. GUERRIER: C. R. Acad. Sci. (Paris) **260**, 2344 (1965).

Diskussion

zu den Vorträgen D. J. L. Luck, H. Tuppy und E. Wintersberger

Vorsitz: *Karlson*

Jakob: I would like to mention the work of J. C. MONOULOU. He adapted anaerobically grown yeast cells ("grande"), which contained promitochondria without cristae, in a special buffer where six hours later mitochondria with cristae appeared. After administration of ^{3}H-thymidine at zero time, the lavel is to be found in the mitochondrial fraction already one hour later.

Parthier: I wonder, how Dr. TUPPY has found DNA-stimulated incorporation of leucine without any addition of the nucleoside triphosphates. Is it possible, perhaps, that in your experiments a firm attachment of the radioactivity to the structural proteins is present and no real protein synthesis?

Wintersberger: The amount of radioactive amino acids incorporated into mitochondrial protein is rather low. It is quite possible therefore, that there are still enough nucleoside triphosphates present inside the mitochondria. Furthermore, I isolated the radioactive product and could show (by digestion), that the radioactive amino-acid was incorporated inside the protein.

Karlson: Were these mitochondria incubated aerobically, so that they would be able to produce ATP?

Wintersberger: Yes. Under anaerobic conditions you get less than 15% of the incorporation.

Klima: Dr. LUCK, is the change of the density of mitochondria, by changing choline concentration, due to the larger amount of the outer envelope or to a change in the composition of the cristae?

Luck: The change in density after adding choline can be correlated directly with the change in phospho-lipid protein-composition. It appears from examining electron micrographs, that the number of cristae per unit area in large mitochondria is not different from that in the small mitochondria typical of high choline cultures. The deficiency in terms of lipid-composition is more serious with respect to lecithin: There is a ninefold difference (per unit protein) in lecithin content of the two kinds of mitochondria (high-choline and low-choline), but only a twofold difference in phospho-lipid.

It may be, that reduction in available amount of lecithin may specifically affect the synthesis of outer membrane. But we have no evidence for this. What changes quickly when you add choline, is the number of mitochondria.

Grundmann: Dr. TUPPY, you used actinomycin C in your leucine-incorporation experiments. This actinomycin C is a mixture of some actinomycins, as you know. In spite of this, are you sure that you really have a DNA-dependence of this incorporation?

Wintersberger: We used actinomycin C, simply because we did not have any actinomycin D available. But there are data on aminoacid incorporation into isolated mitochondria in the literature where inhibition by actinomycin D was observed (e. g. A. M. KROON, Biochim. Biophys. Acta *76*, 165 (1963)).

Sachsenmaier: As far as I know, Neubert working with liver mitochondria does not find inhibition of RNA-synthesis by actinomycin, unless the mitochondria are damaged.

Wintersberger: Our mitochondria were purified by gradient flotation. They, therefore, are probably more permeable than more gently isolated mitochondria.

But there are also contradictory data in the literature concerning the inhibition by actinomycin.

Grundmann: Dr. Luck, have you any idea from your dilution experiments about the distribution of the "old" (the formerly administered) material in the newly built mitochondria ?

Luck: Actually what you would have to do for such an experiment would be an extremely short pulse labeling experiment. And with the specific activity of choline that we had, it was impossible to get exposures for electron microscopic radio-autography. But since one has randomization of the labeled material within a single division , it seems to be unlikely, that there is a "growing point" (in that case one would expect to get non-random distribution of new material and old).

Beermann: As Dr. Luck has shown when he injects these mutant forms of mitochondria one really sees a morphological transformation of mitochondria only at the very last moment before the phenotypic change occurs. I would like to ask Madame Jakob, whether anything similar has been observed in mixing "petite" and "grande" cells of yeast.

Jakob: We don't know, what is happening, at a mitochondrial level, with yeast.

We know vegetative "petites" of two types: neutral ones (which fortunately were discovered first, because otherwise "petite"-genetics would not have happened), and suppressive ones. Crossing neutral-"petite" × "grande" gets a diploid grande which produces by sporulation 4 haploid "grandes" (not-mendelian). On the other hand, crossing suppressive "petite" × "grande" gets a diploid which is "petite". By sporulation of the zygotes immediately after their formation, in this case, come out 4 "petites". But we don't know at all, if this has basis on mitochondria.

Beermann: But the possibility, that in this case there is a selection between "petite" mitochondria and normal ones is not to be excluded.

Jakob: It is not excluded.

Luck: In *Neurospora* systems there is a correlate to non-suppressive petites, namely "poky". In heterocaryon situations, where "poky" is introduced into another cytoplasm by fusion, the "poky" trait is not expressed. ABN-1, a cyto-plasmically inherited alteration resembling "poky", described recently by Tatum (J. Cell Biol. *26*: 413, 1965) does express itself in such an experiment. It could be considered suppressive "poky", if you like.

Karlson: With respect to the multiplication of mitochondria I would like to know, whether their DNA represents only one set of information or several sets of information. If the total DNA of one mitochondrion codes for 10000 amino-acids and this pool of DNA represents the set of, let us say, 10 informative units, then we are back to 1000 amino-acids, which may be just enough for the structural protein, but certainly not enough to account for all the enzymes, which are present in the mitochondria.

Luck: The pieces, which we get of mitochondrial DNA from *Neurospora* as analysed by Dr. Stoeckenius using Kleinschmidt-techniques have a molecular weight of about 13 million. If one makes a calculation based on a number of mito-chondria per mg protein, and assumes uniform distribution of DNA, then one can have 10 such molecules per mitochondrion.

In the case of yeast, there is good evidence that the structural genes for cyto-chrome C, at least, are located in the nucleus. I think that in *Neurospora* the same situation exists. It seems, therefore, unlikely that the synthesis of all mitochondrial proteins is determined by the mitochondrial DNA.

Karlson: How is the information for the cytochromes or the other enzymes, which are coded for in the nucleus, transferred to the mitochondrion ?

Luck: I think, that these enzymes, which could exist in a soluble form, could be synthesized in the cytoplasm and put into the mitochondrion. Because one sees few ribosomes in *Neurospora* mitochondria, I am prejudiced against consider-ing the mitochondrion as a major site for protein synthesis. Structural protein as an insoluble protein would be quite difficult to transport, therefore there might be an advantage in having it made *in situ*.

Beermann: There are not only structural genes, but also regulatory genes, on the chromosomal level at least. There may be some in the mitochondria too.

Luck: Yes, of course. It seems likely that in *Neurospora* the nuclear determinants of mitochondrial function so far described are regulatory. Furthermore it is possible to consider, that the gene-product of mitochondrial DNA is not protein at all, or at least not a protein which is a component of mitochondria. The gene product of mitochondrial DNA may be an RNA or protein which has a regulatory function on structural genes in the nucleus.

Karlson: Is it likely to assume the existence of an additional piece of information for the organization of all the macromolecular structures (especially in the electron-transport chain), or is it possible that all these enzymes become properly arranged by physico-chemical forces only?

Luck: It is possible that like the phage proteins these units are self assembling. But in the case of *Neurospora*, where mitochondria come from mitochondria, we have the possibility that the pre-existing structure may be a primer in Dr. TAYLOR's sense of the word.

Tuppy: On electron micrographs of anaerobic yeast cells which lack typical mitochondria and do not possess a full electron-transport chain one sees vesicles (YOTSUYANAGI), and the density-gradient experiments have also shown, that particles are present, which may be structural and functional precursors of fully developed mitochondria. However , as GREEN and his colleagues have shown it is possible to put the components of the respiratory chain into an array even without a pre-existing matrix, and they will then work at least in electron-transport, though not in oxidative phosphorylation, possibly because the "coupling factors" are lacking.

Parthier: Dr. LUCK, what do you think about the entering of such a big protein molecule like cytochrome oxidase, if you accept that small protein molecules (like RNase or DNase) cannot enter?

Luck: I have no good answer to this. It is possible that cytochromes become part of the mitochondrion as a complex with phospholipid.

Stoeckenius: I would like to offer a definition for a hypothetical structural protein. I would say: It is a protein, which combines with the membrane lipids and holds the enzymes of the membrane in the proper position and configuration for their specific function. It could, of course, in addition, function as an enzyme itself.

As to the reconstruction of a functional chain from soluble components, I don't think that it has been established beyond doubt by GREEN and his collaborators. I have looked at some of GREEN's complexes, prepared by Dr. FLEISCHER, and they appeared as fragments with unit membrane structure in the electron microscope.

The only clear case of functional and morphological reconstruction from non-membranous components, I know of, is the work of RACKER and his collaborators. He isolated the particles from the mitochondrial cristae, and they showed ATPase activity. When the particles are on the membrane, the ATPase activity is oligomycine-sensitive. When they are removed from the membranes, it is not. RACKER has further isolated from the membranes, devoid of the ATPase, a protein fraction, which contains less than 5% lipid. In the electron microscope it is rather amorphous. It certainly shows no membrane structure. This fraction binds the ATPase, and all activity is abolished completely. The same protein fraction, with phospholipid added, forms vesicles, which have a unit membrane structure. There is no enzymatic activity found in these. However, when the phospholipid is added to the membrane protein fraction plus the ATPase, oligomycine-sensitive ATPase activity is regained. In the electron microscope, vesicles are seen, which carry the particles on the outside and in sections show the unit membrane structure.

This is the only reasonably well documented case, I know of, where specific activity is regained by reforming a membrane. It is possible, that the membrane protein fraction contains the so-called structural protein.

Actually Racker has also isolated a so-called structural protein, which is present in this fraction (his factor F 4) and he has shown that it is necessary for reconstitution of parts of the oxydative phosphorylation mechanism.

Wintersberger: This factor F 4 is quite similar to the "structural protein" of Green, except that it was isolated by a different method and, therefore, might be more native than Dr. Green's "structural protein".

Stoeckenius: Green's original preparation was actually inactive in Dr. Racker's system. However, a modified preparation [Richardson, S. H., H. O. Hultin, and S. Fleischer, Arch. Biochem. **105**, 254 (1964)] seems to be active. Mixed with lipids, it reforms vesicles bounded by a unit membrane, but so do other proteins.

Taylor: Are there any reversible mutations in any of the mitochondrial systems?

Luck: No, not in any of the reported cytoplasmic mutants of *Neurospora*. The situation is different for nuclear mutants affecting mitochondrial function.

Jakob: There is no reversal in yeast also. In segregational "petite" mutants like in vegetative "petites" the respiratory system is not functional. In the latter cristae are not found in the mitochondria while in the former there are cristae. If in segregational ones a gene is back-mutating, respiration is regained. It is possible to get a reconstitution by fusion of a segregational "petite" with a vegetative one, but this is no reversal.

Brdicka: Da das „Strukturprotein" bisher jeder Analyse widersteht, könnte man dem Problem vielleicht näher kommen, wenn man untersucht, welche Aminosäuren für die Mitochondrien zur Synthese dieses Proteins essentiell sind.

Neupert: Hefemitochondrien scheinen genügend endogene Aminosäuren zu haben, so daß eine Zugabe von Aminosäuren nicht erforderlich ist.

Etwas anderes ist es bei Rattenlebermitochondrien. Wir haben den Aminosäurebedarf untersucht und ein Maximum der Syntheserate bei einer bestimmten Konzentration eines kompletten Aminosäuregemisches festgestellt. Um den Effekt der einzelnen Aminosäuren zu untersuchen, ließen wir jeweils eine Aminosäure aus dem Gesamtgemisch weg. Dabei zeigte sich, daß die Syntheserate nur im Fall von 5—6 Aminosäuren vermindert wird, und zwar bis zu 20—30%. Zu diesen Aminosäuren zählen hauptsächlich unpolare, wie Alanin, Valin, Serin und auch Prolin.

Karlson: Können Sie abschätzen, wieviel Protein in der Zeiteinheit produziert wird?

Neupert: Das ist für die Rattenlebermitochondrien schwer zu sagen, weil die Radioaktivität sehr verdünnt ist. Aber im Falle des Einbaus in isolierte Locusten-Mitochondrien kann man abschätzen, daß die Syntheserate ungefähr 10% der theoretischen Einbaurate (berechnet aus dem Mitochondrienwachstum in vivo) ist, wenn man die spezifische Aktivität auf das gesamte unlösliche Protein bezieht. Wenn tatsächlich das „Strukturprotein" synthetisiert wird, sind noch höhere Einbauraten zu erwarten, da das „Strukturprotein" nur einen Teil des gesamten unlöslichen Proteins darstellt.

Taylor: Dr. Tuppy showed some profiles of RNA that is formed in mitochondria. The rapidly labeled RNA in most systems seems to be a large 45 S particle. Did you find the same thing in this system?

Tuppy: Dr. Wintersberger found the fast labeled mitochondrial RNA to be rather heterogeneous. The sedimentation constants, however, did not appear to be as high as 45 S. Nor have we evidence yet whether its base composition corresponds to the base composition of the mitochondrial DNA.

Taylor: How is its base-composition different from that in other organelles in the whole cell?

Tuppy: This has not yet been determined.

Sachsenmaier: Dr. Luck, have you made any hybridization experiments, to find out, whether mitochondrial DNA has regions corresponding to nuclear DNA?

Luck: No. But Shipp et al. (PNAS. *54*: 207, 1965) have carried out such experiments with plastid and nuclear DNA of tobacco, where they find no detectable homology.

Jakob: I would like to make some comments concerning the paper that has been recently published by Drs. Tuppy and Wildner entitled "Cytoplasmic transformation": mitochondria of wild-type baker's yeast restoring respiratory capacity in the respiratory deficient "petite" mutant [Biochem. Biophys. Res. Commun. *20*, 733 (1965)].

I think that one should be very critical about this type of experiment. There are several reasons for being particularly cautious in addition to the obvious necessity of excluding rigourously the possibility of contamination. One reason stems from unpublished observations of WRIGHT and SLONIMSKI who tried this type of experiment several years ago. "Grandes" "pseudo-transformants" were found, but after careful analysis their appearance was explained not by a so-called "transformation" but by a fusion (or copulation) of "grande" cells present in the donor mitochondrial preparation in a very small proportion. These cells being so damaged as to be unable to give rise to a colony on plating, but apparently able to fuse. To detect this process one has to use good nuclear markers and I do not think that thiamine and pink color, used in the experiments of TUPPY and WILDNER are completely reliable gene markers. On the other hand, we have found in our laboratory, a few times, respiratory sufficient cells in pure cultures of cytoplasmic "petites". This result is very unexpected. Since not all of the nuclear markers of the "petite" were recovered in the respiratory sufficient clones, it is difficult to rule out completely an unusual contamination in these cases.

Proposals Concerning Replication of the
Golgi Apparatus

By

W. Gordon Whaley, Austin

With 15 Figures

Introduction

Biologists have made more borderline observations, created more confusion, and dealt in more polemics over the Golgi apparatus than they have with any other cell organelle. The story of inadequacies, misrepresentations, straightforward observations, and entertaining speculations has been variously treated by many investigators [6,7,2,47,48,56,16,44, 30, 31, 5]. The literature is enormous and through it the question is frequently raised — how does the Golgi apparatus replicate? Specific concern with replication of this organelle dates back at least as far as the first decade of the twentieth century when Perroncito [49, 50] applied the term "dictyokinesis" to what he assumed to be a process of division of the apparatus and implied that this division more or less parallels division of the nuclear material. Unfortunately the first half century of work and discussion have failed to answer the question conclusively.

The early literature was all based on studies by optical microscopy that could barely define the structure of the Golgi apparatus, let alone give a clear idea of its replication; and the subsequent literature, based on electron microscopy, has reference to static pictures which suggest methods of replication only in terms of positional relationships, apparent modification of structure, and entities interpreted as progenitors. Modern cytology has no better example of the confusion that can result when dynamic events are interpreted on the basis of electron micrographs alone. Furthermore, direct comparability between what is now called the Golgi apparatus in electron micrographs and what Camillo Golgi [22] referred to as the "appareil réticulaire interne" is subject to considerable doubt. The "internal reticular apparatus" of the Purkinje cells of the barn owl as diagrammed by Golgi most likely contained what we now describe as the Golgi apparatus, but it also must have included a number of other cellular components capable of precipitating osmium or silver.

In both optical microscopy and electron microscopy of fixed materials and in phase microscopy of living materials we have in our laboratory examined, over a period of many years, a great number of preparations for the occurrence or nonoccurrence of the images on which the ideas of origin and/or division of the Golgi apparatus are based, as they are recorded in the literature. Our experimental materials have, in one investigation or another, included cells of cryptogams and phanerogams, invertebrates and vertebrates. Although this paper cannot answer the question of how replication of the Golgi apparatus takes place, it will attempt to assess the tenability of some of the hypotheses that have been advanced and evaluate alternative hypotheses that might explain some of the images that have been seen.

The Morphology and Function of the Organelle

Comprehension of the possibilities for replication depends upon knowledge of the general characteristics of the apparatus. Like other organelles, the Golgi apparatus has certain common structural features which distinguish it no matter what changes in extent and form take place with differentiation and the development of particular functions. Over a range of experimental materials the Golgi apparatus may seem to have greater variability than do other organelles. It is doubtful that it does vary more than other organelles. The extent of structural differentiation of organelles is often very great, as, for example, among mitochondria as discussed by LEHNINGER [35]. Often, too, organelles may appear as discrete entities or be very intimately grouped as are mitochondria in the constitution of the *nebenkern*. The absence of a membrane delimiting the Golgi apparatus as a discrete organelle makes its boundaries difficult to define and may enhance the appearance of variability.

The Golgi apparatus is a specialized component of the cytoplasm, distinguished by its ability to carry out specific activities; it has certain structural characteristics that are related to these activities. The most studied of its activities relate to its role in secretion but as several investigators have recognized [see 74], there is a distinct possibility that the apparatus has as yet unexplored functions in nonsecreting cells. The same basic structure appears to pertain, however, in the unspecialized and specialized apparatus.

The most general and ordinarily the most conspicuous component is the "stack" of flattened, membrane-bound sacs referred to as "cisternae," "saccules," or "lamellar units." The application of any of these terms can be confusing, for all of them have been applied to other cellular components. I shall use the term cisternae. The size and number of these cisternae are subject to considerable variation. There may be only a very few, each a few microns in diameter, or there may be several times this many of much greater extent (Fig. 1a, b). The membranes of the cisternae are not studded with ribosomes. The arrangement is often far more complex than a simple stack and is subject to considerable variation.

The thickness of the unit membrane appears to differ somewhat from one fixation procedure to another. As is discussed below, the apparatus usually has a more or less obvious polarity. Sometimes the membranes at one pole may appear different from those toward the other pole, quite possibly reflecting differences in the contents of the cisternae and/or

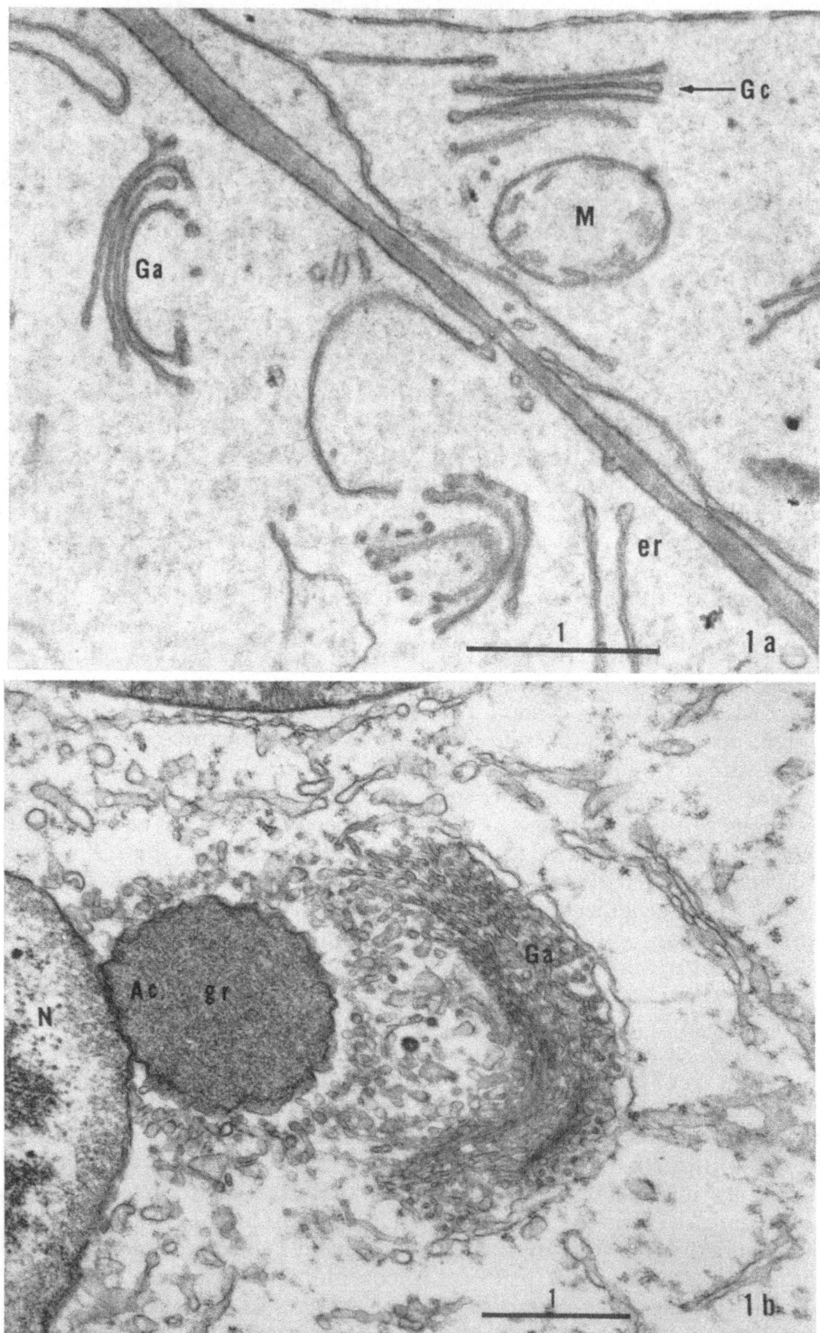

Fig. 1 (a). Simple Golgi apparatus (Ga) in undifferentiated cells of a *Zea mays* root tip. [KMnO₄] (b). Specialized Golgi apparatus involved in acrosomal granule (Ac gr) formation in an Hemipteran spermatid. [glut-OsO₄]. *er* = endoplasmic reticulum, *Gc* = Golgi cisterna, *M* = mitochondrion, *N* = nucleus, *glut* = glutaraldehyde. The same symbols are used in subsequent figures with additional ones introduced as necessary. Fixation is noted in brackets

functional states. What appear to be membrane differences are difficult to evaluate, for the relation between the functional activities of membranes, their precise chemical composition, and their physical appearance in electron microscopy has not been adequately worked out. The distances between adjacent cisternae appear to vary, but whether this variation is related only to the plane of cut or also to functional factors is not known.

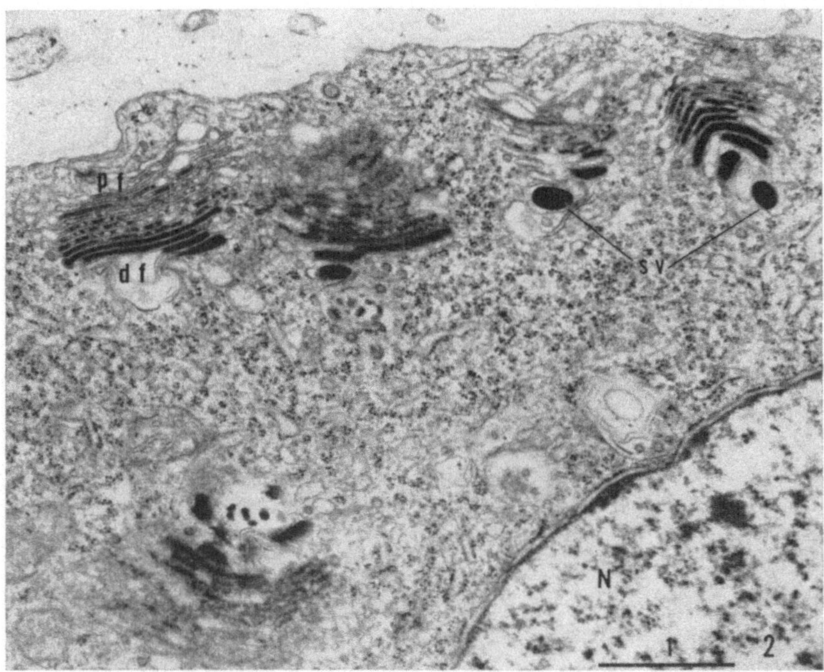

Fig. 2. Golgi apparatus showing density in the cisternal contents which increases from the proximal face (*pf*) to the distal face (*df*). *Helix aspersa* embryo cell forming an electron dense product. [glut-OsO₄] *sv* = secretory vesicles

As is also noted below, there is sometimes detectable intercisternal structure.

The apparatus has what was termed by GRASSÉ and his co-workers a distal face and, opposite it, a proximal face. The distal face was interpreted as characterized by vesiculation. In the zooflagellates studied, the proximal face had a positional association with the parabasal filament. Of greater significance was the extension of this interpretation to hold that the apparatus undergoes vesiculation at the distal extremity and a continuous formation of cisternae at the proximal extremity [23]. This concept calls for a stepwise displacement of cisternae across the apparatus, continuing disappearance from one face and continuing replacement on the other. From studies of secreting plant cell Golgi apparatus MOLLEN-HAUER and WHALEY [42] introduced the terms "forming" and "maturing" faces to make essentially the same distinction GRASSÉ had made. Since

Grassé's terminology has priority, it seems desirable to retain it, even though the apparatus often is not in a definite positional relationship to another cellular component as were the ones he studied in zooflagellates. The important consideration is that the apparatus has two faces which may have very different functions, although there may be circumstances in which simple morphological distinctions between the two faces are not

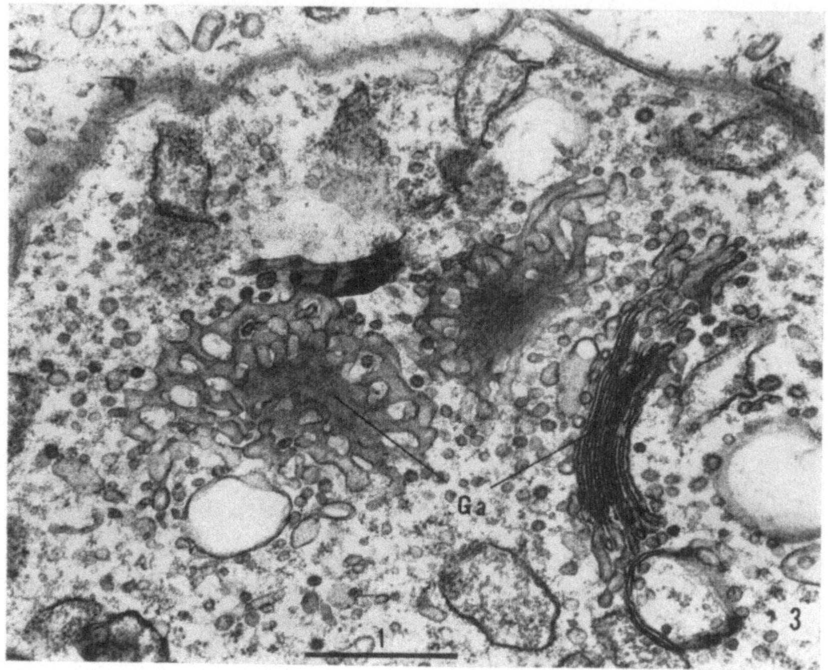

Fig. 3. Golgi apparatus from a fresh-water clam gill showing the fenestrated nature of the cisternae. [OsO_4]

easy to discern. In fact, the morphological evidence distinguishing between the two faces is complicated by differences in fixation reactions, stage of development, intensity of activity and the functions of the organelle. Whatever the variables, the proximal face is apparently characterized by an inflow of molecular components from the cytoplasmic matrix. The distal face is in many instances characterized by the evolution of membrane-bound material.

The relative length of the cisternae within an apparatus may differ from species to species, with stage of ontogeny and very clearly sometimes in relation to activity. If there is a detectable difference in the contents of the cisternae, it seems in most instances to be in the direction of an increase in density from the proximal to the distal face (Fig. 2). Often the cisternae are somewhat fenestrated and sometimes they are practically reticulate (Fig. 3). This reticulation was recognized in the first electron micrographs of the Golgi apparatus [see *32*], but its significance with respect to activity, or even to position along the gradient from the proximal to the distal face, has not been satisfactorily defined. An example of extreme fenestration is seen in neurosecretory cells of the tunicate *Styella* [*33*] where the Golgi apparatus is apparently involved in production of a pigmented secretory material. MOLLENHAUER [*40*] and his co-workers have also shown extreme fenestration in extracted Golgi apparatus after "stabilization" with glutaraldehyde. It would seem, however, that caution should be observed in postulating relationships between this type of structure and particular activities or making interpretations from *in vitro* preparations, because such cisternal morphology is frequent in many types of cells.

Associated with the cisternae are numbers of vesicles. These vesicles may differ in their presumed relationship to the cisternae. They may also differ from one type of cell to another or with respect to size and staining characteristics and thus in their composition [*58, 69*]. In some cases we cannot yet separate these vesicles in terms of their origin, function, or fate. The problem of dealing with these Golgi vesicles is further complicated because some categories of them are frequently referred to as "vacuoles" and some as "granules." With one exception the term "vesicle" is used here as a general designation without reference to the characteristics of the structures. The exception is that those vesicles clearly participating in secretion will be designated as secretion vesicles. This is not to indicate that all secretion necessarily involves the participation of the Golgi apparatus. As other investigators, such as SCHNEPF [*55*] and FAWCETT [*20*], have recognized, some secretory processes involve materials that seem not to be passed through the Golgi apparatus.

Secretion vesicles may be variously separated from the cisternae. In some instances only the very peripheral regions of the cisternae appear to be involved. In other instances there is, by some interpretations, total vesiculation of the cisternae at the distal face of the apparatus. The evidence suggests there is more than one mode of separation of these vesicles from the Golgi apparatus, and HAGADORN et al. [*25*] have supported such a suggestion. Whatever the mode of separation, their form and other characteristics may vary greatly. We have observed different types of secretion vesicles in variously differentiated cells of the maize root tip (Fig. 4a, b, c). HAGADORN et al. [*25*] have also noted that at least four types are produced by the Golgi apparatus in different cells of the nervous system of the leech. In any event, the production of secretion vesicles increases from the proximal to the distal pole.

In the course of their development and separation secretion vesicles may come to contain elaborate structures such as the iron-containing crystalline arrays found by STRUNK [*61*] in the midgut glands of *Limnoria lignorum*, the various crystals found by MANTON and co-workers in vesicles from the Golgi apparatus of algae [*36, 37, 38*], or the complex rod-shaped structures found by TURNER in the scorpion (Fig. 5). It seems apparent that there may be continuing modification of the structural contents of the secretion vesicles after they have been separated from

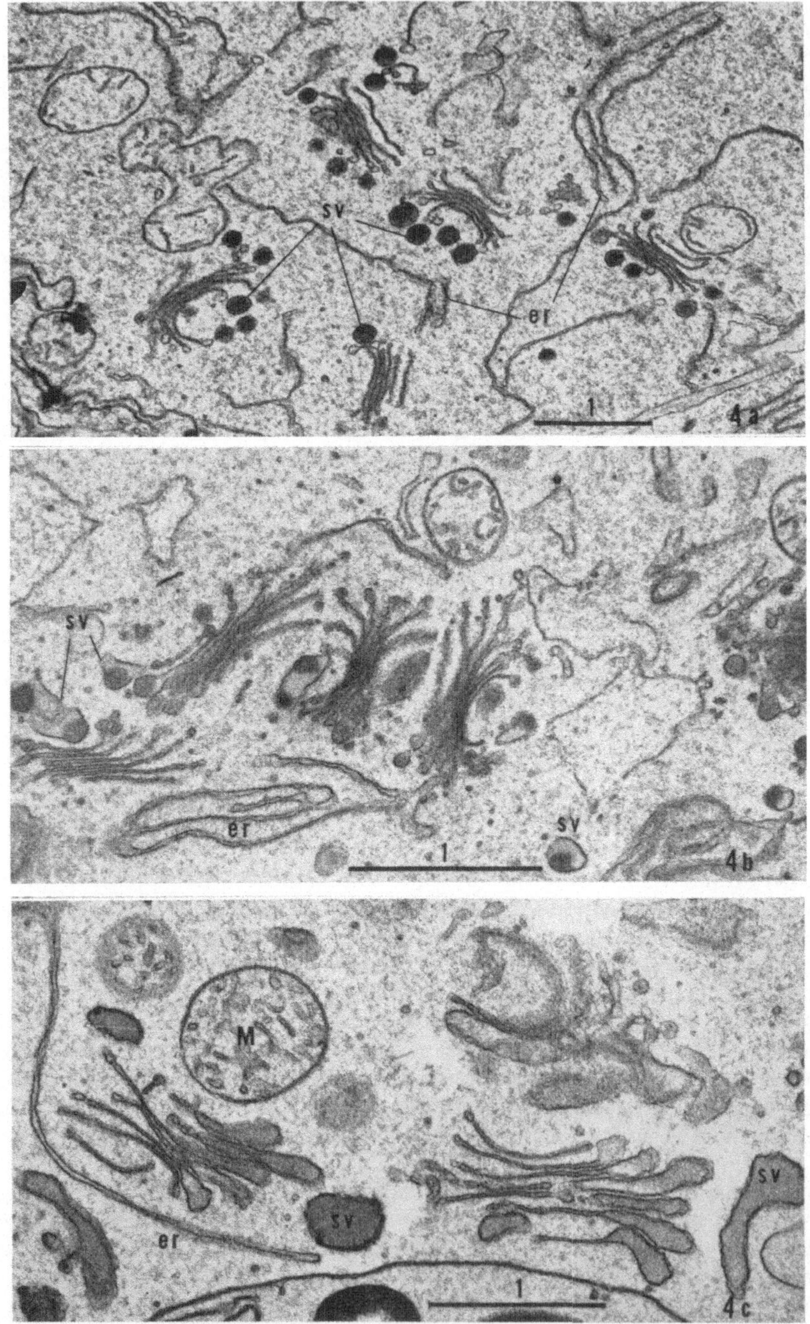

Fig. 4a—c. Different types of secretory vesicles in *Zea mays* root tip cells. [KMnO₄] (a) epidermal cell, (c) secreting outer root cap cell, (b) cells of a region of overlap of differentiated epidermal cells and differentiated root cap cells. The vesicles show characteristics of both the types seen in (a) and (c)

the cisternae — a concept strongly supported by SJÖSTRAND [57]. They
may change in size and density [74] by processes which could include
fusion of the vesicles, dehydration or concentration, further synthetic
activity, or the incorporation of additional materials from the cytoplasm.
The outer surfaces of secretion vesicles sometimes become ribosome-
studded (see Fig. 5).

Close to the proximal pole of the apparatus there are frequently
aggregations of small vesicles. In certain cells in which protein secretion

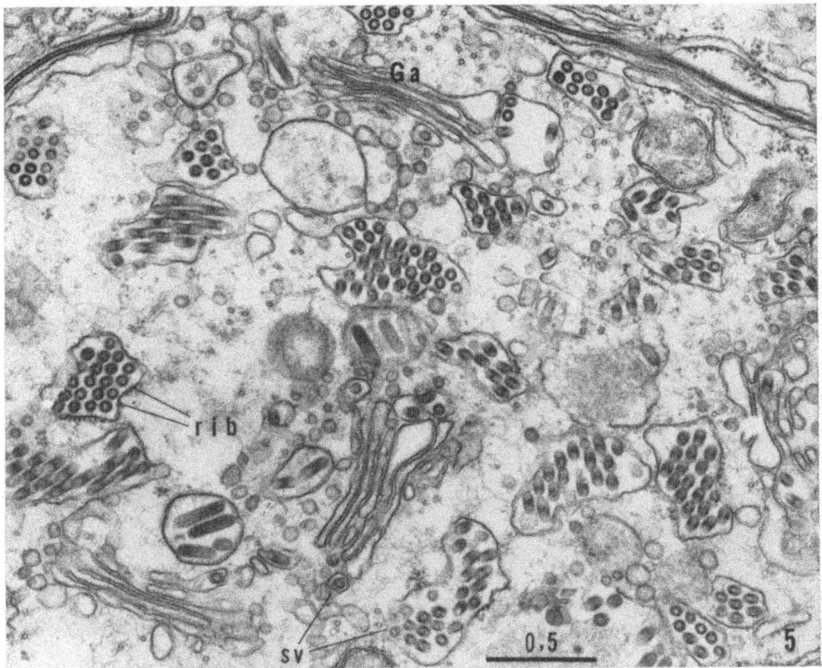

Fig. 5. Golgi apparatus and secretory vesicles containing rod-shaped structures. Some of the
secretory vesicles are studded with ribosomes (rib). Scorpion tissue. [OsO₄]

is an important process there is a conspicuous budding-off of small
vesicles from the endoplasmic reticulum. The interpretation discussed
below that small vesicles from the endoplasmic reticulum represent
vehicles for transfer between this organelle and the Golgi apparatus
would explain the presence of these aggregations of small vesicles at
the proximal pole.

In some Golgi apparatus there are definable intercisternal structures
[62, 41]. In the Golgi apparatus of *Nitella* cells in spermatogenesis,
certain cells of the leech *(Hirudo sp.)*, and maize root tip cells, there is a
system of elongated elements in a patterned arrangement (Fig. 6). The
structural and functional significance of these elements is unknown, and
there is not yet information as to how generally they occur.

The older literature frequently alludes to some differences in the ground substance surrounding the demonstrated components of the Golgi apparatus from that of the remainder of the cytoplasm [58, 16]. Hirsch [see 31] has carefully defined this Golgi ground substance as a special field and has implicated it in the exchange of materials between the general cytoplasmic ground substance and the cisternae of the Golgi apparatus. What is known of the physiological activity of the apparatus makes it

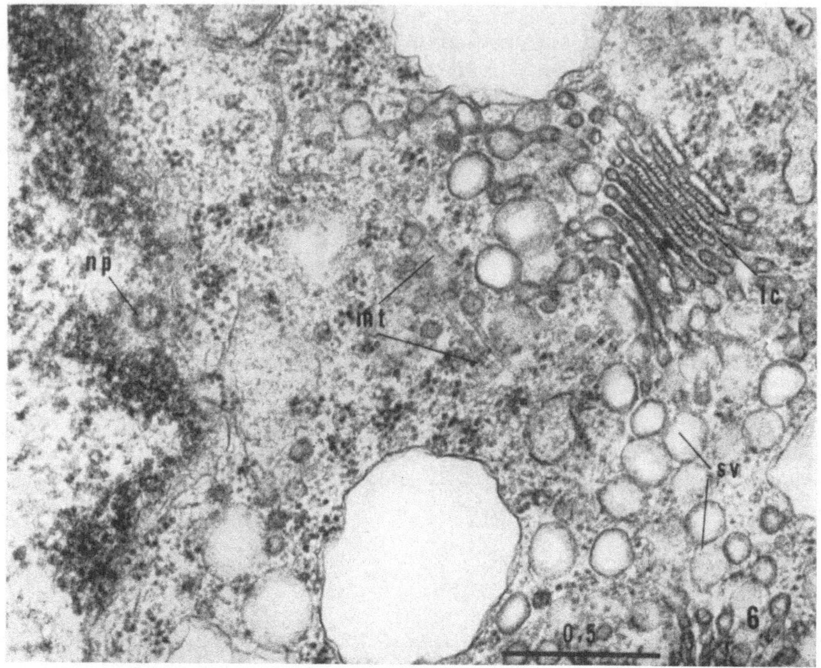

Fig. 6. Golgi apparatus showing intercisternal elements *(ic)*. *Nitella sp.* [glut-OsO₄]
mt = microtubules, *np* = nuclear pore

certain that there are some differences between the ground substance surrounding the cisternae and that of the rest of the cell, but there is no morphological evidence on this point.

The emerging concept is one of an organelle consisting of a stack of cisternae across which there is a polarity, certain activities being more intense at one pole and certain activities more intense at the other. The form, extent and specific characteristics of the organelle may differ considerably from cell type to cell type and ontogenetically during the development of each cell. Additionally the number of organelles within a cell, their distribution, and the extent to which they are aggregated or interrelated is subject to wide variation. Years ago, Wilson [71] described two principal forms of the Golgi apparatus, one localized, the other diffuse, and suggested that they might reflect changing phases of cellular

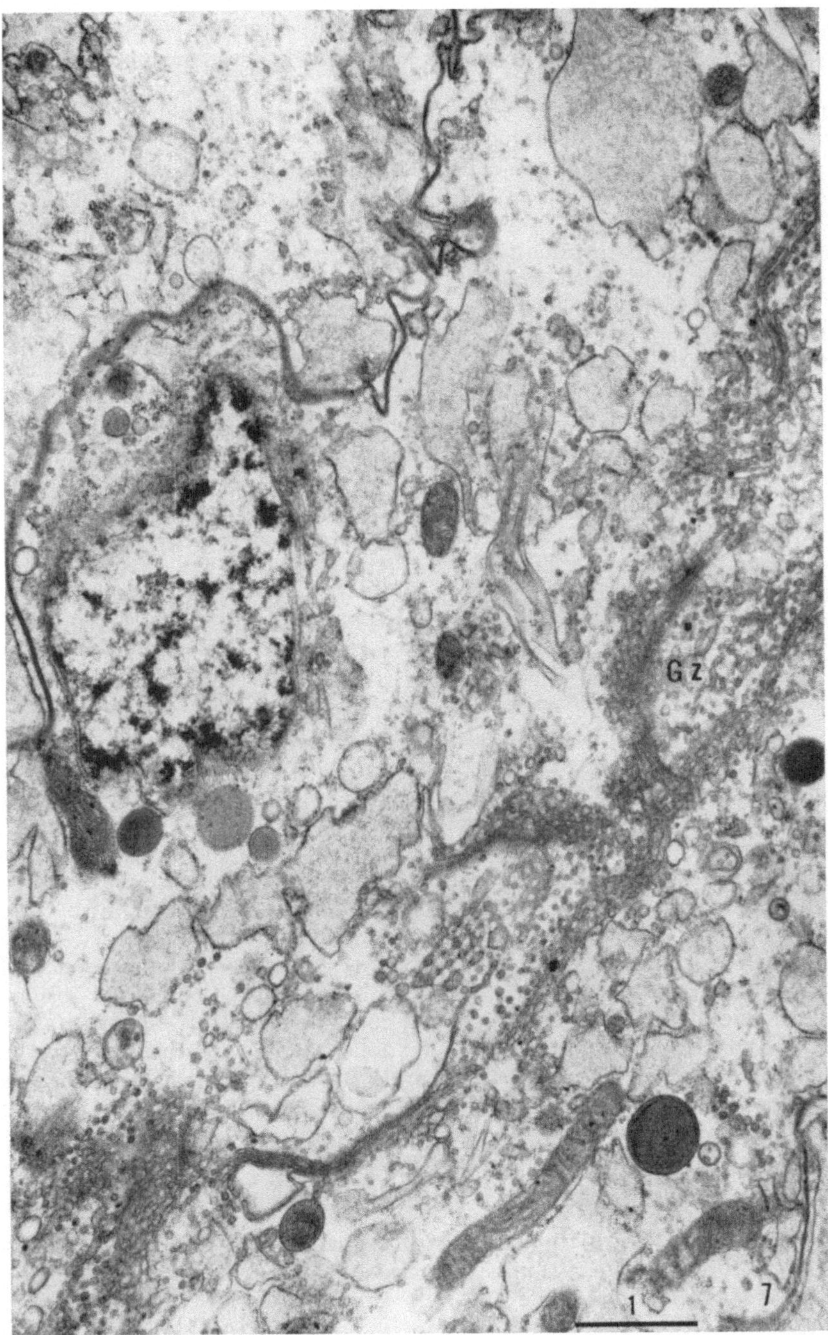

Fig. 7. Extensive and elaborate "Golgi zone" (Gz) from scorpion testis. [glut-OsO$_4$]

activity. Changes in the appearance of the organelle correlated with changes in cellular activity and development are a conspicuous feature of all but the most superficial studies of it by electron microscopy. For example, Hay [27], using larval salamanders, *Amblystoma punctatum*, recognized significant differences between the Golgi apparatus of early blastema cells and those of cells following regeneration which were involved in cartilage matrix formation. Chardard [14] from his studies on the development of the pollen mother cells in orchids has suggested that there exist more or less distinct phases of the Golgi apparatus; a secretory phase and a multiplication phase, each with specific structural characteristics. A further complexity is introduced by an observation of Bonneville and Voeller [3] that could be interpreted to indicate that there may be two quite different types of Golgi apparatus simultaneously within a single cell of the dodder, *Cuscuta campestris* Yuncker.

Some of the controversy regarding the Golgi apparatus has arisen from the fact that, depending on the state of cellular activity and the stage of differentiation, the total cellular complement of Golgi material may appear as a number of individual units that are not interconnected by membranes, highly elaborate "Golgi zones" (Fig. 7) or any of a number of intermediate types. The fact that the literature deals with investigations of many different species, various stages of development, and widely differing cell types has done more to confuse the issue than to contribute to a unified concept. The literature includes varied and sometimes conflicting uses of a number of terms. A case in point is the term "dictyosome."

Perroncito [49] introduced this term to refer to structures resulting from the division of what he assumed to be the equivalent of Golgi's reticular apparatus, but the term has since been variously applied in part in relation to classes of organisms. Because of the varied applications of the term and because, as noted above, the form of the Golgi apparatus may differ depending on the stage of differentiation and the state of cellular activity, the problem of relating the terms "Golgi apparatus" and "dictyosome" has become both very difficult and surrounded by sterile arguments. It is hardly worthwhile either debating the equivalence of these terms or attempting to restrict the term "dictyosome" to a specific type of Golgi apparatus or a particular component of the Golgi apparatus. The terms "Golgi region" and "Golgi zone" have some relevance to the arrangement and position of the complex structures noted above, where they have generally been used in a more inclusive sense than has the term Golgi apparatus.

Interpretation of Function and a Concept

Evidence for a functional polarity of the Golgi apparatus, at least in relation to cellular secretion, seems clearly established, regardless of the morphology of the organelle. The very fact that in many cases a proximal and distal face can be defined on purely morphological grounds supports the concept of functional diversity at the two poles of the apparatus. The proximal face may sometimes be distinguished by certain characteristics of its cisternae, by the absence of any vesicles that are clearly secretory, by close association with the endoplasmic reticulum, or by the presence of ribosomes. The distal face may be characterized by particular patterns of vesiculation, by the abundance of secretory vesicles, or by the absence of ribosomes (Fig. 8, 9). In some cases where the contents of the cisternae are electron-dense or can be stained, an actual gradient can be observed

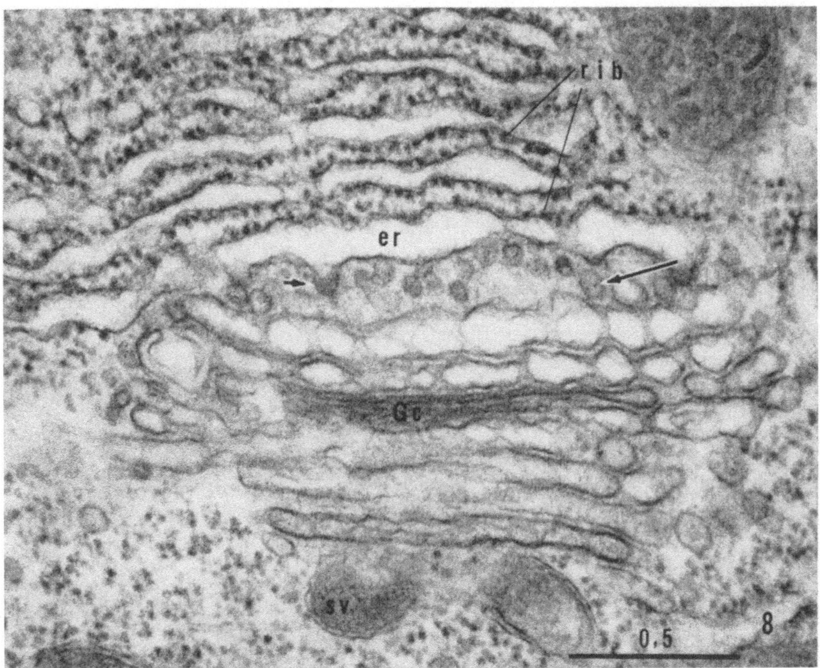

Fig. 8. Golgi apparatus showing the close association of the endoplasmic reticulum (*er*). Arrows show vesicles apparently derived from the endoplasmic reticulum and clustered adjacent to the proximal cisterna of the apparatus. Mealybug testis. [glut-OsO$_4$]

(see Fig. 2). The concept of a gradient is also supported by the demonstration of different enzyme activities in localized regions of the Golgi apparatus of certain neurons as shown by NOVIKOFF and his co-workers [*45, 46*].

The role of the organelle in secretion has received the most study. While synthetic capacities had been previously assumed for the apparatus by certain investigators, direct evidence for the synthesis of complex compounds from simple precursors was lacking. Its role was often looked upon as that of concentrating and modifying materials that became part of a secretory product to be separated from the apparatus in membrane-bound form. However, there is accumulating evidence that certain synthetic capacities may distinguish the Golgi apparatus. FRIEND's studies of Brunner's glands in the mouse [*21*] and some of the work in

352 W. Gordon Whaley:

Leblond's laboratory [51, 52] indicate that the complex carbohydrates that are often part of the secretion products are actually synthesized in the Golgi apparatus. Whatever the activities of the Golgi apparatus are, there would currently appear to be at least two possibilities for interpreting the functional pattern of this polarized organelle.

One hypothesis holds that in relation to secretory activity or otherwise there is an actual loss of cisternae from the distal face and a con-

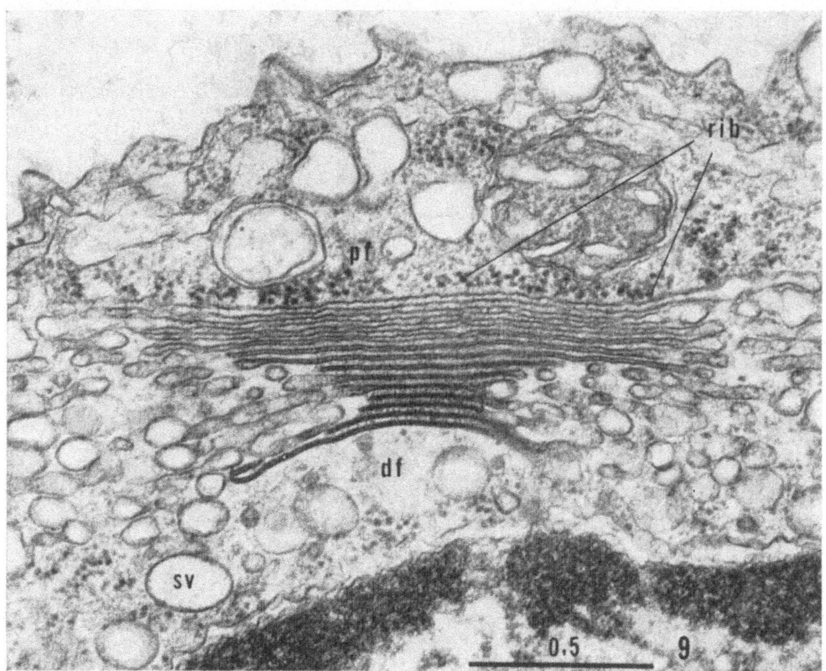

Fig. 9. Golgi apparatus showing ribosomes (*rib*) at the proximal face (*pf*) and none at the distal face (*df*). There is an apparent change in the density of the cisternal contents from the proximal to the distal pole of the apparatus. Intercisternal elements can be seen in the distal region of the apparatus. *Nitella sp.* [glut-OsO₄]

tinuing replacement at the proximal face, with stepwise movement of the cisternal components across the apparatus from the proximal to the distal pole. Grassé [23] assumed this to take place in the Golgi apparatus in zooflagellates. Policard and Baud [53] supposed that this might take place in both normal and pathologic tissues they studied. The apparent total vesiculation that often appears to occur in some secretory cells would seem to support this hypothesis, for maintenance of the apparatus would require the addition of cisternae on the proximal face. The reduction of cisternae by starvation [24, 64] by actinomycin treatment [65] and long-term ethionine treatment [29] is additional evidence for that part of the hypothesis relating to activities at the distal face.

The alternative hypothesis holds that individual cisternae remain intact, though their number is subject to some developmental change and experimental modification. According to this hypothesis, membrane components and other materials would move into the proximal cisternae and then along the gradient of activity buildup within the organelle. This hypothesis does not assume a continuing displacement of cisternae by separation from one side of the apparatus and buildup on the other side, but a flow of materials along the activity gradient and thus across the membranes of the organelle although probably supplemented by materials entering from the periphery. Such Golgi apparatus as those of epidermal root cells that are involved in secretory product evolution do not show any disintegration of cisternae on the distal side. This tends to support the idea that the organelle is, so far as the cisternae are concerned, a more-or-less fixed structure and that there is not a continuing turnover of its unit components.

The second hypothesis seems somewhat more in accord with the idea that the Golgi apparatus is a discrete organelle. This idea, although long in the literature, was qualified somewhat when the Golgi apparatus was considered as a localized area of the endoplasmic reticulum. There is now, however, ample evidence of many sorts to support the view that the Golgi apparatus is a discrete organelle.

Both hypotheses will have to be evaluated in terms of Golgi apparatus development during cellular differentiation, in reference to its functions including those in secretion, and in terms of responses to experimental treatments. In normal development and under the influence of certain experimental agents to be discussed below there may be an increase in the number of cisternae per organelle, and in starvation there may be a reduction in their number. These changes will have to be accounted for.

Both hypotheses assume the existence of certain organizational characteristics at the proximal pole of the apparatus which could relate to the addition of cisternal units. As already noted, the first hypothesis is more readily acceptable with respect to the decrease in the number of cisternae. Both hypotheses assume that membrane components can be added not only to the proximal cisternae but also to the successive cisternae across the apparatus, since the cisternae may either become larger from the proximal to the distal pole, or fail to become smaller when membranes surrounding secretory product are separated from them; and both provide for movement of materials into the individual cisternae throughout the organelle.

There is, unfortunately, no clearcut specific evidence upon which to base a choice between the two hypotheses, and it may be that in a given phase of cellular activity either might pertain. Techniques which would demonstrate the dynamics of functioning in the organelle would be required to select between the alternatives, and these are not currently available.

Despite the fact that we have not determined whether or not continual displacement of cisternae across the apparatus is a characteristic of Golgi apparatus activity, we can profitably examine certain general features of the organelle. The essence of the ideas presented here is not

new, nor is it original with us. We have, however, attempted to clarify
and broaden these ideas into a unified concept designed to provide a basis
for interpretation of the various capacities exhibited by the organelle.

Our present concept is that the proximal field is specifically characterized by the capacity to organize into cisternal units molecular components that move into it. In highly differentiated Golgi apparatus participating in the production of secretory materials certain activities in the
distal region may be dominant. In such Golgi apparatus the organizational

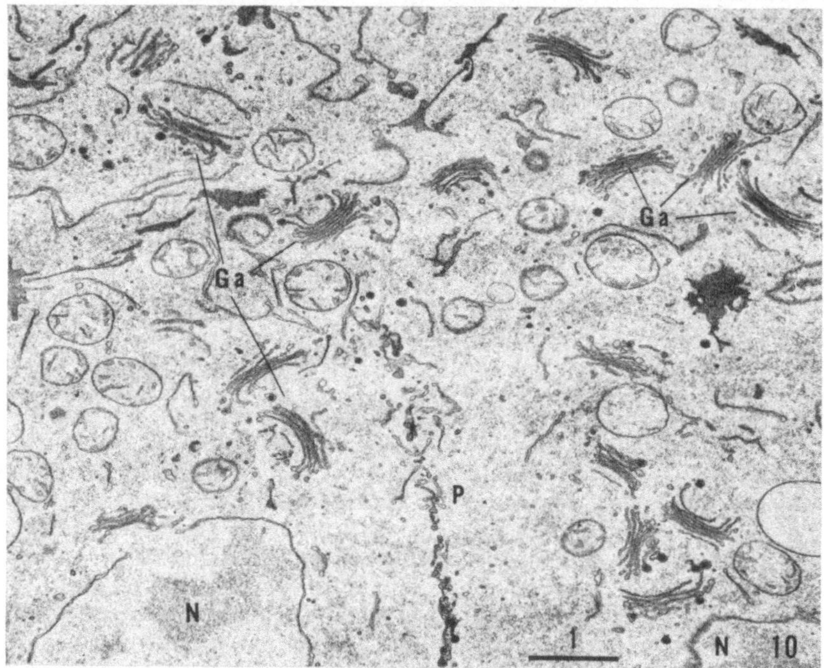

Fig. 10. Golgi apparatus showing secretory activity in an epidermal cell during mitosis.
Some of the secretory vesicles are contributing to the forming plate (p). *Zea mays* root tip.
[KMnO₄]

capacity at the proximal pole would perhaps be restricted to the replacement of cisternal units removed by vesiculation, if the first hypothesis
pertains, or to the mobilization of specific materials into the proximal
cisterna, if the second hypothesis pertains. The concept will provide an
explanation for the normal increase in the number of cisternae which takes
place during differentiation of certain cell types, or experimentally
induced increases in cisternae. The important consideration is that the
capacity for replication of the organelle would, according to this concept,
depend on characteristics or potentialities resident in the organizational
field. As suggested above, in highly differentiated Golgi apparatus these
potentialities may be limited. In fact, at least some such Golgi apparatus
may not be capable of replication. However, in certain instances, as in

the epidermal cells of the root (Fig. 10), Golgi apparatus may be differentiated for participation in secretory activity prior to the time the cells lose their capacity for division, and thus presumably while replication of the Golgi apparatus may still take place. In apparently unspecialized Golgi apparatus, perhaps indeed in the form of the apparatus most subject to replication, the organizational field could extend over a much greater portion, if not all, of the apparatus. The term "field" is not currently popular in biological literature. It has been used here purposely to refer to the organizational region around one pole and in recognition of the fact that it may extend over different proportions of the organelle, depending upon the state of development. Under some conditions molecular components of that portion of the apparatus toward the distal pole may be returned to the general metabolic pool, but not to such an extent that the integrity of the organizational field is lost. It is, of course, from this general metabolic pool that the molecular components move into the organizational field. The composition of this pool must change with differentiation and in relation to specific cellular activities. Morphological study by electron microscopy cannot be expected to resolve the basic problems here. We can, however, profitably review ideas concerning the formation of cisternae, an event which may be essential to replication.

The Formation of Cisternae

Proposals concerning formation of the Golgi cisternae suggest both the direct transfer of membranes from other membranous structures in the cell and formation from nonlamellar components. Evidence of varying weight has been adduced to support both views.

There is now evidence for the sequestering of protein precursors or proteins from the matrix of the cytoplasm into the lumen of the endoplasmic reticulum and their subsequent appearance in the cisternae of the Golgi apparatus [*12, 13, 28, 63, 51, 52*]. This evidence, based on autoradiographic studies, is frequently considered together with the budding off of small vesicles from the endoplasmic reticulum and the presence of numbers of small vesicles adjacent to the Golgi apparatus to support the conclusion that cisternae are formed from the fusion of such vesicles (see Fig. 8). In some instances at least, the contents of these vesicles closely resemble those of the lumen of the endoplasmic reticulum in staining characteristics, but are quite different from those of the proximal cisterna. FRIEND [*21*] from his studies of Brunner's glands in the mouse has pointed out that if there is transfer from the endoplasmic reticulum to the Golgi apparatus, there must be many transition zones. In a number of instances where the Golgi apparatus is actively producing membrane-bound secretory materials no such endoplasmic reticulum vesicles are present, nor is there any consistent association of either rough or smooth endoplasmic reticulum with the Golgi apparatus.

The nonuniversality of endoplasmic reticulum-Golgi apparatus association and the distinct variability in many characteristics of this association, when it does exist, cast doubt on the direct transfer of membranes as an exclusive mechanism for cisternae construction or extension. One possible explanation for the images that have given rise to this idea is that, in certain instances, perhaps where the Golgi apparatus participates in processing a secretory product that is predominantly protein, there is such a transfer, whereas in other cases there is not. If this is so, then the primary importance of these images may relate to transfer of precursors of the secretory product and membrane transfer might be incidental.

Hirsch [31], from observations of pancreatic cells forming zymogen, has proposed a transfer from the endoplasmic reticulum to the Golgi apparatus that is based on the same concept but includes what might be a significant modification. He assumes that after vesicles are budded off from the granular endoplasmic reticulum they become what he terms X-bodies. These transitional protein-containing bodies then undergo modification in what he refers to as the "lamellar-vacuolar field" (a term he uses instead of Golgi field). In this modification the X-bodies change in size and shape and ultimately become aligned with the existing cisternae, each one in turn becoming a new cisterna of the Golgi apparatus. Hirsch does not deal with the concept presented here of a proximal field with certain organizational capacities. It seems resonable to assume that the changes in the X-bodies he suggests might well take place in this organizational field.

Functional relationships among the granular endoplasmic reticulum, the agranular endoplasmic reticulum and the Golgi apparatus appear to be a practical certainty, but as Fawcett [20], in his excellent review of cytoplasmic membranes, has pointed out the mechanisms and pathways are still to be established.

Attention has frequently been called to associations between extensions of the nuclear envelope and the Golgi apparatus. In some developmental stages and particularly in some organisms, including cryptogamic plants, such extensions appear to be very common. Lang [34] has gone so far as to assign a particular name to these extensions, which she has called "amplexi," an amplexus being an embracing element. She notes that the membrane on the side of an amplexus adjacent to a Golgi apparatus is frequently smooth and suggests that the arrangement may provide for direct transfer of material from the nucleus to the Golgi apparatus. Moore and McAlear [43] have supposed that in the fungus *Neobulgaria pura* the Golgi cisternae arise from the nuclear envelope. Weston [66] and his colleagues have provided more convincing evidence for such a derivation. In the chick embryo cells and tumor cells they studied, there is sometimes seen an association of vesicles from the nuclear envelope with the adjacent Golgi apparatus. These vesicles and the projections of material from the nuclear envelope they assume may contribute to the Golgi apparatus are smooth, unlike other extensions of the nuclear envelope also observed in their material (Fig. 11). They suggest, very cautiously, that the nuclear envelope could conceivably give rise to the whole Golgi apparatus. Earlier cytologists commented frequently on the association of the Golgi apparatus and the nuclear surface. Studies by electron microscopy have verified the frequency of this association, notably in embryonic stages and in certain particular specialized cells.

Bouck [4], studying brown algae, observed that small vesicles associated with one face of the Golgi apparatus and possibly involved in fusion might be derived either from the nuclear envelope or from the endoplasmic reticulum depending upon the species. Such an observation raises the

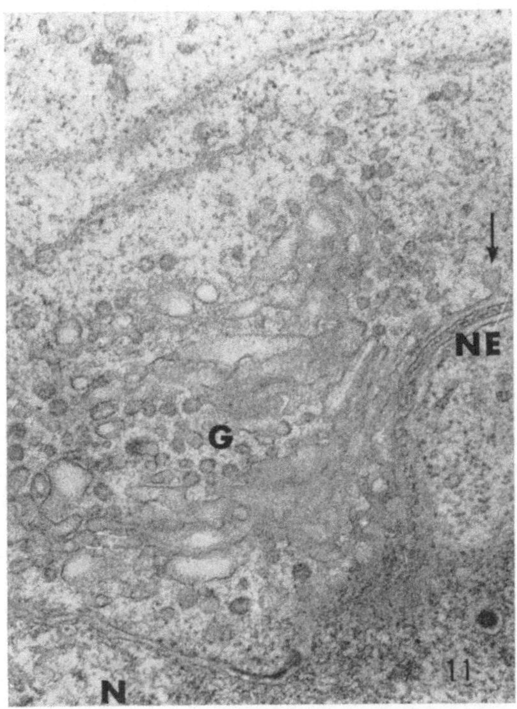

Fig. 11. Golgi apparatus (G) in an ascites tumor cell, showing continuity of vesicles and cisternae with the nuclear envelope (NE). A bleb of the outer membrane of the nuclear envelope (arrow) is similar to vesicles neighboring the Golgi cisternae. (Courtesy of M. H. Greider [66])

fundamental question of the relation of the nuclear envelope to the endoplasmic reticulum [70]. Proposals that Golgi cisternae are derived from the nuclear envelope are not necessarily wholly distinct from proposals that they are derived from the endoplasmic reticulum.

Schnepf [54] has proposed a scheme for increase in the number of cisternae during the buildup of secretory activity in which cisternae may extend, fold over, and then divide at the folds. This would represent derivation of cisternae from cisternae.

Daniels [17] has contributed evidence for another origin of the Golgi apparatus cisternae in the Giant amoeba *Pelomyxa illinoisensis*. The plasma membrane of this organism is fringed. During pinocytosis and phagocytosis infoldings of the plasma membrane result in the formation of plasmalemma vesicles in the cytoplasm. These vesicles are readily distinguished by what appears to be fringe on the inner surface of their membranes. Daniels contends that these fringed vesicles may aggregate at the proximal face of the Golgi apparatus. Some of his micrographs show fringe on the inner surfaces of segments of Golgi cisternae about halfway across the apparatus. The contention is that these vesicles combine with existing cisternal elements, where they undergo some modification and that the plasma membrane thus represents the source of at least part of the Golgi apparatus membranes. Neither the direct comparability of the "fringed" membrane segments nor the postulate concerning their incorporation in relation to the polarity of the Golgi apparatus seems well enough established to support the conclusion that fusion of plasmalemma vesicles gives rise to Golgi cisternae.

It may well be that under certain conditions there is a direct transfer of formed membranes from other cell components to the Golgi cisternae but conclusive proof is lacking. An alternative or supplement to the concept of direct transfer of membranes is the organization of molecular components into membranes in the organizational field.

Studies of cyclic changes in cells and the progress of differentiation have made it clear that the amount of membrane made visible at any one time may differ greatly [68]. Further, such work as that of Stoeckenius [59, 60] has shown that the slightest of changes may bring about alterations between dispersed and lamellar phases of molecular components of concern in membrane structure. Such lability makes it unlikely that direct transfer of membranes would always provide the answer, or even that this is necessarily the most logical thing to look for. It is wise to remain aware that the problem is one of dynamics in a complex molecular system.

In fact, it is not possible to distinguish satisfactorily between the organizational processes involved in cisterna formation and those involved in the extension of cisternal membranes. Membrane extension is an important characteristic of Golgi apparatus development and functioning, at least in relation to the secretory process. Both cisterna organization and membrane extension may take place independently of the differentiation of the apparatus for participation in secretory activity. There is now some limited experimental evidence that bears upon this question of cisterna formation and extension.

When maize roots are exposed to conditions that modify oxidative respiration, for example CO_2 or CO atmosphere, or the presence of KCN in the medium, the processes involved in the aggregation or synthesis of secretion product in the Golgi apparatus are interfered with and the cisternae which would normally be distended with this product do not, after a few hours, show any of the distension characteristic of secretory vesicle development. The already-formed secretion product is no longer visible. Although the secretory processes are blocked the extension of the cisternal membranes may continue. Characteristically in materials so treated the profiles of the cisternae are much longer, and in many instances there is a substantial increase in the number of cisternae per Golgi apparatus (Fig. 12). They frequently curl to form nearly closed or closed structures. This extension of cisternal membrane may well be a result of normal membrane extension with no removal of membrane as a consequence of evolution of secretion vesicles.

HALL and WITKUS [26] have shown increases in the numbers and extent of Golgi cisternae in root meristem cells of *Allium cepa* treated

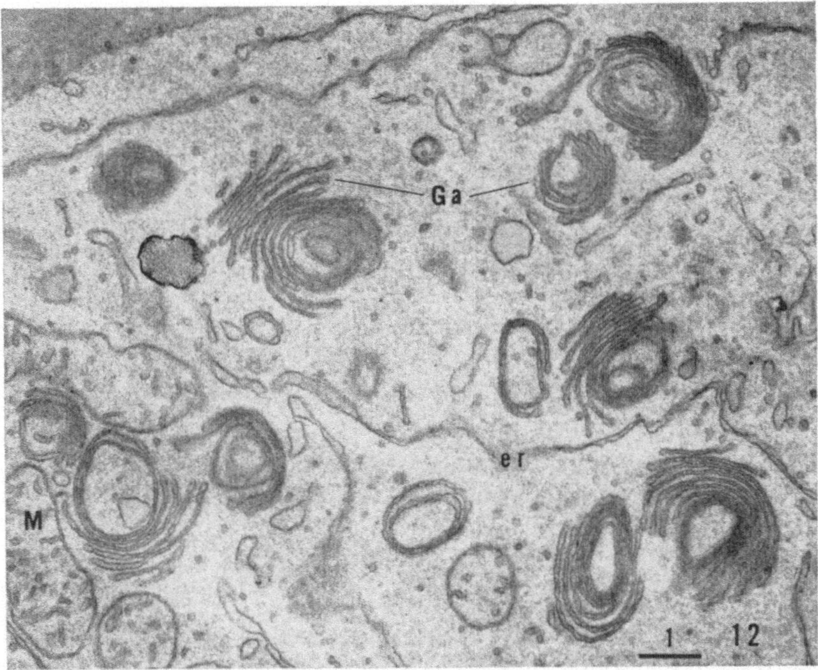

Fig. 12. Golgi apparatus showing extension of the cisternae and an increase in the number of cisternae present as a result of a three-hour exposure to a carbon dioxide atmosphere. Normally, these Golgi apparatus would show accumulation of secretory product (compare with Fig. 4c). [$KMnO_4$]

with 6-hydrouracil, an inhibitor of RNA synthesis which they note acts by blocking biosynthesis of pyrimidines. It is not clear whether HALL and WITKUS were dealing with Golgi apparatus involved in secretory activity, but they apparently disturbed normal activity and produced an increase in the number of cisternae.

WRISCHER [73] has shown that in shoot apices and root apices many treatments that she calls "enzyme-blocking" — X-rays, UV, H_2O_2, NaN_3, KCN, H_2S, $CHCl_3$, and changes in temperature or pH may modify the form of the Golgi apparatus. While WRISCHER does not make a point of it, her figures suggest that extension of the cisternae may continue in these instances too. All that can be said from the sparse evidence at hand is that certain secretory activities can be blocked by such procedures as the inhibition of oxidative respiration without halting the formation of cisternae as components of the Golgi apparatus or the extension of pre-existing cisternae. This situation may provide for the possibility of studying the mechanisms involved in cisterna formation and extension.

The fact that cisternae are formed in the absence of participation in secretory activity seems to us to make the organization of cisternae a fundamental characteristic of the Golgi apparatus. Whether this organization involves the transfer and fusion of actual membrane from other cellular components or organization into lamellar structures of phospholipid and protein molecules or both is by no means clear.

Possible Replication Mechanisms

Hypotheses concerning replication of the Golgi apparatus as an organelle fall into three general categories: (1) so-called *de novo* origin, (2) development of the apparatus from a precursor structure, and (3) some method of replication from an existing Golgi apparatus. In discussing these hypotheses I shall deal with them in the terms used by their proponents to avoid further confusion of the terminology.

Buvat [*8, 9*] has contended that dictyosomes arise *de novo* in the material of the phragmoplast. Carasso and Favard [*11*] have also made a point of the likelihood of *de novo* appearance of dictyosomes in relation to the phragmoplast, and similar suggestions have been made with respect to the origin of the Golgi apparatus in the vicinity of the centrioles. According to Buvat, such *de novo* origin involves a stage in the organization of the apparatus in which the membranes become progressively distinct and in which there is as yet no evidence of vesicle production[*10*]. This stage somewhat resembles the pro-Golgi body stage of Werz [*65*]. At the periphery of the phragmoplast there are aggregations of Golgi apparatus, but so far as we have been able to determine the organelles are complete and apparently active structures [*18*]. This appears to be true from recognizable prophase through telophase. As noted below, we know that in some instances replication of the Golgi apparatus occurs independently of cell division. It is our impression that when it is associated with cell division it may take place prior to any detectable stages of mitosis.

Although *de novo* origin, if it occurs, might take place in any phase of cellular activity, the concepts concerning it generally seem to imply a relationship between this *de novo* origin and cell division. In such a meristematic system as the root apex there is some general association of Golgi apparatus replication and distribution with cell division, for there is not, prior to apparent differentiation, either a conspicuous dilution or a conspicuous increase in the number of Golgi apparatus per cell. The problem is further complicated by our lack of knowledge of the relation between replication of the Golgi apparatus and its differentiation. All that can be said is that in many instances where the Golgi apparatus becomes very complex, such as in spermatogenesis and oogenesis, at least a part of its differentiation follows the cell division stages.

Some aspects of development are indeed a consequence of differentials established in cell division. Asymmetric distribution of Golgi apparatus at the time of cell division could result in differences in the numbers of Golgi apparatus in the daughter cells. In some instances, however, as in the transition region between the initial cells of the root cap and the outer secreting cells in *Zea mays*, there is a marked increase in the number

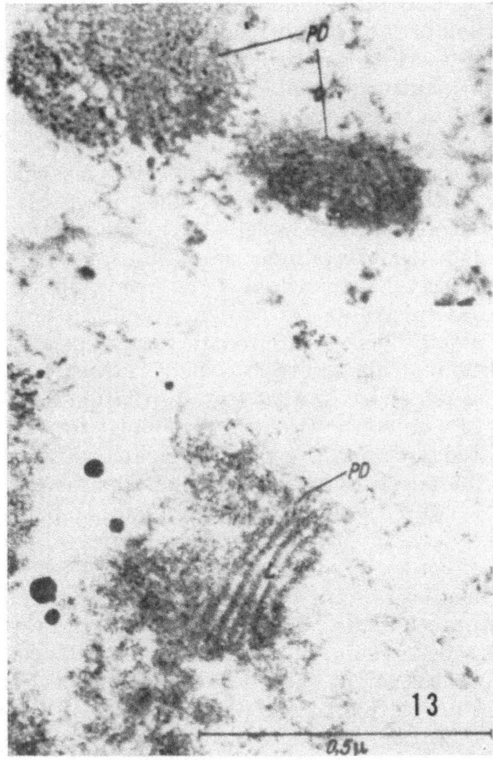

Fig. 13. According to WERZ, the figure shows perinuclear bodies (PD) in succesive stages of development into Golgi apparatus, the lower one showing lamellae (L). *Polyphysa cliftonii*. (Courtesy of G. WERZ [65])

of Golgi apparatus per cell, as also noted by CLOWES and JUNIPER [15]. This represents Golgi apparatus replication in nondividing cells. Cellular differentiation includes not only differentiation of the Golgi apparatus, but also changes in the number per cell and changes in the association of the apparatus with other cellular components. Changes incident to differentiation, and the absence of a consistent relationship to any components of the mitotic apparatus combine to make *de novo* origin in the material of the phragmoplast, in relation to the organized centriole, or any other relationship to the mitotic apparatus, an inadequate explanation for replication of the Golgi apparatus.

Werz [65], Mercer [39], and Hall and Witkus [26] have provided examples of the postulates that the Golgi apparatus has its origin from some other cellular components. Werz has concluded from studies of *Acetabularia* that the Golgi apparatus has its origin in "perinuclear bodies" in which the typical lamella system is progressively organized (Fig. 13). From studies of actinomycin- and puromycin-treated cells and anucleate cells he concludes that the nucleus exercises a direct control over the ontogeny and functioning of the apparatus. Actinomycin prevents the formation of the "perinuclear bodies"; puromycin prevents the transformation of "perinuclear bodies" into Golgi apparatus. In both instances, and in anucleate cells, replication fails to take place. It is most likely that the nucleus controls Golgi apparatus replication more or less directly, but in material specifically selected for study because of Golgi apparatus replication we have seen no conclusive evidence that the apparatus develops progressively from progenitor bodies, perinuclear or otherwise.

Mercer has suggested that in cells of the ovotestis of *Helix pomatia* or *H. aspersa* vacuoles containing amorphous, relatively unhydrated phospholipid materials "('Golgi body' (?))" develop by hydration and crystallization into spherulites of compact, concentric membrane shells which can then swell, open, and burst to form typical Golgi apparatus. The transformation of an amorphous body into a parallel membrane array would seem somewhat similar to the situation described by Werz. Although we have seen concentric arrays of Golgi apparatus membranes, they are associated with adverse physioligical conditions and we have never observed the bursting forth of a Golgi apparatus.

Hall and Witkus have made a considerable point of the aggregations of Golgi apparatus in different stages of development at the poles in cells of *Allium cepa* roots in mitosis. They have suggested that during mitosis double membrane units develop around some of the cytoplasm neighboring the mitotic figure. They propose, very tentatively, that such membrane units might result in initial cisternae in association with which entire Golgi apparatus might then develop. The objections that pertain to *de novo* origin in relation to the phragmoplast also pertain here.

Wischnitzer [72] suggested, with reservations, that in the cells of young amphibian oocytes the Golgi apparatus might originate from the nuclear envelope, whereas in older oocytes replication might take place through bisection. It will be recalled that Weston et al. [67] have proposed that in still-differentiating embryonic stages the apparatus might originate from the nuclear envelope. Such suggestions raise the possibility that the nuclear envelope may be involved in the formation or functioning of the Golgi apparatus in certain developmental stages but not in others. It is possible that the participation of the nuclear envelope reflects nuclear control. If this is so, the character or directness of such control may vary with development. There is indeed experimental evidence that replication of the Golgi apparatus is subject to nuclear control.

In cells of *Chlamydomonas eugametos* exposed to colchicine for extended periods of time there appears to be a distinct increase in the number of Golgi apparatus (Fig. 14). After prolonged treatment, the

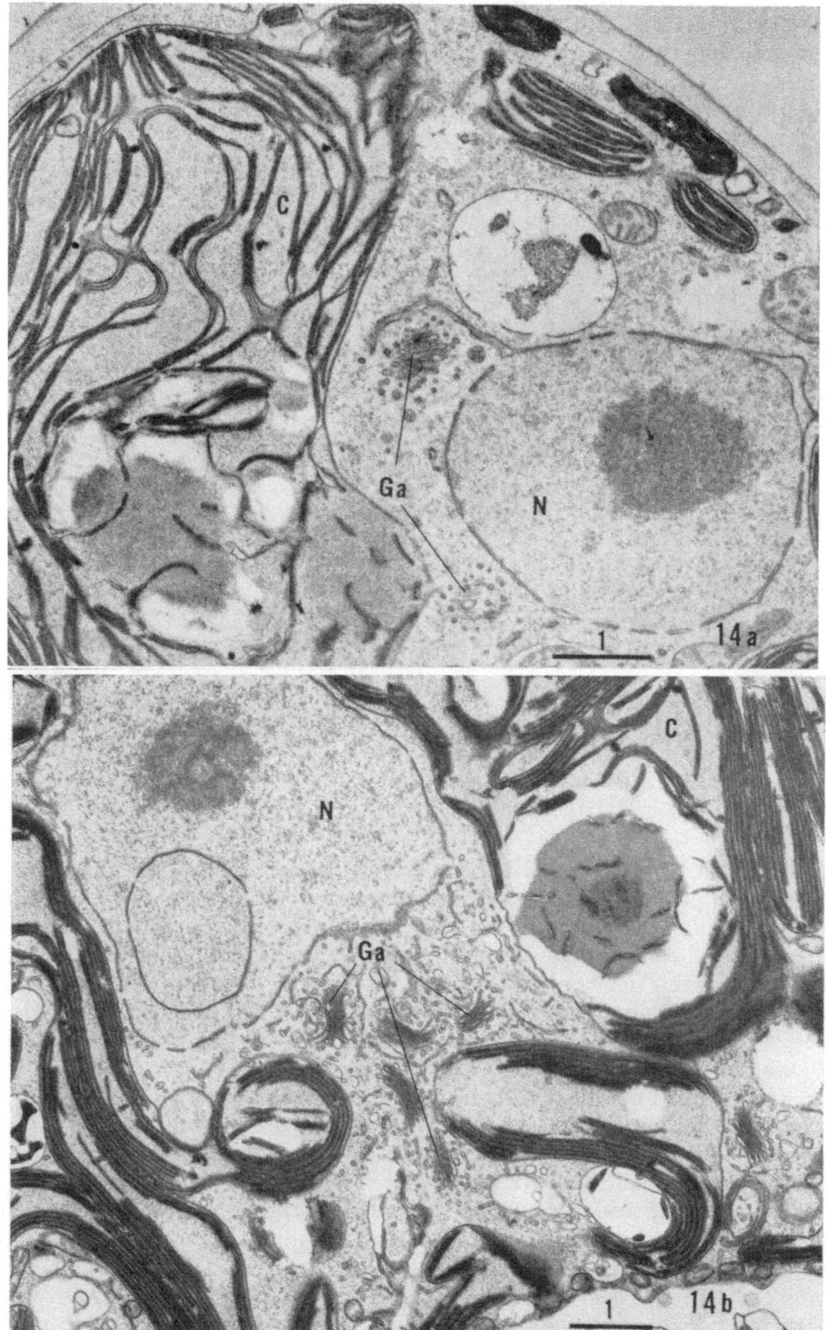

Fig. 14 (a). Normal cell of *Chlamydomonas eugametos* showing two Golgi apparatus adjacent to the nucleus. (b) A comparable cell following exposure to $5 \cdot 10^{-3}$ M colchicine for one week and showing an increase in the number of Golgi apparatus. [$KMnO_4$] C = chloroplast

nuclei of such cells, in which endopolyploidy presumably occurs, may become structures of quite irregular shape, but the envelope remains intact. The buildup of endopolyploidy may determine the number of Golgi apparatus, but under these conditions no evidence for pro-Golgi apparatus bodies or transitional phases has been observed.

The remaining hypotheses have to do with replication from existing Golgi apparatus. The apparatus could fragment along a plane parallel to the long axes of the cisternae. Unless, however, such fragmentation took place so as to include a portion of the organizational field in each component, it might not result in replication. To have replication in

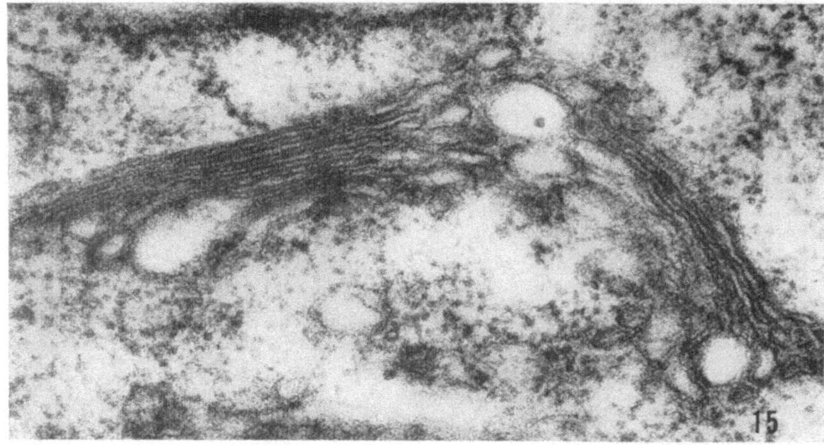

Fig. 15. Meristematic cell of *Elodea canadensis* with two dictyosomes. Their position and length have been interpreted to suggest division of a previously existing dictyosome. (Courtesy of R. Buvat [8])

these circumstances, one fragment would have to develop according to the normal polarity and the other would have to undergo a reconstitution of the missing portion. The differences in numbers of cisternae in Golgi apparatus of various types of cells, and the fact that sections may reveal only part of the cisternae in an apparatus makes it difficult to assess the meaning of the appearance of small numbers of cisternae on which this idea is based. Of equal concern are the physiological factors which reduce the number of cisternae under certain circumstances.

Fragmentation from one pole to the other of the apparatus has also been proposed. In 1956, Afzelius [1] postulated a "break" of the cisternae of the dictyosomes in the growing oocyte of sea urchins which bisected all the cisternae simultaneously. He presumed this to represent a duplicating stage and added in support of the idea the observation that all of the dictyosomes had approximately the same number of cisternae, and all were of the same thickness, $0.2\,\mu$, though their diameters might range from 0.2 to $2.4\,\mu$. Drawert and Mix [19] noted, tentatively, that some of their evidence would support this idea. Buvat [8] had suggested such a pattern of division (Fig. 15), and Chardard [14] supported this particular concept of BUVAT's. Actually, this suggestion was made several times during the period of study of the Golgi apparatus by light microscopy.

This idea has probably had more acceptance than the others. It seems most tenable in Golgi apparatus that have not been highly differentiated in relation to secretory activity. It avoids the organizational difficulties that are involved in all the *de novo* concepts and in proposed transformations from other cellular components.

Constriction of organelles is a common pattern of division and the occasional constriction of cisternae and juxtaposition of Golgi apparatus seems to provide some suggestive evidence for this idea. The significance of this evidence is difficult to evaluate for in many of the materials used to support the idea numbers of the organelles are aggregated into a delineated region. In such cases the end-to-end juxtaposition may simply reflect concentration in a particular arrangement.

As noted, GRASSÉ [*23*] has proposed the continuing formation of cisternae at the proximal face of the apparatus. He further assumes that at certain stages, forces at this face may result in the formation of discontinuous cisternae. If, then, more cisternae are added to each resulting unit, two or more Golgi apparatus may be formed. This proposal relates replication of the Golgi apparatus to forces active in what we have termed the organizational field.

These concepts are examples of the major ideas on how the Golgi apparatus may be replicated. There are variations of each of them based upon studies of different species, different stages of development, and Golgi apparatus with varied activities. Although the ideas concerning replication of the Golgi apparatus seem confused, they are no more so than other problems where studies have ranged from relatively primitive cells to highly specialized ones, or dealing with cells in various stages of differentiation, or with cells with a range of metabolic capacities.

Conclusion

The Golgi apparatus is, apparently, a component of all cells above those at the most primitive level of organization. There is now evidence that it should be considered an organelle with the same degree of integrity as such structures as the mitochondria and the plastids.

Most attention has been given to the activity of the apparatus in the elaboration of secretory materials. These secretory materials include a great range of compounds: in plants — at least part of the substances of the primary wall, some of the components of secondary walls, the slimes that come to cover the surface of many types of cells and at least some of the secretions of insectivorous species; in animals — the zymogen granules of pancreatic cells, mucous secretions, lactoprotein from mammary gland cells, at least components of thyroxin, in certain instances melanin granules and perhaps other pigmented substances, tropocollagen and maybe even collagen. These are but indications of the variety of products. There are many more.

Their nature implicates the Golgi apparatus in specific cellular activities involving both proteins and carbohydrates. There is substantial evidence for the elaboration of at least some of the proteins in the endoplasmic reticulum. There is increasing evidence that the Golgi apparatus itself synthesizes, and does not merely sequester, complex carbohydrates.

The presence of the Golgi apparatus, often apparently in an active state in cells not known to have any secretory functions, suggests that the apparatus may be important in other ways in the economy of the cell. It may be that compounds related to ones that are secreted are processed through the Golgi apparatus and then utilized or retained in the protoplast. It may be, too, that its membrane-organizing capacity is a significant aspect of the maintenance and activity of the cell.

The model of the organelle we have projected assumes it to have an activity gradient which may vary with the functional state of the cell. The important point is that around one pole of the apparatus, the organizational pole, there is a capacity for building the cisternae that are the conspicuous components of the structure in electron micrographs. The source of the additional membrane is not at all clear. Perhaps it could vary with the activity of the apparatus. So far as the extension of existing cisternae is concerned, it would appear to be a mobilization of molecular components, but the possibility of direct membrane transfer has not been definitely ruled out. Whether by vesiculation from one face and building up of new cisternae at the other face, there is a continuing displacement of the cisternae across the structure, as has been postulated, or whether the cisternal components are relatively stable are questions for which there are no definite answers. It is clear that under certain conditions cisternae may disappear and under certain conditions new cisternae are built up. The flow of materials across the structure would seem to present no more problems than the flow of materials into and within other membrane-bound or membrane-containing organelles.

The possibilities for replication include *de novo* origin, modification of or transfer from other cellular components, some sort of fragmentation, or morphogenetic activities in a limited part of the apparatus. Theories of *de novo* origin, first of cells and then of organelles, have fascinated biologists for a long time. They are not subject to ready proof or disproof. But all theories of *de novo* origin involve certain difficult organizational problems. The modification of other cellular components to constitute the beginnings of the Golgi apparatus involves, if anything, even more difficult conceptual problems.

We do not know how the Golgi apparatus is replicated. It seems to us most likely, that either there is a fragmentation parallel to the activity gradient, or that discrete entities are built up in what we have termed the organizational field of the apparatus. Currently, we have no means of distinguishing between these possibilities.

The structure has remarkable capacity for both physiological and morphological differentiation. It is, in most instances, and perhaps always, polarized. It may be that its manner of replication can vary.

Evidence from many sources seems to support the concept of an organizational field around one pole of the apparatus. The term field is permissible only in the absence of specific knowledge, but it will suffice at this stage as a loose definition of a region in which there exists the capacity to organize molecules into cisternal membranes. We are not certain of the source of these molecules and we know nothing about the mechanism by which the cisternae are organized, but the existence of a region with this capacity may provide a starting point for investigation of how the apparatus is replicated, as well as further investigation of the manner of its functioning.

In some pattern which we do not understand information is transmitted, presumably from the nucleus, to this region with the capacity for cisterna formation. In some cells this capacity for cisternal organization may extend over a considerable portion, perhaps even all of the apparatus.

I am only inclined to contend that we are dealing with an organelle with specific characteristics, and the probability of its replication *de novo* or by development from some other specialized cellular component seems to me remote. We do not know enough about the differentiation and functions of the Golgi apparatus, we do not know enough about its polarity, and we do not know even just where to look for its replication. Most important, because we do not know these things, we are unable to design the essential critical experiments.

Addendum

Since the preparation of this paper, BAINTON and FARQUHAR (BAINTON, DOROTHY F., and FARQUHAR, MARILYN G., 1965, The Origin of Polymorphomolecular Leukocyte Granules, J. Cell Biol. 27, 6A—7A) have noted the production of two distinct types of PMN granules in immature granulocytes from opposite poles of the Golgi complex at different developmental stages. This observation has a significant bearing upon any concept of a polarity gradient in Golgi apparatus involved in secretory activity. NOVIKOFF and his associates (see Laboratory Investigations Supplement, January, 1966) have laid emphasis upon exchanges between the endoplasmic reticulum and the distal as well as the proximal face of the Golgi apparatus. This interpretation also has significance for the polarity concept.

Acknowledgement

This paper is actually a result of the collaborative efforts of the author and his colleagues, Dr. MARIANNE DAUWALDER and JOYCE E. KEPHART. While it is presented in this Symposium under single authorship, it would have been impossible of conception or execution without the full and wholehearted participation and the contributions to interpretations of these colleagues. The author must, of course, remain responsible for its inaccuracies and the inadequacies of interpretation.

368 W. Gordon Whaley:

The micrographs presented come from the work of several investigators
in The Cell Research Institute and a few investigators elsewhere as
follows: Figs. 1a, 4b, and 12, H. H. Mollenhauer; Figs. 1b, 3, 5, 6 ,7, 8,
and 9, R. Turner; Figs. 2, 4a, and 4c, Marianne Dauwalder; Fig. 10,
Joan Hunter; Fig. 14, Patricia Walne; Fig. 11, Dr. Marie H. Greider, Department of Anatomy and Pathology, College of Medicine,
Ohio State University, Columbus, Ohio; Fig. 13, Herr Professor Dr.
Günther Werz, Max Planck-Institut für Meeresbiologie, Wilhelms-
haven, Abteilung Hämmerling; Fig. 15, Professor R. Buvat, Université
de Paris, École Normale Supérieure, Laboratoire de Botanique, 24,
Rue Lhomond, Paris V^e. This work has been supported in part by NSF
GB-301 and USPH GM-07289.

References

[1] Afzelius, B. A.: Exp. Cell Res. 11, 67 (1956).
[2] Baker, J. R.: In: Tenth Symposium of the Society for Experimental Biology,
    1955. "Mitochondria and Other Cytoplasmic Inclusions", p. 1. Cambridge
    University Press 1957.
[3] Bonneville, M. A., and B. R. Voeller: J. Cell Biol. 18, 703 (1963).
[4] Bouck, C. B.: J. Cell Biol. 25, 523 (1965).
[5] Bourne, C. H., and H. B. Tewari: In: C. H. Bourne (ed.): Cytology and Cell
    Physiology, p. 377. New York: Academic Press 1964.
[6] Bowen, R. H.: Quart. Rev. Biol. 4, 299 (1929).
[7] — Quart. Rev. Biol. 4, 484 (1929).
[8] Buvat, R.: Annales des Sciences Naturelles, Ser. 11. Bot. et Biol. 19, 121 (1958).
[9] — In: C. H. Bourne and J. E. Danielli (eds.): International Review of
    Cytology, Vol. 14, p. 41. New York: Academic Press 1963.
[10] — Personal communication (1965).
[11] Carasso, N., et P. Favard: In: C. Magnan (ed.): Traité de microscopie
    électronique, Vol. II, p. 901. Paris: Hermann 1961.
[12] Caro, L. G.: J. biophys. biochem. Cytol. 10, 37 (1961).
[13] —, and G. E. Palade: J. Cell Biol. 20, 473 (1964).
[14] Chardard, R.: Rev. Cytol. et Biol. Végétales 24, 1 (1962).
[15] Clowes, F. A. L., and B. E. Juniper: J. exp. Bot. 15, 622 (1964).
[16] Dalton, A. J.: In: J. Brachet and A. E. Mirsky (eds.): The Cell, Vol. II,
    p. 603. New York: Academic Press 1961.
[17] Daniels, E. W.: Z. Zellforsch. 64, 38 (1964).
[18] Dauwalder, M., and W. G. Whaley: J. Cell Biol. (abst.) (in press) (1965).
[19] Drawert, H., and M. Mix: Port. Acta. Biol. Ser. A 7, 17 (1963).
[20] Fawcett, D. W.: In: Symposia of the Society for Cellular Chemistry, Vol. 14,
    Suppl.: S. Seno and E. V. Cowdry (eds.): "Intracellular Membraneous
    Structure", p. 15. Okayama: Chugoku Press Ltd. 1965.
[21] Friend, D. S.: J. Cell Biol. 25, 3/1, 563 (1965).
[22] Golgi, C.: Arch. ital. Biol. 30, 60 (1898).
[23] Grassé, P.-P.: C. R. Acad. Sci. (Paris) 245, 1278 (1957).
[24] Grimstone, A. V.: J. biophys. biochem. Cytol. 6, 369 (1959).
[25] Hagadorn, I. R., H. A. Bern, and R. S. Nishioka: Z. Zellforsch. 58, 714
    (1962/63).
[26] Hall, W. T., and E. R. Witkus: Exp. Cell Res. 36, 494 (1964).
[27] Hay, E. D.: J. biophys. biochem. Cytol. 4, 583 (1958).
[28] —, and J. P. Revel: Devel. Biol. 7, 152 (1963).
[29] Herman, L., and P. J. Fitzgerald: J. Cell Biol. 12, 277 (1962).
[30] Hirsch, G. C.: In: First IUB/IUBS Symposium, T. W. Goodwin and O.
    Lindberg (eds.): "Biological Structure and Function", Vol. 1, p. 195.
    New York and London: Academic Press 1961.

[31] HIRSCH, G. C.: In: Symposia of the Society for Cellular Chemistry, Vol. 14, Suppl.: S. SENO and E. V. COWDRY (eds.): "Intracellular Membraneous Structure", p. 197. Okayama: Chugoku Press Ltd. 1965.

[32] HODGE, A. J., J. D. McLEAN, and F. V. MERCER: J. biophys. biochem. Cytol. 2, 597 (1956).

[33] KESSEL, R. G., and H. W. BEAMS: J. Cell Biol. 25, 1/1, 55 (1965).

[34] LANG, N. J.: Amer. J. Bot. 50, 280 (1963).

[35] LEHNINGER, A. L.: The Mitochondrion. New York: W. A. Benjamin, Inc. 1964.

[36] MANTON, I.: Abstracts of the Tenth International Botanical Congress (Edinburgh), Abs. No. 245, p. 107 (1964).

[37] — and G. F. LEEDALE: Phycol. 1, 37 (1961).

[38] —, and M. PARKE: J. Marine Biol. Assoc. U. K. 42, 565 (1962).

[39] MERCER, E. H.: In: Symposia of the International Society for Cell Biology, R. J. C. HARRIS (ed.): "Interpretation of Ultrastructure", Volume 1, p. 369. New York and London: Academic Press 1962.

[40] MOLLENHAUER, H. H.: Personal communication (1965).

[41] — J. Cell Biol. 24, 504 (1965).

[42] —, and W. G. WHALEY: J. Cell Biol. 17, 222 (1963).

[43] MOORE, R. T., and J. H. McALEAR: J. Cell Biol. 16, 131 (1963).

[44] NOVIKOFF, A. B.: In: J. BRACHET and A. MIRSKY (eds.): The Cell, Vol. II, p. 423. New York: Academic Press 1961.

[45] —, E. ESSNER, S. GOLDFISCHER, and M. HEUS: In: Symposia of the International Society for Cell Biology, R. J. C. HARRIS (ed.): "Interpretation of Ultrastructure", Vol. 1, p. 149. New York and London: Academic Press 1962.

[46] —, N. QUINTANA, H. VILLAVERDE, and R. FORSCHIRM: J. Cell. Biol. 23, 68A (1964).

[47] PALAY, S. L.: In S. PALAY (ed.): Frontiers in Cytology, p. 305. New Haven: Yale University Press 1958.

[48] PERNER, E. S.: Protoplasma (Wien) 49, 407 (1958).

[49] PERRONCITO, A.: Instituto Lombardo di Sci. e lettere, Milan, Rendiconti, 2: ser. 42, 602 (1909).

[50] — Arch. ital. biol. Pisa 54, 307 (1910).

[51] PETERSON, M. R., and C. P. LEBLOND: Exp. Cell Res. 34, 420 (1964).

[52] — — J. Cell Biol. 21, 143 (1964).

[53] POLICARD, A., and C. A. BAUD: Les Structures Inframicroscopiques Normales et Pathologiques des Cellules et des Tissus. Paris: Libraires de L'Académie de Médecine 1958.

[54] SCHNEPF, E.: Z. Naturforsch. 16b, 605 (1961).

[55] — Abstracts of the Tenth International Botanical Congress (Edinburgh), Abs. No. 470, p. 202 (1964).

[56] SITTE, P.: Protoplasma 49, 447 (1958).

[57] SJÖSTRAND, F. S.: In: Ciba Foundation Symposium, "The Exocrine Pancreas", p. 1. Boston: Little, Brown & Co. 1962.

[58] —, and V. HANZON: Exp. Cell Res. 7, 415 (1954).

[59] STOECKENIUS, W.: In: Symposia of the International Society for Cell Biology, R. J. C. HARRIS (ed.): "Interpretation of Ultrastructure", Vol. 1, p. 349. New York and London: Academic Press 1962.

[60] — J. Cell Biol. 12, 221 (1962).

[61] STRUNK, S. W.: J. biophys. biochem. Cytol. 5, 385 (1959).

[62] TURNER, F. R., and W. G. WHALEY: Science 147, 1303 (1965).

[63] WARSHAWSKY, H., C. P. LEBLOND, and B. DROZ: J. Cell Biol. 16, 1 (1962)·

[64] WEISBLUM, B., L. HERMAN, and P. J. FITZGERALD: J. Cell. Biol. 12, 313 (1962).

[65] WERZ, G.: Planta 63, 366 (1964).

[66] WESTON, J. C.: Personal communication concerning manuscript in press: G. ACKERMAN, M. H. GREIDER, and J. C. WESTON (1965).

[67] —, M. H. GREIDER, A. ACKERMAN, and R. F. NIKOLEWSKI: J. appl. Physics 36, 2627 (1965).

[*68*] WHALEY, W. G., J. E. KEPHART, and H. H. MOLLENHAUER: In: Twenty-Second Symposium of the Society for the Study of Development and Growth, M. LOCKE (ed.): "Cellular Membranes in Development", p. 135. New York: Academic Press 1964.

[*69*] —, H. H. MOLLENHAUER, and J. E. KEPHART: J. biophys. biochem. Cytol. **5**, 501 (1959).

[*70*] — —, and J. H. LEECH: J. biophys. biochem. Cytol. **8**, 233 (1960).

[*71*] WILSON, E. B.: In: The Cell in Development and Heredity. New York: Macmillan Co. 1928.

[*72*] WISCHNITZER, S.: Z. Zellforsch. **57**, 202 (1962).

[*73*] WRISCHER, M.: Naturwissenschaften **22**, 522 (1960).

[*74*] ZEIGEL, R. F., and A. J. DALTON: J. Cell Biol. **15**, 45 (1962).

Discussion

Chairman: *Taylor*

Taylor: The Golgi apparatus is a very highly organized structure, that apparently replicates without any nucleic acid.

Stubbe: If the Golgi apparatus could be a structure of a genetic function, one has to try to find species with differences in it. After hybridization we have to look, how the difference is inherited. Are there known any experiments in this direction?

Whaley: Not so far, as I know. One perhaps puzzling observation is that the lowest number of Golgi apparatus we have found in the simpliest sorts of organisms, that still have organized nuclei, is two. One can go quite some distance in an evolutionary sense among the cryptogamic plants before one finds the build up of very large numbers.

I would only add, that up today no one has demonstrated DNA in here. I would not know, how you get to a demonstration of DNA in this sort of a structure.

Schnepf: As I know from the work of Miss MANTON in Leeds there are little flagellates *(Chrysochromulina)* which have only one Golgi apparatus. In the light microscope, she could see the division of it in connexion with cell division.

Whaley: Where we have looked at several sections of algal cells we have usually found two. Though the structure is sometimes visible in phase microscopy we have never seen division activity.

Schnepf: If I am correct, MOORE and McALEAR have shown that in the hymenium of *Neobulgaria*, the Golgi apparatus is found only in special types of cells.

Whaley: There are a number of papers pointing out that there are stages in the life cycle in which there is no Golgi apparatus. However, when we began embryological studies in plant material, we often saw no Golgi apparatus. But the more we have worked at adapting techniques the narrower has become this gap. And so finally in the embryogenesis of *Zea mays*, where we thought we had a period of development of about six days in which there was no Golgi apparatus, we now find beautiful Golgi apparatus.

Karlson: Can it be possible, that the fibrils between the cisternae are contractile elements?

Whaley: We actually know nothing about the nature of these elements between the cisternae; hence I cannot tell you whether they could be contractile elements.

Stoeckenius: I think it would be an even more unifying hypothesis if the Golgi apparatus would be considered as a source of membrane material for all membranous organelles (ER, cell membrane proper, nuclear envelope).

Whaley: I think that it is a sort of membrane machine. The evidence is also conclusive that there is considerable turnover of membrane components from one component of the cell to another. But this is not the only function of the apparatus. Within the cisternae of the apparatus there is formation of sometimes very complex compounds. Further there are many types of cells not known to have secretory activity in which there is obviously a very actively functioning Golgi apparatus.

Wehrmeyer: As you are able to isolate Golgi apparatus without loosing cisternae what do you think about the forces which keep them together?

Whaley: I have not the remotest idea. In the isolation experiments one can sometimes detach two or three cisternae from the rest of the apparatus, but they remain applied one to the other. There is a structural integrity to the apparatus that we do not understand at this stage.

Miller: It has been stated by several workers that there are differences in the thickness of the membranes of the Golgi apparatus and of the ER. Have you made any measurements?

Whaley: One can get a picture of progressive change in the Golgi membranes, as one goes across the Golgi apparatus, in dimensions as well as in certain staining characteristics. The membranes of the Golgi cisternae, the membranes of the endoplasmic reticulum, and the plasma membrane in most of the cells we have studied fall in one general distribution.

Schaller: I would like to ask, whether anybody has done material transfer experiments in the Golgi apparatus during multiplication by labeling, like Dr. Luck's experiments on mitochondria?

Whaley: I don't even know whether it would be possible or not. We have isolated Golgi apparatus but not clean.

Luck: For this type of kinetic experiment you must have clean fractions because of the need for a denominator (e.g. protein). On the other hand, even with a dirty preparation it would be possible to carry out such a study with radioautographs at the electron microscope level where Golgi profiles are easily identified. But to undertake such experiments without knowing about the kinetics of the label would be perhaps a waste of time.

I wonder, if pinocytotic vesicles truly fuse with the Golgi, could you not possibly introduce a marker like ferritin or colloidal gold into the system?

Wohlfarth-Bottermann: W. Stockem (Tagung Dtsch. Ges. E. M. Aachen 1965; Z. Zellforschung, in press) has done such experiments. He has marked these vesicles with amorphous silica. But these particles have never been found in the Golgi apparatus. So I don't believe, that Golgi apparatus is reconstituted by incorporation of pinocytotic vesicles.

Stoeckenius: How well is it established, that these vesicles, which you find on the cell surface actually move from the surface to the Golgi apparatus and not in the opposite way?

Whaley: It is not established at all.

Bajer: I have a question concerning formation of the cell plate. From studies on movements of small particles within the phragmoplast it seems to be rather impossible that Golgi apparati can migrate inside the phragmoplast. It is therefore difficult to visualize the formation of the cell plate by the Golgi apparatus, although, on the other hand, all available evidence seems to give very convincing support to Dr. Whaley's suggestions.

Whaley: At either late prophase or early metaphase the Golgi apparatus undergoes some very special cyclic differentiation. It begins to produce very small vesicles, which Porter and his people have called pectin-vesicles. These invade the phragmoplast in late metaphase. During anaphase one sees these small vesicles usually along the fibers. They move to the plate forming little islands in the plate region. They are a result of a very ephemeral phase of Golgi apparatus activity.

At late anaphase the production of larger vesicles begins. On constitution of the cell plate these large vesicles apparently fuse with the small ones. So far as I know there is no invasion of the phragmoplast by the Golgi apparatus, but the vesicles produced by the Golgi apparatus very clearly can invade this structure, and they very clearly move in an architectural pattern that is determined, at least to some extent, by the spindle fibers.

So the Golgi apparatus not only differentiates from cell to cell in the development of a tissue, but also shows a cyclic differentiation within a given cell.

Organellen-Reduplikation
und Zellkompartimentierung

Von

Eberhard Schnepf, Göttingen

Mit 11 Abbildungen

Durch die Elektronenmikroskopie haben sich die Vorstellungen von der Organisation der Zelle erweitert und vertieft. Was gefunden wurde, bezeugt die Einheit des Lebendigen [65]. Es hat sich dabei gezeigt, daß die Zelle viel komplizierter aufgebaut ist, als man nach den Befunden der Lichtmikroskopie annehmen konnte. Strukturen wurden neu entdeckt oder als allgemein verbreitete Elemente erkannt.

A. Membranen, Kompartimente, Phasen

Es hat nicht daran gefehlt, das neue Wissen zu ordnen, ein neues Schema vom Aufbau der Zelle zu geben. In der Konzeption der *unit membrane* (besser: Elementarmembran, [65]) stellt Robertson (u. a. [54 55]) heraus, daß die verschiedenartigen Membranen, die die Zelle und ihre Organellen umgeben und aufbauen, ähnlich strukturiert sind und Ähnliches abgrenzen. Diese Membranen teilen den Protoplasten in zahlreiche, in sich abgeschlossene Räume auf. Diese „Kompartimente" haben nicht nur für die morphologische, sondern auch für die funktionelle Organisation eine große Bedeutung. Ihre Membranen verhindern, teils passiv durch ihre Permeabilitätseigenschaften, teils durch aktive Transportleistungen, eine gleichmäßige Durchsetzung der Substanzen in der Zelle. Dadurch ist die Zelle in *verschiedene* „wäßrige Mischphasen" [57, 58] aufgeteilt. Diese Kompartimentierung ist nur in Zusammenhang mit der Organellen-Reduplikation zu verstehen.

H. Ruska [58] sowie C. Ruska [57] (vgl. auch Wohlfarth-Bottermann [77]) erkennen allen diesen Phasen eine plasmatische Natur zu. Sie unterscheiden in der tierischen Zelle Grundplasma, Karyoplasma, Reticulumplasma (= der Inhalt der Kompartimente des endoplasmatischen Reticulums), Golgiplasma (= der Inhalt der Kompartimente des Golgiapparates), äußeres Chondrioplasma (im Raum zwischen äußerer und innerer Mitochondrienmembran), inneres Chondrioplasma (die Mitochondrienmatrix) und perinucleäres Plasma (im Raum zwischen den beiden Membranen der Kernhülle). Wenn man dieses Schema konsequent auf die Pflanzenzelle ausdehnt, müßte es noch durch das äußere und das innere Plastidenplasma sowie durch das „Vacuolenplasma" ergänzt werden. „Vacuolenplasma" ist jedoch ein Widerspruch in sich. Das zeigt, daß dieses Ordnungssystem den natürlichen Verhältnissen nicht ganz gerecht wird. Ein anderer Vorschlag (von Robertson [55]) betont mehr die Ähnlichkeit gewisser Kompartimente untereinander und die Verschiedenheit zu anderen (vgl. auch [6]). Aber auch dieses Modell kann meines Erachtens nicht in allen Einzelheiten befriedigen. Eine andere Konzeption [61] wird im folgenden dargelegt.

Alle Zellschemata sind nur schwierig zu diskutieren, weil es an geeigneten Bezeichnungen für die einzelnen Bestandteile fehlt. Das trifft schon für den Begriff „Phase" zu. Dem Inhalt eines Kompartimentes fehlt, wie allen biologischen Systemen, gewöhnlich die Homogenität [15]. Nach H. Ruska [58] stellt er eine „wäßrige Mischphase "dar. Die abgrenzenden Membranen verhalten sich in vieler Hinsicht wie eine Phasengrenze [15]. Sie sind stets in sich geschlossen, haben keine freien Enden oder größere Poren und verzweigen sich nicht. Sie trennen *verschiedene* Mischphasen, und diese sind immer durch Membranen gegeneinander und nach außen hin abgegrenzt. So ist es beispielsweise nicht gelungen, auch bei einem gerade entstandenen Plasmatröpfchen aus einem Myxomyceten-Plasmodium nacktes Protoplasma nachzuweisen [78]. Die Neubildung des Plasmalemmas um solche Tröpfchen beansprucht keine für uns meßbare Zeit. Auch wenn Stoffe aufgenommen oder ausgeschieden werden, bekommt das Plasmalemma keine Diskontinuitäten. Die Umgebung des Plasmakörpers einer Zelle kann man als eine „wäßrige Mischphase" auffassen: Bei nackten Zellen das Medium, sonst die wasserdurchtränkten Zellwände oder die Zwischenzellsubstanz. Die Membranen haben jedoch, anders als eine normale Phasengrenze, eine gewisse Dicke und, das ist noch wichtiger, einen komplexen Aufbau. Man kann sie daher nicht nur als Phasengrenzen, sondern auch als gesonderte Phasen ansprechen. Wenn das hier nicht geschieht, so nur, um die Darstellung zu vereinfachen.

B. Karyoplasma und Cytoplasma

H. Ruska [58] und Wohlfarth-Bottermann [77] nehmen bei der tierischen Zelle 7 verschiedene Mischphasen an, ohne die begrenzenden Membranen selbst dazuzurechnen (bei der pflanzlichen wären es 10, hinzu käme jeweils noch die umgebende Phase). Diese Phasen sind jedoch teilweise nur recht unvollkommen voneinander getrennt. Das zeigt, daß entweder die Phasenregel bei der Organisation der Zelle nur sehr beschränkt gültig ist oder daß diese Regel nicht ganz konsequent angewendet wurde. H. Ruska [58] erwähnt selbst die unvollkommene Trennung von Karyoplasma und Grundcytoplasma. Die Kernhülle ist von mehr oder weniger offenen, jedenfalls nicht durch eine Elementarmembran verschlossenen Poren durchbrochen (vgl. u. a. [3] u. [14] für tierische, [35] u. [73] für pflanzliche Zellen). Außerdem zerfällt die Kernhülle in der Prophase in einzelne Zisternen und restituiert sich gegen Ende der Mitose wieder (u. a. [51]). Karyoplasma und Grundcytoplasma vermischen sich bei der Kernteilung (vgl. [16]).

Wenn man ein Kompartiment als einen Zellraum definiert, der von anderen Kompartimenten immer durch eine Plasmamembran getrennt ist, und eine „Mischphase" (im cytologischen Sinne) als den Inhalt solch eines Raumes, dann gehören Karyoplasma und Grundcytoplasma dem gleichen Kompartiment, der gleichen Mischphase an. Diese kann man am besten als nucleocytoplasmatische Matrix bezeichnen.

C. Endoplasmatisches Reticulum, Golgi-Apparat, Vacuolen

Bei der Mitose zerfällt die Kernhülle in einzelne Zisternen, die sich durch nichts von Zisternen des endoplasmatischen Reticulums unterscheiden (u. a. [51]). Nicht selten kommuniziert auch während der Interphase der perinucleare Raum mit cytoplasmatischen Zisternen des endoplasmatischen Reticulums (u. a. [70, 73], vgl. auch das Zellschema von Wohlfarth-Bottermann [77]). Das zeigt, daß das „perinucleäre Plasma" nicht vom „Reticulumplasma" getrennt ist. Beide gehören demselben Kompartiment, derselben Phase an. Die Kernhülle ist nichts anderes als eine spezialisierte Zisterne des endoplasmatischen Reticulums. Gegen die nucleocytoplasmatische Matrix sind die Kavernen des endoplasmatischen Reticulums dagegen stets durch Elementarmembranen abgegrenzt.

Anders sind die Beziehungen zwischen dem endoplasmatischen Reticulum, dem Golgi-Apparat und den Vacuolen. In den meisten Fällen sind alle drei Systeme von einander durch Elementarmembranen und durch die nucleocytoplasmatische Matrix getrennt. Sie können aber auch in Verbindung treten, so die Zisternen des endoplasmatischen Reticulums bei *Paramaecium* mit der kontraktilen Vacuole [60]. Nach Buvat [8], Poux [52], Maruyama, Gay und Kaufmann [40] u. a. m. bilden sich die Vacuolen aus sich erweiternden Zisternen des endoplasmatischen Reticulums. Möglicherweise können sich auch Golgi-Vesikel zu Vacuolen umwandeln [39].

Ähnliche Zusammenhänge sind auch zwischen den Kompartimenten des endoplasmatischen Reticulums und des Golgi-Apparates beschrieben. Jensen [29] zeigte kürzlich eine direkte Kommunikation zwischen beiden. In vielen tierischen Drüsen werden Sekretvorstufen in den Zisternen des endoplasmatischen Reticulums gebildet. Von diesen lösen sich Vesikel ab, die mit Golgi-Zisternen verschmelzen (u. a. [13], Übersicht bei Helander [27]). Auch für den umgekehrten Prozeß, eine Umwandlung von Golgi-Komponenten in endoplasmatisches Reticulum, gibt es Hinweise [48], [79].

Wenn die Kompartimente des endoplasmatischen Reticulums, des Golgi-Apparates und des Vacuoms auch nur zeitweilig miteinander fusionieren oder auseinander hervorgehen, so sind die in ihnen enthaltenen Phasen bei aller stofflicher Verschiedenheit einander doch so ähnlich, daß sie sich mischen. Sie stellen offenbar nur verschiedene Variationen derselben Phase dar (vgl. das Zellschema von Robertson [55] und Brachet [6]).

Nun können diese drei Systeme nicht nur untereinander, sondern vorübergehend auch mit dem Außenmedium kommunizieren. Der erste Prozeß spielt beim intracellulären Stofftransport, der zweite bei der Stoffaufnahme und Stoffausscheidung eine wichtige Rolle (vgl. z. B. [28]). Vacuolen entstehen beispielsweise als Nahrungsvacuolen bei Phagocytose durch Einstülpen des Plasmalemmas. Pinocytose-Bläschen werden auf ähnliche Weise gebildet. Pulsierende Vacuolen öffnen sich nach außen. Die Experimente von PLOWE [49] mit normalen Zellsafträumen zeigen Entsprechendes.

Die Zisternen des endoplasmatischen Reticulums öffnen sich mitunter gleichfalls direkt (vgl. [8, 23, 27, 41, 64], sowie die erwähnten Zellschemata) oder indirekt über pulsierende Vacuolen [60] nach außen. Eine intermittierende Verbindung mit dem Medium scheint über Pinocytose-Vesikel hergestellt werden zu können [76, 47, 68]. Bei einer granulocrinen Sekretion extruieren in vielen Fällen Golgi-Vesikel ihren Inhalt [27, 44]. Bei der Neubildung von Zellwänden höherer Pflanzen entsteht die Zellplatte durch die Fusion von Golgi-Vesikeln [72], das neue Plasmalemma durch die Aggregation ihrer Membranen [17]. In allen diesen Fällen — aus der Fülle der mehr oder weniger gut gesicherten (vgl. [50]) Angaben wurden nur wenige ausgewählt — verschmilzt die Membran des endoplasmatischen Kompartimentes vorübergehend oder ständig mit dem Plasmalemma. Der von ihr umschlossene Raum öffnet sich frei nach außen. Sein Inhalt vermischt sich mit der Umgebung.

Nach unserer Definition (s. S. 373) bedeutet das, daß die Kompartimente des endoplasmatischen Reticulums, des Golgi-Apparates und des Vacuoms ein Material enthalten, das dem umgebenden Medium phasengleich ist. Es ist eine nichtprotoplasmatische, wäßrige Mischphase, wird also besser nicht als „Plasma" bezeichnet, denn es unterscheidet sich grundlegend von der protoplasmatischen Mischphase, der nucleocytoplasmatischen Matrix. Diese ist ja gerade dadurch charakterisiert, daß sie *nicht* mit dem umgebenden Medium und *nicht* mit den anderen Kompartimenten kommuniziert. Das mag mit ihren lipophilen Eigenschaften zusammenhängen [10].

Das Plasmalemma und seine Einstülpungen, die Membranen des endoplasmatischen Reticulums, die Golgimembranen und die Vacuolenmembranen trennen also stets die protoplasmatische Mischphase (mit Eigenschaften eines hydrophilen Lipoids [2]) von einer wäßrigen Mischphase (dabei kann diese letzte jedoch ebenfalls Proteine und Lipoide enthalten).

D. Protokaryonten

Das gilt auch für die Protokaryonten. Einfache Bakterien bestehen nur aus einem Kompartiment: die nucleocytoplasmatische Matrix (in der die Kernäquivalente völlig frei im Cytoplasma liegen) ist vom Plasmalemma umschlossen.

Bei anderen Bakterien stülpt sich das Plasmalemma ein. Die so entstandenen Räume stehen entweder in offener Verbindung mit der umgebenden wäßrigen Mischphase (z. B. [11, 22]) oder sie lösen sich ab und bilden geschlossene Kompartimente (z. B. [34]). Die inneren Menbranen von Blaualgen (das Thylakoidsystem) können sich ebenfalls mit dem Plasmalemma vereinigen [61], (vgl. S. 379 u. Abb. 4). Dadurch grenzen die Elementarmembranen auch in diesen Fällen protoplasmatische Mischphasen von wäßrigen Mischphasen ab (das gilt natürlich nicht für Lagen der Zellwand, die im elektronenmikroskopischen Dünnschnitt-Bild wie Elementarmembranen aussehen können [30]).

E. Ausnahmen: Mischung verschiedenartiger Phasen

Nur in wenigen Ausnahmefällen scheint diese Trennung durchbrochen werden zu können. Durch grobe mechanische Eingriffe kann das Plasmalemma zerstört werden. Die nucleocytoplasmatische Matrix verquillt dann und mischt sich mit Wasser. Das bedeutet den Tod der Zelle. Wahrscheinlich kommuniziert die Elektrolyt-Lösung einer zur Potentialmessung mikrurgisch eingeführten Elektrode ebenfalls frei mit dem Grundplasma. In der Regel bleibt das Plasmalemma jedoch auch bei solchen Eingriffen erhalten, die eine starke Oberflächenvergrößerung des Protoplasten bewirken [66].

Membranauflösungen als natürliche Prozesse sind von Bennet [4] postuliert, aber selten exakt nachgewiesen. Die Möglichkeit einer Artefaktbildung ist bei entsprechenden Befunden nur schwer auszuschließen [28]. Mit am besten fundiert ist die Auflösung des Tonoplasten in Siebröhren ([33], dort weitere Angaben) zwischen dem strak hydratisierten Cytoplasma [32] und der Vacuole. Dadurch entsteht in diesen Zellen ein stark verdünntes, aufgelockertes Protoplasma (Esau und Cheadle [12]: „Mictoplasma"). Dieses ist jedoch noch immer durch das Plasmalemma nach außen und z. B. durch Membranen des endoplasmatischen Reticulums gegen innere wäßrige Mischphasen abgegrenzt. Es stellt also eine echte protoplasmatische Mischphase dar.

Als Ausnahme kann man auch die Bildung von "compound membranes" [54] ansehen, wie sie u. a. in desmosomalen Schlußleisten und Synapsen [56] und ähnlich auch zwischen den Thylakoiden von Chloroplasten [26, 43] anzutreffen sind. Hier ist die Phase zwischen den Membranen auf ein Minimum reduziert.

Die Konzeption, daß die bisher betrachteten Membranen „Phasengrenzen" zwischen wäßrigen und protoplasmatischen Mischphasen sind, schließt nicht aus, daß Membranen auch de novo entstehen können. Das wird unter anderem von Wohlfarth-Bottermann [75] für möglich gehalten. Mühlethaler [45] postuliert gleichfalls, daß sich die Vacuolen (und damit auch ihre Membranen, die Tonoplasten) durch Entmischung de novo bilden. Die Mehrzahl der Befunde spricht jedoch dafür, daß die Kompartimente des endoplasmatischen Reticulums, des Golgi-Apparates und des Vacuoms durch Teilung redupliziert werden. Dem geht ein Intussuszeptionswachstum ihrer Membranen oder eine Fusion mit anderen Kompartimenten voraus; dabei sind diese drei Systeme wenigstens teilweise ineinander konvertierbar.

F. Mitochondrien und Plastiden

Wie lassen sich nun auch die Membranen und Kompartimente der Mitochondrien und Plastiden in dieses Schema einordnen ? Diese Organellen, die „Plasten", unterscheiden sich von den anderen Zellkomponenten dadurch, daß sie von einer lückenlosen Hülle aus *zwei* Membranen umgeben sind. Die innere kann sich in die Matrix einstülpen und bildet so die Cristae der Mitochondrien und — direkt oder indirekt — die Thylakoide der Plastiden (Übersicht bei [43]).

Außerdem, daran ist kaum noch zu zweifeln, entstehen sie nur durch Teilung auseinander und sind in gewissem Umfang genetisch selbständig. Damit hängt zusammen, daß sie in ihrer Matrix über eine *eigene*, vom Kern wahrscheinlich qualitativ verschiedene DNS verfügen (vgl. die betreffenden Referate dieses Symposiums). Alle Versuche, die Mitochondrien und Plastiden in das Zellschema einzufügen, müssen also diese Gegebenheiten berücksichtigen.

Zwischen den beiden Hüllmembranen und dementsprechend auch im Intracristaraum (bzw. im Thylakoidbinnenraum) befindet sich nach H. Ruska [58] das äußere Mitochondrien- (bzw. Plastiden-) Plasma, nach Frey-Wyssling [16] jedoch eine serumartig verdünnte, wäßrige Lösung. Die Frage nach der Natur dieser Kompartimente hängt eng mit dem Problem der Herkunft der Organellen zusammen (vgl. [55] mit [61]). Phylogenetische Dokumente liegen freilich nicht mehr vor. Man kann jedoch versuchen, aus rezenten Organismen die Entwicklung zu rekonstruieren. Die Ergebnisse solch einer Spekulation sind jedoch — wie vorweg betont sei — nicht beweisbar und bleiben daher Hypothese.

I. Geosiphon

Mitochondrien kommen bei jedem Eukaryonten vor. Sie sind deshalb für solch eine Betrachtung weniger geeignet als Plastiden. In ihren oben aufgezählten Charakteristika gleichen Plastiden einer weiteren, allerdings nur bei wenigen Organismen auftretenden „Organelle", nämlich einem grünen, endosymbiontischen Protokaryonten. Das habe ich [61] an *Geosiphon pyriforme* gezeigt.

Geosiphon ist ein Phycomycet, der auf lehmiger Erde wächst. Die Hyphen nehmen in einer Art von Phagocytose Blaualgen aus der Gattung *Nostoc* auf und bilden dann etwa 1 mm große Blasen, die die Cyanellen enthalten [71, 31]. Abb. 1 zeigt den Rand solch einer Blase mit mehreren Blaualgenzellen. In Abb. 2 erkennt man bei stärkerer Vergrößerung, daß das Plasma des Pilzes gegen den Endosymbionten durch eine Membran getrennt ist, die wie das Plasmalemma zur Zellwand hin dargestellt wird. Die Cyanelle hat eine mehrschichtige Zellwand (Abb. 2, 3). Ihr Plasma wird ebenfalls durch ein Plasmalemma begrenzt.

Abb. 3 stellt *Nostoc*-Zellen aus der Mitte der Blase dar. Hier bildet das Pilzplasma nur einen schmalen Saum um die Blaualgenzelle. Es wird zur Vacuole hin durch den Tonoplast, zum Endosymbionten hin durch das Plasmalemma abgeschlossen. Die Cyanellen liegen also nicht

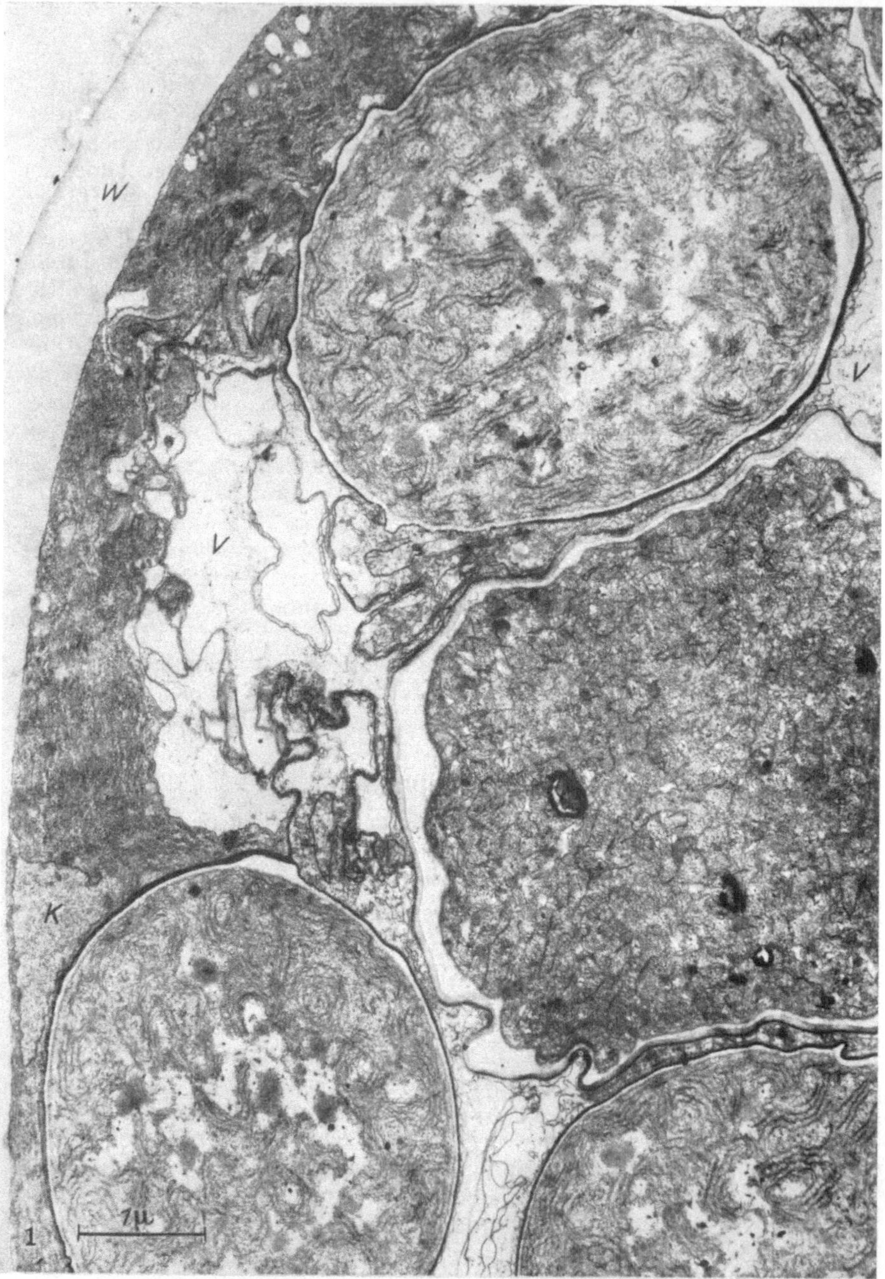

Abb. 1. *Geosiphon pyriforme*, äußerer Teil einer Blase mit 4 endosymbiontischen Blaualgenzellen. Die drei Zellen rechts gehören zu einem *Nostoc*-Faden, die mittlere ist eine Heterocyste. K = Zellkern des Pilzes, V = Vacuole des Pilzes, W = Zellwand des Pilzes. Fix.: OsO_4

frei in der Vacuole, sondern sind stets vom Wirtsplasma eingehüllt und befinden sich in einer Art von Phagocytosebläschen.

Das Blaualgen-Plasmalemma ist kein durchgehender Abschluß. Es geht an vielen Stellen in die Membranen der Thylakoide der Cyanellen

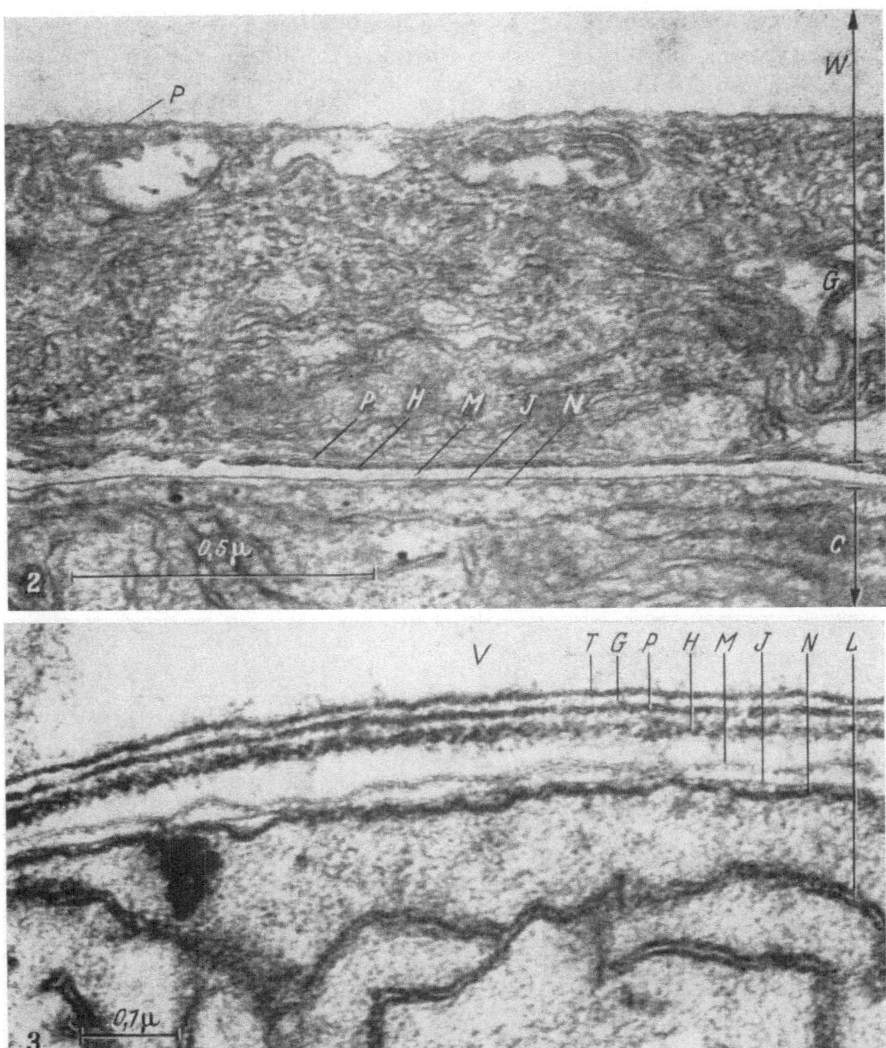

Abb. 2. *Geosiphon pyriforme*, äußerer Teil einer Blase. C = Plasma der Endocyanelle, G = Plasma des Pilzes, W = Zellwand des Pilzes, P = Plasmalemma des Pilzes, H, M, I = Hüll-, Mittel- und Innenschicht der Blaualgenzellwand, N = Plasmalemma der Blaualge. Fix.: OsO₄

Abb. 3. *Geosiphon pyriforme*, Blaseninneres. Strukturen des Pilzes: V = Vacuole, T = Tonoplast, G = Cytoplasma, P = Plasmalemma; Strukturen der Endocyanelle: H, M, I = Hüll-, Mittel- und Innenschicht der Zellwand, N = Plasmalemma, L = Thylakoid. Fix.: KMnO₄

über (Abb. 4). Die Thylakoide kommunizieren also mit dem Raum des Phagocytose-Bläschens. Sie enthalten demzufolge eine wäßrige Mischphase.

Vergleicht man die *Geosiphon*-Endosymbionten mit Chloroplasten, so entsprechen einander:

Geosiphon-Endosymbiose	*Chloroplast*
Plasmalemma des Wirtes (= Membran des Phagocytosebläschens)	äußere Plastidenmembran

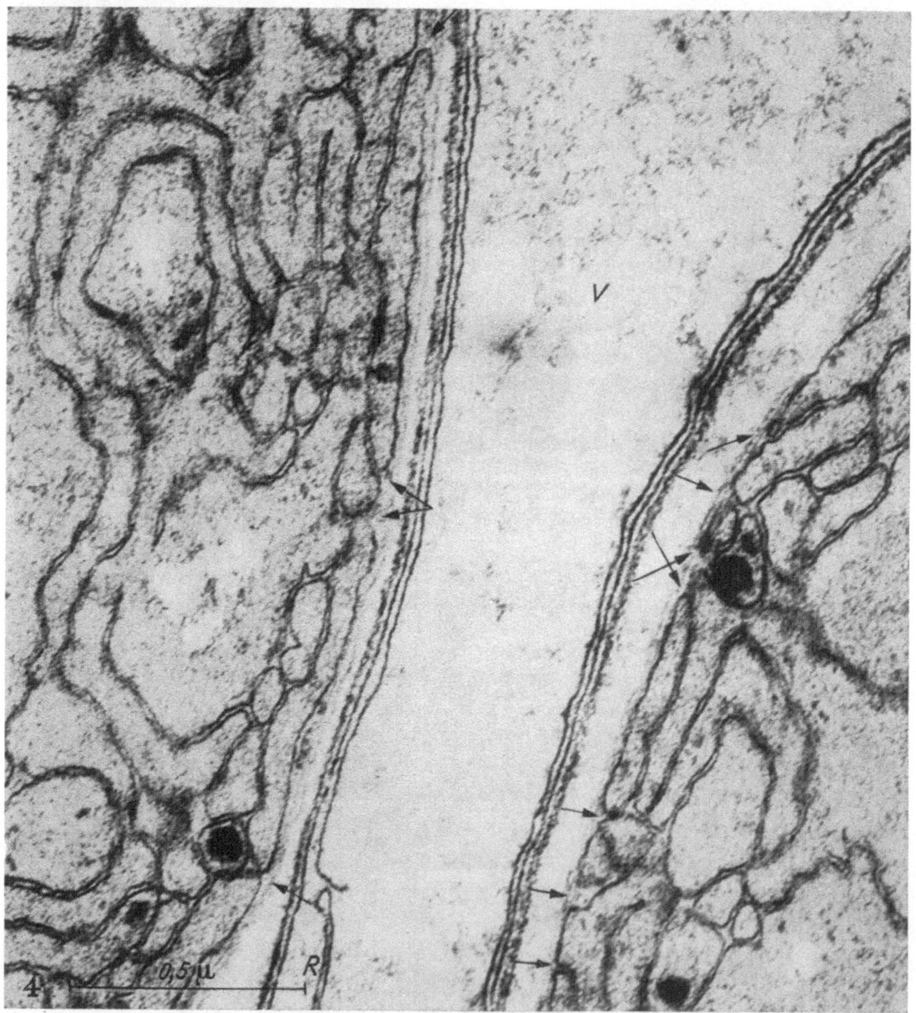

Abb. 4. *Geosiphon pyriforme*, Blaseninneres. V = Vacuole des Pilzes (Tonoplast unten links artifiziell zerrissen). R = Zisterne des endoplasmatischen Reticulums im Pilzplasma. Pfeile: Verbindungen zwischen den Thylakoidmembranen und dem Plasmalemma der Cyanellen, Öffnungen der Thylakoide in die Zellwand. Fix.: $KMnO_4$

Plasmalemma der Cyanelle	innere Plastidenmembran
Nucleocytoplasmatische Matrix des Endosymbionten	Plastidenmatrix (mit DNS, vgl. u. a. [24])
Einstülpungen des Blaualgen-Plasmalemmas = Thylakoide	Einstülpungen der inneren Plastidenmembran = Thylakoide
Rest des Phagocytosebläschens mit Zellwand der Cyanelle (= wäßrige Mischphase)	Raum zwischen äußerer und innerer Plastidenmembran, keine „Plastidenwand"

Der Raum zwischen den beiden Plastidenmembranen enthält hiernach eine wäßrige Mischphase.

II. Glaucocystis

Bei der *Geosiphon*-Endosymbiose können beide Partner unabhängig voneinander existieren. Dadurch und durch den Besitz einer wohl ausgebildeten Zellwand unterscheiden sich die Endocyanellen von Plastiden. Es gibt jedoch weiter entwickelte Symbiosen. Bei diesen ist es noch nicht gelungen, die Organismen unabhängig voneinander zu kultivieren. Die Zellwand des Endosymbionten kann in solchen Fällen weitgehend reduziert sein. Ein solcher ist *Glaucocystis*. GEITLER [19, 20] beschrieb ihren lichtmikroskopischen Zellbau; ihre Feinstruktur wird von uns untersucht [62, 63], vgl. [69]. Der Wirt soll eine chlorococcale Alge sein, verwandt mit *Oocystis*; dagegen spricht jedoch, daß er mit zwei rudimentären Geißeln versehen ist, die aber nicht durch die Zellulosewand treten sowie mit einer pulsierenden Vacuole, die mit dem Golgi-Apparat in Zusammenhang steht. Er besitzt keine Plastiden, dafür aber Blaualgen unbekannter Stellung.

Im Verlauf des Miteinanderlebens hat die Cyanelle ihre Zellwand bis auf die Innenschicht verloren. Diese ist in der Regel mit dem Plasmalemma verschmolzen und wird nur gelegentlich nach OsO_4-Fixierung und bei stärkerer Schrumpfung der Blaualgen sichtbar (Abb. 6, vgl. den Wandbau der *Geosiphon-Nostoc* in Abb. 3). Nach OsO_4-Fixierungen sind solche Schrumpfungen nicht selten; dann liegen die Endosymbionten scheinbar in einem größeren „Phagocytosebläschen" (Abb. 5, 6). Meistens aber (Abb. 7; bei $KMnO_4$-Fixierung immer: Abb. 9) folgt jedoch das Plasmalemma (+innerste Wandschicht) unmittelbar auf die Membran des „Phagocytosebläschens", wie die innere auf die äußere Plastidenmembran folgt. Das Chromatoplasma ist sehr ausgedehnt, das Zentroplasma dadurch nur auf einen kleinen Bereich beschränkt (Abb. 5, 7, 9). Damit unterscheiden sich die Blaualgenzellen im elektronenmikroskopischen Bild kaum von Rotalgenplastiden. In diesen liegen wie bei den Cyanophyceen die Thylakoide einzeln in der Matrix (vgl. die Angaben in [43] sowie z. B. [5, 69]). In beiden findet man osmiophile Globuli.

Der Vergleich läßt sich noch weiter führen. Die *Glaucocystis*-Blaualgen (Abb. 8, 9) sind unregelmäßig langgestreckt. An einem Pol haben sie ein chlorophyll- und thylakoidfreies, stark lichtbrechendes Körperchen. Es

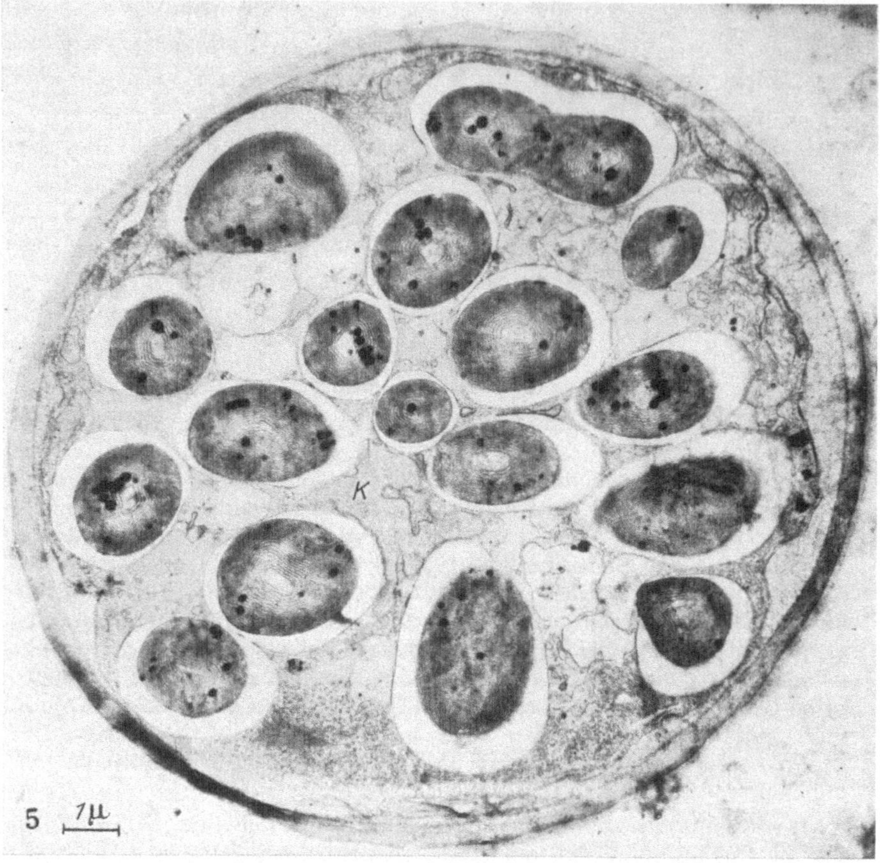

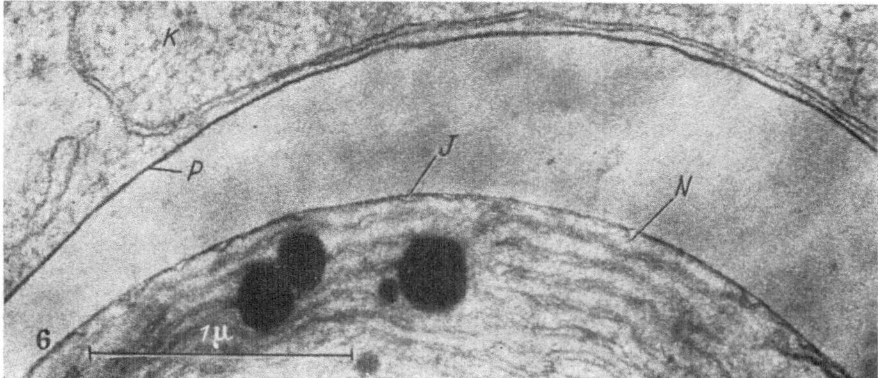

Abb. 5. *Glaucocystis incrassata*, Querschnitt im Bereich des Zellkernes (K, nur angeschnitten). Endocyanellen geschrumpft. Fix.: OsO$_4$

Abb. 6. *Glaucocystis incrassata*, geschrumpfte Endocyanelle in einem „Phagocytosebläschen". K = Zellkern des Wirtes, P = Membran des „Phagocytosebläschen". I = Rest der Blaualgenzellwand (entspricht der Innenschicht in Abb. 2 und 3), N = Plasmalemma der Cyanelle. Fix.: OsO$_4$

liegt innerhalb der Plastidenmembran (Abb. 10) und gibt Proteinreaktionen (vgl. dagegen [19]). Die Cyanellen strahlen von zwei Zentren in der Zellmitte aus, sie liegen mit ihren farblosen Polen zusammen (Abb. 10). Vorzugsweise um diese Körperchen bilden sich im Grundplasma des

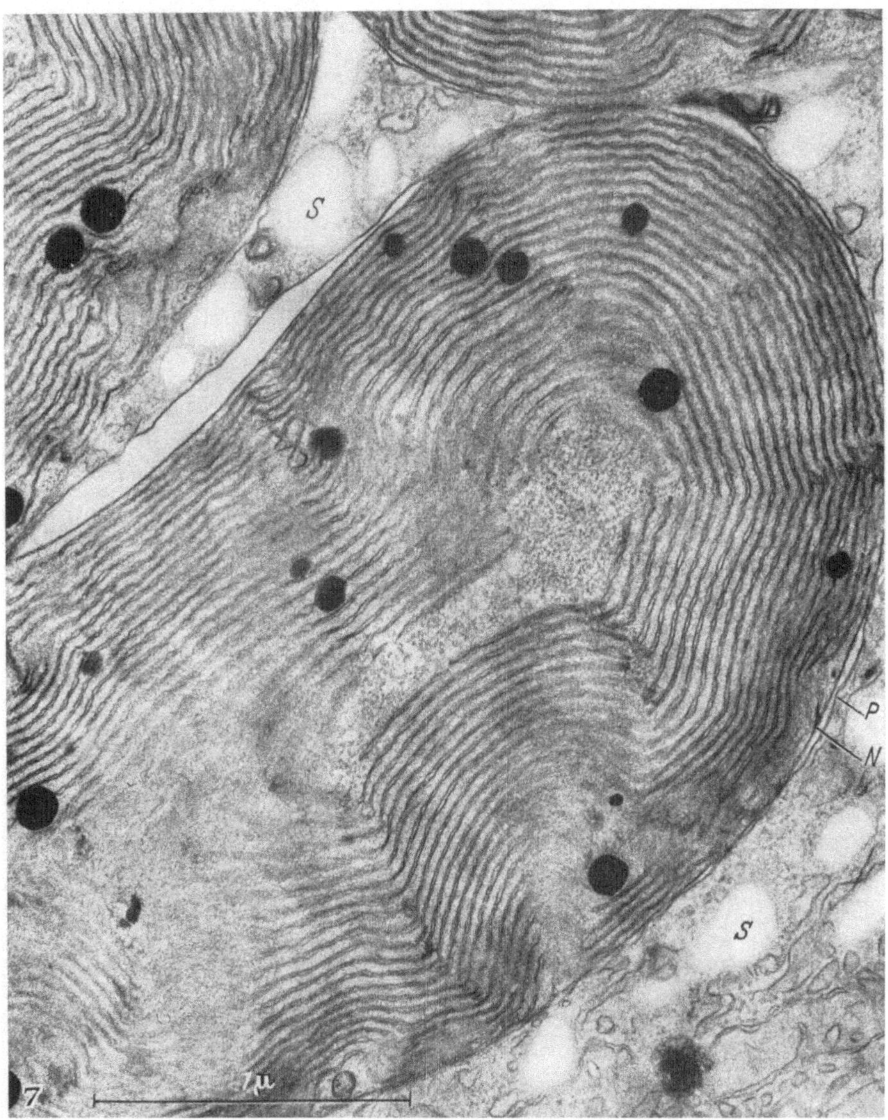

Abb. 7. *Glaucocystis incrassata*, Endocyanelle. N = Plasmalemma (+ Innenschicht der Zellwand) der Cyanelle, P = Membran des „Phagocytosebläschens", S = Körner eines stärkeähnlichen Polysaccharids im Grundplasma des Wirtes. Fix.: OsO_4

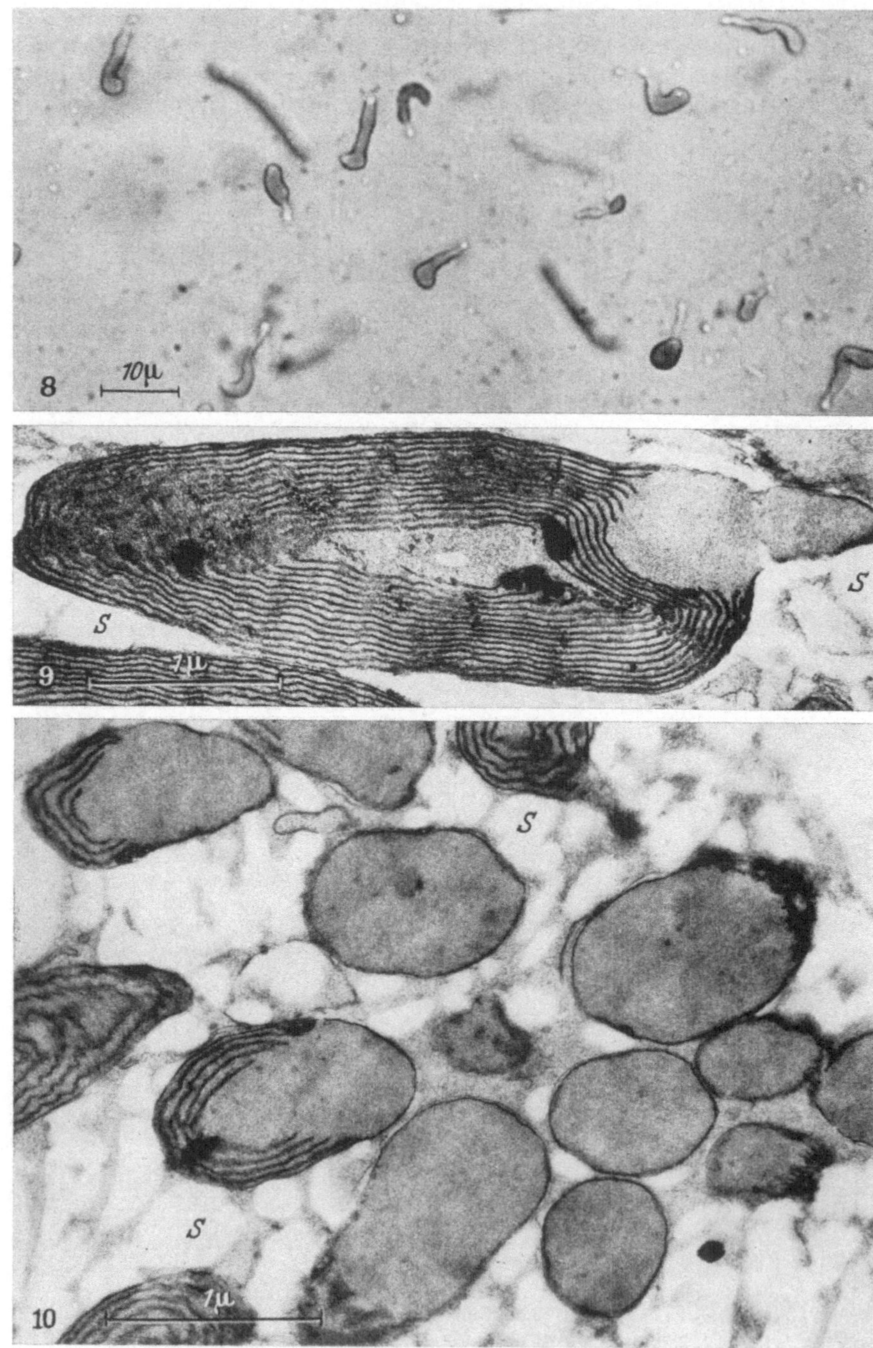

8
10μ
9
S
S
1μ
10
S
S
1μ

Wirtes Körner eines stärkeähnlichen Polysaccharids. — Bei niederen
Rotalgen liegt die Florideenstärke ebenfalls außerhalb der Plastiden, und
zwar vorzugsweise um einen chlorophyllfreien, stark lichtbrechenden
Plastidenabschnitt, der bei einzelligen *Bangiales* häufig in der Mitte der
Zelle liegt und von dem die grünen Plastidenteile strahlig ausgehen: das
Pyrenoid. Es bestehen also gewisse strukturelle und funktionelle Ähn-
lichkeiten zwischen dem Blaualgen-Körperchen und den Rotalgen-
Pyrenoiden. Es läßt sich freilich nicht entscheiden, ob diese Entspre-
chungen nur zufällig sind. Erwähnt sei ferner die sehr ähnliche Aus-
stattung der Cyanophyceenzelle und Rhodoplasten mit Phycocyan und
Phycoerythrin (Übersicht in [46]).

III. Endocyanellen — Rhodoplasten

All das zeigt, wie man sich die Entstehung der Rotalgenplastiden
vorstellen *kann*. Aus einer losen Endosymbiose mit Blaualgen (wie bei
Geosiphon) wird eine innigere Verbindung (wie bei *Glaucocystis*); die
Cyanophyceenzellen verlieren mehr und mehr ihre Selbständigkeit, die
Zellwand wird reduziert, schließlich aufgegeben (vgl. auch [25]). Es erscheint daher schwierig, eine klare cytologische Grenze zwischen Endocyanellen und Rhodoplasten zu ziehen.

Cyanellen können Bestandteile sehr verschiedener Zellen geworden sein. Die
„Glaucophyten" [67] scheinen eher eine Organisationsstufe der Endocyanose darzustellen als einen phylogenetisch selbständigen Stamm [53]. Wenn man den entwickelten Gedanken weiter verfolgt, könnte man auch für die Rhodophyten er-
wägen, ihre Zellen seien Konsortien zwischen Blaualgen und anderen Zellen. Eine
Verwandtschaft zwischen beiden Gruppen ist oft behauptet und bestritten worden
(vgl. zuletzt [21]). Die Annahme, die Rotalgenplastiden hätten sich aus endosymbiontischen Blaualgenzellen entwickelt, könnte eine Brücke zwischen den gegen-
sätzlichen Anschauungen sein. Sie würde vielen cytologischen und biochemischen
Fakten gerecht. Außerdem ließe sie noch Beziehungen zwischen Rotalgen und
anderen Stämmen zu, beispielsweise zwischen höheren Florideen und Ascomyceten,
wie sie schon lange diskutiert werden (Abgaben bei [18]) und eine Stütze in neueren
cytologischen Untersuchungen finden [38].

Zu betonen ist jedoch noch einmal, daß die hier aufgezeigte Ent-
wicklungsmöglichkeit dem phylogenetischen Ablauf entsprochen haben
kann, daß dieser jedoch heute und wohl auch in Zukunft nicht zu beweisen ist. Die Möglichkeit, daß es sich um Analogien und nicht um
Homologien handelt, kann nicht ausgeschlossen werden.

IV. Endosymbiont — Chloroplast

Cyanophyceenzellen kommen aus biochemischen und cytologischen
Gründen als direkte Ausgangsformen wohl nur für die Rotalgenplastiden
in Frage. Für den unmittelbaren Anschluß der Plastiden anderer Gruppen

---

Abb. 8. *Glaucocystis geitleri*, ausgequetschte Endocyanellen, lebend

Abb. 9. *Glaucocystis geitleri*, Endocyanelle längs. Rechts: Thylakoidfreier „Polkörper",
S = Körner eines stärkeähnlichen Polysaccharids im Grundplasma des Wirtes. Fix.: KMnO$_4$

Abb. 10. *Glaucocystis geitleri*, Schnitt durch ein „Polkörperzentrum": zahlreiche Cyanellen
liegen mit ihren thylakoidfreien „Polkörpern" in der Zellmitte zusammen. S = Körner
eines stärkeähnlichen Polysaccharids im Grundplasma des Wirtes. Fix.: KMnO$_4$

an rezente Organismen gibt es keine Hinweise. Dennoch kann man auch bei ihnen annehmen, sie seien aus einer Endosymbiose hervorgegangen, vielleicht sogar polyphyletisch. Die Einwanderung hat sicherlich auf einer sehr frühen Stufe der Phylogenie stattgefunden. Von der Cytologie her gesehen müßten sie sich ebenfalls aus Protokaryonten entwickelt haben.

V. Endosymbiont — Mitochondrium

Für die Herkunft der Mitochondrien gilt das gleiche. LEHNINGER [*36*] vergleicht diese Organellen mit Rickettsien, also intracellulären Parasiten, die bislang in einem synthetischen Medium nicht erfolgreich kultiviert werden konnten. Sie haben die Fähigkeit, wie Mitochondrien die Oxydationen des Citratcyclus und die oxydative Phosphorylierung zu katalysieren, und es gibt Hinweise darauf, daß sie die Glycolyse nicht durchführen können.

In diesem Zusammenhang sei erwähnt, daß am Plasmalemma von Bakterien und seinen Einstülpungen Partikel nachgewiesen wurden, die mit den Elementarpartikeln an der inneren Mitochondrien-Membran identisch zu sein scheinen [*1, 37*].

G. Protocyte und Eucyte

ROBERTSON (u. a. [*55*]) versucht, die Kompartimentierung der Mitochondrien mit der Annahme zu erklären, diese Organellen wären durch die Phagocytose eines zelleigenen Körpers entstanden. Das führt zwar zu ähnlichen Folgerungen über die Natur der Phasen und Elementarmembranen und ihre Konvertierbarkeit, erscheint aber doch recht unwahrscheinlich und kann beispielsweise den DNS-Gehalt der Plasten nicht erklären.

Der Gedanke, daß sich Plastiden und Mitochondrien aus Endosymbionten entwickelt haben könnten, ist nicht neu. Er wird schon von SCHIMPER [*59*] geäußert. BUCHNER [*7*] gibt eine Übersicht über die Wege und Irrwege dieser Vorstellungen. Er hält sie jedoch für so spekulativ, daß sie auf den Gang der Forschung keinen Einfluß gehabt haben.

Die neue, cytologische Konzeption dieser Idee (Abb. 11) gibt nun aber die Möglichkeit, die Mannigfaltigkeit der Membranen und Kompartimente zu überschauen und zu ordnen und das Organisationsprinzip der Zelle zu verstehen: *Eine Elementarmembran trennt stets eine protoplasmatische Mischphase von einer wäßrigen Mischphase.* In einer Protocyte, der Zelle eines Protokaryonten, gibt es nur *eine* protoplasmatische Phase, in einer Eucyte (der Zelle eines Eukaryonten) dagegen zwei oder drei: die eigentliche nucleocytoplasmatische Matrix, die Mitochondrienmatrix und, bei grünen Pflanzen, die Plastidenmatrix.

Dann unterscheidet sich eine Protocyte, ein Bakterium oder eine Cyanophyceenzelle, *grundsätzlich* von einer Eucyte: Eine Eucyte ist ein System höherer Ordnung, eine Verbindung verschiedener „Protocyten", die eine neue, komplexere Einheit gebildet haben. Das biologische Prinzip, daß sich verschiedenartige Teile zu einem neuen Ganzen zusammenfügen, scheint auch auf cellulärem Niveau verwirklicht zu sein.

Durch diese Deutung der Membranen und Kompartimente lassen sich viele zellmorphologische und -physiologische Befunde erklären. Celluläre Parasiten liegen immer in einem „Phagocytosebläschen", die acellulären Viren entstehen dagegen nur in protoplasmatischen Mischphasen; stärkeähnliche Polysaccharide werden ebenfalls nur in solchen gebildet. Das hier entwickelte Zellschema kann außerdem eine Hilfe bei der Identifizierung von Zellstrukturen geben (vgl. z. B. die unterschiedliche Interpretation des *Anthoceros*-Chloroplasten durch MENKE [*42*] und WILSENACH [*74*]). Es erlaubt also eine *deduktive* Analyse.

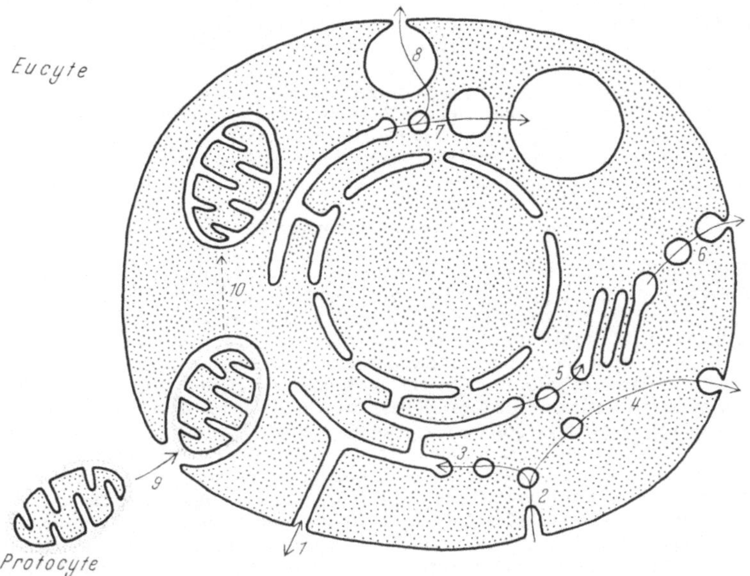

Abb. 11. Stark vereinfachtes Schema vom Aufbau von Protocyte und Eucyte sowie von der Kompartimentierung der Zelle und den Beziehungen zwischen den Membranen und Kompartimenten. Punktiert: Nucleocytoplasmatische Matrix = protoplasmatische Mischphasen; weiß = wäßrige Mischphasen; gestrichelt = Zellwand der Protocyte, Teil der wäßrigen Mischphase. Die durchgehenden Pfeile 1—9 zeigen Möglichkeiten von Membran- und Kompartimentverschmelzungen und damit die Mischbarkeit der wäßrigen Mischphasen. Sie stellen ontogenetische Prozesse dar. 1. Öffnung einer Zisterne des endoplasmatischen Reticulums in das Medium, Verschmelzung des Plasmalemmas mit den Membranen des endoplasmatischen Reticulums. Mischung des Zisterneninhaltes mit der umgebenden wäßrigen Mischphase. 2. Pinocytose, die Membran eines Pinocytosevesikels ist ein Abkömmling des Plasmalemmas. 3. Fusion von Pinocytosevesikeln mit Zisternen des endoplasmatischen Reticulums, vgl. 1. 4. Cytopempsis. 5. Fusion eines Vesikels, das von einer Zisterne des endoplasmatischen Reticulums abgeschnürt wurde, mit einer Golgi-Zisterne. Konvertierbarkeit der Membranen des Golgiapparates und der Membranen des endoplasmatischen Reticulums. Mischbarkeit des Kompartiment-Inhaltes. 6. Extrusion von Golgi-Vesikeln. Einbau von Golgi-Membranen in das endoplasmatische Reticulum, Mischbarkeit des Golgi-Kompartiment-Inhaltes mit der umgebenden wäßrigen Mischphase. 7. Entwicklung von Vacuolen aus Vesikeln des endoplasmatischen Reticulums. 8. pulsierende Vacuole. Verschmelzung von Tonoplast und Plasmalemma. Mischbarkeit des Vacuoleninhaltes mit der umgebenden wäßrigen Mischphase. 9. Phagocytose einer Protocyte. Die Membran des Phagocytosebläschens stammt vom Plasmalemma ab (vgl. *Geosiphon*). Der gestrichelte Pfeil 10 stellt die — hypothetische — phylogenetische Entwicklung eines Plasten (Mitochondrion oder Plastide) aus einem Endosymbionten dar (vgl. *Glaucocystis*)

Summary

The reduplication of organelles and the problem of compartmentation of the cell

The cells of eukaryontic organisms are composed of several compartments. The content of this compartments can be looked at as "mixed phases". Dissimilar mixed phases are separated from each other by unit membranes. Between similar phases there is a permanent or an intermittent connection.

The karyoplasm and the cytoplasm are connected by pores in the nuclear envelope and intermingle in the mitosis. Therefore karyoplasm and cytoplasm belongs to *one* phase, the "protoplasmatic mixed phase" which fills the nucleocytoplasmatic compartment.

The question of the nature of the other compartments and their contents can be answered only in connection with the reduplication of the organelles consisting of theses elements. The compartments of the endoplasmic reticulum, of the Golgi apparatus, and of the vacuom partly originate from each other and can communicate intermittently with each other. They also communicate with the medium which surrounds the cell. This medium including the cell wall is a "watery mixed phase". Therefore, the three systems mentioned above contain a material of the same quality, a watery mixed phase and not a protoplasmatic mixed phase.

In contrast the plastids and mitochondria consist of two compartments; they multiply by identical reduplication and do never arise from other cell elements. They contain a special DNA in the matrix which fills the inner compartment. To understand these characteristics we can compare them with an endosymbiontic protokaryont.

The structure of the cells of the protokaryontic organisms, bacteria and blue green algae is much simpler than that of the eukaryontic cells. The protokaryontic cells consist of the nucleocytoplasmatic matrix (nuclear and cytoplasmic material are unintermittes intermingled), the plasmalemma (limiting the protoplasmatic mixed phase against the watery mixed phase = surrounding medium and cell wall), and in many cases of infoldings of the plasmalemma which can detach and loose their connection with the medium. Therefore they contain a watery mixed phase. That is also right of the thylakoids of the blue green algae: In the endosymbiontic *Nostoc* of *Geosiphon* the membranes of the thylakoids fuse with the plasmalemma.

These *Nostoc* cells are found to be in phagocytotic vesicles, the membrane of which derives from the plasmalemma of the fungus. The algae themselves are surrounded by their own complete cell walls. In the blue green alga which lives in the cells of *Glaucocystis* the cell wall is almost completely reduced. The membrane of the phagocytotic vesicle is almost directly attached to the plasmalemma of the endosymbiont. Thus the endocyanellae resemble in many details the plastids or mitochondria, especially the rhodoplasts of the red algae. There is a further morphological and physiological similarity; the endosymbionts of *Glaucocystis* have a colorless polar body comparable with a pyrenoid.

Certainly it is impossible to give direct proof that plastids and mitochondria originate phylogenetically from endosymbiontic organisms. The resemblance may depend on homology or on analogy. But this concept fits the facts best and allows to understand their structure, their function, and their reduplication. It allows also to comprehend the principle of the compartmentation of the cell: a unit membrane separates always, in these organelles as in the other parts of the cell, a protoplasmatic mixed phase from a watery mixed phase (Fig. 11). Accepting these ideas, an eucyte, a cell of an eukaryontic organism, would differ from a protocyte, a cell of a protocaryontic organism, by being a composition of several "protocytes".

Literatur

[1] ABRAM, D.: J. Cell Biol. **23**, 3A (1964).
[2] BANCHER, E., u. J. HÖLZL: Protoplasma (Wien) **57**, 33 (1963).
[3] BARNES, B. G., and I. M. DAVIS: J. Ultrastruct. Res. 3, 131 (1959).
[4] BENNET, H. S.: J. biophys. biochem. Cytol. 2. Suppl., 99 (1956).
[5] BOUCK, G. B.: J. Cell Biol. **12**, 553 (1962).

[6] Brachet, J.: Sci. American **205**, (3) (1961).
[7] Buchner, P.: Endosymbiose der Tiere mit pflanzlichen Mikroorganismen. Basel-Stuttgart: Birkhäuser 1953.
[8] Buvat, R.: Proc. 5. Int. Congr. Electron Microsc. Philadelphia, S. W 1 (1962).
[9] — — Intern. Rev. Cytol. **14**, 41 (1963).
[10] Drawert, H.: Z. Naturforsch. 3 b, 111 (1948).
[11] Edwards, M. R., and R. W. Stevens: J. Bact. **86**, 414 (1963).
[12] Esau, K., and V. I. Cheadle: Univ. Calif. Publ. Bot. **36**, 253 (1965).
[13] Essner, E., and A. B. Novikoff: J. Cell Biol. **15**, 289 (1962).
[14] Feldherr, C. M.: J. Cell Biol. **14**, 65 (1962).
[15] Frey-Wyssling, A.: Submikroskopische Morphologie des Protoplasmas und seiner Derivate. Berlin: Bornträger 1938.
[16] — — Nova Acta Leopold. NF 22, Heft 147 (1960).
[17] — —, J. F. López-Sáez, and K. Mühlethaler: J. Ultrastruct. Res. **10**, 422 (1964).
[18] Fritsch, F. E.: The Structure and Reproduction of the Algae. Vol. II. Cambridge: University Press 1952.
[19] Geitler, L.: Arch. Protistenk. **47**, 1 (1923).
[20] — — In Hdb. d. Pflanzenphysiologie, Bd. 11, S. 530. Berlin-Göttingen-Heidelberg: Springer 1959.
[21] — — Ber. Dtsch. Bot. Ges. **78**, 101 (1965).
[22] Giesbrecht, P., u. G. Drews: Arch. Mikrobiol. **43**, 152 (1962).
[23] Gresson, R. A. R.: Exp. Cell Res. **26**, 212 (1962).
[24] Gunning, B. E. S.: J. Cell Biol. **24**, 79 (1965).
[25] Hall, W. T., and G. Claus: J. Cell Biol. **19**, 551 (1963).
[26] Heitz, E.: Z. Zellforsch. **53**, 444 (1961).
[27] Helander, H. F.: In K. E. Wohlfarth-Bottermann: Funktionelle und morphologische Organisation der Zelle. Sekretion und Exkretion. 2. wiss. Konf. Ges. Dtsch. Naturf. u. Ärzte 1964, S. 2. Berlin-Heidelberg-New York: Springer 1965.
[28] Holter, H.: In K. E. Wohlfarth-Bottermann: Funktionelle und morphologische Organisation der Zelle. Sekretion und Exkretion. 2. wiss. Konf. Ges. Dtsch. Naturf. u. Ärzte 1964, S. 119. Berlin-Heidelberg-New York: Springer 1965.
[29] Jensen, W. A.: Amer. J. Bot. **52**, 238 (1965).
[30] Kellenberger, E., and A. Ryter: J. biophys. biochem. Cytol. **4**, 323 (1958).
[31] Knapp, E.: Ber. Dtsch. Bot. Ges. **51**, 210 (1933).
[32] Kollmann, R.: Phytomorphol. **14**, 247 (1964).
[33] — u. W. Schumacher: Planta (Berl.) **63**, 155 (1964).
[34] Kran, G., F. W. Schlote, u. H. G. Schlegel: Naturwissenschaften **50**, 728 (1963).
[35] Lance, A.: C. R. Acad. Sci. (Paris) **245**, 352 (1957).
[36] Lehninger, A. L.: The Mitochondrion. New York-Amsterdam: Benjamin 1964.
[37] Löw, H., and B. A. Afzelius: Exp. Cell Res. **35**, 431 (1964).
[38] Maltzahn, K. von, and M. L. Cameron: Naturwissenschaften **51**, 642 (1964).
[39] Marinos, N. G.: J. Ultrastr. Res. **9**, 177 (1963).
[40] Maruyama, K., H. Gay, and B. P. Kaufmann: Amer. J. Bot. **49**, 662 (1962).
[41] McAlear, J. H., and G. A. Edwards: Exp. Cell Res. **16**, 689 (1959).
[42] Menke, W.: Z. Naturforsch. **16 b**, 334 (1961).
[43] — Ber. Dtsch. Bot. Ges. **77**, 340 (1964).
[44] Mollenhauer, H. H., W. G. Whaley, and J. H. Leech: J. Ultrastr. Res. **5**, 193 (1961).
[45] Mühlethaler, K.: 4. Int. Kongr. Elektronenmikr. Berlin Bd. II, S. 491, 1958. Berlin-Göttingen-Heidelberg: Springer 1960.
[46] O hEocha, C.: In R. A. Lewin: Physiology and Biochemistry of Algae. S. 421. New York-London: Academic Press 1962.
[47] Palay, S. L., and L. J. Karlin: J. biophys. biochem. Cytol. **5**, 373 (1959).
[48] Pipan, N.: Z. Zellforsch. **52**, 291 (1960).
[49] Plowe, J. Q.: Protoplasma (Wien) **12**, 196 (1931).

[50] PORTER, K. R.: In: J. BRACHET and A. E. MIRSKY: The Cell, Vol. II. S. 621.
New York-London: Academic Press 1961.
[51] —, and R. D. MACHADO: J. biophys. biochem. Cytol. 7, 167 (1960).
[52] POUX, N.: J. Microsc. 1, 55 (1962).
[53] PRINGSHEIM, E. G.: Farblose Algen. Stuttgart: Fischer 1963.
[54] ROBERTSON, J. D.: J. biophys. biochem. Cytol. 4, 349 (1958).
[55] — In M. LOCKE: Cellular Membranes in Development. S. 1. New York-London:
Academic Press 1964.
[56] —, T. S. BODENHEIMER and D. E. STAGE: J. Cell Biol. 19, 159 (1963).
[57] RUSKA, C.: Proc. 5. Internat. Congr. Electron Microsc. Philadelphia S TT 1
(1962).
[58] RUSKA, H.: S. B. Ges. Bef. ges. Naturwiss. Marburg 82, 3 (1960).
[59] SCHIMPER, A. F. W.: Bot. Z. 41, 105 (1883).
[60] SCHNEIDER, L.: J. Protozool. 7, 75 (1960).
[61] SCHNEPF, E.: Arch. Mikrobiol. 49, 112 (1964).
[62] — Planta (Berl.) (im Druck).
[63] —, u. W. KOCH: in Vorbereitung.
[64] SHATKIN, A. J., and E. L. TATUM: J. biophys. biochem. Cytol. 6, 423 (1959).
[65] SITTE, P.: Ber. Dtsch. Bot. Ges. 74, 177 (1961).
[66] — Protoplasma (Wien) 57, 304 (1963).
[67] SKUJA, H.: In A. ENGLER: Syllabus der Pflanzenfamilien. 12. Aufl. von H.
MELCHIOR und E. WERDERMANN. Bd. I, S. 56. Berlin: Borntraeger 1954.
[68] STAUBESAND, J.: In K. E. WOHLFARTH-BOTTERMANN: Funktionelle und
morphologische Organisation der Zelle. Sekretion und Exkretion. 2. wiss.
Konf. Ges. Dtsch. Naturf. u. Ärzte 1964, S. 162. Berlin-Heidelberg-New
York: Springer 1965.
[69] UEDA, K.: Cytologia (Tokyo) 26, 344 (1961).
[70] WATSON, M. L.: J. biophys. biochem. Cytol. 1, 257 (1955).
[71] WETTSTEIN, F. VON: Öst. Bot. Z. 65, 145 (1915).
[72] WHALEY, W. G., and H. H. MOLLENHAUER: J. Cell Biol. 17, 216 (1963).
[73] — — and J. H. LEECH: J. biophys. biochem. Cytol. 8, 233 (1960).
[74] WILSENACH, R.: J. Cell Biol. 18, 419 (1963).
[75] WOHLFARTH-BOTTERMANN, K. E.: Z. Zellforsch. 50, 1 (1959).
[76] — Protoplasma (Wien) 52, 58 (1960).
[77] — Naturwissenschaften 50, 237 (1963).
[78] — in K. E. WOHLFARTH-BOTTERMANN: Funktionelle und morphologische
Organisation der Zelle. Sekretion und Exkretion. 2. wiss. Konf. Ges. Dtsch.
Naturf. u. Ärzte 1964, S. 118. Berlin-Göttingen-New York: Springer 1965.
[79] —, u. V. MOERICKE: Z. Naturforsch. 14b, 446 (1959).

Mit Unterstützung der Deutschen Forschungsgemeinschaft.

Diskussion

Vorsitz: *Wohlfarth-Bottermann*

Karlson: Was weiß man über den Inhalt des endoplasmatischen Reticulum in
biochemischer Hinsicht?

Schnepf: Im Pankreas z. B. findet man darin Verdauungsenzyme. Diese Enzyme
werden freilich nicht hier, sondern im protoplasmatischen Bereich gebildet. Sie
kommen offensichtlich leicht durch die Membran in die Zisternen.

Parthier: Das ist — nach einem Vortrag, den Prof. SIEKEVITZ kürzlich in Stockholm gehalten hat — nachweislich der Fall. Amylase wird in der Leber an den
Mikrosomen gebildet, findet sich dann aber sehr rasch im nichtplasmatischen
Raum der ER-Zisternen. Er konnte auch zeigen, daß das Enzymprotein noch während seiner Synthese die ER-Membran durchdringt.

David: Inwiefern ist man berechtigt, von einem nucleo-cytoplasmatischen
Raum zu sprechen? In den Poren der Kernmembran befinden sich doch Diaphrag-
men, die beide Phasen voneinander trennen. Und biochemische Erfahrungen sprechen doch auch dafür, daß da ganz unterschiedliche Vorgänge und unterschiedliche
Bestandteile vorhanden sind.

Schnepf: Zugegeben. Aber dieses Diaphragma ist keine Elementarmembran, und während der Mitose mischt sich der Kernraum mit dem cytoplasmatischen Raum.

Grundmann: Ich möchte aber doch daran erinnern, daß man den Kern in einer lebenden Zelle anstechen, ihn zur Seite zwängen kann, usw. In effectu bildet die Kernmembran auch vital einen sicheren Abschluß. Ich glaube auch nicht, daß ihr Verhalten während der Mitose wirklich dagegen spricht. Das ist das einzige, was ich an Ihrer Konzeption auszusetzen habe.

Schnepf: Sicher sind Kernplasma und Cytoplasma etwas Verschiedenes. Sie sind ja auch voneinander getrennt. Ebenso ist sicher auch der Binnenraum der Golgi-Zisternen verschieden vom Binnenraum der endoplasmatischen Zisternen. Aber die Möglichkeit der Vermischung besteht auch hier. Gerade das erscheint mir für meine Konzeption entscheidend.

Bier: Im Hinblick auf die Ionen-Ungleichgewichte und die aktiven Transport-prozesse an der Kernmembran scheint mir Ihr Schema etwas irreführend für den Zustand in der Interphase. Wenn Sie es nur formal auffassen („Zwei Räume können zusammengefaßt werden, wenn sie sich irgendwann im Zellcyclus mischen"), dann ist nichts dagegen einzuwenden.

Weissenfels: Ich glaube, Herr SCHNEPF hat völlig recht, wenn er die *Möglichkeit* einer Vermischung bestimmter Kompartimente in den Vordergrund stellt. Wir haben viele Beispiele dafür, daß solche Möglichkeiten nicht ständig realisiert sind (z. B. Verbindung perinucleare Zisterne — ER; Verbindung ER — Plasmalemma). Im Hinblick darauf mußte eine Gliederung der Zelle in etwa 10 verschiedene Phasen etwas künstlich erscheinen.

Beermann: Es ist aber doch nicht so, daß die Kernmembran vorübergehend existiert, sondern vielmehr so, daß sie vorübergehend verschwindet (nämlich dann, wenn die Chromosomen funktionslos sind). Physiologisch gesehen ist also die Trennung zwischen Cytoplasma und Kern ganz eindeutig da.

Wohlfarth-Bottermann: Herr SCHNEPF, Sie unterscheiden also zwischen einer protoplasmatischen Mischphase und einer wäßrigen Mischphase. Im strengen Sinne ist das Grundplasma aber sicher auch eine wäßrige Mischphase.

Schnepf: Aber von einer grundsätzlich anderen Qualität als in den Zisternen des endoplasmatischen Reticulum, denn es ist nicht mit ihnen mischbar.

Karlson: Können Sie irgendwelche Leitsubstanzen angeben für die „plasma-tischen" und „wäßrigen" Räume?

Schnepf: Stärkeähnliche Polysaccharide wie Glykogen und Stärke werden bei Tieren und bei Pflanzen in der protoplasmatischen Phase gebildet; auch Viren werden nur hier vermehrt.

P. Sitte: Mir scheint vor allem bedeutsam, daß die Hypothese von Herrn SCHNEPF viele Dinge so einfach erklärt, daß man fast sagen möchte: es fällt einem wie Schuppen von den Augen. Daß z. B. die Plastiden und die Mitochondrien sich auszeichnen durch eigene DNA, durch die Fähigkeit zur Autoreduplikation und zugleich durch eine doppelte Membranumgrenzung — und daß sich das alles nun in einem kausalen Zusammenhang sehen läßt —, das finde ich wirklich sehr beeindruckend.

A year ago I discussed this with Dr. VOGELL at Marburg. He says, one should have more biochemical evidence in favour (or against) Dr. SCHNEPFs hypothesis. I think, that such evidence really exists and is not confined to the existance of DNA in plastids and in mitochondria, for we have in chloroplasts, in mitochondria and in bacteria some typical cytochromes and some typical quinones — plastoquinone and ubiquinone —, and we have contractile proteins.

Klima: There seems to be a DNA in the cytoplasm connected with the basalbody of the flagellum, so that in this case a DNA would be in the cytoplasmic matrix without an envelope and without endosymbiosis.

Schnepf: Quite right! This confirms me that I can speak of a nucleocytoplasmic matrix, because there is a DNA in the centriole, too.

Klima: But then it would not be necessary to suppose, that the mitochondria and the chloroplasts would be endosymbionts. They could catch their DNA by invagination.

Schnepf: Yes, that is the opinion of ROBERTSON. He thinks, that the mitochondria originate from a cell-owned protuberance, which is phagocytosed. But it seems most unlikely to me.

P. Sitte: Wenn man sich auf Grund der Hypothese von Herrn SCHNEPF die Entstehung der Eucyte (= Eukaryontenzelle) in der Phylogenese vorzustellen versucht, kommt man zunächst wohl zur Annahme, daß etwa heterotrophe Protocyten (= Zellen protokaryontischer Organismen) andere, autotrophe Protocyten in ihren Zellkörpern aufgenommen und darin weiterhin behalten haben. Es überrascht dann eigentlich nicht, wenn auch im Grundplasma rezenter Eucyten kontraktiles Protein und gelegentlich auch DNA gefunden wird.

Weissenfels: Bei den Bakterien ist das Kernäquivalent nie von einer eigenen Membran umhüllt; das scheint mir in unserem Zusammenhang sehr wichtig.

Klima: Dann muß freilich auch auf jene Fälle hingewiesen werden, wo es nie zu einer Vermischung von Karyo- und Cytoplasma kommt, auch nicht während der Mitose (z. B. bei bestimmten Rhizopoden).

Beermann: I cannot agree that, apart from putative symbionts like plastids and mitochondria, there should be only a little difference between protocaryonts and eucaryonts. I would say: Three cheers for the "little" difference! In fact, the genetic system of eucaryonts seems to differ from that of the protocaryonts in so many essential points that it is hard to see how one could have evolved from the other: Just think of the numerous mechanisms to ensure and control genetic recombination in higher organisms, e. g. mitosis and meiosis and, indeed, the structure of the chromosomes themselves, or think of the ways in which genetic activity may be controlled, think of the nucleolus and last, but not least, of the nuclear membrane.

Schnepf: Das ist sicher eine Frage von großer Wichtigkeit. In diesem Zusammenhang wäre es wertvoll zu wissen, ob in Mitochondrien und Plastiden ein DNA-Histon-Komplex vorliegt, oder ob die Verhältnisse jenen bei Bakterien entsprechen, wo das nicht der Fall ist.

Stoeckenius: Dr. SCHNEPF, I wonder if you would not have to assume, that the nucleus also has been a symbiont originally. It basically has the same structure as the plastids: it is bounded by a double membrane.

Schnepf: There are no unit membranes in the pores of the nuclear envelope.

Stoeckenius: But how has it been formed?

Schnepf: You may have a bacterium, which has a plasmalemm. The plasmalemm folds in, and probably these membranes are also in bacteria different from the normal plasmalemm (they have in some bacteria the function of mitochondria). This is a differentiation into two parts of the membrane — why should this differentiated part of the plasmalemm not differentiate further and build up such a membrane, as it is found around the nucleus in cells of higher organisms?

Stoeckenius: But I don't see, why you could not form your mitochondria and plastids in exactly the same way.

Schnepf: It is a fundamental difference: here pores are not closed by unit membranes. But in the case of mitochondria there are always two closed unit membranes.

P. Sitte: Vielleicht ist eine Klarstellung am Platze: Kein Verfechter von Herrn SCHNEPFs Ansichten — am wenigsten wohl Herr SCHNEPF selbst — wird sich zur Auffassung versteigen, Mitochondrien oder Plastiden seien im Grunde nichts anderes als Bakterien. *Wenn* sie es je — vor Milliarden Jahren — wirklich waren (was naturgemäß völlig im Dunkeln liegt) — heute sind sie es keinesfalls mehr. Insoweit ist eher das verwunderlich, daß man das Kompartimentierungs-Schema von Endosymbionten auf die Plastiden und die Mitochondrien übertragen kann. Ich denke, es ist Herrn SCHNEPFs Verdienst, vor allem darauf hingewiesen zu haben. — Im übrigen stimmen wir wohl alle Herrn BEERMANN zu: Eine Eucyte ist sicher mehr, als ein compositum mixtum von Protocyten, und es muß entscheidende Evolutionsschritte gegeben haben, auf welche die Konzeption von Herrn SCHNEPF keinen Bezug hat.

Grundmann: Ohne Zweifel ist die Einteilung, die uns Herr SCHNEPF gegeben hat, sehr wertvoll. Die Aufgabe der Zukunft wird es jetzt sein, den Inhalt der einzelnen Kompartimente nicht nur biochemisch, sondern (was vielleicht schwieriger ist) auch morphologisch auszufüllen. Ich glaube, dieser Zustand ist geradezu typisch für den derzeitigen Status der cytologischen Forschung. Ihre Konzeption ist dabei eine gute Grundlage.

Wohlfarth-Bottermann: Und auch ein Stimulans!

Schnepf: Ich glaube, daß dabei die Morphologen noch mehr als bisher auf die Hilfe der Biochemiker angewiesen sind.

Karlson: Nur müssen wir uns vor allzu großen Vereinfachungen hüten. Ich möchte z. B. daran erinnern, daß die verschiedenen „protoplasmatischen" Phasen in Herrn SCHNEPFs Schema für ganz verschiedene Funktionen spezialisiert sind (Grundplasma — Mitochondrienplasma, usw.).

Schnepf: Ich stimme da ganz mit Ihnen überein; vielleicht ist das aber in meinem Referat nicht klar genug herausgekommen.

Grundprinzipien der Reduplikation komplexer Systeme

Von

Jörg Klima, Innsbruck

Mit 8 Abbildungen

Im folgenden soll versucht werden, von einem ganz generellen und formalen Standpunkt aus die allgemeinen Gesetzmäßigkeiten der Verdoppelung komplexer Systeme zu formulieren. Diese Musterkarte möglicher Ereignisse soll ein Hilfsmittel werden zur Erforschung des wirklichen Geschehens, wie es in einem bestimmten Fall vorliegt. Bei einer solchen Aufgabenstellung, die im Prinzip viel Ähnlichkeit hat mit der Auffassung der generellen Morphologie von Zwicky [24], ist es selbstverständlich, daß man auf Informationstheorie, Kybernetik, Systemtheorie und symmetrologische Betrachtungen zurückgreifen muß. Bewußt soll in diesem Beitrag der Gehalt an mathematischen Formeln niedrig gehalten werden — auch werden diese eingehend erläutert —, um den Zugang zu den Überlegungen möglichst zu erleichtern.

Definition der Begriffe: Verdoppelung, System, komlexes System

Zu Beginn sollen die Begriffe Verdoppelung, System und komplexes System definiert werden. Vorweg ist festzustellen, daß jede Konfiguration raumzeitlicher Dinge und Ereignisse als *Nachricht* abbildbar ist[1]. Jede Nachricht kann ihrerseits in eine Binärzahl verwandelt werden. Dies soll nun so geschehen, daß sie in einem Code ohne jede Redundanz[2] formuliert ist, ohne jedoch die dem abgebildeten System eigentümliche Redundanz zu verändern. So kann z. B. die Aminosäuresequenz Ala — Phe — Ala, die durch die Basentriplets GCA, UUU und GCU codiert ist, durch folgende Binärzahl wiedergegeben werden 100100111111100111.

Von einer *Verdoppelung* wollen wir dann sprechen[3], wenn folgende Bedingung gilt: In einem bestimmten abgeschlossenen Raum sei zur Zeit t_0 *ein* Gebilde einer bestimmten Art vorhanden, im Zeitpunkt t_1 seien dagegen zwei oder mehrere dieser Gebilde vorhanden. Da der Raum

---

[1] Einführung in die Grundbegriffe bei Zemanek [23].

[2] Als Redundanz wird jener Betrag der Information bezeichnet, um den eine Nachricht verstümmelt werden kann, ohne an effektiver Information zu verlieren.

[3] Der deutsche Ausdruck Verdoppelung, wie auch der englische *reduplication* ist in diesem Zusammenhang unglücklich gewählt; man sollte besser von identischer Vermehrung sprechen.

abgeschlossen ist, können die überzähligen Gebilde nicht durch Zuzug in den Raum gelangt sein.

Wären alle Gebilde miteinander völlig identisch, müßten sie durch dieselbe Binärzahl darstellbar sein. In Wirklichkeit aber kommt eine solche völlig identische Vermehrung bei Organismen nicht vor, denn zwischen den einzelnen Gebilden finden sich geringfügige Abweichungen. Es sei daran erinnert, daß es bei genetisch reinen Linien zu Variationen kommen kann. Auch ist der genetische Code degeneriert, es können daher Basenänderungen auftreten, die sich nicht auf die Aminosäuresequenz auswirken. Da wir alle Gebilde nach einem bestimmten System binär codieren, bedeutet das, daß einzelne Stellen unterschiedlich mit Null und Eins besetzt sein werden. Bildet man nun die Differenz der Zahlen, so bleibt ein Restbetrag stehen, dessen Größe stark schwanken kann, da die Größe des Differenzbetrages abhängig ist vom Stellenwert der Stelle, die unterschiedlich besetzt ist. In unserem Zusammenhang ist aber dieser Stellenwert belanglos, da es ja nur auf die Anzahl der geänderten Stellen ankommt. Solange diese Anzahl genügend klein ist, können wir das System als *identisch verdoppelt* betrachten.

Die folgenden Formeln 1—4 geben in mathematisch strenger Schreibweise wieder, was bisher erörtert wurde. Auf das Problem der identischen Vermehrung wird später noch einmal zurückzukommen sein. Es wird dann eine funktionelle Definition gegeben werden. Dadurch muß die hier vorliegende Formulierung geändert und eine Bewertung der einzelnen Stellen eingeführt werden. Im jetzigen Zusammenhang soll das aber nur erwähnt und zunächst nicht erläutert werden.

$$N_i - N_j = 0; \quad i \neq j \quad \text{ideale identische Vermehrung} \qquad [1]$$

$$1 \leq N_i - N_j \leq N_i - 1 \qquad \qquad \text{tatsächlich} \quad [2]$$

$$N_i = \alpha_{i\,1}\, 2^{n-1} + \alpha_{i\,2}\, 2^{n-2} \ldots \alpha_{i\,(n-1)}\, 2^1 + \alpha_{i\,n}\, 2^0 \qquad \text{vor-} \quad [3]$$

$$1 \leq \sum_{\substack{1 \\ k}}^{n} (\alpha_{ik} - \alpha_{jk})^2 \ll \mathrm{ld}\, N; \quad \alpha \text{ kann nur 0 und 1 betragen} \qquad \text{kommend.} \quad [4]$$

N = Nachricht,

α_{ik} = Argument der Nachricht i an der Stelle k,

ld = Logarithmus dualis.

HALL und FRAGEN [6] definierten 1956 ein *System* als eine Anzahl von Objekten zusammen mit den Beziehungen zwischen diesen Objekten und zwischen ihren Attributen. Diese sehr allgemein und weit gehaltene Definition ergänzen die Autoren noch kurz, indem sie erläutern, was sie unter Objekten, Beziehungen und Attributen verstehen. *Objekte* sind einfach Teile oder Komponenten eines Systems und ihre Auswahl unterliegt keinerlei Beschränkung. Es können Klassen konkreter Gebilde sein wie Nucleonen, Atome, Sterne, Stahlfedern, Gene, Neurone, Muskeln, Organismen, oder abstrakte Objekte wie mathematische Variable, Gleichungen, Regeln, Gesetze etc. Die *Attribute* der Objekte sind ihre Eigenschaften, oder besser gesagt, eine Auswahl ihrer Eigenschaften. Die Beziehungen, von denen in der Definition gesprochen wird, sind jene Beziehungen, die das System zusammenhalten, doch soll die Auswahl der Beziehungen dem Bearbeiter völlig frei überlassen werden und es wird keinerlei Versuch gemacht, zwischen trivialen und nützlichen Beziehungen zu unterscheiden. Auf diese sehr weite und dadurch vage gehaltene Definition wurde bewußt zurückgegriffen, damit nicht schon durch die Definition eines grundlegenden Begriffes bestimmte Positionen bezogen sind.

Als *komplexes System* soll ein System bezeichnet werden, das aus mehreren Teilsystemen aufgebaut ist. Soll der kybernetische Gesichts-

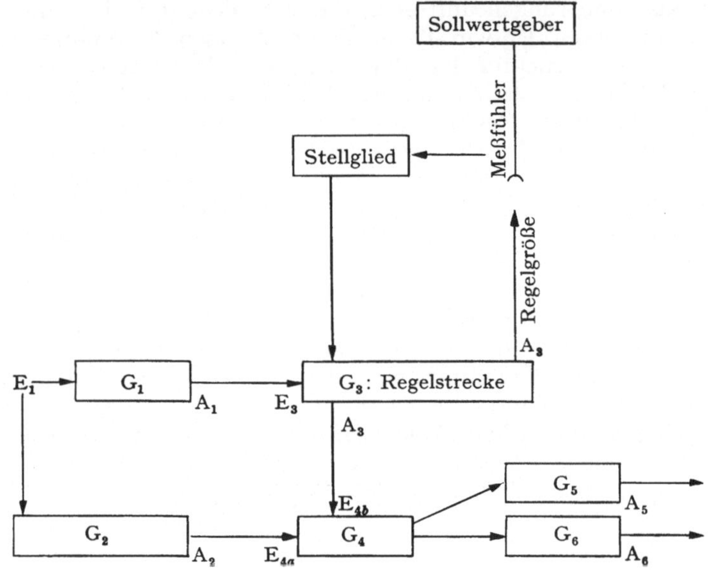

Abb. 1. Schema eines komplexen Systemes, das einen Regelkreis
und eine Vermaschung enthält

punkt stärker im Vordergrund des Interesses liegen, so scheint es nützlich, die zusätzliche Randbedingung einzuführen, daß als komplexes System nur ein solches betrachtet wird, das mindestens einen Regelkreis und eine Vermaschung[1] enthält (Abb. 1). Steht mehr die symmetrologische Betrachtung im Vordergrund, können wir zusätzlich fordern, daß verschiedene Symmetrien in räumlich typisierbarer Lagerung im System vorliegen.

Informationstheoretische Implikationen einer idealen Verdoppelung

Bevor noch auf die Mechanismen der Verdoppelung eingegangen werden kann, sollen kurz die informationstheoretischen Bedingungen für

---

[1] Der Begriff der Vermaschung ist in der Kybernetik noch nicht eindeutig definiert. Im Sprachgebrauch steht er für die Tatsache, daß ein Glied von mindestens zwei anderen Gliedern der Maschine Eingänge erhält und an zwei weitere abgibt.

eine ideale Verdoppelung angegeben werden, da sie auch für die quasi-ideale Verdoppelung von Interesse sind. Jede Informationsübertragung unterliegt der Störmöglichkeit. Nur durch zusätzlichen Aufwand an Information kann die Störung auf jedes beliebig kleine Maß herabgedrückt werden. Überschreitet nun der Informationsgehalt einer Nachricht nicht den Störpegel des übertragenden Systems, kann ideale identische Verdoppelung eintreten. Prinzipiell sind folgende drei Möglichkeiten denkbar:

1. Eine Anordnung, die alte und neue Nachricht miteinander *vergleicht* und bei Abweichungen die neue korrigiert. Für jedes Symbol der alten Nachricht müssen drei Vergleiche mit mindestens je 1 bit Information[1] gezogen werden. A: im Takt, außer Takt; B: gleich oder ungleich; C: welche ist die alte Nachricht? Daß diese drei Vergleiche tatsächlich notwendig sind, kann man aus folgender Überlegung ersehen. Der Vergleich zweier Symbole der beiden Nachrichten auf Gleichheit oder Ungleichheit kann erst dann sinnvoll gezogen werden, wenn feststeht, daß dieselben Stellenwerte der Nachricht verglichen werden (Vergleich „im Takt" oder „außer Takt"). Tritt trotz des Im-Takt-Sein eine Ungleichheit der Symbole ein, muß zur Korrektur feststehen, auf welcher Seite die alte Nachricht steht, damit man die neue ihr angleicht und nicht etwa die alte zerstört. Ist für den Abgleich seinerseits ein bestimmter Störpegel gegeben, müssen die Vergleiche entsprechend oft (oder durch entsprechend viele Kanäle) vollzogen werden, und die Mehrheit entscheidet. Der Aufwand an zusätzlicher Information ist sehr hoch. Es ist zweckmäßig, A als erstes, B als zweites und C als Eventualglied zu schalten. A soll mit geringster primärer Störung ausgelegt werden. Darüber hinaus scheint es unrealistisch, anzunehmen, daß die Störung von A unabhängig von der Nachrichtengröße sei. Es ist eher anzunehmen, daß die Kontrolle des Im-Takt-Sein mit zunehmender Nachrichtengröße störungsanfälliger wird. Für B muß man annehmen, daß der Störpegel für den Vergleich (gleich oder ungleich) nicht unabhängig von der Art der verwendeten Symbole ist. Je mehr Symbole pro Stelle zu unterscheiden sind, desto störanfälliger wird der Vergleich werden, bis schließlich eine Entscheidung „gleich" oder „ungleich" überhaupt nicht mehr gelingt.

2. Der Entwurf eines *selbstkorrigierenden Codes*. Man kann Codierungen formulieren, die es erlauben, eine Störung pro Codewort nicht nur zu erkennen, sondern auch zu eliminieren. Der Aufwand pro bit Nachricht ist groß, außerdem muß die primäre Störung so gering sein, daß sich Doppelfehler praktisch nicht ereignen.

3. Statistisch oder homogen verteilte *Redundanz der Nachricht*. Diese drückt die Störanfälligkeit vom Ausgangswert entsprechend der Redundanz herab. Wirtschaftlichste Version bei an und für sich geringer Störanfälligkeit und großem Nachrichtenumfang [*3*]. Dieser Mechanismus ist im *genetischen Code* verwirklicht.

---

[1] bit ist die Maßeinheit der Information und bedeutet eine ja-nein-Entscheidung; z. B. können durch die Zahlen 0 bis 31 5 bit Information gespeichert werden: $32 = 2^5$.

Praktische Anwendung

Schon die reine Informationstheorie — und damit die grundlegenden Voraussetzungen jeder Verdoppelung — haben ihre Konsequenzen für die Biologie. Dazu einige Beispiele.

Man hat versucht, den *Nachrichtengehalt der Zelle* abzuschätzen. Dabei sind zum Teil sehr unterschiedliche Werte angegeben worden, je nachdem, was man als bedeutungsvolle Variationsmöglichkeiten ansieht. Ein relativ niederer Wert wird ermittelt, wenn man nur die funktionell verschiedenen Genvariationen zählt, ein höherer, wenn man alle Nucleotidbasen der genetischen Substanz berücksichtigt, und ein noch höherer, wenn man auch dem Cytoplasma (und zwar der Anordnung der Zellorganellen im Zellplasma) einen Informationsbetrag zuschreibt. Der Nachrichtengehalt einzelner Zellen liegt nach einer Abschätzung von Quastler [*13*] zwischen 10^5 und 10^{12} bits. Das rechte Gewicht erhält diese Zahl erst, wenn man die Zahl der „möglichen Organismen" $(2^{100\,000} — 2^{1\,000\,000\,000\,000})$, nämlich die Zahl der durch diesen Informationsbetrag möglichen verschiedenen Nachrichten in Beziehung setzt mit der Anzahl der Elementarteilchen im Kosmos, nämlich 2^{240}. Auch bei einer denkbar dichten Besiedlung des Kosmos dürfte die Gesamtzahl aller lebenden und gelebt habenden Organismen 2^{100} nicht übersteigen. Die *Redundanz der genetischen Information* der Zelle kann jetzt annähernd abgeschätzt werden und stammt aus zwei Quellen.

1. Degeneration des Nucleinsäurecodes. Verwendet man die neuesten Daten über die Codierung, so zeigt sich, daß jeweils nur jede dritte Base redundant ist. Die Redundanz für die dritte Base ist aber dann sehr hoch, in manchen Fällen 100%, in den übrigen Fällen 50%. Dies gibt mit Berücksichtigung der sog. *nonsense triplets*, die ebenfalls als redundant angenommen wurden, einen Betrag der Redundanz für den Code von etwa 27%.

2. Die Ersatzmöglichkeit von Aminosäuren in Polypeptidketten ohne Funktionsänderung. Hierbei ist vorerst nur eine grobe Abschätzung möglich: etwa 25%.

Es ist aber denkbar, daß es noch unentdeckte Auswirkungen dieser Änderungen gibt, nämlich im funktionellen Zusammenspiel des gesamten Zellmetabolismus, in der Geschwindigkeit der Neusynthese usw. Auf alle Fälle sind redundante Änderungen für die Evolution von Bedeutung, da sie geänderte Ausgangspositionen schaffen.

Eine genügend lange Nachricht mit m unterschiedlichen Symbolen und einer gleichmäßigen Störungsquelle (d. i. eine solche, die alle Symbole mit gleicher Wahrscheinlichkeit ineinander überführt) muß mit zunehmender Häufigkeit der Nachrichtenübertragung sich immer mehr einer Nachricht mit einer gleichen Häufigkeit von $\frac{1}{m}$ pro Symbol nähern. Der genetische Code besteht aus 4 Symbolen. Als Störmöglichkeit tritt die Mutation auf, die früher oft als gleichmäßige Störungsquelle angesehen wurde. Da aber von Bakterien bekannt ist, daß bei den einzelnen Arten von 1 : 1 stark abweichende Werte des Basen-Verhältnisses vorkommen, kann *keine* spontane Störungsquelle vorliegen, die mit gleicher Häufigkeit die vier Basen ineinander überführt[1]. Die selten vorkommenden Basen müssen mit größerer Wahrscheinlichkeit in die häufigeren umgewandelt werden als umgekehrt [4].

Begriff der funktionell identischen Verdoppelung

Bei der Erstellung der Formel für die reale identische Verdoppelung wurde verlangt, daß die Zahl der Änderungen genügend klein ist. In diesem „genügend klein" steckt natürlich eine gewisse Willkür, denn für die Festlegung dieser Grenze gibt es keine objektiven Normen. Sicher ist, daß *eine* Änderung immer noch toleriert werden muß, wenn wir aus den schon erwähnten Gründen an der Forderung einer idealen identischen Vermehrung nicht festhalten können. Es gibt aber sicher Mutationen in Organismen, die zu weitreichenden funktionellen Änderungen führen, obwohl nur eine einzige Base geändert würde. Wie könnte man nun diese Mutationen, die wir nach allem bisher gesagten als „identisch" verdoppelt betrachten müssen, erfassen? Bei der Formulierung der identischen Verdoppelung wurde nur die redundanzfreie Codierung der Nachricht gefordert, nicht aber die Eliminierung der dem System inhärenten Redundanz. Nunmehr soll auch dies zusätzlich verlangt werden; damit wird aber die identische Verdoppelung rein funktionell bestimmt, da sich dann eine Änderung der Information zumindest unter *einer* Umgebungskonstellation äußern müßte. Als „identisch" würde man dann ansehen, was völlig *gleich funktioniert*, ohne aber das morphologische Substrat zu berücksichtigen — eine für die Biologie unzweckmäßige Definition. Wollte man auf das morphologische Substrat bei gleich funktionierenden Organismen zurückgreifen, müßte man die einzelnen Stellen der nach den Regeln von Seite 395 formulierten Nachricht bewerten und drei verschiedene Kategorien einführen:

a) Stellen, die unter keinen Umständen geändert werden dürfen, da sich sonst eine funktionelle Änderung des Organismus ergibt.

b) Stellen, für die beliebige Änderungen möglich sind, ohne daß sich eine funktionelle Änderung ergibt.

c) Stellen, die reziprok vertauscht werden dürfen, damit sich keine funktionelle Änderung ergibt.

$$\sum_k{}_1^n \left(\alpha_{if_1(k)} - \alpha_{jf_1(k)}\right)^2 = 0 \qquad [5\,\mathrm{a}]$$

$$0 \leqq \sum_k{}_1^n \left(\alpha_{if_2(k)} - \alpha_{jf_2(k)}\right)^2 \geqq \sum_k{}_1^n \beta_{f_2(k)} \qquad [5\,\mathrm{b}]$$

$$\sum_k{}_1^n \left(\alpha_{if_3(k)} - \alpha_{jf_3(k)}\right)^2 = \sum_k{}_1^n \beta_{f_3(k)}; \beta = 1 \; . \qquad [5\,\mathrm{c}]$$

Anstelle der einfachen Zuordnung a_{ik} zu a_{jk} treten Funktionen von k.

Formel 5 formuliert diese Bedingungen in exakter Weise. Die Umständlichkeit des Ausdruckes hat aber zur Folge, daß diese Formulierung

---

[1] Falls im Primärvorgang der Mutation tatsächlich „weißes Rauschen" auftritt, muß durch nachfolgende Einwirkung eine systematische Veränderung (Bevorzugung bestimmter Basen) gesetzt werden.

sehr unhandlich ist und daß es angebrachter erscheint, sich mit der Formulierung von Seite **395** zu begnügen. Es muß aber darauf hingewiesen werden, daß die Mutation nur als funktionelle Abweichung zu fassen ist.

Norbert Wiener [*22*] hat ein prinzipielles Verfahren zur Verdoppelung funktioneller Strukturen angegeben, das in Abb. 2 schematisch dargestellt ist. Die Funktion des weißen Kastens ist innerhalb beliebiger Genauigkeitsgrenzen der des schwarzen Kastens anzugleichen, trotzdem ist der schwarze Kasten dem weißen nur isomorph[1], sonst aber prinzipiell

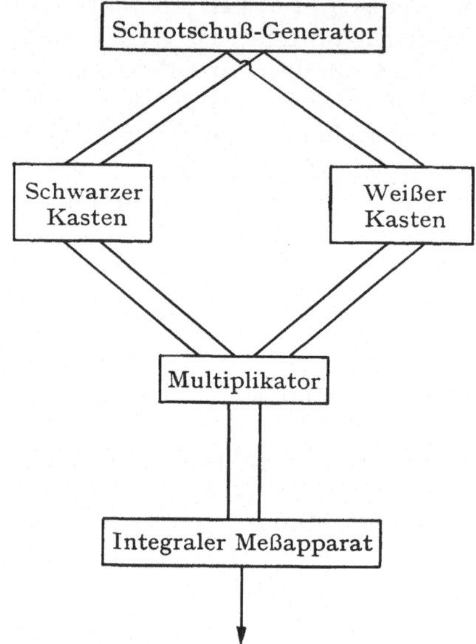

Abb. 2 (nach N. Wiener). Prinzipielle Schaltung zum Abgleich eines weißen Kastens, damit er dieselbe Funktion wie ein beliebiger schwarzer Kasten erfüllt

verschieden gestaltbar. Dieses Modell kann daher auch nicht als Vorbild für eine Verdoppelung von Zellen oder auch nur Zellorganellen angesehen werden, da hierbei nicht nur funktionelle Gleichheit, sondern auch gestaltliche Ähnlichkeit gefordert werden muß.

Mechanismen der Verdoppelung

Nach diesen vorbereitenden Erläuterungen sollen nun im folgenden alle prinzipiell verschiedenen Möglichkeiten der identischen Vermehrung aufgezählt werden. Dabei muß man berücksichtigen, daß eine identische Vermehrung bestimmter Gebilde sich nur in einer bestimmten Umgebung abspielen kann.

---

[1] Isomorphie ist ein Fachausdruck der Mathematik und besagt für Automaten nur, daß zwei Automaten bei denselben Eingangswerten dieselben Ausgangswerte zeigen.

1. Der gesetzmäßige Aufbau identischer Gebilde durch *Wechselwirkungen von Komponenten* einer geeignet zusammengesetzten Umgebung. (Beispiele hierfür sind die Bildung bestimmter Moleküle aus anderen mit oder ohne Beteiligung von Katalysatoren usw.)

2. Der *Zufall*. Er kann erfolgreich wirksam sein in Systemen, die eine geringe Anzahl von Komponenten und eine noch (viel) geringere Anzahl

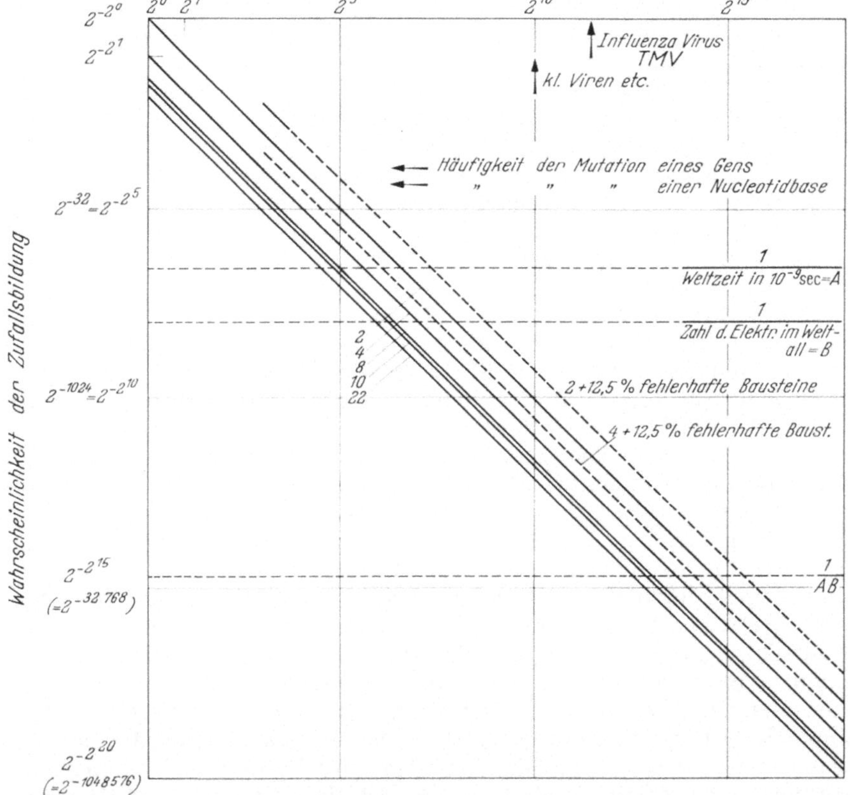

Abb. 3. Auf der Abszisse ist in logarithmischem Maßstab (Zweierpotenzen) die Gesamtzahl aller Komponenten, die eine bestimmte Nachricht aufbauen, eingetragen. Auf der Ordinate ist im doppellogarithmischen Maßstab (2^{-2^n}, $n = 0 - 20$) die Wahrscheinlichkeit, daß eine Nachricht durch eine Zufallskombination entsteht, aufgetragen. Die Kurven sind für 2, 4, 8, 10 und 22 verschiedene Symbole eingezeichnet, für 2 und 4 Symbole sind auch die Kurven für einen Fehlerbereich von 12,5% falscher Bausteine eingezeichnet. Dazu noch einige charakteristische Zahlenwerte

verschiedener Symbole besitzen. Abb. 3 gibt die Zufallswahrscheinlichkeit an, wenn alle Kombinationen die gleiche Bildungswahrscheinlichkeit haben. Für die Kurve mit 2 und 4 Komponenten ist auch eine Korrektur angegeben, wenn man den Einbau falscher Bausteine erlaubt, und zwar bis zu einer Größenordnung von 12,5%. Trotz dieser starken Fehlerdichte ist die Bildungswahrscheinlichkeit nur geringfügig erhöht.

3. *Nachbau durch eine Vorrichtung.* Das ist die uns aus den verschiedenen Fabrikationsprozessen vertrauteste Möglichkeit. Wir müssen in diesem Fall drei Untergruppen unterscheiden, je nachdem wie die Anteile der Steuerung und der Regelung verteilt sind.

a) Das Gebilde wird rein gesteuert aufgebaut: In einem Speicher liegt ein Programm vor, das in eine Vorrichtung zur Erzeugung des Gebildes fließt und von dort wieder zurück in den Speicher. Das ist aber keine Rückkoppelung, sondern nur der Kreisschluß des Steuerprogramms. Das Gebilde selbst zeigt alle zufälligen Störungen, die auf die Vorrichtung eingewirkt haben.

b) Der Nachbau erfolgt geregelt: Statt des Speichers findet sich ein Sollwertgeber, der die zu vervielfältigende Nachricht enthält; ferner ist ein Meßfühler vorhanden, der Sollwert und Regelgröße vergleicht.

c) Die Kombination beider Möglichkeiten: Dabei erfolgt die Rohbearbeitung über ein Steuerprogramm und die feinere Angleichung durch Regelmechanismen.

Bei all diesen Vorgängen des Nachbaues ist wesentlich und wichtig, daß außer dem Steuerprogramm (oder einer Kopie) und den für den Aufbau notwendigen Elementen mindestens ein zusätzliches System vorhanden ist, das als Glied eines Automaten wirkt, Eingang und Ausgang hat, und daß das neu entstehende Gebilde der Ausgang dieses Gliedes ist.

4. Vervielfachung durch *Resonanzphänomene.* In stehenden räumlichen Wellenfeldern können sich Komponenten in den Schwingungsknoten einordnen und so einander entsprechende Strukturen in verschiedenen räumlichen Positionen bilden.

5. *Anlagerung* und *Teilung.* Dies ist ebenfalls eine sehr alte Modellvorstellung für den Nachbau spezifischer biologischer Makromoleküle gewesen, dort ist sie aber nicht verwirklicht. Penrose [12] hat mechanische Modelle entwickeln können, die das geforderte Verhalten zeigen. Er verwendete zwei Sorten von stark asymmetrischen, mit Vorsprüngen und Einbuchtungen geformten Holzklötzchen, die sich in zwei Weisen zusammenlagern können; je zwei komplementäre Einzelteile bilden dann eine ziemlich stabile Einheit. Dabei kann die Bildung der Teilchen, die durch Schütteln aneinandergelagert werden, determiniert werden, indem man einen „Samen" *(Primer)*, nämlich *eine* bestimmte Kombination einführt. Alle anderen Teilchen lagern sich dann entsprechend dem zugefügten Samen aneinander.

Alle bisher besprochenen Mechanismen beruhen darauf, daß die Vervielfältigung des Gebildes in einem Raum abläuft, in dem die Einzelkomponenten schon vorhanden sind, oder in dem das Gebilde durch ein übergeordnetes System geformt wird.

6. *Die Selbstverdoppelung.* R. W. Ashby [1] hat bestritten, daß es eine Selbstverdoppelung gebe, da jede Verdoppelung in einer Umgebung und unter Heranziehung des Materials der Umgebung ablaufen müsse. Dennoch scheint es zweckmäßig, die Vermehrungsweise eines Organismus, bei der man landläufig von Selbstverdoppelung spricht, definitorisch so einzugrenzen, daß man sie von den bisher besprochenen Fällen abgrenzen kann, ohne mit der selbstverständlichen Voraussetzung jeder Ver-

doppelung in einen inneren Widerspruch zu kommen. Von Selbstverdoppelung soll nur dann gesprochen werden, wenn mindestens eine Komponente des Systems *nicht* in der Umgebung vorliegt und *kein* dem System übergeordnetes vorhanden ist, das etwa das System nachbaut. J. VON NEUMANN [*11*] hat gezeigt, daß es logisch möglich ist, einen Automaten zu konstruieren, der diese Fähigkeit hat. Dabei können selbstreproduzierende Systeme nur durch kombinierte Verfahren (1—5) wirken. Es ist sogar ausreichend, nur das Verfahren nach 3 wiederholt zu kombinieren.

Sind in einer bestimmten Umgebung verschiedene sich verdoppelnde Systeme vorhanden, kann und muß Selektion immer dann auftreten, wenn Bildungsgeschwindigkeit und (oder) Zerfallswahrscheinlichkeit der sich verdoppelnden Systeme unterschiedlich sind. Dabei reichern sich jene Systeme an, die die geringste Zerfallsgeschwindigkeit besitzen, wenn alle Systeme in Konkurrenz zumindest um eine Bausteinart stehen. In den Fällen 1, 2, 3b, 3c, 4 und 5 wirkt die Selektion direkt auf die Systeme ein, während im Fall 3a die Selektion nicht direkt einwirken kann. In diesem Fall sind Bildungsgeschwindigkeit und Zerfallswahrscheinlichkeit der aufgebauten Strukturen außer von der Umgebung nur vom Nachbaumechanismus determiniert. Die Selektion kann nur indirekt über Selektion der Nachbaumechanismen vor sich gehen. Im Falle 6 kann äußere Selektion nur auf das Gesamtsystem wirken. Innerhalb der Wechselwirkungen der Teilsysteme kann aber Selektion im System nach den bisher besprochenen Gesichtspunkten wirken.

Die Zelle als „Rechenautomat"

Wenn zu Beginn dieses Artikels dem gesamten komplexen System eine Nachricht zugeordnet wurde, so soll dies nun korrigiert werden. Wir unterscheiden nunmehr zwischen *Programmteil* und *Strukturteil* der Maschine und vergleichen das komplexe System mit einem programmierten Rechenautomaten. Füttern wir nun mit ein und demselben Programm zwei Rechenautomaten, von denen der eine uns in seiner Konstruktion bekannt, der andere aber unbekannt ist, und erhalten wir dasselbe Ergebnis, so dürfen wir dennoch nicht auf die Identität beider Konstruktionen schließen, sondern nur auf deren Isomorphie. Um Aussagen darüber zu machen, ob beide Automaten gleich gebaut sind, müssen wir ihre Konstruktion untersuchen. Rechenautomaten weitgehend unterschiedlicher und daher nicht isomorpher Konstruktion bedürfen zur Lösung ein und derselben Aufgabe verschiedenartig programmierter Programme. Jedes beliebige, überhaupt in einem Rechenautomaten lösbare Programm kann nun einerseits so transformiert werden, daß es zugänglich wird für einen sehr einfachen Rechenautomaten, die sog. Turingmaschine, benannt nach ihrem Entdecker TURING [*21*], einem englischen Mathematiker. Eine Turingmaschine besteht aus sehr einfachen Komponenten: einem Streifen von unbegrenzter, aber nicht unendlicher Länge, der in Quadrate geteilt ist und jeweils um *ein* Feld nach rechts oder links rücken kann; einem Fenster, das jeweils ein

Quadrat freigibt; einer Hilfe, um die aufscheinenden Symbole in dem Fenster niederzuschreiben oder zu löschen; einer Funktionstabelle, welche in eindeutiger Weise angibt, was zu tun ist, wenn man ein bestimmtes Symbol im Fenster in einem bestimmten Stadium erblickt; und dementsprechend auch eine Anzeige dafür, in welchem Stadium sich die Maschine befindet. Mit Hilfe einer solchen Turingmaschine kann jedes algorithmisch lösbare Problem gelöst werden. Es ist bemerkenswert, daß die Wirkungsweise von Enzymmolekülen durch eine Turingmaschine simuliert werden kann [16]. Der Aufwand an Zeit bei der Benutzung einer solchen Turingmaschine kann gegen unendlich hin wachsen, aber doch niemals unendlich werden. Umgekehrt konnte STEINBUCH [17] zeigen, daß jedes Rechenproblem in beliebig kurzer Zeit gelöst werden kann; dafür steigt dann freilich der Aufwand an Schaltelementen, Informationsträgern usw. Es erscheint daher im Prinzip durchaus erfolgversprechend, auch die uns vorliegenden, sich selbst verdoppelnden Systeme der Organismen als „Rechenautomaten" besonderer Art zu betrachten. Vergleicht man ihre Größe, ihren Energiebedarf und die Kürze der Programmdauer, dann muß man sagen, daß sie wahrscheinlich *sehr günstig konstruiert* sind und daß es daher sehr langwieriger Forschung bedarf, ihre logische Struktur aufzudecken.

Wenn auch Lebewesen Automaten sind, deren Bauteile und Funktionsgrößen durchwegs Moleküle sind, so darf man sie doch mit anderen Rechenautomaten, die aus elektronischen Elementen bestehen, vergleichen. Wir müssen dazu einiges über die Konstruktionselemente und Schaltungen von Rechenautomaten erörtern. Einen Bauteil von Rechenmaschinen nennen wir Glied; es zeigt mindestens einen Eingang und einen Ausgang. Falls mehrere Eingänge vorhanden sind, können in jedem dieser Eingänge verschiedene Werte aufscheinen, in den Ausgängen aber jeweils nur derselbe Wert. Die Wirkungsweise eines Gliedes wird dadurch gekennzeichnet, daß jedem möglichen Eingang oder jeder möglichen Eingangskombination ein Ausgangswert (deterministischer Automat) oder eine Wahrscheinlichkeitsverteilung (probabilistischer Automat) der Ausgangswerte zugeordnet werden kann. Allein auf Grund dieser Definition, also aus logischen Gründen, läßt sich über die molekulare Maschine des Organismus folgende Aussage machen:

Mit Ausnahme jener Moleküle, die als Eingang von der Umgebung her aufscheinen, muß jeder Eingang auch Ausgang in der Organisation sein. Jeder Ausgang — mit Ausnahme jener Moleküle, die als Ausgang des gesamten Organismus aufscheinen — muß entweder als Eingang oder als Glied der Maschine wieder auftreten. So ist z. B. die β-Galactosidpermease Ausgang eines Polysoms und Glied für das von außen als Eingang auftretende β-Galactosid. Jedes Glied war einmal Ausgang eines Gliedes in der Maschine und wird, außer im Fall, daß kein *turn-over* besteht, wieder Eingang eines Gliedes. Die jeweils vorhandenen Eingänge, Glieder und Ausgänge sind nicht zeitinvariant, sondern sind Abhängige von der Eigenzeit der Maschine, die sich während ihrer Entwicklung von Teilung zu Teilung rhythmisch verändert. Die Zahl der Glieder wächst mit der Eigenzeit der Maschine an. Nur in erster Annäherung darf bei schnellen Prozessen ein kleiner Zeitabschnitt als im Fließgleichgewicht befindlich und zeitinvariant betrachtet werden. Während bei den meisten technischen Rechenautomaten *ein* Programmstreifen für die gesamte Maschine vorgesehen ist, gilt für die Eucyte sicher, daß *mehrere* Programme vorliegen. Das Hauptprogramm liegt im Zellkern und ist in den einzelnen Chromosomen verankert, also auch im Zellkern in einzelne Abschnitte gegliedert. Daneben gibt es noch weitere Programme durch die Desoxyribonucleinsäure der Plastiden und Mitochondrien. Diese Programme liegen entsprechend der Anzahl der Zellorganellen vielfach vor. Daß die Teilungsfähigkeit der Zelle mit ihrer Funktionsfähigkeit verknüpft ist, ist trivial. Dagegen wäre es nicht trivial, sollte es sich erweisen, daß die Teilungsfähigkeit der sich selbst vermehrenden Zellorganellen — wie Chloroplasten und Mitochondrien — ebenfalls an ihre Funktionsfähigkeit (an die Fähigkeit, ATP bereitzustellen) gebunden ist. Daß nichtgrüne Chloroplasten vermehrungsfähig sind, mag darauf beruhen, daß sie Stärke einerseits speichern, anderseits unter Glykolyse ATP gewinnen. Auch eine nur teilweise Bevorzugung der funktionsfähigen Organellen bei ihrem Wachstum und ihrer Teilung würde dazu führen, daß funktionsuntüchtige Organellen rasch durch intracelluläre Selektion vermindert würden. In diesem Prozeß muß eine *positive* Rückkoppelung auftreten, die jedoch wie jede positive Rückkoppelung gehemmt werden muß, damit es zu keinen Entgleisungen des Stoffwechsels kommt. Der Eintritt aus der Programmierungsphase in die Kopierphase des Programms kann prinzipiell entweder als bedingter Befehl programmiert sein oder als Ende des Programms, das mit einer bestimmten Geschwindigkeit durch den Rechenautomaten läuft. Welche Alternative verwirklicht ist, ist prinzipiell prüfbar und hat Konsequenzen für den gesamten Programmablauf; bei einem bedingten Befehl muß eine Randbedingung erfüllt sein. Zum Beispiel könnte für die Zelle erforderlich sein, daß die gesamte Proteinmenge einen kritischen Wert erreicht. Bei Koppelung an das Programmende muß jede Beschleunigung des Programmablaufes unabhängig von dessen Leistungen zu einem früheren Eintritt in die Kopierphase führen. Diese Einschaltung der Kopierphase des Gesamtprogrammes ist unerläßlich für jeden selbstreproduzierenden Automaten [*2, 9, 10, 11*]. Die unterschiedliche Dauer der G_1-Phase in den Zellen kann durch mehrfaches Durchlaufen einer Programmschleife programmiert werden, dies bedingt dann aber, daß ein Zählwerk vorhanden sein muß, das die Zahl der Cyclen begrenzt.

Bei dem Bemühen, die Zellfunktion modellhaft in der Funktion eines Rechenautomaten darzustellen, sind einige Teilerfolge erzielt worden. Einzelne Stoffwechselfunktionen der Zelle konnten in ihrem quantitativen Verlauf für analoge und digitale Rechenmaschinen programmiert werden [*5, 7*]. SUGITA [*18, 19, 20*] versuchte in mehreren Arbeiten, die Schaltung des Zellautomaten mit Hilfe einer digitalen Maschine, die Zeitglieder enthält, zu simulieren und konnte einige Stoffwechselphänomene nachbilden. Da Zellen molekulare Maschinen sind, ist der Ansatz, ihre Schaltung durch digitale Maschinen darzustellen, durchaus berechtigt, da die einzelnen Glieder der Zellen nur diskrete Werte annehmen können. Es scheint aber wahrscheinlich, daß zahlreichen Enzymmolekülen mehr als zwei diskrete Werte zuzuschreiben sind und daß, wie bei Neuronen, drei Zustände besonders bevorzugt auftreten, nämlich aktiv, aktivierbar und gehemmt. Eine Übertragung der Erfahrungen an Nervennetzen auf die molekulare Maschine der einzelnen Zelle erscheint daher vielversprechend, wurde aber bisher meines Wissens noch nicht unternommen.

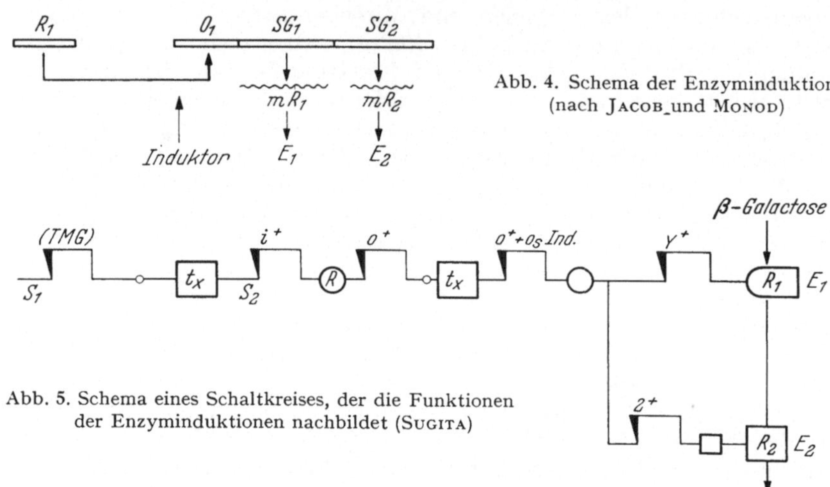

Abb. 4. Schema der Enzyminduktion (nach Jacob und Monod)

Abb. 5. Schema eines Schaltkreises, der die Funktionen der Enzyminduktionen nachbildet (Sugita)

Ich will nun anhand der Arbeit von Sugita [19] kurz einige Überlegungen darüber, wie genphysiologische Muster in Schaltkreismodelle umgesetzt werden, skizzieren. Abb. 4 zeigt das bekannte Schema von Jacob und Monod, Abb. 5 die entsprechende Übersetzung in eine Relaisschaltung mit Zeitverzögerung. Aus diesem Schema kann man die von Monod und Jacob aufgefundenen genetischen Variationen prinzipiell als Schalterdefekte ableiten. Auch sieht man, daß man zur Berücksichtigung von möglichen Zeitgliedern gezwungen wird. Dennoch lassen sich bei diesem einfachen Schaltplan alle Konsequenzen auch bei zeitlicher Verzögerung an ihrem Schema ablesen. Schwieriger ist dies schon bei einem Mechanismus, der zu Schwingungen führt. Abb. 6 zeigt ein Schema der genphysiologischen Koppelung, Abb. 7 dazu den logischen Fluß (a) und die Verwirklichung im Schaltschema (b). Auf Grund des genphysiologischen Schemas kann man noch die Schwingung fordern, doch lassen sich daraus keine zwingenden Angaben über Periodenlänge und Phasenverschiebung zwischen den Produkten P_1 und P_2 angeben. Mit Hilfe des Schaltkreises (8b) kann man beweisen, daß sich folgender Rhythmus einstellt, wenn in beiden Gliedern dieselbe Zeitverzögerung gegeben ist.

t	t_0	t_1	t_2	t_3	t_4	t_5	t_6	t_7	t_8	t_9
P_1	0	0	1	1	0	0	1	1	0	0
P_2	1	0	0	1	1	0	0	1	1	0

In einem Rechenautomaten ist nicht nur im Programm Information
vorhanden, sondern es steckt auch ein Informationsbetrag in der Ver-
drahtung der einzelnen Bauelemente. Auch in den höheren Zellen nehmen
die einzelnen Glieder der Zelle — dabei muß man an die einzelnen Moleküle denken — nicht eine statistisch beliebige Lage ein, sondern bauen
einen geordneten Körper auf. Es erhebt sich nun die Frage, ob die gesamte
Information, welche in den DNA-Molekülen der Zelle enthalten ist und

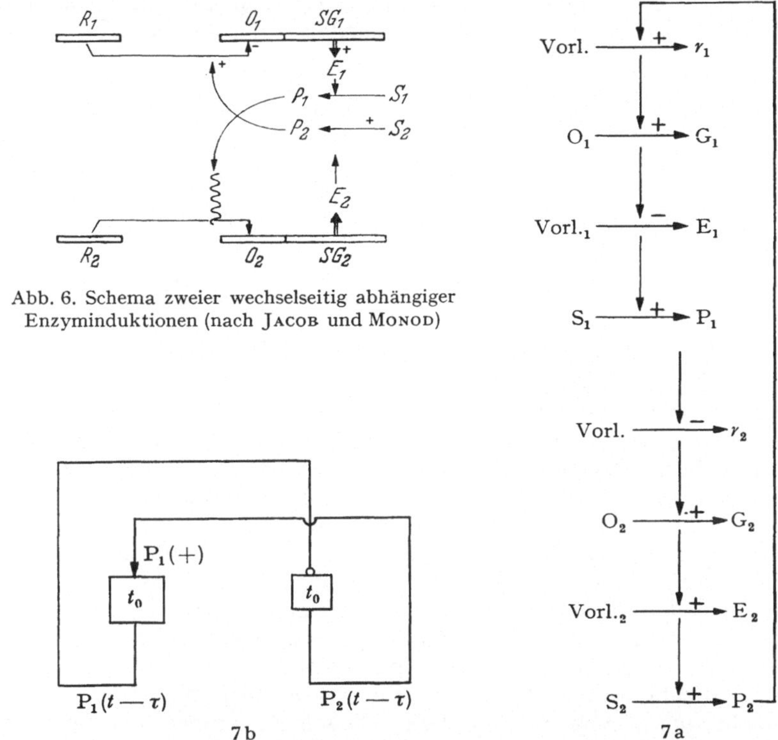

Abb. 6. Schema zweier wechselseitig abhängiger
Enzyminduktionen (nach JACOB und MONOD)

7 b

7 a

Abb. 7 a und b. a Schema des Ablaufes der Steuerung und Regelung von Abb. 6.
b Schaltschema einer digitalen Maschine für Abb. 6 (nach SUGITA)

direkt oder indirekt auf den Bau der Moleküle übertragen wird — direkt
im Falle gewisser Ribonucleinsäuren, indirekt über die Ribonucleinsäuren auf die Eiweißmoleküle oder über das Zusammenwirken verschiedener Enzyme auf niedermolekulare Bestandteile — die gesamte Information darstellt, die eine Zelle enthält, oder ob es darüber hinaus einen
Informationsbetrag gibt, der im Zusammenbau der Bestandteile der Zelle
verankert ist. Die räumliche Konfiguration einer Proteinkette scheint
allein durch die Primärstruktur und die gegebenen physiko-chemischen
Bedingungen der Umgebung bestimmt. Eine Faltungsisomerie ein und
derselben Proteinkette unter physiko-chemisch gleichen Bedingungen
scheint heute äußerst unwahrscheinlich. Es kommt zu gesetzmäßigen

Zusammenlagerungen solcher Einzelproteinmoleküle zu Bausteinen höherer Ordnung, nicht nur in der Zelle, sondern auch *in vitro* unter bestimmten physiko-chemischen Bedingungen [8].

Symmetrie der Zelle

Es erscheint nach dem oben Gesagten daher durchaus erwägenswert, die *räumliche* Struktur der Zelle auf Grund gewisser Symmetrieprinzipien und gewisser Minima- oder Maximaforderungen abzuleiten. Mit Bruno Sander [15] möchte ich Symmetrie definieren als das Vorhandensein

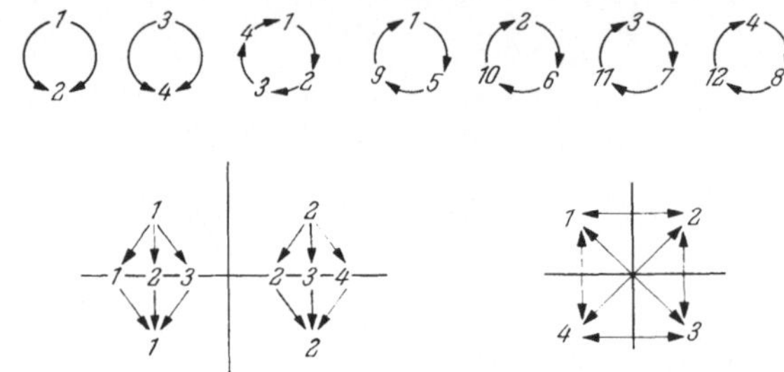

Abb. 8. Beispiele für erlaubte und verbotene (durchkreuzte) Transformationen

von irgendwelchen Gleichheiten in solchen Raumlagen, welche durch die Symmetrieoperationen ineinander übergehen. Die Gleichheiten sind dabei von jeglicher Art und können auch nur statistisch einander entsprechen, so daß man sie als aus derselben Gesamtheit entnommene Probe betrachten darf. Als Symmetrieoperationen sind zu bezeichnen: Spiegelung, Drehung, Translation und — gegenüber der Mineralogie neu, aber für das biologische Gebiet zweckentsprechend eingeführt — die *Transformation*, nämlich die eineindeutige Umwandlung eines Bauelementes in ein anderes; diese Transformationen müssen eine Gruppe bilden, die man auf einem Kreis anordnen kann, so daß sie in sich rückführbar sind, und ihre Anzahl muß endlich sein. Für mathematisch nicht Vorgebildete möchte ich in Abb. 8 einige Beispiele solcher Transformationen geben, die gestattet sind, und einige verbotene (durchgekreuzt).

Auf Grund welcher Bedingungen kann Symmetrie ohne zusätzliche Information entstehen?

1. Als Packungssymmetrie durch die Eigenschaften der Bausteine.

2. Als Abbildungssymmetrie symmetrischer Gebilde oder symmetrischer Kräftefelder.

3. Als räumliche Abbildungen zeitlicher Rhythmen.

Entgegen früheren Anschauungen über die kolloide Natur des Protoplasmas sind verschiedene Symmetriekonfigurationen in den lebenden Zellen aufgefunden worden. Am bemerkenswertesten ist der symmetrische Bau der DNS, deren Symmetrieelement eine zehnzählige Drehspiegelebene ist, die zusätzlich noch transformierenden Charakter trägt. Im ungepaarten Zustand ist das DNS-Molekül eine aperiodische lineare Struktur.

Aperiodische flächige Strukturen wurden bisher in der Zelle noch nicht eindeutig nachgewiesen. Die Eiweißmoleküle kann man als aperiodische räumliche Konfigurationen ansehen; ferner finden sich gleichendige und ungleichendige Drehachsen zweizähliger, drei-, vier-, sechs- und, in der Mineralogie nicht bekannt, sieben [14] und neunzähliger Symmetrie. Außerdem kommt häufig bei manchen Virusgruppen eine fünfzählige Symmetrie vor, und zwar eine Gruppierung von sechs fünfzähligen Drehachsen, die einen Ikosaeder jeweils dort durchstoßen, wo fünf Dreiecke zusammentreffen. Dazu kommen noch unendlich-zählige Drehachsen: in der Einzahl bei Golgi-Apparat und manchen Mitochondrienkonfigurationen (nämlich dem gut ausgebildeten cristae-Typ) oder in der Lamellenstruktur des Sehstäbchens, ferner unendlich viele unendlichzählige Drehachsen in der Kugeloberfläche des Kernes, der statistisch gleichmäßig dicht mit Kernporen besetzt ist.

In welchem Verhältnis stehen nun die eingangs erwähnten Verdoppelsarten mit den Symmetrieeigenschaften gewisser Bauteile? Daß durch Nachbau alle Symmetrien erzielt werden können, ist offensichtlich und bedarf keiner besonderen Erwähnung. Dabei kann die Symmetrieeigenschaft durch andersartig codierte Information gespeichert sein. Aperiodische, lineare und flächige Strukturen lassen sich durch Spiegelung verdoppeln bzw. durch Transformationsspiegelung komplementär abbilden. Dagegen würde ein dreidimensionaler aperiodischer Körper nur unter Gestaltwandel, indem er sich zeitweise in eine Fläche oder Strecke abrollt und wieder faltet, durch Spiegelung bzw. Transformationsspiegelung abbildbar sein. Senkrecht zu einer beliebigzähligen homopolaren Drehachse läßt sich eine Verdoppelung durch Teilung und Wachstum erzielen. Auch für eine n-zählige heteropolare Drehachse gilt dies, wenn sie als zusätzliches Symmetrieelement noch eine Translationsebene senkrecht zur Drehachse aufweist. Auch alle Gleitspiegel- und Transformationsgleitspiegelebenen sind prinzipiell als Teilungsebene ohne Symmetrieverlust denkbar. Ferner läßt sich natürlich an allen Gruppen eine Teilung unter Symmetrieverlust und Restitution durchführen. Bezeichnend ist nur, daß dies neben dem Nachbau die einzige Möglichkeit für die Ikosaedergruppe und die unendlichzähligen Drehachsen der Kernmembran ist.

Eine die ganze Zelle erfassende Symmetrie bleibt nach dem Zusammenbau so vieler verschiedener Symmetrien, die zum Teil nur in Ein- oder Zweizahl vertreten sind, nicht mehr vorhanden. Man muß jeweils die Teilgefüge der Zelle herausgreifen, um an ihnen die betreffenden Symmetriezüge zu erkennen. Dagegen zeigen die einzelnen Zelltypen durchaus typisierbare Lagebeziehungen der verschiedenen Symmetrieelemente und Bestandteile zueinander. Um dies trotz des hohen Wassergehaltes der Zelle durch Packungssymmetrie erklären zu können, muß man annehmen, daß in der Zelle eine *Dichtestpackung der Wirkungsradien* der Bausteine vorliegt, da Lockerpackungen zu beliebiger Raumkonfiguration bei gleicher Raumerfüllung führen[1]. Ein gutes Beispiel

---

[1] Symmetrische Lockerpackungen sind nur durch zusätzliche Information möglich.

dafür sind die Myelinfiguren, die in den Zellen auftreten. Wenn sich die Lipoide entmischt haben, so liegen sie nur in den Flächen dichtgepackt vor, dagegen sind die Abstände senkrecht zur Fläche variabel. Dementsprechend bilden sich ungeordnete zentrische Schläuche und Kugelfiguren aus, die herrschende Kräftefelder abbilden können.

Während bei der Protocyte ein weitgehend statistisch verteiltes Ensemble von molekularen Automaten vorliegt, die nur in den Randbezirken streng geordnete Bereiche aufweisen, nämlich in der Zellmembran und Zellwand und im kreisförmig geschlossenen DNS-Molekül ein zentrales Steuerprogramm als kreisförmiges Band besitzen, ist die höhere Zelle durch eine große Zahl einzelner Kompartimente ausgezeichnet. Diesen können wir nicht völlig gerecht werden, wenn wir nur berücksichtigen, daß diverse Metaboliten in dem Gesamtensemble der einzelnen Kompartimente in unterschiedlicher Konzentration vorliegen; wir können ihnen vielmehr nur dann gerecht werden, wenn wir versuchen, die Zelle als Automaten mit mehreren gleicharbeitenden Subautomaten, zwischen denen behinderte Diffusion besteht, nachzumodellieren oder nachzurechnen. Die einzelnen Zellorganellen können auch bei ihrer Verdoppelung als sich unabhängig vermehrende Gebilde erweisen. So treten z. B. bei *Spirogyra* durch bestimmte Zufälle Zellen mit zwei Chromatophoren auf, die den Ursprung zu einem entsprechenden Zellklon geben. Auch die Umwandlung von diesen zwei Chromatophoren tragenden Zellen zu Zellen, die nur einen Chromatophor tragen, findet statt und diese bilden wieder einen entsprechenden Klon.

Die Erkenntnisse der Informationstheorie, Kybernetik, Symmetrologie und Systemtheorie, die neue Gesichtspunkte in die Lehre von Zellen gebracht haben, erscheinen sehr erfolgversprechend und verlockend. Vor uns liegt eine große Aufgabe, eine Fahrt hinaus auf ein offenes Gewässer, dessen Ufer und Grenze vorerst noch nicht zu ahnen ist.

Zusammenfassung

Es wird eine Definition der Begriffe Verdoppelung, System und komplexes System gegeben. Bei Organismen kommt eine ideale identische Verdoppelung nicht vor. Eine strenge Formulierung der real vorliegenden Verdoppelung wird durch mathematische Formeln gegeben. Weiters werden einige informationstheoretische Bedingungen der idealen und quasi-idealen Verdoppelung besprochen und auf die drei möglichen Formen der Erhaltung der Information hingewiesen. Einige Beispiele der Anwendung in der Biologie werden erörtert und dabei mit Freese festgestellt, daß die Mutation keine gleichmäßige Störungsquelle ist. Der Begriff der funktionellen identischen Verdoppelung wird ebenfalls streng formuliert, aber als unhandlich abgelehnt. Alle Verdoppelungsarten kann man in sechs Kategorien, die angeführt sind, einordnen. Es wird gezeigt, welche Konsequenzen die Auffassung der Zelle als Rechenautomat mit sich bringt. In einem letzten Abschnitt werden die Symmetrieeigentümlichkeiten von Zellen besprochen und der Symmetrie-

begriff gegenüber der Mineralogie um die Transformation als Symmetrie-
operation erweitert. Die Verträglichkeit der einzelnen Symmetriekombi-
nationen mit den Verdoppelungen wird erörtert.

Literatur

[1] ASHBY, W. R.: The aspects of the theory of artificial intelligence. Proc. I.
    Intern. Symp. Biosimulation. Locarno 1960. New York: Plenum Press
    1962.
[2] BURKS, W. A.: Computation, behaviour and structure in fixed and growing
    automata. In: YOVITS, M. C., and S. CAMERON (eds.): Self organizing
    systems. p. 282. New York: Pergamon Press 1959.
[3] DANCOFF, S. M., and II. QUASTLER: The information content and error rate
    of living things. In: H. QUASTLER (ed.): Information theory in biology.
    p. 263. Urbana: University of Illinois Press 1953.
[4] FREESE, E.: J. theor. Biol. 3, 82 (1962).
[5] FUKUDA, N.: J. theor. Biol. 1, 440 (1961).
[6] HALL, A. D., and R. E. FRAGEN: Yearb. Gen. Syst. theor. 1, 18 (1956).
[7] HESS, B., and B. CHANCE: Naturwissenschaften 46, 248 (1959).
[8] HUXLEY, H. E.: J. molec. Biol. 7, 281 (1963).
[9] KALMAR, L.: Über eine Variante des Neumannschen selbstreproduzierenden
    Automaten. In: Mathematische und physikalische Probleme der Kyber-
    netik. Berlin: Akad. Verl. 1963.
[10] KEMENY, J.: Sci. Amer. 192, IV, 58 (1955).
[11] NEUMANN, J. VON: The general and logical theory of automata. In: L. A. JEFF-
    RESS (ed.): Cerebral mechanisms in behavior. New York-London: John
    Wiley & Sons 1951.
[12] PENROSE, L. S.: Sci. Amer. 200, VI, 105 (1959).
[13] QUASTLER, H.: The domain of information theory in biology. In: H. P. YOCKEY
    (ed.): Symp. Information theory in biology. Gatlinburg (Tenn.) 1956.
    p. 187. New York: Pergamon Press 1958.
[14] RUDZINSKA, M. A.: J. Cell Biol. 25, No. 3/Pt. 1/2, 459 (1965).
[15] SANDER, B.: Einführung in die Gefügekunde der geologischen Körper. 1. T.:
    Allgemeine Gefügekunde u. Arbeiten im Bereich Handstück bis Profil.
    Wien-Innsbruck: Springer-Verl. 1948.
[16] STAHL, W. R., and H. E. GOHEEN: J. theor. Biol. 5, 266 (1963).
[17] STEINBUCH, K.: Persönl. Mitteilung (1960).
[18] SUGITA, M.: J. theor. Biol. 1, 415 (1961).
[19] — J. theor. Biol. 4, 179 (1963).
[20] —, and N. FUKUDA: J. theor. Biol. 5, 412 (1963).
[21] TURING, A. M.: Proc. London Math. Soc. 42, 230 (1936); 43, 265 (1937).
[22] WIENER, N.: Naturwissenschaften 48, 174 (1961).
[23] ZEMANEK, H.: Elementare Informationstheorie. München: R. Oldenbourg
    1959.
[24] ZWICKY, F.: Morphology of propulsive power. Monograph No. 1 of the Soc.
    for Morphological Res. Pasadena 1962.

Diskussion

Vorsitz: *P. Sitte*

Stubbe: Bei der Aufzählung möglicher Modelle wurde der Ausdruck „Selektion"
gebraucht — in welchem Sinne?

Klima: Selektion ergibt sich durch unterschiedliche Bildungs- oder Zerfallsgeschwindigkeit der Automaten. Sie ergibt sich immer dann, wenn es eine mittlere
Lebensdauer der Nachricht und gleichzeitig Konkurrenz um einen Baustein gibt.

P. Sitte: Der Begriff Selektion ist hier also im allgemeinsten Sinne verwendet
worden.

H. Sitte: Wenn gewisse Möglichkeiten der Mutation, die sich durch das „weiße Rauschen" ergeben, Letalfaktoren einschließen, dann ist durch diese wieder eine Selektion gegeben.

Klima: Ja.

H. Sitte: Wenn unter den Möglichkeiten, die das „weiße Rauschen" bietet, nur bestimmte Fälle „lebensfähig" sind, dann wird das Mutationsgeschehen nicht notwendig zu gleicher Häufigkeit aller 4 Basen führen, selbst wenn es einem „weißen Rauschen" entspricht.

Klima: Das Problem ist: Kann der Trend, daß eine Nachricht ständig mehr ähnliche Symbole enthält durch die Selektion aufgehalten werden oder nicht? Er kann es nicht; das hat Freese gezeigt in einer Ableitung, die im Moment jedoch schwer zu skizzieren ist. Wir sehen, daß das Basenverhältnis $G + C = 50\%$ möglich ist. Aber es ist nur eines unter anderen; daß es nicht das einzige ist, das gerade ist das Verwunderliche.

Karlson: Ich habe den Eindruck, daß der Formalismus der Informationstheorie *mehr* Möglichkeiten offen läßt als die biologischen und biochemischen Tatsachen, von denen wir in den anderen Vorträgen gesprochen haben.

Klima: Das stimmt. Wir haben vorerst noch einen weitgehend leeren Formalismus, der erst durch Daten gefüllt werden muß. Die Daten kann nur die biologische Forschung liefern.

Umgekehrt könnte aber dieser Formalismus, wenn man ihn weiter ausbaut, auch weitere Auskünfte liefern. Etwa im Beispiel von Jacob und Monod bringt die erste Übertragung einer einfachen Enzyminduktion in ein Schaltersystem keinen Fortschritt, weil sie uns unmittelbar einsichtig ist. Wenn wir aber zwei solche Mechanismen zusammenschalten, dann können wir nicht mehr ohne weiteres sagen, welche Rhythmen auftreten, und wie es sich auswirkt, wenn wir bestimmte Zeitglieder einführen. Das können wir aber dann durchrechnen.

Bei einfachen Systemen mag also vieles trivial erscheinen, aber je mehr man sich den wirklichen Verhältnissen nähert, desto genauer muß der Schaltplan sein, und zwar wirklich in Form eines Schaltplans. Man kann beispielsweise einen Radioapparat nachbauen und analysieren ohne Schaltplan, aber ich würde mich außerstande sehen, ein Elektronenmikroskop ohne Schaltplan zu reparieren. Mit dem Schaltplan finde ich auch in diesem Falle einen Fehler ohne weiteres.

Karlson: Der Schaltplan ist im Falle des Elektronenmikroskops etwas dem Gerät Gemäßes. Hier ist es ein Ersatz, vielleicht ein wertvoller und vereinfachender, aber jedenfalls doch sicher ein *Ersatz* für die Fakten.

P. Sitte: Herr Klima ist im Moment in einer schwierigen Lage; denn die ersten Anwendungen, die gerade jetzt erst versucht werden, sind natürlich noch nicht sehr überzeugend. Aber in der Zukunft kann sich das ändern. Ich erinnere mich, gelesen zu haben, daß Faraday seine Physik ohne Mathematik betrieben hat. Heute ist die theoretische Physik ohne Mathematik vollkommen unvorstellbar: Man kommt nicht mehr darum herum, sich eines unanschaulichen Denkschemas zu bedienen.

Karlson: Nichts gegen die Anwendung der Mathematik! Ich frage nur: Ist es notwendig, vom biologischen Modell auf ein elektrisches (in Form eines Schaltplans) zurückzugehen, oder ist das bloß bequem?

Klima: Wenn es auch bloß bequem ist, ist es schon gerechtfertigt.

Sachsenmaier: Wenn ich Sie recht verstanden habe, könnte man theoretisch durch Verknüpfung zweier oder mehrerer Regelkreise einen Rhythmus konstruieren, der z. B. die Länge des Zellteilungscyclus mit den Teilphasen G_1, S und G_2 determiniert (jeder „Wellenberg" würde etwa die Induktion oder Aktivierung eines Mitose-spezifischen Enzyms anzeigen).

Klima: Wenn man ein Mitose-auslösendes Enzym annehmen darf, könnte man sicher ein derartiges Modell machen. Wie allerdings für diesen Fall das Schaltschema aussehen müßte, kann ich aus dem Stegreif nicht sagen. Immerhin sind derartige Versuche bereits unternommen worden.